COMPUTATIONAL BIOLOGY AND APPLIED BIOINFORMATICS

Edited by **Heitor Silvério Lopes**
and **Leonardo Magalhães Cruz**

INTECHOPEN.COM

Computational Biology and Applied Bioinformatics
Edited by Heitor Silvério Lopes and Leonardo Magalhães Cruz

Published by InTech

Janeza Trdine 9, 51000 Rijeka, Croatia

Edition 2014

Publishing Process Manager Davor Vidic

Technical Editor Teodora Smiljanic

Cover Designer InTech Design Team

Image Copyright Gencay M. Emin, 2010. Used under license from Shutterstock.com

Additional hard copies can be obtained from orders@intechopen.com

Computational Biology and Applied Bioinformatics,
Edited by Heitor Silvério Lopes and Leonardo Magalhães Cruz
p. cm.
ISBN 978-953-307-629-4

INTECH
open science | open minds

Contents

Preface

Nowadays it is difficult to imagine an area of knowledge that can continue developing without the use of computers and informatics. It is not different with biology, that has seen an unpredictable growth in recent decades, with the rise of a new discipline, bioinformatics, bringing together molecular biology, biotechnology and information technology. More recently, the development of high throughput techniques, such as microarray, mass spectrometry and DNA sequencing, has increased the need of computational support to collect, store, retrieve, analyze, and correlate huge data sets of complex information. On the other hand, the growth of the computational power for processing and storage has also increased the necessity for deeper knowledge in the field. The development of bioinformatics has allowed now the emergence of systems biology, the study of the interactions between the components of a biological system, and how these interactions give rise to the function and behavior of a living being.

Bioinformatics is a cross-disciplinary field and its birth in the sixties and seventies depended on discoveries and developments in different fields, such as: the proposed double helix model of DNA by Watson and Crick from X-ray data obtained by Franklin and Wilkins in 1953; the development of a method to solve the phase problem in protein crystallography by Perutz's group in 1954; the sequencing of the first protein by Sanger in 1955; the creation of the ARPANET in 1969 at Stanford UCLA; the publishing of the Needleman-Wunsch algorithm for sequence comparison in 1970; the first recombinant DNA molecule created by Paul Berg and his group in 1972; the announcement of the Brookhaven Protein DataBank in 1973; the establishment of the Ethernet by Robert Metcalfe in the same year; the concept of computers network and the development of the Transmission Control Protocol (TCP) by Vint Cerf and Robert Khan in 1974, just to cite some of the landmarks that allowed the rise of bioinformatics. Later, the Human Genome Project (HGP), started in 1990, was also very important for pushing the development of bioinformatics and related methods of analysis of large amount of data.

This book presents some theoretical issues, reviews, and a variety of bioinformatics applications. For better understanding, the chapters were grouped in two parts. It was not an easy task to select chapters for these parts, since most chapters provide a mix of review and case study. From another point of view, all chapters also have extensive biological and computational information. Therefore, the book is divided into two parts. In Part I, the chapters are more oriented towards literature review and theoretical issues. Part II consists of application-oriented chapters that report case studies in which a specific biological problem is treated with bioinformatics tools.

Molecular phylogeny analysis has become a routine technique not only to understand the sequence-structure-function relationship of biomolecules but also to assist in their classification. The first chapter of Part I, by Kolekar et al., presents the theoretical basis, discusses the fundamental of phylogenetic analysis, and a particular view of steps and methods used in the analysis.

Methods for protein function and gene expression are briefly reviewed in Hupert-Kocurek and Kaguni's chapter, and contrasted with the traditional approach of mapping a gene via the phenotype of a mutation and deducing the function of the gene product, based on its biochemical analysis in concert with physiological studies. An example of experimental approach is provided that expands the current understanding of the role of ATP binding and its hydrolysis by DnaC during the initiation of DNA replication. This is contrasted with approaches that yield large sets of data, providing a different perspective on understanding the functions of sets of genes or proteins and how they act in a network of biochemical pathways of the cell.

Due to the importance of transcriptional regulation, one of the main goals in the post-genomic era is to predict how the expression of a given gene is regulated based on the presence of transcription factor binding sites in the adjacent genomic regions. Nain et al. review different computational approaches for modeling and identification of regulatory elements, as well as recent advances and the current challenges.

In Hwa et al., an approach is proposed to group proteins into putative functional groups by designing a workflow with appropriate bioinformatics analysis tools, to search for sequences with biological characteristics belonging to the selected protein family. To illustrate the approach, the workflow was applied to LARGE-like protein family.

Microarray technology has become one of the most important technologies for unveiling gene expression profiles, thus fostering the development of new bioinformatics methods and tools. In the chapter by Lowery et al. a thorough review of microarray technology is provided, with special focus on MRNA and microRNA profiling of breast cancer.

MicroRNAs are a class of small RNAs of approximately 22 nucleotides in length that regulate eukaryotic gene expression at the post-transcriptional level. Santos and Soares present several tools and computational pipelines for miRNA identification, discovery and expression from sequencing data.

Currently, the mass spectroscopy-based methods represent very important and flexible tools for studying the dynamic features of proteins and their complexes. Such high-resolution methods are especially used for characterizing critical regions of the systems under investigation. Vitale et al. present a thorough review of mass spectrometry and the related computational methods for studying the three-dimensional structure of proteins.

Rodrigues and Kluskens review synthetic biology approaches for the development of alternatives for cancer diagnosis and drug development, providing several application examples and pointing challenging directions of research.

Biological sequence alignment is an important and widely used task in bioinformatics. It is essential to provide valuable and accurate information in the basic research, as well as in daily use of the molecular biologist. The well-known Smith and Waterman algorithm is an optimal sequence alignment method, but it is computationally expensive for large instances. This fact fostered the research and development of specialized hardware platforms to accelerate biological data analysis that use that algorithm. Hasan and Al-Ars provide a thorough discussion and comparison of available methods and hardware implementations for sequence alignment on different platforms.

Exciting and updated issues are presented in Part II, where theoretical bases are complemented with case studies, showing how bioinformatics analysis pipelines were applied to answer a variety of biological issues.

During the last years we have witnessed an exponential growth of the biological data and scientific articles. Consequently, retrieving and categorizing documents has become a challenging task. The second part of the book starts with the chapter by Addis et al. that propose a multiagent system for retrieving and categorizing bioinformatics publications, with special focus on the information extraction task and adopted hierarchical text categorization technique.

Computational immunology is a field of science that encompasses high-throughput genomic and bioinformatic approaches to immunology. On the other hand, grid computing is a powerful alternative for solving problems that are computationally intensive. Pappalardo and Chiachio present two different studies of using computational immunology approaches implemented in a grid infrastructure: modeling atherosclerosis and optimal protocol searching for vaccine against mammary carcinoma.

Despite the growing number of proteins discovered as sub-product of the many genome sequencing projects, only a very few number of them have a known three-dimensional structure. A possible way to infer the full structure of an unknown protein is to identify potential secondary structures in it. Chidambaram et al. compare the performance of several machine learning and evolutionary computing methods for the classification of secondary structure of proteins, starting from their primary structure.

Cancer is one of the most important public health problems worldwide. Breast and cervical cancer are the most frequent in female population. Salcedo et al. present a study about the functional analysis of the cervical carcinoma transcriptome, with focus on the methods for unveiling networks and finding new genes associated to cervical cancer.

In Sawada and Mitaku, the number distribution of transmembrane helices is investigated to show that it is a feature under natural selection in prokaryotes and how membrane proteins with high number of transmembrane helices disappeared in random mutations by simulation data.

In Chu et al., an alignment approach using the pure best hit strategy is proposed to classify TIM barrel protein domain structures in terms of the superfamily and family categories with high accuracy.

Jamil and Sabeena use classic bioinformatic tools, such as ClustalW for Multiple Sequence Alignment, SCI-PHY server for superfamily determination, ExPASy tools for pattern matching, and visualization softwares for residue recognition and functional elucidation to determine the functional diversity of the enolase enzyme superfamily.

Quality assessment of structure predictions is an important problem in bioinformatics because quality determines the application range of predictions. Piedra et al. briefly review some applications used in protein structure prediction field, were they are used to evaluate overall prediction quality, and show how structure comparison methods can also be used to identify the more reliable parts in "de novo" analysis and how this information can help to refine/improve these models.

In Fu, a new method is presented that explores potential genes in intergenic regions of an annotated genome on the basis of their gene expression activity. The method was applied to the M. tuberculosis genome where potential protein-coding genes were found, based on bioinformatics analysis in conjunction with transcriptional evidence obtained using the Affymetrix GeneChip. The study revealed potential genes in the intergenic regions, such as DNA-binding protein in the CopG family and a nickel binding GTPase, as well as hypothetical proteins.

Cai et al. present a new method for developmental studies. It combines experimental studies and computational analysis to predict the trans-acting factors and transcriptional regulatory networks for mouse embryonic retinal development.

The chapter by Roberts shows how advances in bioinformatics can be applied to the development of improved therapeutic strategies. The chapter describes how functional genomics experimentation and bioinformatics tools could be applied to the design of synthetic promoters for therapeutic and diagnostic applications or adapted across the biotech industry. Designed synthetic gene promoters can than be incorporated in novel gene transfer vectors to promote safer and more efficient expression of therapeutic genes for the treatment of various pathological conditions. Tools used to analyze data obtained from large-scale gene expression analyses, which are subsequently used in the smart design of synthetic promoters are also presented.

Bruskin et al. describe how candidate genes commonly involved in psoriasis and Crohn's disease were detected using lists of differentially expressed genes from microarrays experiments with different numbers of probes. These gene codes for proteins are particular targets for elaborating new approaches to treating these pathologies. A comprehensive meta-analysis of proteomics and transcriptomics of psoriatic lesions from independent studies is performed. Network-based analysis revealed similarities in regulation at both proteomics and transcriptomics level.

Some eukaryotic mRNAs have multiple ORFs, which are recognized as polycistronic mRNAs. One of the well-known extra ORFs is the upstream ORF (uORF), that functions as a regulator of mRNA translation. In Ao-Kondo et al., this issue is addressed and an introduction to the mechanism of translation initiation and functional roles of uORF in translational regulation is given, followed by a review of how the authors identified novel small proteins with Mass Spectrometry and a discussion on the progress of bioinformatics analyses for elucidating the diversification of short coding regions defined by the transcriptome.

Acrylamide might feature toxic properties, including neurotoxicity and carcinogenicity in both mice and rats, but no consistent effect on cancer incidence in humans could be identified. In the chapter written by Lima and Carloni, the authors report the use of bioinformatics tools, by means of molecular docking and molecular simulation procedures, to predict and explore the structural determinants of acrylamide and its derivative in complex with all of their known cellular target proteins in human and mice.

Professor Heitor Silvério Lopes
Bioinformatics Laboratory, Federal University of Technology – Paraná,
Brazil

Professor Leonardo Magalhães Cruz
Biochemistry Department, Federal University of Paraná,
Brazil

Part 1

Reviews

1

Molecular Evolution & Phylogeny: What, When, Why & How?

Pandurang Kolekar[1], Mohan Kale[2] and Urmila Kulkarni-Kale[1]
[1]Bioinformatics Centre, University of Pune
[2]Department of Statistics, University of Pune
India

1. Introduction

The endeavour for the classification and study of evolution of organisms, pioneered by Linneaus and Darwin on the basis of morphological and behavioural features of organisms, is now being propelled by the availability of molecular data. The field of evolutionary biology has experienced a paradigm shift with the advent of sequencing technologies and availability of molecular sequence data in the public domain databases. The post-genomic era provides unprecedented opportunities to study the process of molecular evolution, which is marked with the changes organisms acquire and inherit. The species are continuously subjected to evolutionary pressures and evolve suitably. These changes are observed in terms of variations in the sequence data that are collected over a period of time. Thus, the molecular sequence data archived in various databases are the snapshots of the evolutionary process and help to decipher the evolutionary relationships of genes/proteins and genomes/proteomes for a group of organisms. It is known that the individual genes may evolve with varying rates and the evolutionary history of a gene may or may not coincide with the evolution of the species as a whole. One should always refrain from discussing the evolutionary relationship between organisms when analyses are performed using limited/partial data. Thorough understanding of the principles and methods of phylogeny help the users not only to use the available software packages in an efficient manner, but also to make appropriate choices of methods of analysis and parameters so that attempts can be made to maximize the gain on huge amount of available sequence data.

As compared to classical phylogeny based on morphological data, molecular phylogeny has distinct advantages, for instance, it is based on sequences (as descrete characters) unlike the morphological data, which is qualitative in nature. While the tree of life is depicted to have three major branches as bacteria, archaea and eukaryotes (it excludes viruses), the trees based on molecular data accounts for the process of evolution of bio-macromolecules (DNA, RNA and protein). The trees generated using molecular data are thus referred to as 'inferred trees', which present a hypothesized version of what might have happened in the process of evolution using the available data and a model. Therefore, many trees can be generated using a dataset and each tree conveys a story of evolution. The two main types of information inherent in any phylogenetic tree are the topology (branching pattern) and the branch lengths.

Before getting into the actual process of molecular phylogeny analysis (MPA), it will be helpful to get familiar with the concepts and terminologies frequently used in MPA.

Phylogenetic tree: A two-dimensional graph depicting nodes and branches that illustrates evolutionary relationships between molecules or organisms.

Nodes: The points that connect branches and usually represent the taxonomic units.

Branches: A branch (also called an edge) connects any two nodes. It is an evolutionary lineage between or at the end of nodes. Branch length represents the number of evolutionary changes that have occurred in between or at the end of nodes. Trees with uniform branch length (cladograms), branch lengths proportional to the changes or distance (phylograms) are derived based on the purpose of analysis.

Operational taxonomic units (OTUs): The known external/terminal nodes in the phylogenetic tree are termed as OTU.

Hypothetical taxonomic units (HTUs): The internal nodes in the phylogenetic tree that are treated as common ancestors to OTUs. An internal node is said to be bifurcating if it has only two immediate descendant lineages or branches. Such trees are also called binary or dichotomous as any dividing branch splits into two daughter branches. A tree is called a 'multifurcating' or 'polytomous' if any of its nodes splits into more than two immediate descendants.

Monophyletic: A group of OTUs that are derived from a single common ancestor containing all the descendents of single common ancestor.

Polyphyletic: A group of OTUs that are derived from more than one common ancestor.

Paraphyletic: A group of OTUs that are derived from a common ancestor but the group doesn't include all the descendents of the most recent common ancestor.

Clade: A monophyletic group of related OTUs containing all the descendants of the common ancestor along with the ancestor itself.

Ingroup: A monophyletic group of all the OTUs that are of primary interest in the phylogenetic study.

Outgroup: One or more OTUs that are phylogenetically outside the ingroup and known to have branched off prior to the taxa included in a study.

Cladogram: The phylogenetic tree with branches having uniform lengths. It only depicts the relationship between OTUs and does not help estimate the extent of divergence.

Phylogram: The phylogenetic tree with branches having variable lengths that are proportional to evolutionary changes.

Species tree: The phylogenetic tree representing the evolutionary pathways of species.

Gene tree: The phylogenetic tree reconstructed using a single gene from each species. The topology of the gene tree may differ from 'species tree' and it may be difficult to reconstruct a species tree from a gene tree.

Unrooted tree: It illustrates the network of relationship of OTUs without the assumption of common ancestry. Most trees generated using molecular data are unrooted and they can be rooted subsequently by identifying an outgroup. Total number of bifurcating unrooted trees can be derived using the equation: $Nu = (2n-5)!/2^{n-3}(n-3)!$

Rooted tree: An unrooted phylogenetic tree can be rooted with outgroup species, as a common ancestor of all ingroup species. It has a defined origin with a unique path to each ingroup species from the root. The total number of bifurcating rooted trees can be calculated using the formula, $Nr = (2n-3)!/2^{n-2}(n-2)!$ (Cavalli-Sforza & Edwards, 1967). Concept of unrooted and rooted trees is illustrated in Fig. 1.

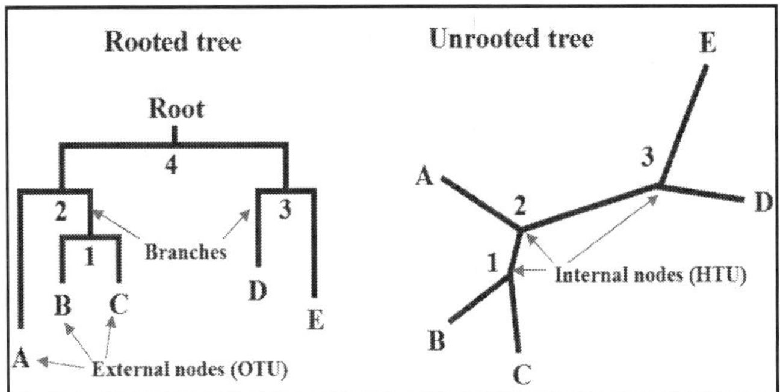

Fig. 1. Sample rooted and unrooted phylogenetic trees drawn using 5 OTUs . The external and internal nodes are labelled with alphabets and Arabic numbers respectively. Note that the rooted and unrooted trees shown here are one of the many possible trees (105 rooted and 15 unrooted) that can be obtained for 5 OTUs.

The MPA typically involves following steps

- Definition of problem and motivation to carry out MPA
- Compilation and curation of homologous sequences of nucleic acids or proteins
- Multiple sequence alignments (MSA)
- Selection of suitable model(s) of evolution
- Reconstruction of phylogenetic tree(s)
- Evaluation of tree topology

A brief account of each of these steps is provided below.

2. Definition of problem and motivation to carry out MPA

Just like any scientific experiment, it is necessary to define the objective of MPA to be carried out using a set of molecular sequences. MPA has found diverse applications, which include classification of organisms, DNA barcoding, subtyping of viruses, study the co-evolution of genes and proteins, estimation of divergence time of species, study of the development of pandemics and pattern of disease transmission, parasite-vector-host relationships etc. The biological investigations where MPA constitute a major part of analyses are listed here. A virus is isolated during an epidemic. Is it a new virus or an isolate of a known one? Can a genotype/serotype be assigned to this isolate just by using the molecular sequence data? A few strains of a bacterium are resistant to a drug and a few are sensitive. What and where are the changes that are responsible for such a property? How do I choose the attenuated strains, amongst available, such that protection will be offered against most of the wild type strains of a given virus? Thus, in short, the objective of the MPA plays a vital role in deciding the strategy for the selection of candidate sequences and adoption of the appropriate phylogenetic methods.

3. Compilation and curation of homologous sequences

The compilation of nucleic acid or protein sequences, appropriate to undertake validation of hypothesis using MPA, from the available resources of sequences is the next step in MPA.

At this stage, it is necessary to collate the dataset consisting of homologous sequences with the appropriate coverage of OTUs and outgroup sequences, if needed. Care should be taken to select the equivalent regions of sequences having comparable lengths (± 30 bases or amino acids) to avoid the subsequent errors associated with incorrect alignments leading to incorrect sampling of dataset, which may result in erroneous tree topology. Length differences of >30 might result in insertion of gaps by the alignment programs, unless the gap opening penalty is suitably modified. Many comprehensive primary and derived databases of nucleic acid and protein sequences are available in public domain, some of which are listed in Table 1. The database issue published by the journal 'Nucleic Acids research' (NAR) in the month of January every year is a useful resource for existing as well as upcoming databases. These databases can be queried using the 'text-based' or 'sequence-based' database searches.

Database	URL	Reference
	Nucleotide	
GenBank	http://www.ncbi.nlm.nih.gov/genbank/	Benson et al., 2011
EMBL	http://www.ebi.ac.uk/embl/	Leinonen et al., 2011
DDBJ	http://www.ddbj.nig.ac.jp/	Kaminuma et al., 2011
	Protein	
GenPept	http://www.ncbi.nlm.nih.gov/protein	Sayers et al., 2011
Swiss-Prot	http://expasy.org/sprot/	The UniProt Consortium (2011)
UniProt	http://www.uniprot.org/	The UniProt Consortium (2011)
	Derived	
RDP	http://rdp.cme.msu.edu/	Cole et al., 2009
HIV	http://www.hiv.lanl.gov/content/index	Kuiken et al., 2009
HCV	http://www.hcvdb.org/	http://www.hcvdb.org/

Table 1. List of some of the commonly used nucleotide, protein and molecule-/species-specific databases.

Text-based queries are supported using search engines viz., Entrez and SRS, which are available at NCBI and EBI respectively. The list of hits returned after the searches needs to be curated very carefully to ensure that the data corresponds to the gene/protein of interest and is devoid of partial sequences. It is advisable to refer to the feature-table section of every entry to ensure that the data is extracted correctly and corresponds to the region of interest. The sequence-based searches involve querying the databases using sequence as a probe and are routinely used to compile a set of homologous sequences. Once the sequences are compiled in FASTA or another format, as per the input requirements of MPA software, the sequences are usually assigned with unique identifiers to facilitate their identification and comparison in the phylogenetic trees. If the sequences posses any ambiguous characters or low complexity regions, they could be carefully removed from sequences as they don't contribute to evolutionary analysis. The presence of such regions might create problems in alignment, as it could lead to equiprobable alternate solutions to 'local alignment' as part of a global alignment. Such regions possess 'low' information content to favour a tree topology over the other. The inferiority of input dataset interferes with the analysis and interpretation of the MPA. Thus, compilation of well-curated sequences, for the problem at hand, plays a crucial role in MPA.

The concept of homology is central to MPA. Sequences are said to be homologous if they share a common ancestor and are evolutionarily related. Thus, homology is a qualitative description of the relationship and the term %homology has no meaning. However, supporting data for deducing homology comes from the extent of sequence identity and similarity, both of which are quantitative terms and are expressed in terms of percentage.

The homologous sequences are grouped into three types, viz., orthologs (same gene in different species), paralogs (the genes that originated from duplication of an ancestral gene within a species) and xenologs (the genes that have horizontally transferred between the species). The orthologous protein sequences are known to fold into similar three-dimensional shapes and are known to carry out similar functions. For example, haemoglobin alpha in horse and human. The paralogous sequences are copies of the ancestral genes evolving within the species such that nature can implement a modified function. For example haemoglobin alpha and beta in horse. The xenologs and horizontal transfer events are extremely difficult to be proved only on the basis of sequence comparison and additional experimental evidence to support and validate the hypothesis is needed. The concepts of sequence alignments, similarity and homology are extensively reviewed by Phillips (2006).

4. Multiple sequence alignments (MSA)

MSA is one of the most common and critical steps of classical MPA. The objective of MSA is to juxtapose the nucleotide or amino acid residues in the selected dataset of homologous sequences such that residues in the column of MSA could be used to derive the sequence of the common ancestor. The MSA algorithms try to maximize the matching residues in the given set of sequences with a pre-defined scoring scheme. The MSA produces a matrix of characters with species in the rows and character sites in columns. It also introduces the gaps, simulating the events of insertions and deletions (also called as indels). Insertion of gaps also helps in making the lengths of all sequences same for the sake of comparison. All the MSA algorithms are guaranteed to produce optimal alignment above a threshold value of detectable sequence similarity. The alignment accuracy is observed to decrease when sequence similarity drops below 35% towards the twilight (<35% but > 25%) and moonlight zones (<25%) of similarity. The character matrix obtained in MSA reveals the pattern of conservation and variability across the species, which in turn reveals the motifs and the signature sequences shared by species to retain the fold and function. The analysis of variations can be gainfully used to identify the changes that explain functional and phenotypic variability, if any, across OTUs.

Many algorithms have been specially developed for MSA and subsequently improved to achieve higher accuracy. One of the popular heuristics-based MSA approach follows progressive alignment procedure, in which sequences are compared in a pair wise fashion to build a distance matrix containing percent identity values. A clustering algorithm is then applied to distance matrix to generate a guide tree. The algorithm then follows a guide tree to add the pair wise alignments together starting from the leaf to root. This ensures the sequences with higher similarity are aligned initially and distantly related sequences are progressively added to the alignment of aligned sequences. Thus, the gaps inserted are always retained. A suitable scoring function, sum-of-pairs, consensus, consistency-based etc.

is employed to derive the optimum MSA (Nicholas et al., 2002; Batzoglou, 2005). Most of the MSA packages use Needleman and Wunsch (1970) algorithm to compute pair wise sequence similarity. The ClustalW is the widely used MSA package (Thompson et al., 1994). Recently many alternative MSA algorithms are also being developed, which are enlisted in Table 2. The standard benchmark datasets are used for comparative assessment of the alternative approaches (Aniba et al., 2010; Thompson et al., 2011). Irrespective of the proven performance of MSA methods for individual genes and proteins, some of the challenges and issues regarding computational aspects involved in handling genomic data are still the causes of concern (Kemena & Notredame, 2009).

Alignment programs	Algorithm description	Available at / Reference
ClustalW	Progressive	http://www.ebi.ac.uk/Tools/msa/clustalw2/ ; Thompson et al., 1994
MUSCLE	Progressive/iterative	http://www.ebi.ac.uk/Tools/msa/muscle/ ; Edgar, 2004
T-COFFEE	Progressive	http://www.ebi.ac.uk/Tools/msa/tcoffee/ ; Notredame et al., 2000
DIALIGN2	Segment-based	http://bibiserv.techfak.uni-bielefeld.de/dialign/ ; Morgenstern et al., 1998
MAFFT	Progressive/iterative	http://mafft.cbrc.jp/alignment/software/ ; Katoh et al., 2005
Alignment visualization programs		
*BioEdit		http://www.mbio.ncsu.edu/bioedit/bioedit.html ; Hall, 1999
MEGA5		http://www.megasoftware.net/ ; Kumar et al., 2008
DAMBE		http://dambe.bio.uottawa.ca/dambe.asp ; Xia & Xie, 2001
CINEMA5		http://aig.cs.man.ac.uk/research/utopia/cinema ; Parry-Smith et al., 1998

*: Not updated since 2008, but the last version is available for use.

Table 2. List of commonly used multiple sequence alignment programs and visualization tools.

The MSA output can also be visualized and edited, if required, with the software like BioEdit, DAMBE etc. Multiple alignment output shows the conserved and variable sites, usually residues are colour coded for the ease of visualisation, identification and analysis. The character sites in MSA can be divided as conserved (all the sequences have same residue or base), variable-non-informative (singleton site) and variable-informative sites. The sites containing gaps in all or majority of the species are of no importance from the evolutionary point of view and are usually removed from MSA while converting MSA data to input data for MPA. A sample MSA is shown in Fig. 2. The sequences of surface hydrophobic (SH) protein from various genotypes (A to M) of Mumps virus, are aligned. A careful visual inspection of MSA allows us to locate the patterns and motifs (LLLXIL) in a given set of sequences. Apart from MPA, the MSA data in turn can be used for the construction of position specific scoring matrix (PSSM), generation of consensus sequence,

sequence logos, identification and prioritisation of potential B- and T-cell epitopes etc. Nowadays the databases of curated, pre-computed alignments of reference species are also being made available, which can be used for the benchmark comparison, evaluation purpose (Thompson et al., 2011) and it also helps to keep the track of changes that get accumulated in the species over a period of time. For example, in case of viruses, observed changes are correlated with emergence of new genotypes (Kulkarni-Kale et al., 2004; Kuiken et al., 2005).

```
                    10        20        30        40        50
            ....|....|....|....|....|....|....|....|....|....|....|..
AD90231     MPAIQPPLYLTFLLLILLYLIITLYVWTILTINHKTAVRYAALYQRSCSRWGFDQSL
AD00663     MPANQPPLYLTFLLLILLYLIITLYVWTILTINHKTAVRYAALYQRSCSRWGFDQSL
AD90232     MPAIQPPLYLTFLLLTLLYLIITLYVWTILTINHNTAVRYAALYQRSFSRWGFDQSL
AX72944     MPAIQPPLYLTFLLLILLYLIITLYVWTILTINHKTAVRYAALYQRSCSRWGFDQSL
BD90236     MPAIQPPLYPTFLLLILLSLIVTLYVWIISTITYKTVVRHAALYQRSFFRWSFDHSL
BD90233     MPAIQPPLYPTFLLLILLSLIITLYVWIISTITYKTAVRHASLYQRSFSRWSFDHSL
BD90234     MPAIQPPLYPTFLLLILLSLIITLYVWIISTITYKTAVRHAALHQRSFSRWSLDHSL
BAB003414   MPAIQPPLYPTFLLLILLSLIITLYVWIISTITYKTAVRHAALYQRSFFRWSFDHSL
CX63709     MPAIQPPLYLTFLLLILLYRIITLYVWVVSTITYKTAVRHAALYQRSLFRWSFDHSL
CY08214     MPAIQPLLYLTFLLLILLYLIITLYVWVVSTITYKTAVRHAALYQRSLFRWSFDHSL
DAF142769   MPVIQPPLYLTSLLLILLYLIITLYVWIILTITYKTAVRHEALYQRSFSRWSFDHSP
DAF142766   MPVIQPPLYLTSLLLFLLYLVITLYVWIILTITYKTAVRHEALYQRSFSRWSFDHSL
DY08212     MPAIQPPLYLTSLLLILLYLIITLYVWIILTITYKTAVRHEALYQRSFSRWSFDHSL
EX63711     MPLIQPPLYLTFLLLMLLYRIITLYVWSLSTITYKTSVRHASLYQRSFFRWSVDHSL
EX63710     MPAIQPPLYLTFLLLMLLYRIITLYVWSLSTITYKTSVRHASLYQRSFFRWSVDHSL
FZ77158     MSAIHPPLYLTFLLLILLHLIITLYVWIMLTITHKTAVQHAGLYQRSLFRWSFDHSL
FZ77160     MSAIHPPLYLTFLLLILLHLIITLYVWIMLTITHKTAVQHVALYQRSRSRWGLDHSL
GAF142764   MPAIQPPLYLTFLLLILLYLIITLYVWIILTITYKTAVRHAALYQRSFFHWSFDHSL
GAY380075   MPAIQPPLYLTFLLLILLYLIITSYVWIILTITYKTAVRHAALYQRSFFHWSFDHSL
HAF142771   MPAIQPPLHLTFLLLILLYLIITLYVWITLTITYKTAVRHATLYQRSFFRWSFDHPL
HAF142762   MPAIQPPLHLTFLLLILLYLIITFYVWITSTITYKAAVRHATLYQRSFFRWSFDHPL
ID86174     MPAIQPPLYLTFLLLILLYMIITLYVWIILTITYKTAVRHAALYQGSFFRWSFDHSL
IAF180374   MPAIQPPLYLTFLLLILLYMIITLYAWIILTITYKTAVRHAALYQGSFFRWSFEHSL
JAB003417   MPAIQPSLYLTFLLLVLLYLIIALYVWIILAITYNTAAQHAALHRGSLSRWSFDHSF
JAF142770   MPAIQPPLYLTFLLLVLLYLIITLYVWIILAITYKTAVQHAALYQGSLFRWSFDHSF
KAF365891   MPAIQLPLYPTFLLLTLLYLITTLYVWTISTITYKTTVRHAALHQRSFFRWSFDHSL
LAB105483   MPAIQPPLYLTLLLLILLYLIITLYVCIILTITYKTAERHATLYQRSFFRWSFDHSL
LAB105480   MPAIQPPLYLTLLLLILLYLIITLYVCIILTITYKTAERHATLYQRSFFRWSFDHSL
MEU069917   MPAIQPPLYLTFLLLILLYLIITLYVWIILTITYKTSVRHSVLHQRSFFRWSFDHSL
MEU069918   MPAIQPPLYLTFLLLILLYLIITLYVWIILTITYKTSVRHSVLHQRSFFRWSFDHSL
MEU069922   MPAIQPPLYLTFLLLILLYLIITLYVWIILTITYKTSVRHSVLHQRSFFRWSFDHSL
MEU069923   MPAIQPPLYLTFLLLILLYLIITLYVWIILTITYKTSVRHSVLHQRSFFRWSFDHSL
            *      *  * *** **      *      *          *   *   *
```

Fig. 2. The complete multiple sequence alignment of the surface hydrophobic (SH) proteins of Mumps virus genotypes (A to M) carried out using ClustalW. The MSA is viewed using BioEdit. The species labels in the leftmost column begin with genotype letter (A-M) followed by GenBank accession numbers. The scale for the position in alignment is given at the top of the alignment. The columns with conserved residues are marked with an "*" in the last row.

5. Selection of a suitable model of evolution

The existing MPA methods utilize the mathematical models to describe the evolution of sequence by incorporating the biological, biochemical and evolutionary considerations. These mathematical models are used to compute genetic distances between sequences. The use of appropriate model of evolution and statistical tests help us to infer maximum evolutionary information out of sequence data. Thus, the selection of the right model of sequence evolution becomes important as a part of effective MPA. Two types of approaches are adapted for the building of models, first one is empirical i.e. using the properties revealed through comparative studies of large datasets of observed sequences, and the other is parametrical, which uses biological and biochemical knowledge about the nucleic acid and protein sequences, for example the favoured substitution patterns of residues. Parametric models obtain the parameters from the MSA dataset under study. Both types of approaches result in the models based on the Markov process, in the form of matrix representing the rate of all possible transitions between the types of residues (4 nucleotides in nucleic acids and 20 amino acids in proteins). According to the type of sequence (nucleic acid or protein), two categories of models have been developed.

5.1 Models of nucleotide substitution

The nucleotide substitution models are based on the parametric approach with the use of mainly three parameters i) nucleotides frequencies, ii) rate of nucleotide substitutions and iii) rate heterogeneity. Nucleotide frequencies, account for the compositional sequence constraints such as GC content. These are subsequently used in a model to allow the substitutions of a certain type to occur more likely than others. The nucleotide substitution parameter is used to represent a measure of biochemical similarity. Higher the similarity between the nucleotide bases, the more is the rate of substitution between them, for example, the transitions are more frequent than transversions. A parameter of rate heterogeneity accounts for the unequal rates of substitution across the variable sites, which can be correlated with the constraints of genetic code, selection for the gene function etc. The site variability is modelled by gamma distribution of rates across sites. The shape parameter of gamma distribution determines amount of heterogeneity among sites, larger values of shape parameter gives a bell shaped distribution suggesting little or no rate variation across the sites whereas small values of it gives J-shaped distribution indicating high rate variation among sites along with low rates of evolution at many sites.
Varieties of nucleotide substitution models have been developed with a set of assumptions and parameters described as above. Some of the well-known models of nucleotide substitutions include Jukes-Cantor (JC) one-parameter model (Jukes & Cantor, 1969), Kimura two-parameter model (K2P) (Kimura, 1980), Tamura's model (Tamura, 1992), Tamura and Nei model (Tamura & Nei, 1993) etc. These models make use of different biological properties such as, transitions, transversions, G+C content etc. to compute distances between nucleotide sequences. The substitution patterns of nucleotides for some of these models are shown in Fig. 3.

5.2 Models of amino acid replacement

In contrast to nucleotide substitution models, amino acid replacement models are developed using empirical approach. Schwarz and Dayhoff (1979) developed the most widely used model of protein evolution in which, the replacement matrix was obtained from the alignment of globular protein sequences with 15% divergence. The Dayhoff matrices, known as PAM matrices, are also used by database searching methods. The similar methodology was adopted by other model developers but with specialized databases. Jones et al., (1994) have derived a replacement matrix specifically for membrane proteins, which has values significantly different from Dayhoff matrix suggesting the remarkably different pattern of amino acid replacements observed in the membrane proteins. Thus, such a matrix will be more appropriate for the phylogenetic study of membrane proteins. On the other hand, Adachi and Hasegawa (1996) obtained a replacement matrix using mitochondrial proteins across 20 vertebrate species and can be effectively used for mitochondrial protein phylogeny. Henikoff and Henikoff (1992) derived the series of BLOSUM matrices using local, ungapped alignments of distantly related sequences. The BLOSUM matrices are widely used in similarity searches against databases than for phylogenetic analyses.

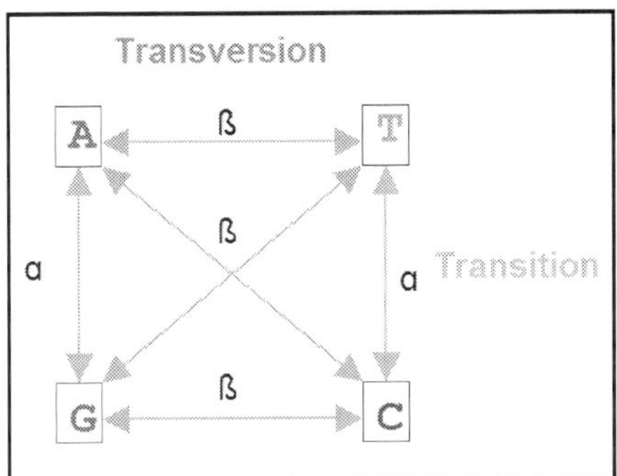

Fig. 3. The types of substitutions in nucleotides. α denotes the rate of transitions and β denotes the rate of transversions. For example, in the case of JC model α=β while in the case of K2P model α>β.

Recently, structural constraints of the nucleic acids and proteins are also being incorporated in the building of models of evolution. For example, Rzhetsky (1995) contributed a model to estimate the substitution patterns in ribosomal RNA genes with the account of secondary structure elements like stem-loops in ribosomal RNAs. Another approach introduced a model with the combination of protein secondary structures and amino acid replacement (Lio & Goldman, 1998; Thorne et al., 1996). The overview of different models of evolution and the criteria for the selection of models is also provided by Lio & Goldman (1998); Luo et al. (2010).

6. Reconstruction of a phylogenetic tree

The phylogeny reconstruction methods result in a phylogenetic tree, which may or may not corroborate with the true phylogenetic tree. There are various methods of phylogeny reconstruction that are divided into two major groups viz. character-based and distance-based.

Character-based methods use a set of discrete characters, for example, in case of MSA data of nucleotide sequences, each position in alignment is reffered as "character" and nucleotide (A, T, G or C) present at that position is called as the "state" of that "character". All such characters are assumed to evolve independent of each other and analysed separately. Distance-based methods on other hand use some form of distance measure to compute the dissimilarity between pairs of OTUs, which subsequently results in derivation of distance matrix that is given as an input to clustering methods like Neighbor-Joining (N-J) and Unweighted Pair Group Method with Arithmetic mean (UPGMA) to infer phylogenetic tree. The character-based and distance-based methods follow exhaustive search and/or stepwise clustering approach to arrive at an optimum phylogenetic tree, which explains the evolutionary pattern of the OTUs under study. The exhaustive search method examines theoretically all possible tree topologies for a chosen number of species and derives the best tree topology using a set of certain criteria. Table 3 shows the possible number of rooted and unrooted trees for n number of species/OTUs.

Number of OTUs	Number of unrooted trees	Number of rooted trees
2	1	1
3	1	3
4	3	15
5	15	105
6	105	945
10	2027025	34459425

Table 3. The number of possible rooted and unrooted trees for a given number of OTUs. The number of possible unrooted trees for n OTUs is given by $(2n-5)!/[2^{n-3}(n-3)!]$; and rooted trees is given by $(2n-3)!/[2^{n-2}(n-2)!]$

Whereas, stepwise clustering methods employ an algorithm, which begins with the clustering of highly similar OTUs. It then combines the clustered OTUs such that it can be treated as a single OTU representing the ancestor of combined OTUs. This step reduces the complexity of data by one OTU. This process is repeated and in a stepwise manner adding the remaining OTUs until all OTUs are clustered together. The stepwise clustering approach is faster and computationally less intensive than the exhaustive search method.
The most widely used distance-based methods include N-J & UPGMA and character-based methods include Maximum Parsimony (MP) and Maximum Likelihood (ML) methods (Felsenstein, 1996). All of these methods make particular assumptions regarding evolutionary process, which may or may not be applicable to the actual data. Thus, before selection of a phylogeny reconstruction method, it is recommended to take into account the assumptions made by the method to infer the best phylogenetic tree. The list of widely used phylogeny inference packages is given in Table 4.

Package	Available from / Reference
PHYLIP	http://evolution.genetics.washington.edu/phylip.html ; Felsenstein, 1989
PAUP	http://paup.csit.fsu.edu/ ; Wilgenbusch & Swofford, 2003
MEGA5	http://www.megasoftware.net/ ; Kumar et al., 2008
MrBayes	http://mrbayes.csit.fsu.edu/ ; Ronquist & Huelsenbeck, 2003
TREE-PUZZLE	http://www.tree-puzzle.de/ ; Schmidt et al., 2002

Table 4. The list of widely used packages for molecular phylogeny.

6.1 Distance-based methods of phylogeny reconstruction

The distance-based phylogeny reconstruction begins with the computation of pair wise genetic distances between molecular sequences with the use of appropriate substitution model, which is built on the basis of evolutionary assumptions, discussed in section 4. This step results in derivation of a distance matrix, which is subsequently used to infer a tree topology using the clustering method. Fig. 4 shows the distance matrix computed for a sample sequence dataset of 5 OTUs with 6 sites using Jukes-Cantor distance measure. A distance measure possesses three properties, (a) a distance of OTU from itself is zero, D(i, i) = 0; (b) the distance of OTU i from another OTU j must be equal to the distance of OTU j from OTU i, D(i, j) = D(j, i); and (c) the distance measure should follow the triangle inequality rule i.e. D(i, j) ≤ D(i, k) + D(k, j). The accurate estimation of genetic distances is a crucial requirement for the inference of correct phylogenetic tree, thus choice of the right model of evolution is as important as the choice of clustering method. The popular methods used for clustering are UPGMA and N-J.

```
5 6
A    AACAAC
B    AACCAC
C    ACCAAC
D    CACCAT
E    ACACAT
```

	A	B	C	D	E
A	0.000000				
B	0.188486	0.000000			
C	0.188486	0.440840	0.000000		
D	0.823959	0.440840	1.647918	0.000000	
E	1.647918	0.823959	0.823959	0.823959	0.000000

Fig. 4. The distance matrix obtained for a sample nucleotide sequence dataset using Jukes-Cantor model. Dataset contains 5 OTUs (A-E) and 6 sites shown in Phylip format. Dnadist program in PHYLIP package is used to compute distance matrix.

6.1.1 UPGMA method for tree building

The UPGMA method was developed by Sokal and Michener (1958) and is the most widely used clustering methodology. The method is based on the assumptions that the rate of substitution for all branches in the tree is constant (which may not hold true for all data) and branch lengths are additive. It employs hierarchical agglomerative clustering algorithm, which produces ultrametric tree in such a way that every OTU is equidistant from the root.

The clustering process begins with the identification of the highly similar pair of OTUs (i & j) as decided from the distance value D(i, j) in distance matrix. The OTUs i and j are clustered together and combined to form a composite OTU ij. This gives rise to new distance matrix shorter by one row and column than initial distance matrix. The distances of un-clustered OTUs remain unchanged. The distances of remaining OTUs (for e.g. k) from composite OTUs are represented as the average of the initial distances of that OTU from the individual members of composite OTU (i.e. D(ij, k) = [D(i, k) + D(j, k)]/2). In this way a new distance matrix is calculated and in the next round, the OTUs with least dissimilarity are clustered together to form another composite OTU. The remaining steps are same as discussed in the first round. This process of clustering is repeated until all the OTUs are clustered.

The sample calculations and steps involved in UPGMA clustering algorithm using distance matrix shown in Fig. 4 are given below.

Iteration 1: OTU A is minimally equidistant from OTUs B and C. Randomly we select the OTUs A and B to form one composite OTU (AB). A and B are clustered together. Compute new distances of OTUs C, D and E from composite OTU (AB). The distances between unclustered OTUs will be retained. See Fig. 4 for initial distance matrix and Fig. 5 for updated matrix after first iteration of UPGMA.
d(AB,C) = [d(A,C) + d(B,C)]/2 = [0.188486 + 0.440840]/2 = 0.314633
d(AB,D) = [d(A,D) + d(B,D)]/2 = [0.823959 + 0.440840]/2 = 0.632399
d(AB,E) = [d(A,E) + d(B,E)]/2 = [1.647918 + 0.823959]/2 = 1.235938

	AB	C	D	E
AB	0.000000			
C	0.314633	0.000000		
D	0.632399	1.647918	0.000000	
E	1.235938	0.823959	1.647918	0.000000

Fig. 5. The updated distance matrix and clustering of A and B after the 1st iteration of UPGMA.

Iteration 2: OTUs (AB) and C are minimally distant. We select these OTUs to form one composite OTU (ABC). AB and C are clustered together. We then compute new distances of OTUs D and E from composite OTU (ABC). See Fig. 5 for distance matrix obtained in iteration 1 and Fig. 6 for updated matrix after the second iteration of UPGMA.
d(ABC,D) = [d(AB,D) + d(C,D)]/2 = [0.632399 + 1.647918]/2 = 1.140158
d(ABC,E) = [d(AB,E) + d(C,E)]/2 = [1.235938 + 0.823959]/2 = 1.029948

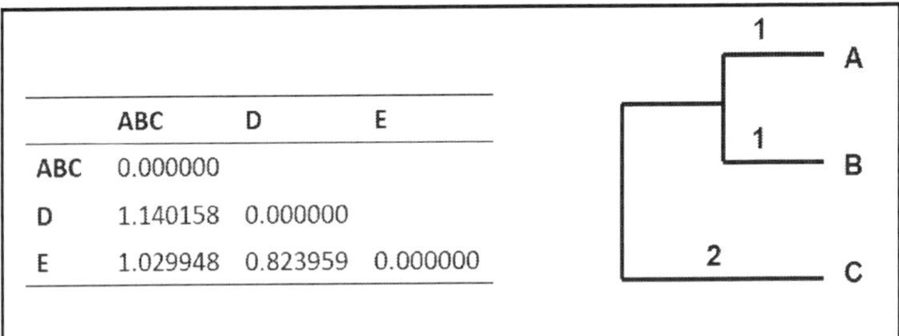

	ABC	D	E
ABC	0.000000		
D	1.140158	0.000000	
E	1.029948	0.823959	0.000000

Fig. 6. The updated distance matrix and clustering of A, B and C after the 2nd iteration of UPGMA.

Iteration 3: OTUs D and E are minimally distant. We select these OTUs to form one composite OTU (DE). D and E are clustered together. Compute new distances of OTUs (ABC) and (DE) from each other. Finally the remaining two OTUs are clustered together. See Fig. 6 for distance matrix obtained in iteration 2 and Fig. 7 for updated matrix after third iteration of UPGMA.
d(ABC,DE) = [d(ABC,D) + d(ABC,E)]/2 = [1.140158 + 1.029948]/2 = 1.085053

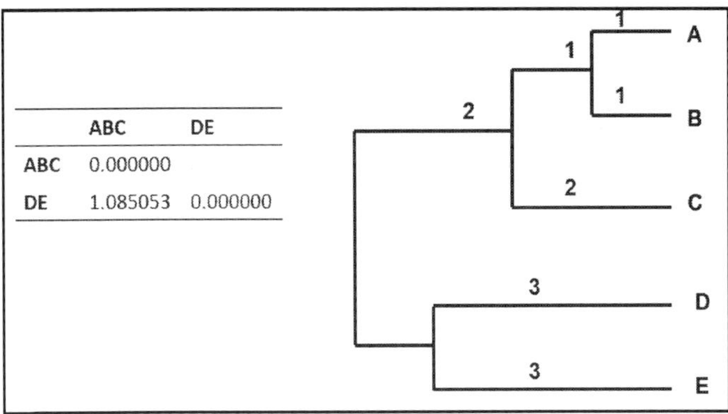

	ABC	DE
ABC	0.000000	
DE	1.085053	0.000000

Fig. 7. The updated distance matrix and clustering of OTUs after the 3rd iteration of UPGMA. Numbers on the branches indicate branch lengths, which are additive.

6.1.2 N-J method for tree building

The N-J method for clustering was developed by Saitou and Nei (1987). It reconstructs the unrooted phylogenetic tree with branch lengths using minimum evolution criterion that minimizes the lengths of tree. It does not assume the constancy of substitution rates across sites and does not require the data to be ultrametric, unlike UPGMA. Hence, this method is more appropriate for the sites with variable rates of evolution.

N-J method is known to be a special case of the star decomposition method. The initial tree topology is a star. The input distance matrix is modified such that the distance between every pair of OTUs is adjusted using their average divergence from all remaining OTUs. The least dissimilar pair of OTUs is identified from the modified distance matrix and is combined together to form single composite OTU. The branch lengths of individual members, clustered in composite OTU, are computed from internal node of composite OTU. Now the distances of remaining OTUs from composite OTU are redefined to give a new distance matrix shorter by one OTU than the initial matrix. This process is repeated till all the OTUs are grouped together, while keeping track of nodes, which results in a final unrooted tree topology with minimized branch lengths. The unrooted phylogenetic tree, thus obtained can be rooted using an outgroup species. The BIONJ (Gascuel 1997), generalized N-J (Pearson et al., 1999) and Weighbor (Bruno et al., 2000) are some of the recently proposed alternative versions of N-J algorithm. The sample calculation and steps involved in N-J clustering algorithm, using distance matrix shown in Fig. 4, are given below.

Iteration 1: Before starting the actual process of clustering the vector r is calculated as following with N=5, refer to the initial distance matrix given in Fig. 4 for reference values.

r(A) = [d(A,B)+ d(A,C)+ d(A,D)+ d(A,E)]/(N-2) = 0.949616

r(B) = [d(B,A)+ d(B,C)+ d(B,D)+ d(B,E)]/(N-2) = 0.631375

r(C) = [d(C,A)+ d(C,B)+ d(C,D)+ d(C,E)]/(N-2) = 1.033755

r(D) = [d(D,A)+ d(D,B)+ d(D,C)+ d(D,E)]/(N-2) = 1.245558

r(E) = [d(E,A)+ d(E,B)+ d(E,C)+ d(E,D)]/(N-2) = 1.373265

Using these r values, we construct a modified distance matrix, Md, such that

MD(i,j) = d(i,j) - (ri + rj).

See Fig. 8 for Md.

Md	A	B	C	D	E
A	0.000000				
B	-1.392505	0.000000			
C	-1.794885	-1.224290	0.000000		
D	-1.371215	-1.436093	-0.631395	0.000000	
E	-0.674963	-1.180681	-1.583061	-1.794864	0.000000

Fig. 8. The modified distance matrix Md and clustering for iteration 1 of N-J.

As can be seen from Md in Fig. 8, OTUs A and C are minimally distant. We select the OTUs A and C to form one composite OTU (AC). A and C are clustered together.

Iteration 2: Compute new distances of OTUs B, D and E from composite OTU (AC). Distances between unclustered OTUs will be retained from the previous step.

d(AC,B) = [d(A,B) + d(C,B)-d(A,C)]/2 = 0.22042

d(AC,D) = [d(A,D) + d(C,D) -d(A,C)]/2 = 1.141695

d(AC,E) = [d(A,E) + d(C,E) -d(A,C)]/2 = 1.141695

Compute r as in the previous step with N=4. See Fig. 9 for new distance matrix and r vector.

d	AC	B	D	E
AC	0.000000			
B	0.220420	0.000000		
D	1.141695	1.647918	0.000000	
E	1.141695	0.823959	1.647918	0.000000

r	
AC	1.251905
B	1.346148
D	2.218765
E	1.806786

Fig. 9. The new distance matrix D and vector r obtained for NJ algorithm iteration 2.

Now, we compute the modified distance matrix, Md as in the previous step and cluster the minimally distant OTUs. See Fig. 10

Md	AC	B	D	E
AC	0.000000			
B	-2.377633	0.000000		
D	-2.327975	-1.916995	0.000000	
E	-1.916996	-2.328975	-2.377633	0.000000

Fig. 10. The modified distance matrix Md, obtained during N-J algorithm iteration 2.

In this step, AC & B and D & E are minimally distant, so we cluster AC with B and D with E. Repeating the above steps we will finally get the following phylogenetic tree, Fig. 11.

Both the distance-based methods, UPGMA and N-J, are computationally faster and hence suited for the phylogeny of large datasets. N-J is the most widely used distance-based method for phylogenetic analysis. The results of these methods are highly dependent on the model of evolution selected a priori.

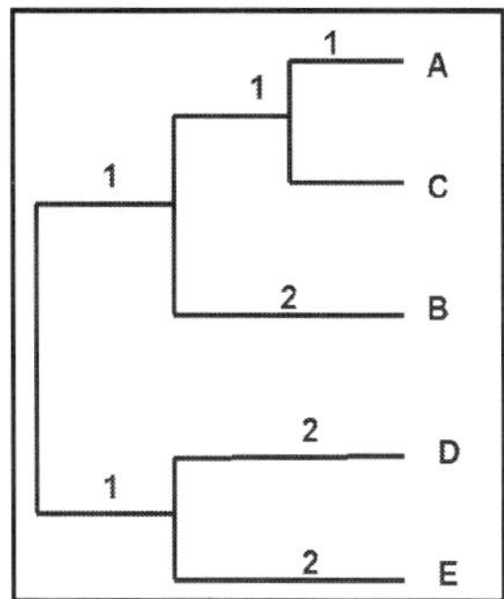

Fig. 11. The phylogenetic tree obtained using N-J algorithm for distance matrix in Fig 4. Numbers on the branches indicate branch length.

6.2 Character-based methods of phylogeny reconstruction

The most commonly used character-based methods in molecular phylogenetics are Maximum parsimony and Maximum likelihood. Unlike the distance-based MPA, character-based methods use character information in alignment data as an input for tree building. The aligned data is in the form of character-state matrix where the nucleotide or amino acid symbols represent the states of characters. These character-based methods employ optimality criterion with the explicit definition of objective function to score the tree topology in order to infer the optimum tree. Hence, these methods are comparatively slower than distance-based clustering algorithms, which are simply based on a set of rules and operations for clustering. But character based methods are advantageous in the sense that they provide a precise mathematical background to prefer one tree over another unlike in distance-based clustering algorithms.

6.2.1 Maximum parsimony

The Maximum parsimony (MP) method is based on the simple principle of searching the tree or collection of trees that minimizes the number of evolutionary changes in the form of change of one character state into other, which are able to describe observed differences in the informative sites of OTUs. There are two problems under the parsimony criterion, a) determining the length of the tree i.e. estimating the number of changes in character states, b) searching overall possible tree topologies to find the tree that involves minimum number of changes. Finally all the trees with minimum number of changes are identified for each of the informative sites. Fitch's algorithm is used for the calculation of changes for a fixed tree topology (Fitch, 1971). If the number of OTUs, N is moderate, this algorithm can be used to calculate the changes for all possible tree topologies and then the most parsimonious rooted tree with minimum number of changes is inferred. However, if N is very large it becomes computationally expensive to calculate the changes for the large number of possible rooted trees. In such cases, a branch and bound algorithm is used to restrict the search space of tree topologies in accordance with Fitch's algorithm to arrive at parsimonious tree (Hendy & Penny, 1982). However, this approach may miss some parsimonious topologies in order to reduce the search space.

An illustrative example of phylogeny analysis using Maximum parsimony is shown in Table 5 and Fig. 12. Table 5 shows a snapshot of MSA of 4 sequences where 5 columns show the aligned nucleotides. Since there are four taxa (A, B, C & D), three possible unrooted trees can be obtained for each site. Out of 5 character sites, only two sites, viz., 4 & 5 are informative i.e. sites having at least two different types of characters (nucleotides/amino acids) with a minimum frequency 2. In the Maximum parsimony method, only informative sites are analysed. Fig. 12 shows the Maximum parsimony phylogenetic analysis of site 5 shown in Table 5. Three possible unrooted trees are shown for site 5 and the tree length is calculated in terms of number of substitutions. Tree II is favoured over trees I and III as it can explain the observed changes in the sequences just with a single substitution. In the same way unrooted trees can be obtained for other informative sites such as site 4. The most parsimonious tree among them will be selected as the final phylogenetic tree. If two or more trees are found and no unique tree can be inferred, trees are said to be equally parsimonious.

OTUs	Character sites				
	1	2	3	4	5
A	A	A	C	A	A
B	A	A	C	C	A
C	A	C	T	A	G
D	C	A	C	C	G

Table 5. Example of phylogenetic analysis from 5 aligned character sites in 4 OTUs using Maximum parsimony method.

Tree I	Tree II	Tree III
Taxon C (G), Taxon D (G), Taxon A (A), Taxon B (A)	Taxon C (G), Taxon A (A), Taxon D (G), Taxon B (A)	Taxon C (G), Taxon D (G), Taxon B (A), Taxon A (A)
Tree length = 2	Tree length = 1	Tree length = 2

• : Substitution

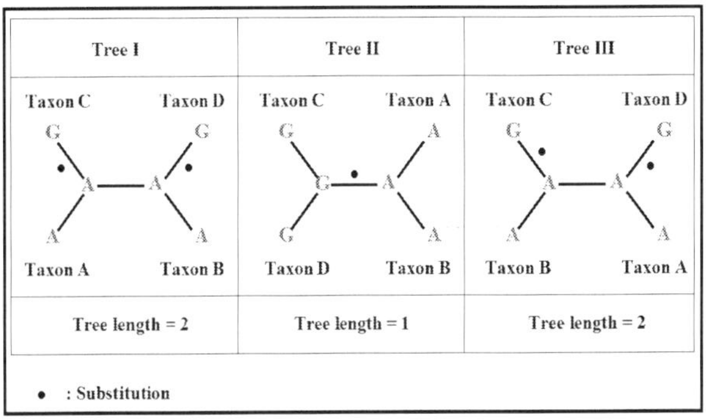

Fig. 12. Example showing various tree topologies based on site 5 in Table 5 using the Maximum parsimony method.

This method is suitable for a small number of sequences with higher similarity and was originally developed for protein sequences. Since this method examines the number of evolutionary changes in all possible trees it is computationally intensive and time consuming. Thus, it is not the method of choice for large sized genome sequences with high variation. The unequal rates of variation in different sites can lead to erroneous parsimony tree with some branches having longer lengths than others as parsimony method assumes the rate of change across all sites to be equal.

6.2.2 Maximum likelihood

As mentioned in the beginning, another character based method for the MPA is the Maximum likelihood method. This method is based on probabilistic approach to phylogeny. This approach is different from the methods discussed earlier. In this method probabilistic models for phylogeny are developed and the tree would be reconstructed using Maximum likelihood method or by sampling method for the given set of sequences. The main difference between this method and some of the available methods discussed before is that it ranks various possible tree topologies according to their likelihood. The same can be obtained by either using the frequentist approach (using the probability (data | tree)) or by using the Baysian approach (likelihood based on the posterior probabilities i.e. by using probability (tree | data)). This method also facilitates computing the likelihood of a sub-tree topology along the branch.

To make the method operative, one must know how to compute $P(x^* \mid T,t^*)$ probability of set of data given tree topology T and set of branch length t*. The tree having maximum probability or the one, which maximizes the likelihood would be chosen as the best tree. The maximization can also be based on the posterior probability P(tree | data) and can be carried out by obtaining required probability using $P(x^* \mid T,t^*)$=P(data | tree) and by applying the Baye's theorem.

The exercise of maximization involves two steps:

a. A search over all possible tree topologies with order of assignment of sequences at the leaves specified.

b. For each topology, a search over all possible lengths of edges in t*

As mentioned in the chapter earlier, the number of rooted trees for given number of sequences (N) grows very rapidly even as N increases to 10. An efficient search procedure for these tasks is required, which was proposed by Felsenstein (1981) and is extensively being used in the MPA. The maximization of likelihood of edge lengths can be carried out using various optimization techniques.

An alternative method is to search stochastically over trees by sampling from posterior distribution $P(T,t^* \mid x^*)$. This method uses techniques such as Monte Carlo method, Gibb's sampling etc. The results of this method are very promising and are often recommended.

Having briefly reviewed the principles, merits and limitations of various methods available for reconstruction of phylogenetic trees using molecular data, it becomes evident that the choice of method for MPA is very crucial. The flowchart shown in Fig. 13 is intended to serve as a guideline to choose a method based on extent of similarity between the sequences. However, it is recommended that one uses multiple methods (at least two) to derive the trees. A few programs have also been developed to superimpose trees to find out similarities in the branching pattern and tree topologies.

7. Assessing the reliability of phylogenetic tree

The assessment of the reliability of phylogenetic tree is an important part of MPA as it helps to decide the relationships of OTUs with a certain degree of confidence assigned by statistical measures. Bootstrap and Jackknife analyses are the major statistical procedures to evaluate the topology of phylogenetic tree (Efron, 1979; Felsenstein, 1985).

In bootstrap technique, the original aligned dataset of sequences is used to generate the finite population of pseudo-datasets by "sampling with replacement" protocol. Each pseudo-dataset is generated by sampling n character sites (columns in the alignment)

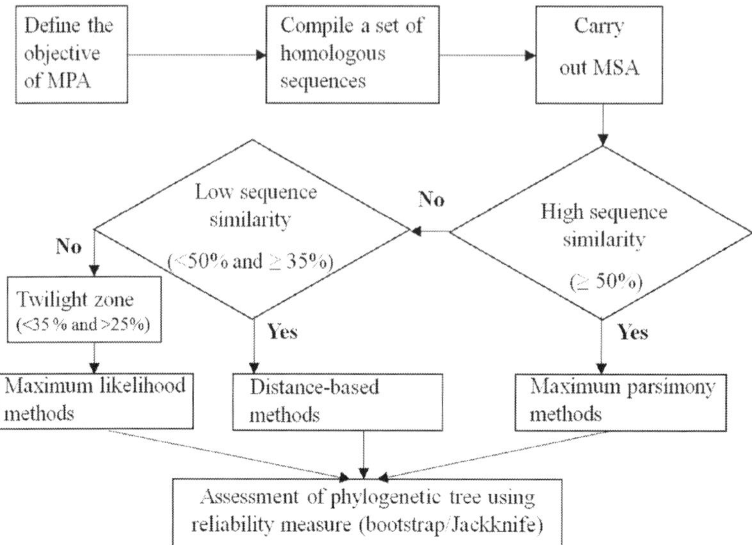

Fig. 13. Flowchart showing the analysis steps involved in phylogenetic reconstruction.

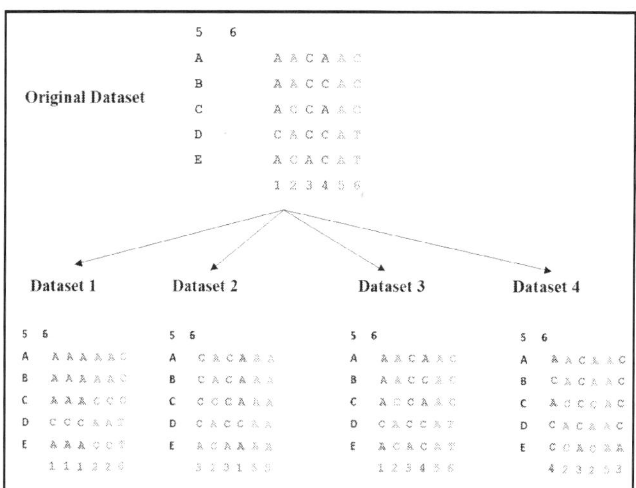

Fig. 14. The procedure to generate pseudo-replicate datasets of original dataset using bootstrap is shown above. The character sites are shown in colour codes at the bottom of datasets to visualize "sampling with replacement protocol".

randomly from original dataset with a possibility of sampling the same site repeatedly, in the process of regular bootstrap. This leads to generation of population of datasets, which are given as an input to tree building methods thus giving rise to population of phylogenetic trees. The consensus phylogenetic tree is then inferred by the majority rule that groups those OTUs, which are found to cluster most of the times in the population of trees. The branches in consensus phylogenetic tree are labelled with bootstrap support values enabling the significance of the relationship of OTUs as depicted using a branching pattern. The procedure for regular bootstrap is illustrated in the Fig. 14. It shows the original dataset along with four pseudo-replicate datasets.

The sites in the original dataset are colour coded to visualize the "sampling with replacement protocol" used in generation of pseudo-replicate datasets 1-4. Seqboot program in PHYLIP package was used for this purpose with choice of regular bootstrap. For example, pseudo-replicate dataset 1 contains the site 1 (red) from original dataset sampled 3 times. In the general practice, usually 100 to 1000 datasets are generated and for each of the datasets phylogenetic tree is obtained. The consensus phylogenetic tree is then obtained by majority rule. The reliability of the consensus tree is assessed from the "branch times" value displayed along the branches of tree.

In Jackknife procedure, the pseudo-datasets are generated by "sampling without replacement" protocol. In this process, sampling (<n) character sites randomly from original dataset generates each pseudo dataset. This leads to generation of population of datasets, which are given as an input to tree building methods thus giving rise to population of phylogenetic trees. The consensus phylogenetic tree is inferred by the majority rule that groups those OTUs, which are found to be clustered most of the times in the population of trees.

8. The case study of Mumps virus phylogeny

We have chosen a case study of Mumps virus (MuV) phylogeny using the amino acid sequences of surface hydrophobic (SH) proteins. There are 12 different known genotypes of MuV, which are designated through A to L, based on the sequence similarity of SH gene sequences. Recently a new genotype of MuV, designated as M, has been identified during parotitis epidemic 2006-2007 in the state of São Paulo, Brazil (Santos et al., 2008). Extensive phylogenetic analysis of newly discovered genotype with existing genotypes of reference strains (A-L) has been used for the confirmation of new genotype using character-based Maximum likelihood method (Santos et al., 2008). In the case study to be presented here, we have used distance-based Neighbor-Joining method with an **objective to re-confirm the presence of new MuV genotype M.** The dataset reported in Santos et al., (2008) is used for the re-confirmation analysis. The steps followed in the MPA are listed below.

a. Compilation and curation of sequences: The sequences of SH protein of the strains of reference genotypes (A to L) as well as newly discovered genotype (M) of MuV were retrieved using GenBank accession numbers as given in Santos et al., (2008). Sequences were saved in Fasta format.
b. Multiple sequence alignment (MSA): SH proteins were aligned using ClustalW (See Fig. 2). MSA was saved in Phylip or interleaved (.phy) format.
c. Bootstrap analysis: 100 pseudo-replicate datasets of the original MSA data (obtained in step b) were generated using regular bootstrap methods in Seqboot program of PHYLIP package.
d. Derivation of distance: The distances between sequences in each dataset were calculated using Dayhoff PAM model assuming uniform rate of variation at all sites. The 'outfile' generated by Seqboot program was used as an input to Protdist program in PHYLIP package.

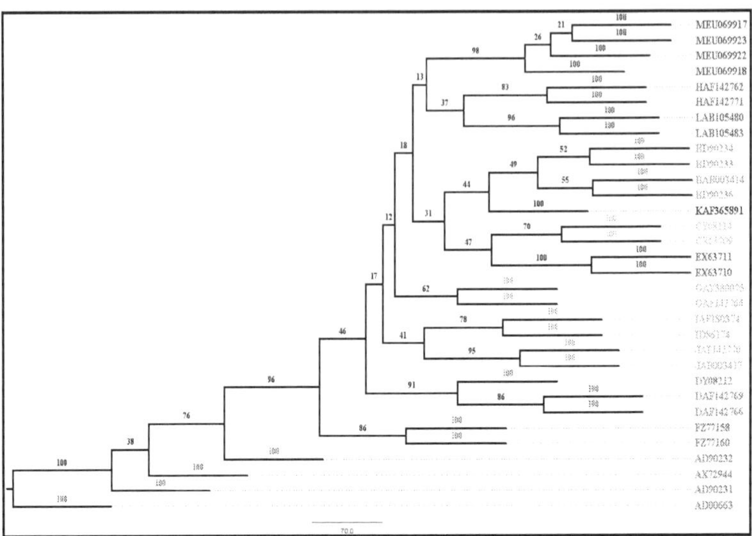

Fig. 15. The unrooted consensus phylogenetic tree obtained for Mumps virus genotypes using Neighbor-Joining method. The first letter in OTU labels indicates the genotype (A-M), which is followed by the GenBank accession numbers for the sequences. The OTUs are also colour coded according to genotypes as following, A: red; B: light blue; C: yellow; D: light green; E: dark blue; F: magenta; G: cyan; H: brick; I: pink; J: orange; K: black; L: dark green; M: purple. All of the genotypes have formed monophyletic clades with high bootstrap support values shown along the branches. The monophyletic clade of M genotypes (with 98 bootstrap support at its base) separated from the individual monophyletic clades of other genotypes (A-L) re-confirms the detection of new genotype M.

e. Building phylogenetic tree: The distance matrices obtained in the previous step were given as an input to N-J method to build phylogenetic trees. The 'outfile' generated by Protdist program containing distance matrices was given as an input to Neighbor program in PHYLIP package.
f. The consensus phylogenetic tree was then obtained using Consense program. For this purpose the 'outtree' file (in Newick format) generated by Neighbor program was given as an input to Consense program.
g. The consensus phylogenetic tree was visualized using FigTree software (available from http://tree.bio.ed.ac.uk/software/figtree/). The consensus unrooted phylogenetic tree is shown in Fig. 15.

The phylogenetic tree for the same dataset was also obtained by using Maximum parsimony method, implemented as the Protpars program in PHYLIP by carrying out MSA and bootstrap as detailed above. The consensus phylogenetic tree is shown in Fig. 16.

Comparison of the trees shown in Fig. 15 & Fig. 16 with that of the published tree re-confirms the emergence of new MuV genotype M during the epidemic in São Paulo, Brazil (Santos et al., 2008), as the members of genotype M have formed a distinct monophyletic clade similar to the known genotypes (A-L). But, a keen observer would note the differences in ordering of clades in the two phylograms obtained using two different methods viz., N-J and MP. For example, the clade of genotype J is close to the clade of genotype I in the N-J phylogram (see Fig. 15) whereas in the MP phylogram (Fig. 16) the clade of genotype J is shown to cluster with the clade of genotype F. Such differences in the ordering of clades are observed some times as these methods (N-J & MP) employ different assumptions and models of evolution. The user can interpret the results with reasonable confidence where the similar clustering pattern of clades is observed in trees drawn using multiple methods. The user, on the other hand, should refrain from over interpretation of sub-tree topolologies, where branching order doesn't match in the trees drawn using different methods. Similarly, a lot of case studies pertaining to the emergence of new species as well as evolution of individual genes/proteins have been published. It is advisable to re-run through a few case studies, which are published, to understand the way in which the respective authors have interpreted the results on the basis of phylogenetic analyses.

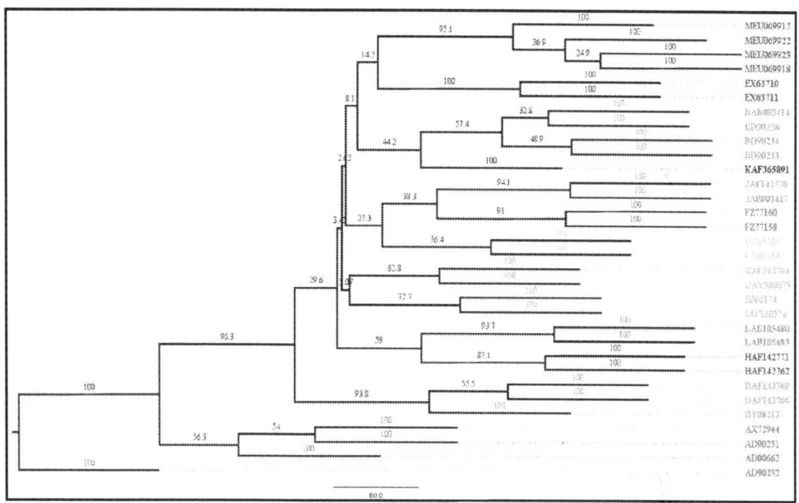

Fig. 16. The unrooted consensus phylogenetic tree obtained for Mumps virus genotypes using Maximum parsimony method. The labelling of OTUs and colour coding is same as in Fig. 15.

9. Challenges and opportunities in phylogenomics

The introduction of next-generation sequencing technology has totally revived the pace of genome sequencing. It has inevitably posed challenges on traditional ways of molecular phylogeny analysis based on single gene, set of genes or markers. Currently the phylogeny based on molecular markers such as 16S rRNA, mitochondrial, nuclear genes etc. provide the taxonomic backbone for Tree of Life (http://tolweb.org/tree/phylogeny.html). But the single gene based phylogeny does not necessarily reflect the phylogenetic history among the genomes of organisms from which these genes are derived. Also the types of evolutionary events such as lateral gene transfer, recombination etc. may not be revealed through the phylogeny of single gene. Thus whole genome based phylogeny analyses become important for deeper understanding of the evolutionary pattern in the organisms (Konstantinidis &

Tiedje, 2007). But whole genome based phylogeny poses many challenges to the traditional methods of MPA, major concerns of them being the size, memory and computational complexity involved in alignment of genomes (Liu et al., 2010).
The methods of MSA developed so far are adequate to handle the requirements of limited amount of data viz. individual gene or protein sequences from various organisms. The increased size of data in terms of the whole genome sequences, however, poses constrains on use and applicability of currently available methods of MSA as they become computationally intensive with requirement of higher memory. The uncertainty associated with alignment procedures, which leads to variations in the inferred phylogeny, has also been pointed out to be the cause of concern (Wong et al., 2008). The benchmark datasets are made available to validate performance of multiple sequence alignment methods (Kemena & Notredame, 2009). These challenges have opened up opportunities for development of alternative approaches for MPA with emergence of alignment-free methods for the same (Kolekar et al., 2010; Sims et al., 2009; Vinga & Almeida, 2003). The field of MPA is also evolving with attempts to develop novel methods based on various data mining techniques viz. Hidden Markov Model (HMM) (Snir & Tuller, 2009), Chaos game theory (Deschavanne et al., 1999), Return Time Distributions (Kolekar et al., 2010) etc. The recent approaches are undergoing refinement and will have to be evaluated with the benchmark datasets before they are routinely used. However, sheer dimensionality of genomic data demands their application. These approaches along with the conventional approaches are extensively reviewed elsewhere (Blair & Murphy, 2010; Wong & Nielsen, 2007).

10. Conclusion

The chapter provides excursion of molecular phylogeny analyses for potential users. It gives an account of available resources and tools. The fundamental principles and salient features of various methods viz. distance-based and character-based are explained with worked out examples. The purpose of the chapter will be served if it enables the reader to develop overall understanding, which is critical to perform such analyses involving real data.

11. Acknowledgment

PSK acknowledges the DBT-BINC junior research fellowship awarded by Department of Biotechnology (DBT), Government of India. UKK acknowledges infrastructural facilities and financial support under the Centre of Excellence (CoE) grant of DBT, Government of India.

12. References

Adachi, J. & Hasegawa, M. (1996) Model of amino acid substitution in proteins encoded by mitochondrial DNA. *Journal of Molecular Evolution* 42(4):459-468.

Aniba, M.; Poch, O. & Thompson J. (2010) Issues in bioinformatics benchmarking: the case study of multiple sequence alignment. *Nucleic Acids Research* 38(21):7353-7363.

Batzoglou, S. (2005) The many faces of sequence alignment. *Briefings in Bioinformatics* 6(1):6-22.

Benson, D.; Karsch-Mizrachi, I.; Lipman, D.; Ostell, J. & Sayers, E. (2011) GenBank. *Nucleic Acids Research* 39(suppl 1):D32-D37.

Blair, C. & Murphy, R. (2010) Recent Trends in Molecular Phylogenetic Analysis: Where to Next? *Journal of Heredity* 102(1):130.

Bruno, W.; Socci, N. & Halpern A. (2000) Weighted neighbor joining: a likelihood-based approach to distance-based phylogeny reconstruction. *Molecular Biology and Evolution* 17(1):189.

Cavalli-Sforza, L. & Edwards, A. (1967) Phylogenetic analysis. Models and estimation procedures. *American Journal of Human Genetics* 19(3 Pt 1):233.

Cole, J.; Wang, Q.; Cardenas, E.; Fish, J.; Chai, B.; Farris, R.; Kulam-Syed-Mohideen, A.; McGarrell, D.; Marsh, T.; Garrity, G. & others. (2009) The Ribosomal Database Project: improved alignments and new tools for rRNA analysis. *Nucleic Acids Research* 37(suppl 1):D141-D145.

Deschavanne, P.; Giron, A.; Vilain, J.; Fagot, G. & Fertil, B. (1999) Genomic signature: characterization and classification of species assessed by chaos game representation of sequences. *Mol Biol Evol* 16(10):1391-9.

Edgar, R.(2004). MUSCLE: a multiple sequence alignment method with reduced time and space complexity. *BMC Bioinformatics* 5:113

Efron, B. (1979) Bootstrap Methods: Another Look at the Jackknife. *Ann. Statist.* 7:1-26.

Felsenstein, J. (1981) Evolutionary trees from DNA sequences: a maximum likelihood approach. *J. Mol Evol* 17:368-376.

Felsenstein, J. (1985) Confidence Limits on Phylogenies: An Approach Using the Bootstrap. *Evolution* 39 783-791.

Felsenstein, J.(1989). PHYLIP-phylogeny inference package (version 3.2). *Cladistics* 5:164-166

Felsenstein, J. (1996) Inferring phylogenies from protein sequences by parsimony, distance, and likelihood methods. *Methods Enzymol* 266:418-27.

Fitch, W. (1971) Toward Defining the Course of Evolution: Minimum Change for a Specific Tree Topology. *Systematic Zoology* 20(4):406-416.

Gascuel, O. (1997) BIONJ: an improved version of the NJ algorithm based on a simple model of sequence data. *Molecular Biology and Evolution* 14(7):685-695.

Hall, T. (1999). BioEdit: A user-friendly biological sequence alignment editor and analysis program for Windows 95/98/NT. *Nucleic Acids Symp Ser* 41:95-98.

Hendy, M. & Penny, D. (1982) Branch and bound algorithms to determine minimal evolutionary trees. *Mathematical Biosciences* 59(2):277-290.

Henikoff, S. & Henikoff, J. (1992) Amino acid substitution matrices from protein blocks. *Proceedings of the National Academy of Sciences of the United States of America* 89(22):10915.

Jones, D.; Taylor, W. & Thornton, J. (1994) A mutation data matrix for transmembrane proteins. *FEBS Letters* 339(3):269-275.

Jukes, T. & Cantor, C. (1969) Evolution of protein molecules. In *"Mammalian Protein Metabolism"*(HN Munro, Ed.). Academic Press, New York.

Kaminuma, E.; Kosuge, T.; Kodama, Y.; Aono, H.; Mashima, J.; Gojobori, T.; Sugawara, H.; Ogasawara, O; Takagi, T.; Okubo, K. & others. (2011). DDBJ progress report. Nucleic Acids Research 39(suppl 1):D22-D27

Katoh, K.; Kuma, K.; Toh, H. & Miyata T (2005) MAFFT version 5: improvement in accuracy of multiple sequence alignment. *Nucleic Acids Res* 33:511 - 518

Kemena, C. & Notredame, C. (2009) Upcoming challenges for multiple sequence alignment methods in the high-throughput era. *Bioinformatics* 25(19):2455-2465.

Kimura, M. (1980) A simple method for estimating evolutionary rates of base substitutions through comparative studies of nucleotide sequences. *Journal of Molecular Evolution* 16(2):111-120.

Kolekar, P.; Kale, M. & Kulkarni-Kale, U. (2010) `Inter-Arrival Time' Inspired Algorithm and its Application in Clustering and Molecular Phylogeny. *AIP Conference Proceedings* 1298(1):307-312.

Konstantinidis, K. & Tiedje, J. (2007) Prokaryotic taxonomy and phylogeny in the genomic era: advancements and challenges ahead. *Current Opinion in Microbiology* 10(5):504-509.

Kuiken, C.; Leitner, T.; Foley, B.; Hahn, B.; Marx, P.; McCutchan, F.; Wolinsky, S.; Korber, B.; Bansal, G. & Abfalterer, W. (2009) HIV sequence compendium 2009. *Document LA-UR*:06-0680

Kuiken, C.; Yusim, K.; Boykin, L. & Richardson, R. (2005) The Los Alamos hepatitis C sequence database. *Bioinformatics* 21(3):379.

Kulkarni-Kale, U.; Bhosle, S.; Manjari, G. & Kolaskar, A. (2004) VirGen: a comprehensive viral genome resource. *Nucleic Acids Research* 32(suppl 1):D289.

Kumar, S.; Nei, M.; Dudley, J. & Tamura, K. (2008) MEGA: a biologist-centric software for evolutionary analysis of DNA and protein sequences. *Brief Bioinform*, 9(4):299-306.

Leinonen, R.; Akhtar, R.; Birney, E.; Bower, L.; Cerdeno-TÃ¡rraga, A.; Cheng, Y.; Cleland, I.; Faruque, N.; Goodgame, N.; Gibson, R. & others. (2011) The European Nucleotide Archive. *Nucleic Acids Research* 39(suppl 1):D28-D31.

Lio, P. & Goldman, N. (1998) Models of molecular evolution and phylogeny. *Genome Res* 8(12):1233-44.

Liu, K.; Linder, C. & Warnow, T. (2010) Multiple sequence alignment: a major challenge to large-scale phylogenetics. *PLoS Currents* 2.

Luo, A.; Qiao, H.; Zhang, Y.; Shi, W.; Ho, S.; Xu, W.; Zhang, A. & Zhu, C. (2010) Performance of criteria for selecting evolutionary models in phylogenetics: a comprehensive study based on simulated datasets. *BMC Evolutionary Biology* 10(1):242.

Morgenstern, B.; French, K.; Dress, A. & Werner, T. (1998). DIALIGN: finding local similarities by multiple sequence alignment. *Bionformatics* 14:290 - 294

Needleman, S. & Wunsch, C. (1970) A general method applicable to the search for similarities in the amino acid sequence of two proteins. *Journal of Molecular Biology* 48(3):443-453.

Nicholas, H.; Ropelewski, A. & Deerfield DW. (2002) Strategies for multiple sequence alignment. *Biotechniques* 32(3):572-591.

Notredame, C.; Higgins, D. & Heringa, J. (2000) T-Coffee: A novel method for fast and accurate multiple sequence alignment. *Journal of Molecular Biology* 302:205 - 217

Parry-Smith, D.; Payne, A.; Michie, A.& Attwood, T. (1998). CINEMA--a novel colour INteractive editor for multiple alignments. *Gene* 221(1):GC57-GC63

Pearson, W.; Robins, G. & Zhang, T. (1999) Generalized neighbor-joining: more reliable phylogenetic tree reconstruction. *Molecular Biology and Evolution* 16(6):806.

Phillips, A. (2006) Homology assessment and molecular sequence alignment. *Journal of Biomedical Informatics* 39(1):18-33.

Ronquist, F. & Huelsenbeck, J. (2003). MrBayes 3: Bayesian phylogenetic inference under mixed models. *Bioinformatics* 19(12):1572-1574

Rzhetsky, A. (1995) Estimating substitution rates in ribosomal RNA genes. *Genetics* 141(2):771.

Saitou, N. & Nei, M. (1987) The neighbor-joining method: a new method for reconstructing phylogenetic trees. *Mol Biol Evol* 4(4):406-25.

Santos, C.; Ishida, M.; Foster, P.; Sallum, M.; Benega, M.; Borges, D.; Corrêa, K.; Constantino, C.; Afzal, M. & Paiva, T. (2008) Detection of a new mumps virus genotype during parotitis epidemic of 2006–2007 in the State of São Paulo, Brazil. *Journal of Medical Virology* 80(2):323-329.

Sayers, E.; Barrett, T.; Benson, D.; Bolton, E.; Bryant, S.; Canese, K.; Chetvernin, V.; Church, D.; DiCuccio, M.; Federhen, S. & others. (2011). Database resources of the National Center for Biotechnology Information. *Nucleic Acids Research* 39(suppl 1):D38-D51.

Schwartz, R. & Dayhoff, M. (1979) Matrices for detecting distant relationships. M. 0. Dayhoff (ed.), *Atlas of protein sequence and structure* 5:353-358.

Schmidt, H.; Strimmer, K.; Vingron, M. & Von Haeseler A. (2002). TREE-PUZZLE: maximum likelihood phylogenetic analysis using quartets and parallel computing. *Bioinformatics* 18(3):502.

Sims, G.; Jun, S.; Wu, G. & Kim, S. (2009) Alignment-free genome comparison with feature frequency profiles (FFP) and optimal resolutions. *Proc Natl Acad Sci U S A* 106(8):2677-82.

Snir, S. & Tuller, T. (2009) The net-hmm approach: phylogenetic network inference by combining maximum likelihood and hidden Markov models. *Journal of bioinformatics and computational biology* 7(4):625-644.

Sokal, R. & Michener, C. (1958) A statistical method for evaluating systematic relationships. *Univ. Kans. Sci. Bull.* 38:1409-1438.

Tamura, K. (1992) Estimation of the number of nucleotide substitutions when there are strong transition-transversion and G+C-content biases. *Molecular Biology and Evolution* 9(4):678-687.

Tamura, K. & Nei, M. (1993) Estimation of the number of nucleotide substitutions in the control region of mitochondrial DNA in humans and chimpanzees. *Molecular Biology and Evolution* 10(3):512-526.

The UniProt Consortium. (2011). Ongoing and future developments at the Universal Protein Resource. *Nucleic Acids Research* 39(suppl 1):D214-D219.

Thompson, J.; Higgins, D. & Gibson, T. (1994) CLUSTAL W: improving the sensitivity of progressive multiple sequence alignment through sequence weighting, position-specific gap penalties and weight matrix choice. *Nucleic Acids Res* 22(22):4673-80.

Thompson, J.; Linard, B.; Lecompte, O. & Poch, O. (2011) A Comprehensive Benchmark Study of Multiple Sequence Alignment Methods: Current Challenges and Future Perspectives. *PLoS ONE* 6(3):e18093.

Thorne, J.; Goldman, N. & Jones, D. (1996) Combining protein evolution and secondary structure. *Molecular Biology and Evolution* 13(5):666-673.

Vinga, S. & Almeida, J. (2003) Alignment-free sequence comparison-a review. *Bioinformatics* 19(4):513-23.

Wilgenbusch, J. & Swofford, D. (2003). Inferring Evolutionary Trees with PAUP*. *Current Protocols in Bioinformatics*. 6.4.1–6.4.28

Wong, K.; Suchard, M. & Huelsenbeck, J. (2008) Alignment Uncertainty and Genomic Analysis. *Science* 319(5862):473-476.

Wong, W. & Nielsen, R. (2007) Finding cis-regulatory modules in Drosophila using phylogenetic hidden Markov models. *Bioinformatics* 23(16):2031-2037.

Xia, X. & Xie, Z. (2001). DAMBE: software package for data analysis in molecular biology and evolution. *Journal of Heredity* 92(4):371

2

Understanding Protein Function - The Disparity Between Bioinformatics and Molecular Methods

Katarzyna Hupert-Kocurek[1] and Jon M. Kaguni[2]
[1]University of Silesia
[2]Michigan State University
[1]Poland
[2]United States of America

1. Introduction

Bioinformatics has its origins in the development of DNA sequencing methods by Alan Maxam and Walter Gilbert (Maxam and Gilbert, 1977), and by Frederick Sanger and coworkers (Sanger et al., 1977). By entirely different approaches, the first genomes determined at the nucleotide sequence level were that of bacteriophage ϕX174, and the recombinant plasmid named pBR322 composed of about 5,400 (Sanger et al., 1977), or 4,400 base pairs (Sutcliffe, 1979), respectively. In contrast, two articles that appeared in February 2001 reported on the preliminary DNA sequence of the human genome, which corresponds to 3 billion nucleotides of DNA sequence information (Lander et al., 2001; Venter et al., 2001). Only two years later, the GenBank sequence database contained more than 29.3 billion nucleotide bases in greater than 23 million sequences. With the development of new technologies, experts predict that the cost to sequence an individual's DNA will be about $1000. This reduction in cost suggests that efforts in the area of comparative genomics will increase substantially, leading to an enormous database that vastly exceeds the existing one.

By way of comparative genomics approaches, computational methods have led to the identification of homologous genes shared among species, and their classification into superfamilies based on amino acid sequence similarity. In combination with their evolutionary relatedness, superfamily members have been clustered into clades. In addition, high throughput sequencing of small RNAs and bioinformatics analyses have contributed to the identification of regions between genes that can code small RNAs (siRNA, microRNA, and long noncoding RNA), which act during the development of an organism to modulate gene expression at the post-transcriptional level (Fire et al., 1998; Hamilton and Baulcombe, 1999) reviewed in Elbashir et al., 2001; Ghildiyal and Zamore, 2009; Christensen et al., 2010). An emerging area is functional genomics whereby gene function is deduced using large-scale methods by identifying the involvement of specific genes in metabolic pathways. More recently, phenotype microarray methods have been used to correlate the functions of genes of microbes with cell phenotypes under a variety of growth conditions (Bochner, 2009).

These methods contrast with the traditional approach of mapping a gene via the phenotype of a mutation, and deducing the function of the gene product based on its biochemical analysis in concert with physiological studies. Such studies have been performed to confirm the functional importance of conserved residues shared by superfamily members, and also to determine the role of specific residues for a given protein. In comparison, comparative genomics methods are unable to distinguish if a nonconserved amino acid among superfamily members is functionally important, or simply reflects sequence divergence due to the absence of selection during evolution. Without functional information, it is not possible to determine if a nonconserved amino acid is important.

2. Bioinformatics analysis of AAA+ proteins

On the basis of bioinformatics analysis, P-loop nucleotide hydrolases compose a very large group of proteins that use an amino acid motif named the phosphate binding loop (P-loop) to hydrolyze the phosphate ester bonds of nucleotides. A positively charged group in the side chain of an amino acid (often lysine) in the P-loop promotes nucleotide hydrolysis by interacting with the phosphate of the bound nucleotide. Additional bioinformatics analysis of this group of proteins led to a category of nucleotidases containing the Walker A and B motifs, as well as additional motifs shared by the AAA (ATPases Associated with diverse cellular Activities) superfamily (Beyer, 1997; Swaffield and Purugganan, 1997). These diverse activities include protein unfolding and degradation, vesicle transport and membrane fusion, transcription and DNA replication. The additional motifs of the AAA superfamily differentiate its members from the larger set of P-loop nucleotidases. Neuwald *et al.*, and Iyer *et al.* then integrated structural information with bioinformatics analysis to classify members of the AAA+ superfamily into clades (Neuwald et al., 1999; Iyer et al., 2004). These clades are the clamp loader clade, the DnaA/CDC6/ORC clade, the classical AAA clade, the HslU/ClpX/Lon/ClpAB-C clade, and the Helix-2 insert clade. The last two clades have been organized into the Pre-sensor 1 hairpin superclade.

Members of the superfamily of AAA+ ATPases carry a nucleotide-binding pocket called the AAA+ domain that ranges from 200 to 250 amino acids, which is formed by an αβα-type Rossmann fold followed by several α helices (Figure 1) (Lupas and Martin, 2002; Iyer et al., 2004; Hanson and Whiteheart, 2005). Such proteins often assemble into ring-shaped or helical oligomers (Davey et al., 2002; Iyer et al., 2004; Erzberger and Berger, 2006). Using the nomenclature of Iyer *et al.*, the Rossmann fold is formed by a β sheet of parallel strands arranged in a β5-β1-β4-β3-β2 series. Its structure resembles a wedge. An α helix preceding the β1 strand and a loop that is situated across the face of the β sheet is a distinguishing feature of the AAA+ superfamily. Another characteristic is the position of several α helices positioned above the wide end of the wedge. The P-loop or the Walker A motif (GX_4GKT/S where X is any amino acid) is located between the β1 strand and the following α helix. The Walker B motif (ϕϕϕϕDE where ϕ is a hydrophobic amino acid) coordinates a magnesium ion complexed with the nucleoside triphosphate via the conserved aspartate residue. The conserved glutamate is thought to interact with a water molecule to make it a better nucleophile for nucleotide hydrolysis.

AAA+ proteins also share conserved motifs named the Sensor 1, Box VII, and Sensor 2 motifs that coordinate ATP hydrolysis with a change in conformation (Figure 1) (Lupas and Martin, 2002; Iyer et al., 2004; Hanson and Whiteheart, 2005). Relative to the primary amino acid sequence, these motifs are on the C-terminal side of the Walker B motif. The Sensor 1 motif contains a polar amino acid at the end of the β4 strand. On the basis of the X-ray crystal structure of N-ethylmaleimide-sensitive factor, an ATPase involved in intracellular vesicle fusion (Beyer, 1997; Swaffield and Purugganan, 1997), this amino acid together with the acidic residue in the Walker B motif interacts with and aligns the activated water molecule during nucleotide hydrolysis. The Box VII motif, which is also called the SRH (Second Region of Homology) motif, contains an arginine named the arginine finger by its analogous function with the corresponding arginine of GTPase activator proteins that interacts with GTP bound to a small G protein partner to promote GTP hydrolysis. The crystal structures of several AAA+ proteins have shown that the Box VII motif in an individual molecule is located some distance away from the nucleotide binding pocket. In AAA+ proteins that assemble into ring-shaped or helical oligomers, the Box VII motif of one protomer directs an arginine residue responsible for interaction with the γ phosphate of ATP toward the ATP binding pocket of the neighboring molecule. It is proposed that this interaction or lack thereof coordinates ATP hydrolysis with a conformational change. The Sensor 2 motif, which resides in one of the α helices that follow the Rossmann fold, also contains a conserved arginine. For proteins whose structures contain the bound nucleoside triphosphate or a nucleotide analogue, this amino acid interacts with the γ phosphate of the nucleotide. As reviewed by Ogura (Ogura et al., 2004), this residue is involved in ATP binding or its hydrolysis in some but not all AAA+ proteins. Like the arginine finger residue, this arginine is thought to coordinate a change in protein conformation with nucleotide hydrolysis.

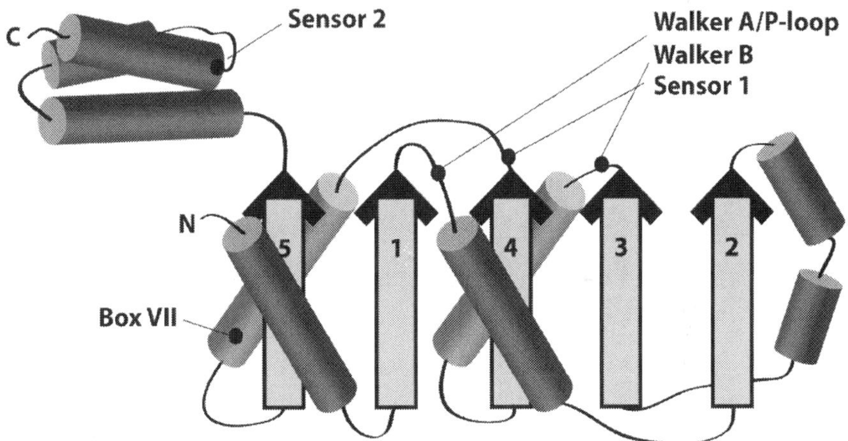

Fig. 1. Structural organization of the AAA+ domain, and the locations of the Walker A/P-loop, Walker B, Sensor 1, Box VII and Sensor 2 motifs are shown (adapted from ref. (Erzberger and Berger, 2006)).

Because this chapter focuses on members of the DnaA/CDC6/ORC or initiator clade, the following summarizes properties of this clade and not others. Like the clamp loader clade, proteins in the initiator clade as represented by DnaA and DnaC have a structure resembling an open spiral on the basis of X-ray crystallography (Erzberger et al., 2006; Mott et al., 2008). In comparison, oligomeric proteins in the remaining clades form closed rings. A characteristic feature of proteins in the initiator clade is the presence of two α helices between the $\beta 2$ and $\beta 3$ strands (Figure 1). Compared with the function of DnaA in the initiation of *E. coli* DNA replication, DnaC plays a separate role. Their functions are described in more detail below. The ORC/CDC6 group of eukaryotic proteins in the initiator clade, like DnaA and DnaC, act to recruit the replicative helicase to replication origins at the stage of initiation of DNA replication (Lee and Bell, 2000; Liu et al., 2000). The origin recognition complex (ORC) is composed of six related proteins named Orc1p through Orc6p, and likely originated along with Cdc6p from a common ancestral gene.

Bioinformatics analysis of DnaC suggests that this protein is a paralog of DnaA, arising by gene duplication and then diverging with time to perform a separate role from DnaA during the initiation of DNA replication (Koonin, 1992). This notion leads to the question of what specific amino acids are responsible for the different functions of DnaA and DnaC despite the shared presence of the AAA+ amino acid sequence motifs. Presumably, specific amino acids that are not conserved between these two proteins have critical roles in determining their different functions, but how are these residues identified and distinguished from those that are not functionally important? In addition, some amino acids that are conserved among homologous DnaC proteins, which were identified by multiple sequence alignment of twenty-eight homologues (Figure 2), are presumable responsible for the unique activities of DnaC, but what are these unique activities? These issues underscore the limitation of deducing the biological function of protein by relying only on bioinformatics analysis.

3. Reverse genetics as an approach to identify the function of an unknown gene

Using various amino acid sequence alignment methods for a particular gene, the postulated function for this gene remains unknown if amino acid sequence homology is not obtained relative to a gene of known function. In such cases, the general approach is to employ reverse genetics to attempt to correlate a phenotype with a mutation in the gene. By way of comparison, forward genetics begins with a phenotype caused by a specific mutation at an unknown site in the genome. The approximate position of the gene can be determined by classical genetic methods that involve its linkage to another mutation that gives rise to a separate phenotype. Refined linkage mapping can localize the gene of interest, followed by PCR (polymerase chain reaction) amplification of the region and DNA sequence analysis to determine the nature of the mutation. As a recent development, whole genome sequencing has been performed to map mutations, dispensing with the classical method of genetic linkage mapping (Lupski et al., 2010; Ng and Kirkness, 2010). The DNA sequence obtained may reveal that the gene and the corresponding gene product have been characterized in the same or different organism, and disclose its physiological function.

In a reverse genetics approach with a haploid organism, the standard strategy is to inactivate the gene with the hope that a phenotype can be measured. Inactivation can be achieved either by deleting the gene or by insertional mutagenesis, usually with a transposon. As examples, transposon mutagenesis has been performed with numerous microbial species, and with *Caenorhabditis elegans* (Vidan and Snyder, 2001; Moerman and Barstead, 2008; Reznikoff and Winterberg, 2008). Using *E. coli* or *S. cerevisiae* as model organisms for gene disruption, one method relies on replacing most of the gene with a drug resistance cassette, or a gene that causes a detectable phenotype. The technique of gene disruption relies on homologous recombination in which the drug resistance gene, for example, has been joined to DNA sequences that are homologous to the ends of the target gene (Figure 3). After introduction of this DNA into the cell, recombination between the ends of the transfected DNA and the homologous regions in the chromosome leads to

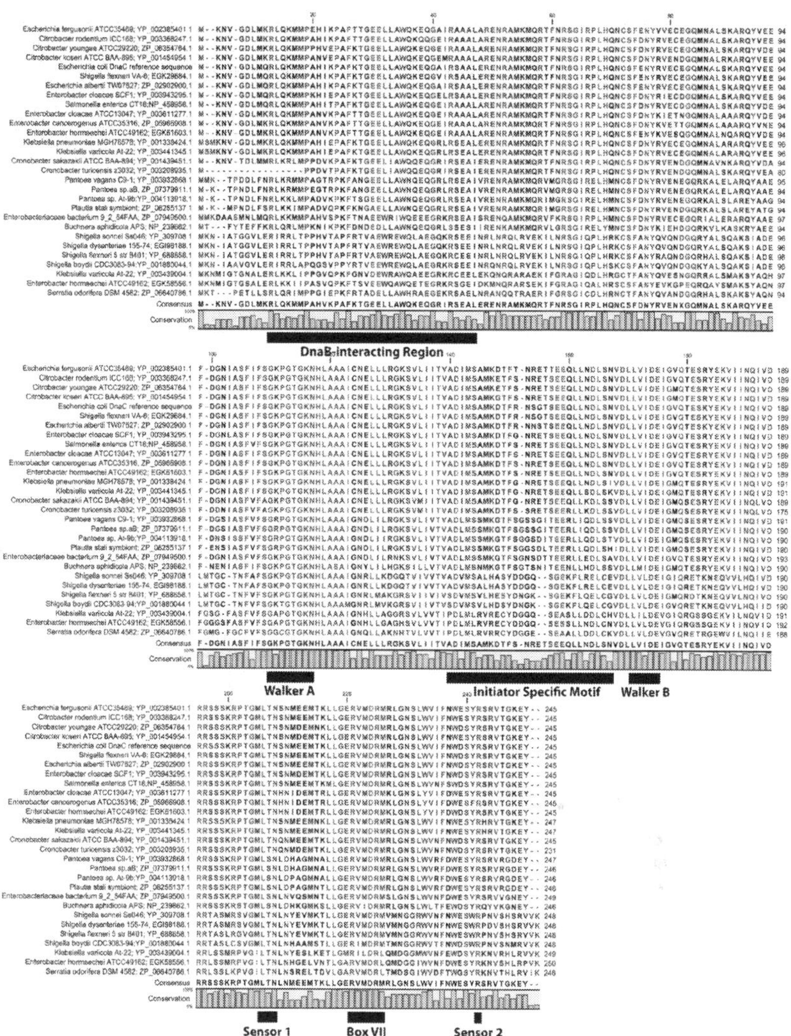

Fig. 2. Multiple sequence alignment of *dnaC* homologues using the Constraint-based Multiple Alignment Tool reveals amino acids that are highly conserved. The Walker A, Walker B, Sensor I, Sensor II and Box VII motifs shared among the AAA+ family of ATPases are shown below the alignment. These motifs are involved in ATP binding, ATP hydrolysis, and coordinating a conformational change with ATP hydrolysis. The figure also shows the region of DnaC near the N-terminus that is involved in interaction with DnaB helicase (Ludlam et al., 2001; Galletto et al., 2003; Mott et al., 2008), and the ISM (Initiator/loader-Specific Motif), which corresponds to two α helices between the β2 and β3 strands, that distinguishes members of the DnaA/CDC6/ORC clade from other AAA+ clades. The ISM causes DnaC assembled as oligomers to form a spiral structure (Mott et al., 2008).

replacement of the chromosomal copy of the gene with the drug resistance cassette, after which the excised copy of the chromosomal gene is lost. In both *E. coli* and *S. cerevisiae*, this approach has been used in seeking to correlate a phenotype with genes of unknown function, and to identify those that are essential for viability (Winzeler et al., 1999; Baba et al., 2006). By either gene disruption or transposon mutagenesis, genetic mapping of the mutation can be performed by inverse PCR where primers complementary to a sequence near the ends of the drug resistance cassette or the transposon are used. This approach first involves partially digesting the chromosomal DNA with a restriction enzyme followed by ligation of the resulting fragments to form a collection of circular DNAs. DNA sequence analysis of the amplified DNA with the primers described above followed by comparison of the nucleotide sequence with the genomic DNA sequence can identify the site of the disrupted gene, or the site of insertion of the transposon.

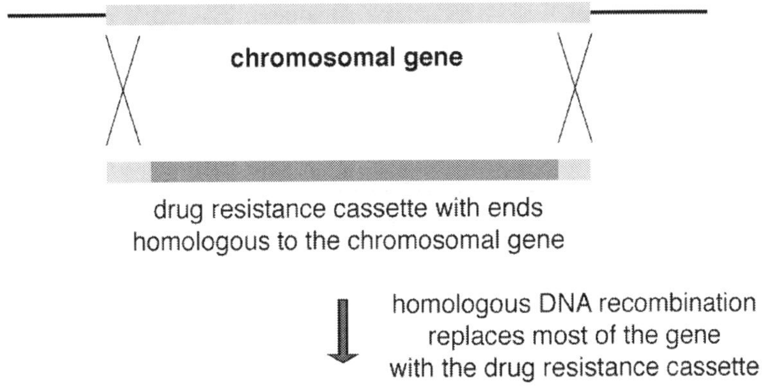

Fig. 3. DNA recombination between the chromosomal gene and homologous DNA at ends of the drug resistance gene leads to replacement of the chromosomal copy of the gene with the drug resistance cassette. The chromosomal gene, and homologous DNA at the ends of the drug resistance cassette are indicated by the lighter shaded rectangles. The drug resistance gene is indicated by the darker rectangles. The thin line represents chromosomal DNA flanking the gene.

With a multicellular organism, a similar strategy that relies on homologous recombination is used to delete a gene. The type of cell to introduce the deletion is usually an embryonic stem cell so that the effect of deletion can be measured in the whole organism. Many eukaryotic organisms have two complete sets of chromosomes. Because the process of homologous recombination introduces the deletion mutation in one of the two pairs of chromosomes, yielding a heterozygote, the presence of the wild type copy on the sister chromosome may conceal the biological effect of the deletion. Thus, the ideal objective is to delete both copies of a gene in order to measure the associated phenotype. To attempt to obtain an organism in which both copies of a gene have been "knocked out," the standard strategy is to mate heterozygous individuals. By Mendelian genetics, one-fourth of the progeny should carry the deletion in both copies of the gene. The drawback with the approach of deleting a gene

is that it may be essential for viability as suggested if a homozygous knockout organism cannot be obtained. Another pitfall is that it may not be possible to construct a heterozygous knockout because the single wild type copy is insufficient to maintain viability. In either case, no other hint of gene function is obtained except for the requirement for life.
Another complication with attempting to determine the role of a eukaryotic gene by deleting it is the existence of gene families where a specific biochemical function is provided by allelic variants. Hence, to correlate a phenotype by introducing a mutation into a specific allelic variant requires inactivation of all other members of the family. A further complication with eukaryotic organisms is that a product formed by an enzyme of interest in one biochemical pathway may be synthesized via an alternate pathway that involves a different set of proteins. In these circumstances, deletion of the gene does not yield a measurable phenotype.
In the event that deletion of a gene is not possible, an alternate approach to characterize the function of an unknown gene is by RNA interference (reviewed in Carthew and Sontheimer, 2009; Fischer, 2010; Krol et al., 2010). This strategy exploits a natural process that acts to repress the expression of genes during development, or as cells progress through the cell cycle (Fire et al., 1998; Ketting et al., 1999; Tabara et al., 1999). Small RNA molecules named microRNA (miRNA) and small interfering RNA (siRNA) become incorporated into a large complex called the RNA-inducing silencing complex (RISC), which reduces the expression of target genes by facilitating the annealing of the RNA with the complementary sequence in a messenger RNA (Liu et al., 2003). The duplex RNA is recognized by a component of the RISC complex, followed by degradation of the messenger RNA to block its expression. The RNA interference pathway has been adapted as a method to reduce or "knockdown" the expression of a specific gene in order to explore its physiological function. Compared with other genetic methods that examine the effect of a specific amino acid substitution on a particular activity of a multifunctional protein, the knockout and knockdown approaches are not as refined in that they measure the physiological effect of either the reduced function, or the loss of function of the entire protein.

4. *E. coli* as a model organism for structure-function studies

Escherichia coli is a rod-shaped bacterium (0.5 micron x 2 microns in the nongrowing state) that harbors a 4.4 x 10^6 base pair genome encoding more than 4,000 genes. By transposon-based insertional mutagenesis and independently by systematic deletion of each open reading frame, these genes have been separated into those that are essential for viability, and those that are considered nonessential (Baba et al., 2006). Of the total, about 300 genes are of undetermined function, including 37 genes that are essential. BLAST analysis indicates that some of the genes of unknown function are conserved among bacteria, suggesting their functional importance.
In comparison, many of the genes of known function have been studied extensively. Among these are the genes required for duplication of the bacterial chromosome, including a subset that acts at the initiation stage of DNA replication. The following section describes a specific example that focuses on DnaC protein. Studies on this protein take advantage of bioinformatics in combination with its X-ray crystallographic structure, molecular genetic analysis, and the biochemical characterization of specific mutant DnaC proteins to obtain new insight into its role in DNA replication.

5. Molecular analysis of *E. coli* DnaC, an essential protein involved in the initiation of DNA replication, and replication fork restart

DNA replication is the basis for life. Occurring only once per cell cycle, DNA replication must be tightly coordinated with other major cellular processes required for cell growth so that each progeny cell receives an accurate copy of the genome at cell division (reviewed in DePamphilis and Bell, 2010). Improper coordination of DNA replication with cell growth leads to aberrant cell division that causes cell death in severe cases. In addition, the failure to control the initiation process leads to excessive initiations, followed by the production of double strand breaks that apparently arise due to head-to-tail fork collisions. In eukaryotes, aneuploidy and chromosome fusions appear if the broken DNA is not fixed that can lead to malignant growth.
In bacteria, chromosomal DNA replication starts at a specific locus called *oriC* (Figure 4).

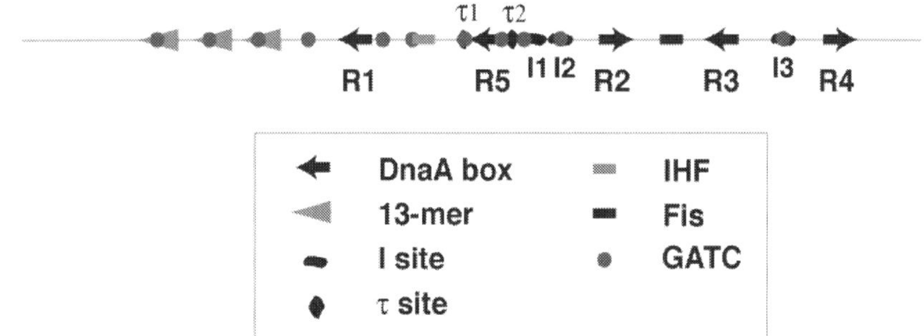

Fig. 4. Organization of DNA sequence motifs in the *E. coli* replication origin (*oriC*). Near the left border are 13-mer motifs that are unwound by DnaA complexed to ATP. Sites recognized by DnaA are the DnaA boxes (arrow), I-sites (warped oval), and τ-sites (warped circle). Sites recognized by IHF (shaded rectangle), Fis (filled rectangle), and DNA adenine methyltransferase (shaded circle) are also indicated.

Recent reviews describe the independent mechanisms that control the frequency of initiation from this site (Nielsen and Lobner-Olesen, 2008; Katayama et al., 2010). In *Escherichia coli*, the minimal *oriC* sequence of 245 base pairs contains DNA-binding sites for many different proteins that either act directly in DNA replication, or modulate the frequency of this process (reviewed in Leonard and Grimwade, 2009). One of them is DnaA, which is the initiator of DNA replication, and has been placed in one of the clades of the AAA+ superfamily via bioinformatics analysis (Koonin, 1992; Erzberger and Berger, 2006). DnaA binds to a consensus 9 base pair sequence known as the DnaA box. There are five DnaA boxes individually named R1 through R5 within *oriC* that are similar in sequence and are recognized by DnaA (Leonard and Grimwade, 2009). In addition to these sites, DnaA complexed to ATP specifically recognizes three I- sites and τ-sites in *oriC*, which leads to the unwinding of three AT-rich 13-mer repeats located in the left half of *oriC*. Binding sites are also present for IHF protein (integration host factor) and FIS protein (factor for inversion stimulation). As these proteins induce bends in DNA, their apparent ability to modulate the binding of DnaA to the respective sites in *oriC* may involve DNA bending. Additionally, *oriC* carries 11 GATC sequences recognized by DNA adenine methyltransferase, and sites recognized by IciA, Rob, and SeqA. The influence of IHF, FIS, IciA, Rob and SeqA proteins on the initiation process is described in more detail in a review (Leonard and Grimwade, 2005).

At the initiation stage of DNA replication, the first step requires the binding of DnaA molecules, each complexed to ATP, to the five DnaA boxes, I- and τ- sites of *oriC*. After binding, DnaA unwinds the duplex DNA in the AT-rich region to form an intermediate named the open complex. HU or IHF stimulates the formation of the open complex. In the next step, the replicative helicase named DnaB becomes stably bound to the separated strands of the open complex to form an intermediate named the prepriming complex. At this juncture, DnaC must be complexed to DnaB for a single DnaB hexamer to load onto each of the separated strands. DnaC protein must then dissociate from the complex in order for DnaB to be active as a helicase. Following the loading of DnaB, this helicase enlarges the unwound region of *oriC*, and then interacts with DnaG primase (Tougu and Marians, 1996). This interaction between DnaB and DnaG primase, which synthesize primer RNAs that are extended by DNA polymerase III holoenzyme during semi-conservative DNA replication, marks the transition between the process of initiation and elongation stage of DNA replication (McHenry, 2003; Corn and Berger, 2006; Langston et al., 2009). Replication fork movement that is supported by DnaB helicase and assisted by a second helicase named Rep proceeds bidirectionally around the chromosome until it reaches the terminus region (Guy et al., 2009). The two progeny DNAs then segregate near opposite poles of the cell before septum formation and cell division.

DnaC protein (27 kDa) is essential for cell viability because it is required during the initiation stage of DNA replication (reviewed in Kornberg and Baker, 1992; Davey and O'Donnell, 2003). DnaC is also required for DNA replication of the single stranded DNA of phage ϕX174, and for many plasmids (e.g. pSC101, P1, R1). DnaC additionally acts to resurrect collapsed replication forks that appear when a replication fork encounters a nick, gap, double-stranded break, or modified bases in the parental DNA (Sandler et al., 1996). This process of restarting a replication fork involves assembly of the replication restart primosome that contains PriA, PriB, PriC, DnaT, DnaB, DnaC, and Rep protein (Sandler, 2005; Gabbai and Marians, 2010). The major roles of DnaC at *oriC*, at the replication origins of the plasmids and bacteriophage described above, or in restarting collapsed replication forks is to form a complex with DnaB, which is required to escort the helicase onto the DNA, and then to depart. Since the discovery of the *dnaC* gene over 40 years ago (Carl, 1970), its ongoing study by various laboratories using a variety of approaches continue to reveal new aspects of the molecular mechanisms of DnaC in DNA replication.

Biochemical analysis combined with the X-ray crystallographic structure of the majority of *Aquifex aeolicus* DnaC (residues 43 to the C-terminal residue at position 235) reveals that DnaC protein consists of a smaller N-terminal domain that is responsible for binding to the C- terminal face of DnaB helicase, and larger ATP-binding region of 190 amino acids (Figure 2; (Ludlam et al., 2001; Galletto et al., 2003; Mott et al., 2008)). Sequence comparison of homologues of the *dnaC* gene classifies DnaC as a member of the AAA+ family of ATPases (Koonin, 1992; Davey et al., 2002; Mott et al., 2008). However unlike other AAA+ proteins, DnaC contains two additional α helices named the ISM motif (Initiator/loader-Specific Motif) that directs the oligomerization of this protein into a right-handed helical filament (Mott et al., 2008). In contrast, the majority AAA+ proteins lacking these α helices assemble into a closed-ring. Phylogenetic analysis of the AAA+ domain reveals that DnaC is most

closely related to DnaA, suggesting that both proteins arose from a common ancestor (Koonin, 1992). In support, the X-ray crystallographic structures of the ATPase region of DnaA and DnaC are very similar (Erzberger et al., 2006; Mott et al., 2008).
For DnaC, ATP increases its affinity for single-stranded DNA, which stimulates its ATPase activity (Davey et al., 2002; Biswas et al., 2004). Other results suggest that ATP stabilizes the interaction of DnaC with DnaB in the DnaB-DnaC complex (Wahle et al., 1989; Allen and Kornberg, 1991), which contradicts studies that support the contrary conclusion that ATP is not necessary for DnaC to form a stable complex with DnaB (Davey et al., 2002; Galletto et al., 2003; Biswas and Biswas-Fiss, 2006). As mutant DnaC proteins bearing amino acid substitutions in the Walker A box are both defective in ATP binding and apparently fail to interact with DnaB, the consequence is that these mutants cannot escort DnaB to *oriC* (Ludlam et al., 2001; Davey et al., 2002). Hence, despite the ability of DnaB by itself to bind to single-stranded DNA in vitro (LeBowitz and McMacken, 1986), DnaC is essential for DnaB to become stably bound to the unwound region of *oriC* (Kobori and Kornberg, 1982; Ludlam et al., 2001). The observation that DnaC complexed to ATP interacts with DnaA raises the possibility that both proteins act jointly in helicase loading (Mott et al., 2008). Together, these observations indicate that the ability of DnaC to bind to ATP is essential for its function in DNA replication, but the paradox about the role of ATP binding and its hydrolysis on the activity of DnaC and about the mechanism that leads to the dissociation of DnaC from DnaB have been long-standing issues.

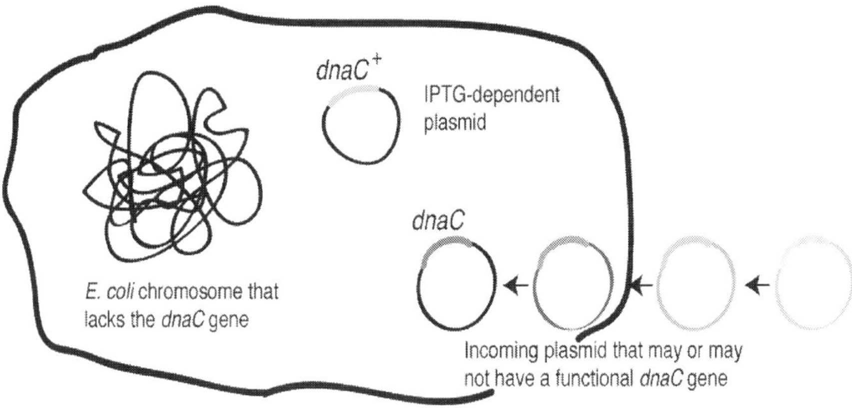

Fig. 5. Plasmid shuffle method. With an *E. coli* strain lacking the chromosomal copy of *dnaC*, a plasmid carrying the wild type *dnaC* gene residing in the bacterial cell can provide for *dnaC* function. Using a plasmid that depends on IPTG for its maintenance, the strain will not survive in the absence of IPTG. An introduced plasmid that does not depend on IPTG for its propagation can sustain viability in the absence of IPTG only if it encodes a functional *dnaC* allele.

As described above, one of the characteristics of AAA+ proteins is the presence of a conserved motif named Box VII, which carries a conserved arginine called the "arginine finger". Structural studies of other AAA+ proteins have led to the proposal that this arginine interacts with the γ phosphate of ATP to promote and coordinate ATP hydrolysis with a conformational change. Recent experiments were performed to examine the role of the arginine finger of DnaC and to attempt to clarify how ATP binding and its hydrolysis by DnaC are involved in the process of initiation of DNA replication (Makowska-Grzyska and Kaguni, 2010). Part of this study relied on an *E. coli* mutant lacking the chromosomal copy of the *dnaC* gene (Hupert-Kocurek et al., 2007). Because the *dnaC* gene is essential, this deficiency of the host strain can be complemented by a plasmid encoding the *dnaC* gene that depends on IPTG (isopropyl β-D-1-thiogalactopyranoside) for plasmid maintenance (Figure 5). If another plasmid is introduced into the null *dnaC* mutant, it maintains viability of the strain in the absence of IPTG only if it carries a functional *dnaC* allele. In contrast, if the second plasmid carries an inactivating mutation in *dnaC*, the host strain cannot survive without IPTG. This plasmid exchange method showed that an alanine substitution for the arginine finger residue inactivated DnaC (Makowska-Grzyska and Kaguni, 2010). Biochemical experiments performed in parallel showed that this conserved arginine plays a role in the signal transduction process that involves ATP hydrolysis by DnaC that then leads to the release of DnaC from DnaB. Finally, the interaction of primase with DnaB that is coupled with primer formation is also apparently necessary for DnaC to dissociate from DnaB.

6. Conclusions

In summary, the combination of various experimental approaches on the study of DnaC have led to insightful experiments that expand our understanding of the role of ATP binding and its hydrolysis by DnaC during the initiation of DNA replication. Evidence suggests that ATP hydrolysis by DnaC that leads to the dissociation of DnaC from DnaB helicase is coupled with primer formation that requires an interaction between DnaG primase and DnaB. Hence, these critical steps are involved in the transition from the process of initiation to the elongation phase of DNA replication in *E. coli*.

This example on the molecular mechanism of DnaC protein is a focused study of one protein and its interaction with other required proteins during the process of initiation of DNA replication. One may consider this a form of vertical thinking. It contrasts with bioinformatics approaches that yield large sets of data for proteins based on the DNA sequences of genomes, and with microarray approaches that, for example, survey the expression of genes and their regulation at the genome level under different conditions, or identify interacting partners for a specific protein. The vast wealth of data from these global approaches provide a different perspective on understanding the functions of sets of genes or proteins and how they act in a network of biochemical pathways of the cell.

7. Acknowledgements

We thank members of our labs for discussions on the content and organization of this chapter. This work was supported by a grant GM090063 from the National Institutes of Health, and the Michigan Agricultural Station to JMK.

8. References

Allen, G. J. and A. Kornberg (1991). Fine balance in the regulation of DnaB helicase by DnaC protein in replication in Escherichia coli. J Biol Chem 266(33): 22096-22101

Baba, T., T. Ara, M. Hasegawa, Y. Takai, Y. Okumura, M. Baba, K. A. Datsenko, M. Tomita, B. L. Wanner and H. Mori (2006). Construction of *Escherichia coli* K-12 in-frame, single-gene knockout mutants: the Keio collection. Mol Syst Biol 2: 1-11

Beyer, A. (1997). Sequence analysis of the AAA protein family. Protein Sci 6(10): 2043-58, 0961-8368 (Print), 0961-8368 (Linking)

Biswas, S. B. and E. E. Biswas-Fiss (2006). Quantitative analysis of binding of single-stranded DNA by *Escherichia coli* DnaB helicase and the DnaB-DnaC complex. Biochemistry 45(38): 11505-11513

Biswas, S. B., S. Flowers and E. E. Biswas-Fiss (2004). Quantitative analysis of nucleotide modulation of DNA binding by the DnaC protein of *Escherichia* coli. Biochem J 379: 553-562

Bochner, B. R. (2009). Global phenotypic characterization of bacteria. FEMS Microbiol Rev 33(1): 191-205, 0168-6445 (Print), 0168-6445 (Linking)

Carl, P. L. (1970). *Escherichia coli* mutants with temperature-sensitive synthesis of DNA. Mol Gen Genet 109(2): 107-122

Carthew, R. W. and E. J. Sontheimer (2009). Origins and Mechanisms of miRNAs and siRNAs. Cell 136(4): 642-55, 1097-4172 (Electronic), 0092-8674 (Linking)

Christensen, N. M., K. J. Oparka and J. Tilsner (2010). Advances in imaging RNA in plants. Trends Plant Sci 15(4): 196-203, 1878-4372 (Electronic), 1360-1385 (Linking)

Corn, J. E. and J. M. Berger (2006). Regulation of bacterial priming and daughter strand synthesis through helicase-primase interactions. Nucleic Acids Res 34(15): 4082-4088

Davey, M. J., L. Fang, P. McInerney, R. E. Georgescu and M. O'Donnell (2002). The DnaC helicase loader is a dual ATP/ADP switch protein. EMBO J 21(12): 3148-3159

Davey, M. J., D. Jeruzalmi, J. Kuriyan and M. O'Donnell (2002). Motors and switches: AAA+ machines within the replisome. Nat Rev Mol Cell Biol 3(11): 826-835

Davey, M. J. and M. O'Donnell (2003). Replicative helicase loaders: ring breakers and ring makers. Curr Biol 13(15): R594-596

DePamphilis, M. L. and S. D. Bell (2010). Genome duplication, Garland Science/Taylor & Francis Group, 9780415442060, London

Elbashir, S. M., J. Harborth, W. Lendeckel, A. Yalcin, K. Weber and T. Tuschl (2001). Duplexes of 21-nucleotide RNAs mediate RNA interference in cultured mammalian cells. Nature 411(6836): 494-8, 0028-0836 (Print), 0028-0836 (Linking)

Erzberger, J. P. and J. M. Berger (2006). Evolutionary relationships and structural mechanisms of AAA+ proteins. Annu Rev Biophys Biomol Struct 35: 93-114

Erzberger, J. P., M. L. Mott and J. M. Berger (2006). Structural basis for ATP-dependent DnaA assembly and replication-origin remodeling. Nat Struct Mol Biol 13(8): 676-683

Fire, A., S. Xu, M. K. Montgomery, S. A. Kostas, S. E. Driver and C. C. Mello (1998). Potent and specific genetic interference by double-stranded RNA in *Caenorhabditis elegans*. Nature 391(6669): 806-11, 0028-0836 (Print), 0028-0836 (Linking)

Fischer, S. E. (2010). Small RNA-mediated gene silencing pathways in *C. elegans*. Int J Biochem Cell Biol 42(8): 1306-15, 1878-5875 (Electronic), 1357-2725 (Linking)

Gabbai, C. B. and K. J. Marians (2010). Recruitment to stalled replication forks of the PriA DNA helicase and replisome-loading activities is essential for survival. DNA Repair (Amst) 9(3): 202-209, 1568-7856 (Electronic), 1568-7856 (Linking)

Galletto, R., M. J. Jezewska and W. Bujalowski (2003). Interactions of the *Escherichia coli* DnaB Helicase Hexamer with the Replication Factor the DnaC Protein. Effect of Nucleotide Cofactors and the ssDNA on Protein-Protein Interactions and the Topology of the Complex. J Mol Biol 329(3): 441-465

Ghildiyal, M. and P. D. Zamore (2009). Small silencing RNAs: an expanding universe. Nat Rev Genet 10(2): 94-108, 1471-0064 (Electronic), 1471-0056 (Linking)

Guy, C. P., J. Atkinson, M. K. Gupta, A. A. Mahdi, E. J. Gwynn, C. J. Rudolph, P. B. Moon, I. C. van Knippenberg, C. J. Cadman, M. S. Dillingham, R. G. Lloyd and P. McGlynn (2009). Rep provides a second motor at the replisome to promote duplication of protein-bound DNA. Mol Cell 36(4): 654-666, 1097-4164 (Electronic), 1097-2765 (Linking)

Hamilton, A. J. and D. C. Baulcombe (1999). A species of small antisense RNA in posttranscriptional gene silencing in plants. Science 286(5441): 950-952, 0036-8075 (Print), 0036-8075 (Linking)

Hanson, P. I. and S. W. Whiteheart (2005). AAA+ proteins: have engine, will work. Nat Rev Mol Cell Biol 6(7): 519-529

Hupert-Kocurek, K., J. M. Sage, M. Makowska-Grzyska and J. M. Kaguni (2007). Genetic method to analyze essential genes of *Escherichia coli*. Appl Environ Microbiol 73(21): 7075-7082

Iyer, L. M., D. D. Leipe, E. V. Koonin and L. Aravind (2004). Evolutionary history and higher order classification of AAA+ ATPases. J Struct Biol 146(1-2): 11-31, 1047-8477 (Print), 1047-8477 (Linking)

Katayama, T., S. Ozaki, K. Keyamura and K. Fujimitsu (2010). Regulation of the replication cycle: conserved and diverse regulatory systems for DnaA and *oriC*. Nat Rev Microbiol 8(3): 163-170, 1740-1534 (Electronic), 1740-1526 (Linking)

Ketting, R. F., T. H. Haverkamp, H. G. van Luenen and R. H. Plasterk (1999). Mut-7 of *C. elegans*, required for transposon silencing and RNA interference, is a homolog of Werner syndrome helicase and RNaseD. Cell 99(2): 133-141, 0092-8674 (Print), 0092-8674 (Linking)

Kobori, J. A. and A. Kornberg (1982). The *Escherichia coli dnaC* gene product. II. Purification, physical properties, and role in replication. J Biol Chem 257(22): 13763-13769

Koonin, E. V. (1992). DnaC protein contains a modified ATP-binding motif and belongs to a novel family of ATPases including also DnaA. Nucleic Acids Res 20(8): 1997

Kornberg, A. and T. A. Baker (1992). DNA Replication Second Edition, W.H. Freeman and Company, 9781891389443, New York

Krol, J., I. Loedige and W. Filipowicz (2010). The widespread regulation of microRNA biogenesis, function and decay. Nat Rev Genet 11(9): 597-610, 1471-0064 (Electronic), 1471-0056 (Linking)

Lander, E. S., L. M. Linton, B. Birren, C. Nusbaum, M. C. Zody, J. Baldwin, K. Devon, K. Dewar, M. Doyle, W. FitzHugh, R. Funke, D. Gage, K. Harris, A. Heaford, J. Howland, L. Kann, J. Lehoczky, R. LeVine, P. McEwan, K. McKernan, J. Meldrim, J. P. Mesirov, C. Miranda, W. Morris, J. Naylor, C. Raymond, M. Rosetti, R. Santos, A. Sheridan, C. Sougnez, N. Stange-Thomann, N. Stojanovic, A. Subramanian, D. Wyman, J. Rogers, J. Sulston, R. Ainscough, S. Beck, D. Bentley, J. Burton, C. Clee, N. Carter, A. Coulson, R. Deadman, P. Deloukas, A. Dunham, I. Dunham, R. Durbin, L. French, D. Grafham, S. Gregory, T. Hubbard, S. Humphray, A. Hunt, M. Jones, C. Lloyd, A. McMurray, L. Matthews, S. Mercer, S. Milne, J. C. Mullikin, A. Mungall, R. Plumb, M. Ross, R. Shownkeen, S. Sims, R. H. Waterston, R. K. Wilson, L. W. Hillier, J. D. McPherson, M. A. Marra, E. R. Mardis, L. A. Fulton, A. T. Chinwalla, K. H. Pepin, W. R. Gish, S. L. Chissoe, M. C. Wendl, K. D. Delehaunty, T. L. Miner, A. Delehaunty, J. B. Kramer, L. L. Cook, R. S. Fulton, D. L. Johnson, P. J. Minx, S. W. Clifton, T. Hawkins, E. Branscomb, P. Predki, P. Richardson, S. Wenning, T. Slezak, N. Doggett, J. F. Cheng, A. Olsen, S. Lucas, C. Elkin, E. Uberbacher, M. Frazier, R. A. Gibbs, D. M. Muzny, S. E. Scherer, J. B. Bouck, E. J. Sodergren, K. C. Worley, C. M. Rives, J. H. Gorrell, M. L. Metzker, S. L. Naylor, R. S. Kucherlapati, D. L. Nelson, G. M. Weinstock, Y. Sakaki, A. Fujiyama, M. Hattori, T. Yada, A. Toyoda, T. Itoh, C. Kawagoe, H. Watanabe, Y. Totoki, T. Taylor, J. Weissenbach, R. Heilig, W. Saurin, F. Artiguenave, P. Brottier, T. Bruls, E. Pelletier, C. Robert, P. Wincker, D. R. Smith, L. Doucette-Stamm, M. Rubenfield, K. Weinstock, H. M. Lee, J. Dubois, A. Rosenthal, M. Platzer, G. Nyakatura, S. Taudien, A. Rump, H. Yang, J. Yu, J. Wang, G. Huang, J. Gu, L. Hood, L. Rowen, A. Madan, S. Qin, R. W. Davis, N. A. Federspiel, A. P. Abola, M. J. Proctor, R. M. Myers, J. Schmutz, M. Dickson, J. Grimwood, D. R. Cox, M. V. Olson, R. Kaul, N. Shimizu, K. Kawasaki, S. Minoshima, G. A. Evans, M. Athanasiou, R. Schultz, B. A. Roe, F. Chen, H. Pan, J. Ramser, H. Lehrach, R. Reinhardt, W. R. McCombie, M. de la Bastide, N. Dedhia, H. Blocker, K. Hornischer, G. Nordsiek, R. Agarwala, L. Aravind, J. A. Bailey, A. Bateman, S. Batzoglou, E. Birney, P. Bork, D. G. Brown, C. B. Burge, L. Cerutti, H. C. Chen, D. Church, M. Clamp, R. R. Copley, T. Doerks, S. R. Eddy, E. E. Eichler, T. S. Furey, J. Galagan, J. G. Gilbert, C. Harmon, Y. Hayashizaki, D. Haussler, H. Hermjakob, K. Hokamp, W. Jang, L. S. Johnson, T. A. Jones, S. Kasif, A. Kaspryzk, S. Kennedy, W. J. Kent, P. Kitts, E. V. Koonin, I. Korf, D. Kulp, D. Lancet, T. M. Lowe, A. McLysaght, T. Mikkelsen, J. V. Moran, N. Mulder, V. J. Pollara, C. P. Ponting, G. Schuler, J. Schultz, G. Slater, A. F. Smit, E. Stupka, J. Szustakowski, D. Thierry-Mieg, J. Thierry-Mieg, L. Wagner, J. Wallis, R. Wheeler, A. Williams, Y. I. Wolf, K. H. Wolfe, S. P. Yang, R. F. Yeh, F. Collins, M. S. Guyer, J. Peterson, A. Felsenfeld, K. A. Wetterstrand, A. Patrinos, M. J. Morgan, P. de Jong, J. J. Catanese, K. Osoegawa, H. Shizuya, S. Choi and Y. J. Chen (2001).

Initial sequencing and analysis of the human genome. Nature 409(6822): 860-921, 0028-0836 (Print), 0028-0836 (Linking)

Langston, L. D., C. Indiani and M. O'Donnell (2009). Whither the replisome: emerging perspectives on the dynamic nature of the DNA replication machinery. Cell Cycle 8(17): 2686-2691, 1551-4005 (Electronic), 1551-4005 (Linking)

LeBowitz, J. H. and R. McMacken (1986). The *Escherichia coli* dnaB replication protein is a DNA helicase. J Biol Chem 261(10): 4738-48,

Lee, D. G. and S. P. Bell (2000). ATPase switches controlling DNA replication initiation. Curr Opin Cell Biol 12(3): 280-285, 0955-0674 (Print), 0955-0674 (Linking)

Leonard, A. C. and J. E. Grimwade (2005). Building a bacterial orisome: emergence of new regulatory features for replication origin unwinding. Mol Microbiol 55(4): 978-985

Leonard, A. C. and J. E. Grimwade (2009). Initiating chromosome replication in *E. coli*: it makes sense to recycle. Genes Dev 23(10): 1145-1150,

Liu, J., C. L. Smith, D. DeRyckere, K. DeAngelis, G. S. Martin and J. M. Berger (2000). Structure and function of Cdc6/Cdc18: implications for origin recognition and checkpoint control. Mol Cell 6(3): 637-648

Liu, Q., T. A. Rand, S. Kalidas, F. Du, H. E. Kim, D. P. Smith and X. Wang (2003). R2D2, a bridge between the initiation and effector steps of the Drosophila RNAi pathway. Science 301(5641): 1921-1925, 1095-9203 (Electronic), 0036-8075 (Linking)

Ludlam, A. V., M. W. McNatt, K. M. Carr and J. M. Kaguni (2001). Essential amino acids of *Escherichia coli* DnaC protein in an N-terminal domain interact with DnaB helicase. J Biol Chem 276(29): 27345-27353

Lupas, A. N. and J. Martin (2002). AAA proteins. Curr Opin Struct Biol 12(6): 746-53, 0959-440X (Print), 0959-440X (Linking)

Lupski, J. R., J. G. Reid, C. Gonzaga-Jauregui, D. Rio Deiros, D. C. Chen, L. Nazareth, M. Bainbridge, H. Dinh, C. Jing, D. A. Wheeler, A. L. McGuire, F. Zhang, P. Stankiewicz, J. J. Halperin, C. Yang, C. Gehman, D. Guo, R. K. Irikat, W. Tom, N. J. Fantin, D. M. Muzny and R. A. Gibbs (2010). Whole-genome sequencing in a patient with Charcot-Marie-Tooth neuropathy. N Engl J Med 362(13): 1181-91, 1533-4406 (Electronic), 0028-4793 (Linking)

Makowska-Grzyska, M. and J. M. Kaguni (2010). Primase Directs the Release of DnaC from DnaB. Mol Cell 37(1): 90-101

Maxam, A. M. and W. Gilbert (1977). A new method for sequencing DNA. Proc Natl Acad Sci U S A 74(2): 560-4, 0027-8424 (Print), 0027-8424 (Linking)

McHenry, C. S. (2003). Chromosomal replicases as asymmetric dimers: studies of subunit arrangement and functional consequences. Mol Microbiol 49(5): 1157-1165

Moerman, D. G. and R. J. Barstead (2008). Towards a mutation in every gene in *Caenorhabditis elegans*. Brief Funct Genomic Proteomic 7(3): 195-204, 1477-4062 (Electronic), 1473-9550 (Linking)

Mott, M. L., J. P. Erzberger, M. M. Coons and J. M. Berger (2008). Structural synergy and molecular crosstalk between bacterial helicase loaders and replication initiators. Cell 135(4): 623-634

Neuwald, A. F., L. Aravind, J. L. Spouge and E. V. Koonin (1999). AAA+: A class of chaperone-like ATPases associated with the assembly, operation, and disassembly of protein complexes. Genome Res 9(1): 27-43

Ng, P. C. and E. F. Kirkness (2010). Whole genome sequencing. Methods Mol Biol 628: 215-226, 1940-6029 (Electronic), 1064-3745 (Linking)

Nielsen, O. and A. Lobner-Olesen (2008). Once in a lifetime: strategies for preventing re-replication in prokaryotic and eukaryotic cells. EMBO Rep 9(2): 151-156

Ogura, T., S. W. Whiteheart and A. J. Wilkinson (2004). Conserved arginine residues implicated in ATP hydrolysis, nucleotide-sensing, and inter-subunit interactions in AAA and AAA+ ATPases. J Struct Biol 146(1-2): 106-112

Reznikoff, W. S. and K. M. Winterberg (2008). Transposon-based strategies for the identification of essential bacterial genes. Methods Mol Biol 416: 13-26, 1064-3745 (Print), 1064-3745 (Linking)

Sandler, S. J. (2005). Requirements for replication restart proteins during constitutive stable DNA replication in *Escherichia coli* K-12. Genetics 169(4): 1799-1806

Sandler, S. J., H. S. Samra and A. J. Clark (1996). Differential suppression of *priA2::kan* phenotypes in *Escherichia coli* K- 12 by mutations in priA, lexA, and dnaC. Genetics 143(1): 5-13

Sanger, F., G. M. Air, B. G. Barrell, N. L. Brown, A. R. Coulson, C. A. Fiddes, C. A. Hutchison, P. M. Slocombe and M. Smith (1977). Nucleotide sequence of bacteriophage phi X174 DNA. Nature 265(5596): 687-95, 0028-0836 (Print), 0028-0836 (Linking)

Sanger, F., S. Nicklen and A. R. Coulson (1977). DNA sequencing with chain-terminating inhibitors. Proc Natl Acad Sci U S A 74(12): 5463-7, 0027-8424 (Print), 0027-8424 (Linking)

Sutcliffe, J. G. (1979). Complete nucleotide sequence of the *Escherichia coli* plasmid pBR322. Cold Spring Harb Symp Quant Biol 43 Pt 1: 77-90, 0091-7451 (Print), 0091-7451 (Linking)

Swaffield, J. C. and M. D. Purugganan (1997). The evolution of the conserved ATPase domain (CAD): reconstructing the history of an ancient protein module. J Mol Evol 45(5): 549-563, 0022-2844 (Print), 0022-2844 (Linking)

Tabara, H., M. Sarkissian, W. G. Kelly, J. Fleenor, A. Grishok, L. Timmons, A. Fire and C. C. Mello (1999). The rde-1 gene, RNA interference, and transposon silencing in *C. elegans*. Cell 99(2): 123-32, 0092-8674 (Print), 0092-8674 (Linking)

Tougu, K. and K. J. Marians (1996). The interaction between helicase and primase sets the replication fork clock. J Biol Chem 271(35): 21398-405,

Venter, J. C., M. D. Adams, E. W. Myers, P. W. Li, R. J. Mural, G. G. Sutton, H. O. Smith, M. Yandell, C. A. Evans, R. A. Holt, J. D. Gocayne, P. Amanatides, R. M. Ballew, D. H. Huson, J. R. Wortman, Q. Zhang, C. D. Kodira, X. H. Zheng, L. Chen, M. Skupski, G. Subramanian, P. D. Thomas, J. Zhang, G. L. Gabor Miklos, C. Nelson, S. Broder, A. G. Clark, J. Nadeau, V. A. McKusick, N. Zinder, A. J. Levine, R. J. Roberts, M. Simon, C. Slayman, M. Hunkapiller, R. Bolanos, A. Delcher, I. Dew, D. Fasulo, M. Flanigan, L. Florea, A. Halpern, S. Hannenhalli, S. Kravitz, S. Levy, C. Mobarry, K. Reinert, K. Remington, J. Abu-Threideh, E. Beasley, K. Biddick, V. Bonazzi, R.

Brandon, M. Cargill, I. Chandramouliswaran, R. Charlab, K. Chaturvedi, Z. Deng, V. Di Francesco, P. Dunn, K. Eilbeck, C. Evangelista, A. E. Gabrielian, W. Gan, W. Ge, F. Gong, Z. Gu, P. Guan, T. J. Heiman, M. E. Higgins, R. R. Ji, Z. Ke, K. A. Ketchum, Z. Lai, Y. Lei, Z. Li, J. Li, Y. Liang, X. Lin, F. Lu, G. V. Merkulov, N. Milshina, H. M. Moore, A. K. Naik, V. A. Narayan, B. Neelam, D. Nusskern, D. B. Rusch, S. Salzberg, W. Shao, B. Shue, J. Sun, Z. Wang, A. Wang, X. Wang, J. Wang, M. Wei, R. Wides, C. Xiao, C. Yan, A. Yao, J. Ye, M. Zhan, W. Zhang, H. Zhang, Q. Zhao, L. Zheng, F. Zhong, W. Zhong, S. Zhu, S. Zhao, D. Gilbert, S. Baumhueter, G. Spier, C. Carter, A. Cravchik, T. Woodage, F. Ali, H. An, A. Awe, D. Baldwin, H. Baden, M. Barnstead, I. Barrow, K. Beeson, D. Busam, A. Carver, A. Center, M. L. Cheng, L. Curry, S. Danaher, L. Davenport, R. Desilets, S. Dietz, K. Dodson, L. Doup, S. Ferriera, N. Garg, A. Gluecksmann, B. Hart, J. Haynes, C. Haynes, C. Heiner, S. Hladun, D. Hostin, J. Houck, T. Howland, C. Ibegwam, J. Johnson, F. Kalush, L. Kline, S. Koduru, A. Love, F. Mann, D. May, S. McCawley, T. McIntosh, I. McMullen, M. Moy, L. Moy, B. Murphy, K. Nelson, C. Pfannkoch, E. Pratts, V. Puri, H. Qureshi, M. Reardon, R. Rodriguez, Y. H. Rogers, D. Romblad, B. Ruhfel, R. Scott, C. Sitter, M. Smallwood, E. Stewart, R. Strong, E. Suh, R. Thomas, N. N. Tint, S. Tse, C. Vech, G. Wang, J. Wetter, S. Williams, M. Williams, S. Windsor, E. Winn-Deen, K. Wolfe, J. Zaveri, K. Zaveri, J. F. Abril, R. Guigo, M. J. Campbell, K. V. Sjolander, B. Karlak, A. Kejariwal, H. Mi, B. Lazareva, T. Hatton, A. Narechania, K. Diemer, A. Muruganujan, N. Guo, S. Sato, V. Bafna, S. Istrail, R. Lippert, R. Schwartz, B. Walenz, S. Yooseph, D. Allen, A. Basu, J. Baxendale, L. Blick, M. Caminha, J. Carnes-Stine, P. Caulk, Y. H. Chiang, M. Coyne, C. Dahlke, A. Mays, M. Dombroski, M. Donnelly, D. Ely, S. Esparham, C. Fosler, H. Gire, S. Glanowski, K. Glasser, A. Glodek, M. Gorokhov, K. Graham, B. Gropman, M. Harris, J. Heil, S. Henderson, J. Hoover, D. Jennings, C. Jordan, J. Jordan, J. Kasha, L. Kagan, C. Kraft, A. Levitsky, M. Lewis, X. Liu, J. Lopez, D. Ma, W. Majoros, J. McDaniel, S. Murphy, M. Newman, T. Nguyen, N. Nguyen, M. Nodell, S. Pan, J. Peck, M. Peterson, W. Rowe, R. Sanders, J. Scott, M. Simpson, T. Smith, A. Sprague, T. Stockwell, R. Turner, E. Venter, M. Wang, M. Wen, D. Wu, M. Wu, A. Xia, A. Zandieh and X. Zhu (2001). The sequence of the human genome. Science 291(5507): 1304-1351, 0036-8075 (Print), 0036-8075 (Linking)

Vidan, S. and M. Snyder (2001). Large-scale mutagenesis: yeast genetics in the genome era. Curr Opin Biotechnol 12(1): 28-34, 0958-1669 (Print), 0958-1669 (Linking)

Wahle, E., R. S. Lasken and A. Kornberg (1989). The dnaB-dnaC replication protein complex of *Escherichia coli*. I. Formation and properties. J Biol Chem 264(5): 2463-2468

Winzeler, E. A., D. D. Shoemaker, A. Astromoff, H. Liang, K. Anderson, B. Andre, R. Bangham, R. Benito, J. D. Boeke, H. Bussey, A. M. Chu, C. Connelly, K. Davis, F. Dietrich, S. W. Dow, M. El Bakkoury, F. Foury, S. H. Friend, E. Gentalen, G. Giaever, J. H. Hegemann, T. Jones, M. Laub, H. Liao, N. Liebundguth, D. J. Lockhart, A. Lucau-Danila, M. Lussier, N. M'Rabet, P. Menard, M. Mittmann, C. Pai, C. Rebischung, J. L. Revuelta, L. Riles, C. J. Roberts, P. Ross-MacDonald, B. Scherens, M. Snyder, S. Sookhai-Mahadeo, R. K. Storms, S. Veronneau, M. Voet, G.

Volckaert, T. R. Ward, R. Wysocki, G. S. Yen, K. Yu, K. Zimmermann, P. Philippsen, M. Johnston and R. W. Davis (1999). Functional characterization of the S. cerevisiae genome by gene deletion and parallel analysis. Science 285(5429): 901-906, 0036-8075 (Print), 0036-8075 (Linking)

3

In Silico Identification of Regulatory Elements in Promoters

Vikrant Nain[1], Shakti Sahi[1] and Polumetla Ananda Kumar[2]
[1]*Gautam Buddha University, Greater Noida*
[2]*National Research Centre on Plant Biotechnology, New Delhi*
India

1. Introduction

In multi-cellular organisms development from zygote to adult and adaptation to different environmental stresses occur as cells acquire specialized roles by synthesizing proteins necessary for each task. In eukaryotes the most commonly used mechanism for maintaining cellular protein environment is transcriptional regulation of gene expression, by recruiting required transcription factors at promoter regions. Owing to the importance of transcriptional regulation, one of the main goals in the post-genomic era is to predict gene expression regulation on the basis of presence of transcription factor (TF) binding sites in the promoter regions. Genome wide knowledge of TF binding sites would be useful to build transcriptional regulatory networks model that result in cell specific differentiation. In eukaryotic genomes only a fraction (< 5%) of total genome codes for functional proteins or RNA, while remaining DNA sequences consist of non-coding regulatory sequences, other regions and sequences still with unknown functions.

Since the discovery of trans-acting factors in gene regulation by Jacob and Monads in lac operon of *E. coli*, scientists had an interest in finding new transcription factors, their specific recognition and binding sequences. In DNAse footprinting (or DNase protection assay); transcription factor bound regions are protected from DNAse digestion, creating a "footprint" in a sequencing gel. This methodology has resulted in identification of hundreds of regulatory sequences. However, limitation of this methodology is that it requires the TF and promoter sequence (100-300 bp) in purified form. Our knowledge of known transcription factors is limited and recognition and binding sites are scattered over the complete genome. Therefore, in spite of high degree of accuracy in prediction of TF binding site, this methodology is not suitable for genome wide or across the genomes scanning.

Detection of TF binding sites through phylogenetic footprinting is gradually becoming popular. It is based on the fact that random mutations are not easily accepted in functional sequences, while they continuously keep on tinkering non functional sequences. Many comparative genomics studies have revealed that during course of evolution regulatory elements remain conserved while the non-coding DNA sequences keep on mutating. With an ever increasing number of complete genome sequence from multiple organisms and mRNA profiling through microarray and deep sequencing technologies, wealth of gene expression data is being generated. This data can be used for identification of regulatory elements through intra and inter species comparative genomics. However, the identification of TF binding sites in promoters still remains one of the major challenges in bioinformatics due to following reasons:

1. Very short (5-15 nt) size of regulatory motifs that also differ in their number of occurrence and position on DNA strands with respect to transcription start site. This wide distribution of short TF binding sites makes their identification with commonly used sequence alignment programmes challenging.
2. A very high degree of conservation between two closely related species generally shows no clear signature of highly conserved motifs.
3. Absence of significant similarities between highly diverse species hinders the alignment of functional sequences.
4. Sometimes, functional conservation of gene expression is not sufficient to assure the evolutionary preservation of corresponding cis-regulatory elements (Pennacchio and Rubin, 2001).
5. Transcription factors binding sites are often degenerate.

In order to overcome these challenges, in the last few years novel approaches have been developed that integrate comparative, structural, and functional genomics with the computational algorithms. Such interdisciplinary efforts have increased the sensitivity of computational programs to find composite regulatory elements.

Here, we review different computational approaches for identification of regulatory elements in promoter region with seed specific legumin gene promoter analysis as an example. Based on the type of DNA sequence information the motif finding algorithms are classified into three major classes: (1) methods that use promoter sequences from co regulated genes from a single genome, (2) methods that use orthologous promoter sequences of a single gene from multiple species, also known as phylogenetic footprinting and (3) methods that use promoter sequences of co regulated genes as well as phylogenetic footprinting (Das and Dai, 2007).

2. Representation of DNA motifs

In order to discover motifs of unknown transcription factors, models to represent motifs are essential (Stormo, 2000).There are three models which are generally used to describe a motif and its binding sites:

1. string representation (Buhler and Tompa, 2002)
2. matrix representation (Bailey and Elkan, 1994) and
3. representation with nucleotide dependency (Chin and Leung, 2008)

2.1 String representation

String representation is the basic representation using string of symbols or nucleotides A, C, G and T of length-l to describe a motif. Wildcard symbols are introduced into the string to represent choice from a subset of symbols at a particular position. The International Union of Pure and Applied Chemistry (IUPAC) nucleic acid codes (Thakurta and Stormo, 2007) are used to represent the information about degeneracy for example: W = A or T ('Weak' base pairing); S= C or G ('Strong' base pairing); R= A or G (Purine); Y= C or T (Pyrimidine); K= G or T (Keto group on base); M= A or C (Amino group on base); B= C, G, or T; D= A, G, or T ; H= A, C, or T ; V= A, C, or G; N= A, C, G, or T.

2.2 Matrix representation

In matrix representation, motifs of length l are represented by position weight matrices (PWMs) or position specific scoring matrices (PSSMs) of size 4x l. This gives the occurrence probabilities of each of the four nucleotides at a position j. The score of any specific sequence is the sum of the position scores from the weight matrix corresponding to that sequence. Using this representation an entire genome can be scanned by a matrix and the score at every position obtained (Stormo, 2000). Any sequence with score that is higher than the predefined cut-off is a potential new binding site. A consensus sequence is deduced from a multiple alignment of input sequences and then converted into a position weight matrix.
A PWM score is the sum of position-specific scores for each symbol in the substring. The matrix has one row for each symbol of the alphabet, and one column for each position in the pattern. The score assigned by a PWM to a substring $S = \left(s_j\right)_{j=1}^{N}$, is defined as $\sum_{j=1}^{N} m_{sj,j}$ where j represents position in the substring, s_j is the symbol at position j in the substring, and $m_{\alpha,j}$ is the score in row α, column j of the matrix.
Although matrix representation appears superior, the solution space for PWMs and PSSMs, which consists of 4l real numbers is infinite in size, and there are many local optimal matrices, thus, algorithms generally either produce a suboptimal motif matrix or take too long to run when the motif is longer than 10 bp (Francis and Henry, 2008).

2.3 Representation with nucleotide dependency

The interdependence between neighboring nucleotides with similar number of parameters as string and matrix representations is described by Scored Position Specific Pattern (SPSP). A set of length-l binding site patterns can be described by a SPSP representation P, which contains c ($c \leq l$) sets of patterns Pi, $1 \leq i \leq c$, where each set of patterns Pi contains length-li patterns $P_{i,j}$ of symbols A, C, G, and T and $\sum_i l_i = l$. Each length- l_i pattern $P_{i,j}$ is associated with a score $s_{i,j}$ that represents the "closeness" of a pattern to be a binding site. The lower the score, the pattern is more likely a binding site (Henry and Fracis, 2006).

3. Methods of finding TF binding sites in a DNA sequence

3.1 Searching known motifs

Development of databases of complete information on experimentally validated TF binding site is indispensable for promoter sequence analysis. Information about TF binding sites remain scattered in literature. In the last one and half decade phenomenal increase in computational power, cheaper electronic storage with faster communication technologies, have resulted in development of a range of web accessible databases having experimentally validated TF binding sites. These TF binding site databases are not only highly useful for identification of putative TF binding sites in new promoter sequences (Table1), but also are valuable for providing positive dataset required for improvement and validation of new TF binding site prediction algorithms.

3.1.1 TRANSFAC

TRANSFAC is the largest repository of transcription factors binding sites. TRANSFAC (TRANSFAC 7.0, 2005) web accessible database consists of 6,133 factors with 7,915 sites, while professional version (TRANSFAC 2008.3) consists of 11,683 factors with 30,227 sites. TRANSFAC database is composed of six tables SITE, GENE, FACTOR, CELL, CLASS and

MATRIX. GENE table gives a short explanation of the gene where a site (or group of sites) belongs to; FACTOR table describes the proteins binding to these sites. CELL gives brief information about the cellular source of proteins that have been shown to interact with the sites. CLASS contains some background information about the transcription factor classes, while the MATRIX table gives nucleotide distribution matrices for the binding sites of transcription factors. This database is most frequently used as reference for TFB sites as well as for development of new algorithms. However, new users find it difficult to access the database because it requires search terms to be entered manually. There is no criterion to select the organism, desired gene or TF from a list, so web interface is not user friendly. Other web tools such as TF search and Signal Scan overcome this limitation to certain extent.

3.1.2 Signal Scan

Signal Scan finds and lists homologies of published TF binding site signal sequences in the input DNA sequence by using TRANSFAC, TFD and IMD databases. It also allows to select from different classes viz mammal, bird, amphibian, insect, plant, other eukaryotes, prokaryote, virus (TRANSFAC only), insect and yeast (TFD only).

3.1.3 TRRD

The transcription regulatory region database (TRRD) is a database of transcription regulatory regions of the eukaryotic genome. The TRRD database contains three interconnected tables: TRRDGENES (description of the genes as a whole), TRRDSITES (description of the sites), and TRRDBIB (references). The current version, TRRD 3.5, comprises of the description of 427 genes, 607 regulatory units (promoters, enhancers, and silencers), and 2147 transcription factor binding sites. The TRRDGENES database compiles the data on human (185 entries), mouse (126), rat (69), chicken (29), and other genes.

Developmental/Environmental stimulus	Transcription factor binding site	Position	Sequence
Core promoter	TATA Box	-33	tcccTATAaataa
	Cat Box	-49	gCCAAc
Stress responsive	G Box	-66	tgACGgtgt
	ABRE	-76	acaccttctttgACTGtccatccttc
	ABI4	-245	CACCg
Pathogen defense	W Box	-72	cttctTTGAcgtgtcca
	TCA		gAGAAgagaa
Light Response	I box	-302	gATATga
Wound specific	WUN	-348	tAATTacac
	TCA	-646	gAGAAgagaa
Seed Specific	Legumin	-118	tccatacCCATgcaagctgaagaatgtc
	Opaque-2	-348	TAATtacacatatttta
	Prolamine box	-385	TTaaaTGTAAAAgtAa
	AAGAA-motif	-294	agaAAGAa

Table 1. *In silico* analysis of pigeonpea legumin gene promoter for identification of regulatory elements. Database search reveals that it consist of regulatory elements that can direct its activation under different envirnmental conditions and developmental stages.

3.1.4 PlantCARE

PlantCARE is database of plant specific cis-Acting regulatory elements in the promoter regions (Lescot et al., 2002). It generates a sequence analysis output on a dynamic webpage, on which TF binding sites are highlighted in the input sequence. The database can be queried on names of transcription factor (TF) sites, motif sequence, function, species, cell type, gene, TF and literature references. Information regarding TF site, organism, motif position, strand, core similarity, matrix similarity, motif sequence and function are listed whereas the potential sites are mapped on the query sequence.

3.1.5 PLACE

PLACE is another database of plant *cis*-acting regulatory elements extracted from published reports (Higo et al., 1999). It also includes variations in the motifs in different genes or plant species. PLACE also includes non-plant cis-elements data that may have homologues with plant. PLACE database also provides brief description of each motif and links to publications.

3.1.6 RegulonDB

RegulonDB is a comprehensive database of gene regulation and interaction in *E. coli*. It consists of data on almost every aspect of gene regulation such as terminators, promoters, TF binding sites, active and inactive transcription factor conformations, matrices alignments, transcription units, operons, regulatory network interactions, ribosome binding sites (rbs), growth conditions, gene product and small RNAs.

3.1.7 ABS

ABS is a database of known TF binding sites identified in promoters of orthologous vertebrate genes. It has 650 annotated and experimental validated binding sites from 68 transcription factors and 100 orthologous target genes in human, mouse, rat and chicken genome sequences. Although it's a simple and easy-to-use web interface for data retrieval but it does not facilitate either analysis of new promoter sequence or mapping user defined motif in the promoter.

3.1.8 MatInspector

MatInspector identifies cis-acting regulatory elements in nucleotide sequences using library of weight matrices (Cartharius et al., 2005). It is based on novel matrix family concept, optimized thresholds, and comparative analysis that overcome the major limitation of large number of redundant binding sites predicted by other programs. Thus it increases the sensitivity of reducing false positive predictions. MatInspector also allows integration of output with other sequence analysis programs e.g. DiAlignTF, FrameWorker, SequenceShaper, for an in-depth promoter analysis and designing regulatory sequences. MatInspector library contains 634 matrices representing one of the largest libraries available for public searches.

3.1.9 JASPAR

JASPAR is the another open access database that compete with the commercial TF binding site databases such as TRANSFAC (Portales-Casamar et al., 2009). The latest release has a collection of 457 non-redundant, curated profiles. It is a collection of smaller databases, viz JASPAR CORE, JASPAR FAM, JASPAR PHYLOFACTS, JASPAR POLII and others, among which JASPAR CORE is most commonly used. The JASPAR CORE database contains a curated, non-redundant set of profiles, derived from published collections of experimentally determined transcription factor binding sites for multicellular eukaryotes (Portales-Casamar et al., 2009). The JASPAR database can also be accessed remotely through external application programming interface (API).

3.1.10 Cister: cis-element cluster finder

Cister is based on the technique of posterior decoding, with Hidden Markov model and predicts regulatory regions in DNA sequences by searching for clusters of cis-elements (Frith et al., 2001). The Cister input page consists of 16 common TF sites to define a cluster and additional user defined PWM or TRANSFAC entries can also be entered. For web based analysis maximum input sequence length is 100 kb, however, the program is downloadable for standalone applications and analysis of longer sequences.

3.1.11 MAPPER

MAPPER stands for Multi-genome Analysis of Positions and Patterns of Elements of Regulation It is a platform for the computational identification of TF binding sites in multiple genomes (Marinescu et al., 2005). The MAPPER consists of three modules, the MAPPER database, the Search Engine, and rSNPs and combines TRANSFAC and JASPAR data. However, MAPPER database is limited to TFBSs found only in the promoter of genes from the human, mouse and *D.melanogaster* genomes.

3.1.12 Stubb

Like Cister, Stubb also uses hidden Markov models (HMM) to obtain a statistically significant score for modules (Sinha et al., 2006). STUBB is more suitable for finding modules over genomic scales with small set of transcription factors whose binding sites are known. Stubb differs from MAPPER in that the application of latter is limited to binding sites of a single given motif in an input sequence.

3.1.13 Clover

Clover is another program for identifying functional sites in DNA sequences. It take a set of DNA sequences that share a common function, compares them to a library of sequence motifs (e.g. transcription factor binding patterns), and identifies which, if any, of the motifs are statistically overrepresented in the sequence set (Frith et al., 2004). It requires two input files one for sequences in fasta format and another for sequence motif. Clover provides JASPAR core collection of TF binding sites that can be converted to clover format. Clover is also available as standalone application for windows, Linux as well as Mac operating systems.

3.1.14 RegSite

Regsite consists of plant specific largest repository of transcription factor binding sites. Current RegSite release contains 1816 entries. It is used by transcription start site prediction programs (Sinha et al., 2006).

3.1.15 JPREdictor

JPREdictor is a JAVA based cis-regulatory TF binding site prediction program (Fiedler and Rehmsmeier, 2006). The JPREdictor can use different types of motifs: Sequence Motifs, Regular Expression Motifs, PSPMs as well as PSSMs and the complex motif type (MultiMotifs). This tool can be used for the prediction of cis-regulatory elements on a genome-wide scale.

3.2 Motif finding programs

3.2.1 Phylogenetic footprinting

Comparative DNA sequence analysis shows local difference in mutation rates and reveals a functional site by virtue of its conservation in a background of non-functional sequences. In the phylogenetic equivalent, regulatory elements are protected from random drift across evolutionary time by selection. Orthologous noncoding DNA sequences from multiple species provide a strong base for identification of regulatory elements by Phylogenetic footprinting (Fig. 1) (Rombauts et al., 2003).
The major advantage of phylogenetic footprinting over the single genome is that multigene approach requires data of co regulated genes. While phylogenetic footprinting can identifying regulatory elements present in single gene, that remain conserved during the course of divergence of two species under investigation. With steep increase in available complete genome sequences, across species comparisons for a wide variety of organisms has become possible (Blanchette and Tompa, 2002; Das and Dai, 2007). A multiple sequence alignment algorithm suited for phylogenetic footprinting should be able to indentify small (5-15 bp) sequence in a background of highly diverse sequences.

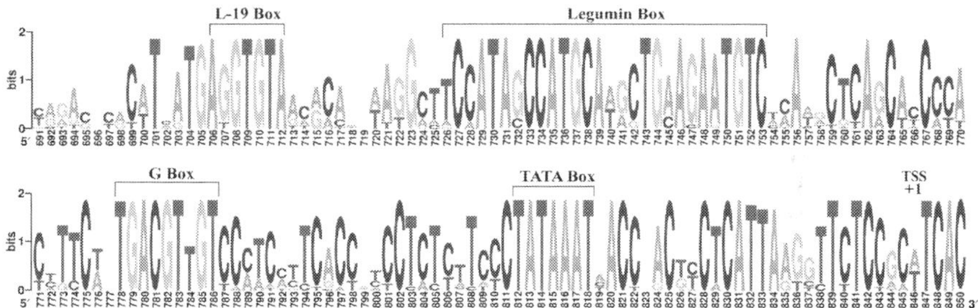

Fig. 1. Identification of new regulatory elements (L-19) in legumin gene promoters by phylogenetic footprinting.

3.2.1.1 Clustal W, LAGAN, AVID

In phylogenetic footprinting primary aim is to construct global multiple alignment of the orthologous promoter sequences and then identify a region conserved across orthologous sequences. Alignment algorithms, such as ClustalW (Thompson et al., 1994), LAGAN (Brudno et al., 2003), AVID (Bray et al., 2003) and Bayes-Block Aligner (Zhu edt al., 1998), have proven useful for phylogenetic footprinting, but the short length of the conserved motif compared to the length of the non-conserved background sequence; and their variable position in a promoter hampers the alignment of conserved motifs. Moreover multiple sequence alignment does not reveal meaningful biological information if the species used for comparison are too closely related. If the species are too distantly related, it is difficult to find an accurate alignment. It requires computational tools that bypass the requirement of sequence alignment completely and have the capabilities to identify short and scattered conserved regions.

3.2.1.2 MEME, Consensus, Gibbs sampler, AlignAce

In cases where multiple alignment algorithms fails, motif finding algorithms such as MEME, Consensus and Gibbs sampler have been used (Fig. 2). The feasibility of using comparative DNA sequence analysis to identify functional sequences in the genome of *S. cerevisiae*, with the goal of identifying regulatory sequences and sequences specifying nonprotein coding RNAs was investigated (Cliften et al., 2001). It was found that most of the DNA sequences of the closely related *Saccharomyces* species aligned to *S.cerevisiae* sequences and known promoter regions were conserved in the alignments. Pattern search algorithms like CONSENSUS (Hertz et al., 1990), Gibbs sampling (Lawrence et al., 1993) and AlignAce (Roth et al., 1998) were useful for identifying known regulatory sequence elements in the promoters, where they are conserved through the most diverged *Saccharomyces* species. Gibbs sampler was used for motif finding using phylogenetic footprinting in proteobacterial genomes (McCue et al., 2001). These programs employ two approaches for motif finding. One approach is to employ a training set of transcription factor binding sites and a scoring scheme to evaluate predictions. The scoring scheme is often based on information theory and the training set is used to empirically determine a score threshold for reporting of the predicted transcription factor binding sites. The second method relies on a rigorous statistical analysis of the predictions, based upon modeled assumptions. The statistical significance of a sequence match to a motif can be accessed through the determination of p-value. P-value is the probability of observing a match with a score as good or better in a randomly generated search space of identical size and nucleotide composition. The smaller the p-value, the lesser the probability that the match is due to chance alone. Since the motif finding algorithms assume the input sequences to be independent, therefore, they are limited by the fact that the data sets containing a mixture of some closely related species will have an unduly high weight in the results of motifs reported.

Multiple genome sequences were compared that are as optimally diverged as possible in *Saccharomyces* genomes. Phylogenetic footprints were searched among the genome sequences of six *Saccharomyces* species using the sequence alignment tool CLUSTAL W and many statistically significant conserved sequence motifs (Cliften et al., 2003) were found.

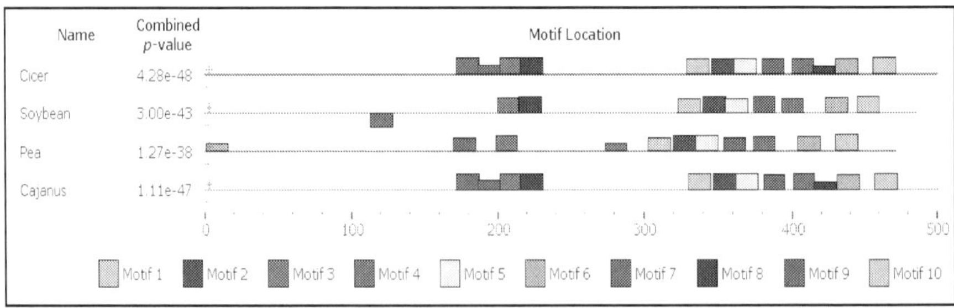

Fig. 2. Combined Block diagram of an MEME output highlighting conserved motifs in promoter regions of legumin seed storage protein genes of four different species.

3.2.1.3 Footprinter

This promising novel algorithm was developed to overcome the limitations imposed by motif finding algorithms. This algorithm identifies the most conserved motifs among the input sequences as measured by a parsimony score on the underlying phylogenetic tree (Blanchette and Tompa, 2002). It uses dynamic programming to find most parsimonious k-mer from each of the input sequences where k is the motif length. In general, the algorithm selects motifs that are characterized by a minimal number of mismatches and are conserved over long evolutionary distances. Furthermore, the motifs should not have undergone independent losses in multiple branches. In other words, the motif should be present in the sequences of subsequent taxa along a branch. The algorithm, based on dynamic programming, proceeds from the leaves of the phylogenetic tree to its root and seeks for motifs of a user-defined length with a minimum number of mismatches. Moreover, the algorithm allows a higher number of mismatches for those sequences that span a greater evolutionary distance. Motifs that are lost along a branch of the tree are assigned an additional cost because it is assumed that multiple independent losses are unlikely in evolution. To compensate for spurious hits, statistical significance is calculated based on a random set of sequences in which no motifs occur.

3.2.1.4 CONREAL

CONREAL (Conserved Regulatory Elements Anchored Alignment Algorithm) is another motif finding algorithm based on phylogenetic footprinting (Berezikov et al., 2005). This algorithm uses potential motifs as represented by positional weight matrices (81 vertebrate matrices form JASPAR database and 546 matrices from TRANSFAC database) to establish anchors between orthologous sequences and to guide promoter sequence alignment. Comparison of the performance of CONREAL with the global alignment programs LAGAN and AVID using a reference data set, shows that CONREAL performs equally well for closely related species like rodents and human, and has a clear added value for aligning promoter elements of more divergent species like human and fish, as it identifies conserved transcription-factor binding sites that are not found by other methods.

3.2.1.5 PHYLONET

The PHYLONET computational approach identifies conserved regulatory motifs directly from whole genome sequences of related species without reliance on additional information was developed by (Wang and Stormo, 2005). The major steps involved are: i) construction of phylogenetic profiles for each promoter , ii) searching through the entire profile space of all the promoters in the genome to identify conserved motifs and the promoters that contain them using algorithm like BLAST, iii) determination of statistical significance of motifs (Karlin and Altschul, 1990). By comparing promoters using phylogenetic profiles (multiple sequence alignments of orthologous promoters) rather than individual sequences, together with the application of modified Karlin- Altschul statistics, they readily distinguished biologically relevant motifs from background noise. When applied to 3524 *Saccharomyces cerevisiae* promoters with *Saccharomyces mikatae, Saccharomyces kudriavzevii,* and *Saccharomyces bayanus* sequences as references PHYLONET identified 296 statistically significant motifs with a sensitivity of >90% for known transcription factor binding sites. The specificity of the predictions appears very high because most predicted gene clusters have additional supporting evidence, such as enrichment for a specific function, in vivo binding by a known TF, or similar expression patterns.

However, the prediction of additional transcription factor binding sites by comparison of a motif to the promoter regions of an entire genome has its own problems due to the large database size and the relatively small width of a typical transcription factor binding site. There is an increased chance of identification of many sites that match the motif and the variability among the transcription factor binding sites permits differences in the level of regulation, due to the altered intrinsic affinities for the transcription factor (Carmack et al., 2007).

3.2.1.6 Phyloscan

PhyloScan combines evidence from matching sites found in orthologous data from several related species with evidence from multiple sites within an intergenic region.The orthologous sequence data may be multiply aligned, unaligned, or a combination of aligned and unaligned. In aligned data, PhyloScan statistically accounts for the phylogenetic dependence of the species contributing data to the alignment and, in unaligned data; the evidence for sites is combined assuming phylogenetic independence of the species. The statistical significance of the gene predictions is calculated directly, without employing training sets (Carmack et al., 2007). The application of the algorithm to real sequence data from seven Enterobacteriales species identifies novel Crp and PurR transcription factor binding sites, thus providing several new potential sites for these transcription factors.

3.3 Software suites for motif discovery

3.3.1 BEST

BEST is a suite of motif-finding programs that include four motif-finding programs: AlignACE (Roth et al., 1998), BioProspector(Liu et al., 2001), Consensus(Hertz and Stormo, 1999), MEME (Bailey et al., 2006) and the optimization program BioOptimizer (Jensen and Liu, 2004). BEST was compiled on Linux, and thus it can only be run on Linux machines (Che et al., 2005).

3.3.2 Seqmotifs

Seqmotifs is a suite of web based programs to find regulatory motifs in coregulated genes of both prokaryotes and eukaryotes. In this suite BioProspector (Liu et al., 2001) is used for finding regulatory motifs in prokaryote or lower eukaryote sequences while CompareProspector(Liu et al., 2002) is used for higher eukaryotes. Another program Mdscan (Liu et al., 2002) is used for finding protein-DNA interaction sites from ChIP-on-chip targets. These programs analyze a group of sequences of coregulated genes so they may share common regulatory motifs and output a list of putative motifs as position-specific probability matrices, the individual sites used to construct the motifs, and the location of each site on the input sequences. CompareProspector has been used for identification of transcription factors Mef2, My, Srf, and Sp1 motifs from a human-muscle-specific co regulated genes. Additionally in a *C. elegans*-*C briggsae* comparison, CompareProspector found the PHA-4 motif and the UNC-86 motif.(Liu et al., 2004) Another *C. Elegans* CompareProspector analysis showed that intestine genes have GATA transcription factor binding motif that was latter experimentally validated (Pauli et al., 2006).

3.3.3 RSAT

The Regulatory Sequence Analysis Tools (RSAT) is an integrated online tool to analyze regulatory sequences in co regulated genes (Thomas-Chollier et al., 2008). The only input required is a list of genes of interest; subsequently upstream sequences of desired distance can be retrieved and appended to the list. Further tasks of putative regulatory signals detection, matching positions for the detected signals in the original dataset can then be performed. The suite includes programs for sequence retrieval, pattern discovery, phylogenetic footprint detection, pattern matching, genome scanning and feature map drawing. Random controls can be performed with random gene selections or by generating random sequences according to a variety of background models (Bernoulli, Markov). The results can be displayed graphically highlighting the desired features. As RSAT web services is implemented using SOAP and WSDL, so either Perl, Python or Java scripts can used for developing custom workflows by combining different tools.

3.3.4 TFM

TFM (Transcription Factor Matrices) is a software suite for identifying and analyzing transcription factor binding sites in DNA sequences. It consists of TFM-Explorer, TFM-Scan, TFM-Pvalue and TFM-Cluster.

TFM-Explorer (Transcription Factor Matrix Explorer) proceeds in two steps: scans sequences for detecting all potential TF binding sites, using JASPAR or TRANSFAC and extracts significant transcription factor.

TFM-Scan is a program dedicated to the location of large sets of putative transcription factor binding sites on a DNA sequence. It uses Position Weight Matrices such as those available in the Transfac or JASPAR databases. The program takes as input a set of matrices and a DNA sequence. It computes all occurrences of matrices on the sequence for a given P-value threshold. The algorithm is very fast and allows for large-scale analysis. TFM-Scan is also able to cluster similar matrices and similar occurrences

TFM-Pvalue is a software suite providing tools for computing the score threshold associated to a given P-value and the P-value associated to a given score threshold. It uses Position Weight Matrices such as those available in the Transfac or JASPAR databases. The program takes as input a matrix, the background probabilities for the letters of the DNA alphabet and a score or a P-value.

3.3.5 rVISTA

rVISTA (regulatory VISTA) combines searching the major transcription binding site database TRANSFAC Professional from Biobase with a comparative sequence analysis and this procedure reduced the number of predicted transcription factor binding sites by several orders of magnitude (Loots and Ovcharenko, 2004). It can be used directly or through links in mVISTA, Genome VISTA, or VISTA Browser. Human and mouse sequences are aligned using the global alignment program AVID (Bray et al., 2003).

3.3.6 Mulan

Mulan brings together several novel algorithms: the TBA multi-aligner program for rapid identification of local sequence conservation and the multiTF program for detecting evolutionarily conserved transcription factor binding sites in multiple alignments. In addition, Mulan supports two-way communication with the GALA database; alignments of multiple species dynamically generated in GALA can be viewed in Mulan, and conserved transcription factor binding sites identified with Mulan/multiTF can be integrated and overlaid with extensive genome annotation data using GALA. Local multiple alignments computed by Mulan ensure reliable representation of short- and large-scale genomic rearrangements in distant organisms. Mulan allows for interactive modification of critical conservation parameters to differentially predict conserved regions in comparisons of both closely and distantly related species. The uses and applications of the Mulan tool through multispecies comparisons of the *GATA3* gene locus and the identification of elements that are conserved in a different way in avians than in other genomes, allowing speculation on the evolution of birds.

3.3.7 MotifVoter

MotifVoter is a variance based ensemble method for discovery of binding sites. It uses 10 most commonly used individual basic motif finders as its component (Wijaya et al., 2008). AlignACE (Hughes et al., 2000), MEME (Bailey and Elkan, 1994; Bailey et al., 2006), ANNSpec, Mitra, BioProspector, MotifSampler, Improbizer, SPACE, MDScan and Weeder. All programs can be selected individually or collectively. Though the existing ensemble methods overall perform better than stand-alone motif finders, the improvement gained is not substantial. These methods do not fully exploit the information obtained from the results of individual finders, resulting in minor improvement in sensitivity and poor precision.

3.3.8 ConSite

ConSite is a, web-based tool for finding cis-regulatory elements in genomic sequences (Sandelin et al., 2004). Two genomic sequences submitted for analysis are aligned by ORCA method. Alternatively, prealigned sequences can be submitted in ClustalW, MSF (GCG), Fasta or Pair wise BLAST format. For analysis Transcription factors can be selected on the basis of species, name, domain or user defined matrix (raw counts matrix or position weight matrix). Predictions are based on the integration of binding site prediction generated with high-quality transcription factor models and cross-species comparison filtering (phylogenetic footprinting). ConSite (Sandelin et al., 2004) is based on the JASPAR database (Portales-Casamar et al., 2009). By incorporating evolutionary constraints, selectivity is increased by an order of magnitude as compared to single sequence analysis. ConSite offers several unique features, including an interactive expert system for retrieving orthologous regulatory sequences.

3.3.9 OPOSSUM

OPOSSUM identifies statistically over-represented, conserved TFBSs in the promoters of co-expressed genes (Ho Sui et al., 2005). OPOSSUM integrates a precomputed database of predicted, conserved TFBSs, derived from phylogenetic footprinting and TFBS detection algorithms, with statistical methods for calculating overrepresentation. The background data set was compiled by identifying all strict one-to-one human/mouse orthologues from the Ensemble database. These orthologues were then aligned using ORCA, a pair-wise DNA alignment program. The conserved non-coding regions were identified. The conserved regions which fell within 5000 nucleotides upstream and downstream of the transcription start site (TSS) were then scanned for TF sites using the position weight matrices (PWMs) from the JASPAR database (Portales-Casamar et al., 2009). These TF sites were stored in the OPOSSUM database and comprise the background set.

3.3.10 TOUCAN2

TOUCAN 2 is an operating system independent, open source, JAVA based workbench for regulatory sequence analysis (Aerts et al., 2004). It can be used for detection of significant transcription factor binding sites from comparative genomics data or for detection of combinations of binding sites in sets of co expressed/co regulated genes. It tightly integrates with Ensemble and EMBL for retrieval of sequences data. TOUCAN provides options to align sequences with different specialized algorithms *viz* AVID (Bray et al., 2003), LAGAN (Brudno et al., 2003), or BLASTZ. MotifScanner algorithm is used to search occurrence of sites of transcription factors by using libraries of position weight matrices from TRANSFAC 6 (Matys et al., 2003), JASPAR, PLANTCARE (Lescot et al., 2002), SCPD and others. Motif Sampler can be used for detection of over-represented motifs. More significantly TOUCAN provides an option to select cis-regulatory modules using the ModuleSearch In essence TOUCAN 2 provides one of the best integration of different algorithms for identification of cis regulatory elements.

3.3.11 WebMOTIFS

WebMOTIFS web server combines TAMO and THEME tools for identification of conserved motifs in co regulated genes (Romer et al., 2007). TAMO combines results from four motif discovery programs viz AlignACE, MDscan, MEME, and Weeder, followed by clustering of results (Gordon et al., 2005). Subsequently Bayesian motif analysis of known motifs is done by THEME. Thus it integrates *de novo* motif discovery programs with Bayesian approaches to identify the most significant motifs. However, current version of Web MOTIFS supports motif discovery only for S. *cerevisiae, M. musculus*, and *H. sapiens* genomes.

3.3.12 Pscan

Pscan is a software tool that scans a set of sequences (e.g. promoters) from co-regulated or co-expressed genes with motifs describing the binding specificity of known transcription factors (Zambelli et al, 2009). It assesses which motifs are significantly over or underrepresented, providing thus hints on which transcription factors could be common regulators of the genes studied, together with the location of their candidate binding sites in the sequences. Pscan does not resort to comparisons with orthologous sequences and experimental results show that it compares favorably to other tools for the same task in terms of false positive predictions and computation time.

4. Identification of regulatory elements in legumin promoter

The plant seeds are rich source of almost all essential supplements of diet *viz.* proteins, carbohydrates, and lipids. Seeds not only provide nutrients to germinating seeding but are major source of energy and other cellular building blocks for human and other heterotrophs as well. Consequently, there is a plethora of pathogens attacking plant seeds. Certain pests such as coleopteran insects of the family Bruchidae, have evolved with leguminous plants (Sales et al., 2000). It is believed that these seeds, most of which are not essential for the establishment of the new plant following germination, contribute to the protection and defense of seeds against pathogens and predators.

Genes encoding seed storage proteins, like zein, phaseolin and legumin, were among the first plant genes studied at gene expression level. Hetrologous expression of reporter genes confirmed that there promoters direct the seed specific expression and subsequently these seed storage gene promoters were used for developing transgenic plants for expressing different genes of interest. Earlier we isolated legumin gene promoter from pigeon pea (*Cajanus cajan*) and identified regulatory elements in its promoter region(Jaiswal et al., 2007). Sequence analysis with PLACE, PLANTCARE, MATINSPECTOR and TRANSFC shows that legumin promoter not only consist of regulatory elements for seed specific expression but also have elements that are present in promoters of other genes involved in pathogen defense, abiotic stress, light response and wounding (Table 1). Our pervious study also confirmed that legumin promoter in expressed in the developing seedling as well (Jaiswal et al., 2007). Recent studies have shown that these promoters are expressed in non seed tissues as well (Zakharov et al., 2004) and play a role in bruchid (insect) resistance (Sales et al., 2000).

In such a scenario where seed storage protein performs an additional task of pathogen defense its prompter must have responsive elements to such stresses. In fact legumin promoter consists of transcription factor binding site for wounding, a signal of insect attack and pathogen defense (Table 1). Since promoter sequences are available for legumin promoter from different species it becomes a good system for identification novel regulatory elements in these promoter. Phylogenetic footprinting analysis reveals presentence of another conserved motif 19 base pair downstream to legumin box (Fig. 1), named L-19 box (Jaiswal et al., 2007). Further MEME analysis shows that in addition to four conserved blocks present for TATA box, G box, Legumin box and L-19 box, there are other conserved, non overlapping sequence blocks are present that were not revealed by multiple sequence alignment based phylogenetic footprinting (Fig. 2).

5. Conclusion

With the critical role of cis-regulatory elements in differentiation of specific cells leading to growth, development and survival of an organism, scientists have a great interest in their identification and characterization. However, due to the limited availability of known transcription factors identification of the corresponding regulatory elements through conventional DNA-protein interaction techniques is challenging. Rapid increase in number of complete genome sequences, identification of co-regulated genes through microarray technology with available electronic storage and computational power has put before the scientific community a challenge to integrate these advancing technology and develop computational program to identify these short but functionally important regulatory sequences. Although some progress has been made in this direction and databases like TRANSFC and others store libraries of transcription factor binding sites. However, there are limitations primarily because publicly available libraries are very small and complete datasets are not freely available. Secondly, because of their very small size there is certain degree of redundancy in binding and therefore the chances of false prediction are very high. These limitations have been overcome to some extent by databases like JASPAR that are freely available and have a collection of regulatory elements large enough to compete with the commercially available datasets. Another concern with cis-acting regulatory elements is that the data pool of these functional non coding transcription factor binding sites is very small (a few thousands), compared with the fact that thousand of genes are expressed in a cell at any point of time and every gene is transcribed by a combination of minimum 5-10 transcription factors. Phylogenetic footprinting, has enormous potential in identifying new regulatory elements and completing the gene network map. Although routine sequence alignment programs such as clustalW fail to align short conserved sequences in a background of hyper variable sequences, more sophisticated sequence alignment programs have been specially developed for identification of conserved regulatory elements. These programs such as CONREAL uses available transcription factor binding site data to align the two sequences that decreases the chances of missing a regulatory site considerably. Moreover, other approaches such as MEME altogether abandons the presence of sequence alignment and directly identifies the rare conserved blocks even if they have jumbled up to the complementary strand. With the increasing sophistication and accuracy of motif finding programs and available genome sequences it can be assumed that knowledge of these regulatory sequences will definitely increase (Table 2). Once we have sufficient data it can be used for development of synthetic promoters with desired expression patterns (Fig. 3).

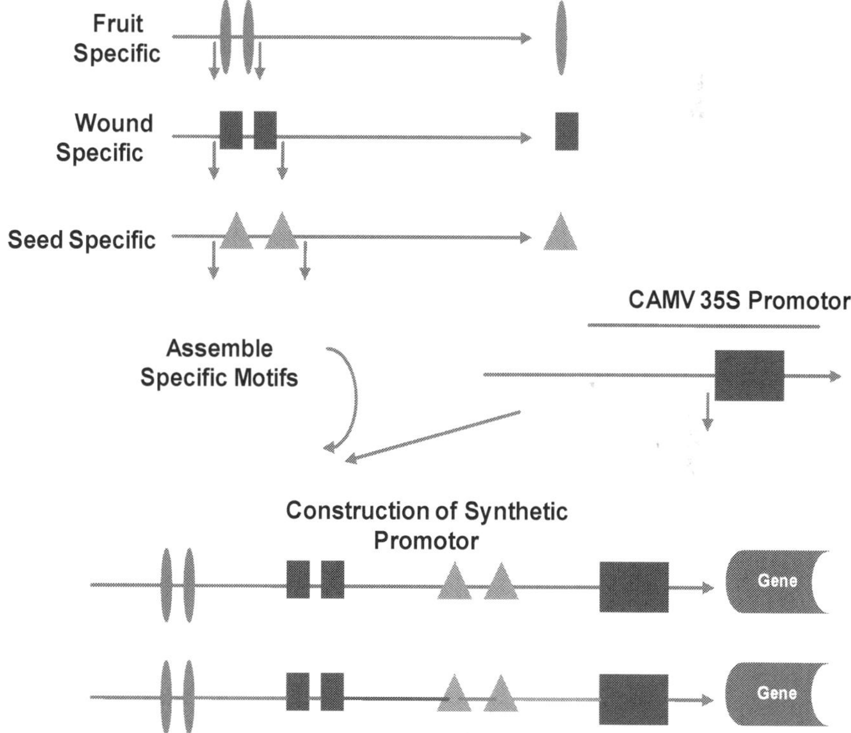

Fig. 3. Future prospects: development of synthetic promoters for expression of gene of interest in desired tissue at defined time and developmental stage. Example: Regulatory elements can be combined from wound, fruit and seed specific promoters and combined with strong CaMV 35S promoter for high level expression of desired gene in all these tissues.

Tools	Web site
Sequence Alignment	
Blast-Z	http://www.bx.psu.edu/miller_lab/
Dialign	http://dialign.gobics.de/chaos-dialign-submission
AVID (Mvista)	http://genome.lbl.gov/vista/mvista/submit.shtml
Lagan	http://lagan.stanford.edu/lagan_web/index.shtml
Clustal W	http://www.ebi.ac.uk/Tools/msa/clustalw2/
TF Binding Site search	
Consite	http://www.phylofoot.org/consite
CONREAL	http://conreal.niob.knaw.nl/
PromH	http://www.softberry.com/berry.phtml?topic=promhg&group=programs&subgroup=promoter
Trafac	http://trafac.cchmc.org/trafac/index.jsp
Footprinter	http://wingless.cs.washington.edu/htbin-post/unrestricted/FootPrinterWeb/FootPrinterInput2.pl
rVISTA	http://rvista.dcode.org/
TFBIND	http://tfbind.hgc.jp/
TESS	http://www.cbil.upenn.edu/cgi-bin/tess/tess
TFSearch	http://www.cbrc.jp/research/db/TFSEARCH.html
Toucan	http://homes.esat.kuleuven.be/~saerts/software/toucan.php
Phyloscan	http://bayesweb.wadsworth.org/cgi-bin/phylo_web.pl
OFTBS	http://www.bioinfo.tsinghua.edu.cn/~zhengjsh/OTFBS/
PROMO	http://alggen.lsi.upc.es/cgi-bin/promo_v3/promo/promoinit.cgi?dirDB=TF_8.3
R-Motif	http://bioportal.weizmann.ac.il/~lapidotm/rMotif/html/
Motif Finding	
MEME	http://meme.sdsc.edu/meme4_6_1/intro.html
AlignAce	http://atlas.med.harvard.edu/cgi-bin/alignace.pl
MotifVoter	http://compbio.ddns.comp.nus.edu.sg/~edward/MotifVoter2/
RSAT	http://rsat.scmbb.ulb.ac.be/rsat/
Gibbs Sampler	http://bayesweb.wadsworth.org/gibbs/gibbs.html
BioProspector	http://ai.stanford.edu/~xsliu/BioProspector/
MatInspector	http://www.genomatix.de/
Improbizer	http://users.soe.ucsc.edu/~kent/improbizer/improbizer.html
WebMOTIFS	http://fraenkel.mit.edu/webmotifs-tryit.html
Psacn	http://159.149.109.9/pscan/
FootPrinter	http://wingless.cs.washington.edu/htbin-post/unrestricted/FootPrinterWeb/FootPrinterInput2.pl

Table 2. Regulatory sequences identification programs.

6. References

Aerts, S., Van Loo, P., Thijs, G., Mayer, H., de Martin, R., Moreau, Y., and De Moor, B. (2004). TOUCAN 2: the all-inclusive open source workbench for regulatory sequence analysis. Nucleic Acids Research 33, W393-W396.

Bailey, T.L., and Elkan, C. (1994). Fitting a mixture model by expectation maximization to discover motifs in biopolymers. Proc Int Conf Intell Syst Mol Biol 2, 28-36.

Bailey, T.L., Williams, N., Misleh, C., and Li, W.W. (2006). MEME: discovering and analyzing DNA and protein sequence motifs. Nucleic Acids Research 34, W369-W373.

Berezikov, E., Guryev, V., and Cuppen, E. (2005). CONREAL web server: identification and visualization of conserved transcription factor binding sites. Nucleic Acids Research 33, W447-W450.

Blanchette, M., and Tompa, M. (2002). Discovery of Regulatory Elements by a Computational Method for Phylogenetic Footprinting. Genome Research 12, 739-748.

Bray, N., Dubchak, I., and Pachter, L. (2003). AVID: A Global Alignment Program. Genome Research 13, 97-102.

Brudno, M., Do, C.B., Cooper, G.M., Kim, M.F., Davydov, E., Program, N.C.S., Green, E.D., Sidow, A., and Batzoglou, S. (2003). LAGAN and Multi-LAGAN: Efficient Tools for Large-Scale Multiple Alignment of Genomic DNA. Genome Research 13, 721-731.

Buhler, J., and Tompa, M. (2002). Finding motifs using random projections. J Comput Biol 9, 225-242.

Carmack, C.S., McCue, L., Newberg, L., and Lawrence, C. (2007). PhyloScan: identification of transcription factor binding sites using cross-species evidence. Algorithms for Molecular Biology 2, 1.

Cartharius, K., Frech, K., Grote, K., Klocke, B., Haltmeier, M., Klingenhoff, A., Frisch, M., Bayerlein, M., and Werner, T. (2005). MatInspector and beyond: promoter analysis based on transcription factor binding sites. Bioinformatics 21, 2933-2942.

Che, D., Jensen, S., Cai, L., and Liu, J.S. (2005). BEST: Binding-site Estimation Suite of Tools. Bioinformatics 21, 2909-2911.

Cliften, P., Sudarsanam, P., Desikan, A., Fulton, L., Fulton, B., Majors, J., Waterston, R., Cohen, B.A., and Johnston, M. (2003). Finding Functional Features in Saccharomyces Genomes by Phylogenetic Footprinting. Science.

Cliften, P.F., Hillier, L.W., Fulton, L., Graves, T., Miner, T., Gish, W.R., Waterston, R.H., and Johnston, M. (2001). Surveying Saccharomyces Genomes to Identify Functional Elements by Comparative DNA Sequence Analysis. Genome Research 11, 1175-1186.

Das, M., and Dai, H.-K. (2007). A survey of DNA motif finding algorithms. BMC Bioinformatics 8, S21.

Fiedler, T., and Rehmsmeier, M. (2006). jPREdictor: a versatile tool for the prediction of cis-regulatory elements. Nucleic Acids Research 34, W546-W550.

Francis, C., and Henry, C.M.L. (2008). DNA Motif Representation with Nucleotide Dependency. IEEE/ACM Trans. Comput. Biol. Bioinformatics 5, 110-119.

Frith, M.C., Hansen, U., and Weng, Z. (2001). Detection of cis -element clusters in higher eukaryotic DNA. Bioinformatics 17, 878-889.

Frith, M.C., Fu, Y., Yu, L., Chen, J.F., Hansen, U., and Weng, Z. (2004). Detection of functional DNA motifs via statistical overâ€ representation. Nucleic Acids Research 32, 1372-1381.

Gordon, D.B., Nekludova, L., McCallum, S., and Fraenkel, E. (2005). TAMO: a flexible, object-oriented framework for analyzing transcriptional regulation using DNA-sequence motifs. Bioinformatics 21, 3164-3165.

Henry, C.M.L., and Fracis, Y.L.C. (2006). Discovering DNA Motifs with Nucleotide Dependency, Y.L.C. Francis, ed, pp. 70-80.

Hertz, G.Z., and Stormo, G.D. (1999). Identifying DNA and protein patterns with statistically significant alignments of multiple sequences. Bioinformatics 15, 563-577.

Hertz, G.Z., Hartzell, G.W., and Stormo, G.D. (1990). Identification of Consensus Patterns in Unaligned DNA Sequences Known to be Functionally Related. Comput Appl Biosci 6, 81 - 92.

Higo, K., Ugawa, Y., Iwamoto, M., and Korenaga, T. (1999). Plant cis-acting regulatory DNA elements (PLACE) database: 1999. Nucleic Acids Research 27, 297-300.

Ho Sui, S.J., Mortimer, J.R., Arenillas, D.J., Brumm, J., Walsh, C.J., Kennedy, B.P., and Wasserman, W.W. (2005). oPOSSUM: identification of over-represented transcription factor binding sites in co-expressed genes. Nucleic Acids Research 33, 3154-3164.

Hughes, J.D., Estep, P.W., Tavazoie, S., and Church, G.M. (2000). Computational identification of Cis-regulatory elements associated with groups of functionally related genes in Saccharomyces cerevisiae. Journal of Molecular Biology 296, 1205-1214.

Jaiswal, R., Nain, V., Abdin, M.Z., and Kumar, P.A. (2007). Isolation of pigeon pea (Cajanus cajan L.) legumin gene promoter and identification of conserved regulatory elements using tools of bioinformatics. Indian Journal of experimental Biology 6, 495-503.

Jensen, S.T., and Liu, J.S. (2004). BioOptimizer: a Bayesian scoring function approach to motif discovery. Bioinformatics 20, 1557-1564.

Karlin, S., and Altschul, S.F. (1990). Methods for assessing the statistical significance of molecular sequence features by using general scoring schemes. Proceedings of the National Academy of Sciences 87, 2264-2268.

Lawrence, C.E., Altschul, S.F., Boguski, M.S., Liu, J.S., Neuwald, A.F., and Wootton, J.C. (1993). Detecting subtle sequence signals: a Gibbs sampling strategy for multiple alignment. Science 262, 208-214.

Lescot, M., DÃ©hais, P., Thijs, G., Marchal, K., Moreau, Y., Van de Peer, Y., RouzÃ, P., and Rombauts, S. (2002). PlantCARE, a database of plant cis-acting regulatory elements and a portal to tools for in silico analysis of promoter sequences. Nucleic Acids Research 30, 325-327.

Liu, X., Brutlag, D.L., and Liu, J.S. (2001). BioProspector: discovering conserved DNA motifs in upstream regulatory regions of co-expressed genes. Pac Symp Biocomput, 127-138.

Liu, X.S., Brutlag, D.L., and Liu, J.S. (2002), An algorithm for finding protein-DNA binding sites with applications to chromatin-immunoprecipitation microarray experiments. Nat Biotech 20, 835-839.

Liu, Y., Liu, X.S., Wei, L., Altman, R.B., and Batzoglou, S. (2004). Eukaryotic Regulatory Element Conservation Analysis and Identification Using Comparative Genomics. Genome Research 14, 451-458.

Loots, G.G., and Ovcharenko, I. (2004). rVISTA 2.0: evolutionary analysis of transcription factor binding sites. Nucleic Acids Research 32, W217-W221.

Loots, G.G., Locksley, R.M., Blankespoor, C.M., Wang, Z.E., Miller, W., Rubin, E.M., and Frazer, K.A. (2000). Identification of a Coordinate Regulator of Interleukins 4, 13, and 5 by Cross-Species Sequence Comparisons. Science 288, 136-140.

Marinescu, V., Kohane, I., and Riva, A. (2005). MAPPER: a search engine for the computational identification of putative transcription factor binding sites in multiple genomes. BMC Bioinformatics 6, 79.

Matys, V., Fricke, E., Geffers, R., GÃ¶ÃŸling, E., Haubrock, M., Hehl, R., Hornischer, K., Karas, D., Kel, A.E., Kel-Margoulis, O.V., Kloos, D.U., Land, S., Lewicki-Potapov, B., Michael, H., MÃ¼nch, R., Reuter, I., Rotert, S., Saxel, H., Scheer, M., Thiele, S., and Wingender, E. (2003). TRANSFACÂ®: transcriptional regulation, from patterns to profiles. Nucleic Acids Research 31, 374-378.

Pauli, F., Liu, Y., Kim, Y.A., Chen, P.-J., and Kim, S.K. (2006). Chromosomal clustering and GATA transcriptional regulation of intestine-expressed genes in C. elegans. Development 133, 287-295.

Pennacchio, L.A., and Rubin, E.M. (2001). Genomic strategies to identify mammalian regulatory sequences. Nat Rev Genet 2, 100-109.

Portales-Casamar, E., Thongjuea, S., Kwon, A.T., Arenillas, D., Zhao, X., Valen, E., Yusuf, D., Lenhard, B., Wasserman, W.W., and Sandelin, A. (2009). JASPAR 2010: the greatly expanded open-access database of transcription factor binding profiles. Nucleic Acids Research.

Rombauts, S., Florquin, K., Lescot, M., Marchal, K., RouzÃ©, P., and Van de Peer, Y. (2003). Computational Approaches to Identify Promoters and cis-Regulatory Elements in Plant Genomes. Plant Physiology 132, 1162-1176.

Romer, K.A., Kayombya, G.-R., and Fraenkel, E. (2007). WebMOTIFS: automated discovery, filtering and scoring of DNA sequence motifs using multiple programs and Bayesian approaches. Nucleic Acids Research 35, W217-W220.

Roth, F.P., Hughes, J.D., Estep, P.W., and Church, G.M. (1998). Finding DNA regulatory motifs within unaligned noncoding sequences clustered by whole-genome mRNA quantitation. Nat Biotech 16, 939-945.

Sales, M.Ì.c.P., Gerhardt, I.R., Grossi-de-SÃ¡, M.F.t., and Xavier-Filho, J. (2000). Do Legume Storage Proteins Play a Role in Defending Seeds against Bruchids? Plant Physiology 124, 515-522.

Sandelin, A., Wasserman, W.W., and Lenhard, B. (2004). ConSite: web-based prediction of regulatory elements using cross-species comparison. Nucleic Acids Research 32, W249-W252.

Sinha, S., Liang, Y., and Siggia, E. (2006). Stubb: a program for discovery and analysis of cis-regulatory modules. Nucleic Acids Research 34, W555-W559.

Stormo, G.D. (2000). DNA Binding Sites: Representation and Discovery. Bioinformatics 16, 16 - 23.

Thomas-Chollier, M., Sand, O., Turatsinze, J.-V., Janky, R.s., Defrance, M., Vervisch, E., Brohee, S., and van Helden, J. (2008). RSAT: regulatory sequence analysis tools. Nucleic Acids Research 36, W119-W127.

Thompson, J.D., Higgins, D.G., and Gibson, T.J. (1994). CLUSTAL W: improving the sensitivity of progressive multiple sequence alignment through sequence weighting, position-specific gap penalties and weight matrix choice. Nucleic Acids Research 22, 4673-4680.

Wang, T., and Stormo, G.D. (2005). Identifying the conserved network of cis-regulatory sites of a eukaryotic genome. Proceedings of the National Academy of Sciences of the United States of America 102, 17400-17405.

Wijaya, E., Yiu, S.-M., Son, N.T., Kanagasabai, R., and Sung, W.-K. (2008). MotifVoter: a novel ensemble method for fine-grained integration of generic motif finders. Bioinformatics 24, 2288-2295.

Zakharov, A., Giersberg, M., Hosein, F., Melzer, M., MÃ¼ntz, K., and Saalbach, I. (2004). Seed-specific promoters direct gene expression in non-seed tissue. Journal of Experimental Botany 55, 1463-1471.

Zhu, J., Liu, J.S., and Lawrence, C.E. (1998). Bayesian adaptive sequence alignment algorithms. Bioinformatics 14, 25-39.

4

In Silico Analysis of Golgi Glycosyltransferases: A Case Study on the LARGE-Like Protein Family

Kuo-Yuan Hwa[1,2], Wan-Man Lin and Boopathi Subramani
Institute of Organic & Polymeric Materials
[1]Department of Molecular Science and Engineering
[2]Centre for Biomedical Industries
National Taipei University of Technology, Taipei, Taiwan, ROC

1. Introduction

Glycosylation is one of the major post-translational modification processes essential for expression and function of many proteins. It has been estimated that 1% of the open reading frames of a genome is dedicated to glycosylation. Many different enzymes are involved in glycosylation, such as glycosyltransferases and glycosidases.

Traditionally, glycosyltransferases are classified based on their enzymatic activities by Enzyme Commission (http://www.chem.qmul.ac.uk/iubmb/enzyme/). Based on the activated donor type, glycosyltransferases are named, for example glucosyltransferase, mannosyltransferase and *N*-acetylglucosaminyltransferases. However, classification of glycosyltransferases based on the biochemical evidence is a difficult task since most of the enzymes are membrane proteins. Reconstruction of enzymatic assay for the membrane proteins are intrinsically more difficult than soluble proteins. Thus the purification of membrane-bound glycosyltransferase is a difficult task. On the other hand, with the recent advancement of genome projects, DNA sequences of an organism are readily available. Furthermore, bioinformatics annotation tools are now commonly used by life science researchers to identify the putative function of a gene. Hence, new approaches based on *in silico* analysis for classifying glycosyltransferase have been used successfully. The best known database for classification of glycosyltransferase by *in silico* approach is the CAZy (Carbohydrate- Active enZymes) database (http://afmb.cnrs-mrs.fr/CAZy/) (Cantarel et al., 2009).

Glycosyltransferases are enzymes involved in synthesizing sugar moieties by transferring activated saccharide donors into various macro-molecules such as DNA, proteins, lipids and glycans. More than 100 glycosyltransferases are localized in the endoplasmic reticulum (ER) and Golgi apparatus and are involved in the glycan synthesis (Narimatsu, H., 2006). The structural studies on the ER and golgi glycosyltransferases has revealed several common domains and motifs present between them. The glycosyltransferases are grouped into functional subfamilies based on similarities of sequence, their enzyme characteristics, donor specificity, acceptor specificity and the specific donor and acceptor linkages (Ishida et al., 2005). The glycosyltransferase sequences comprise of 330-560 amino acids long and share the same type II transmembrane protein structure with four functional domains: a short cytoplasmic domain, a targeting / membrane anchoring domain, a stem region and a catalytic domain (Fukuda et al., 1994). Mammals utilize only 9 sugar nucleotide donors for glycosyltransferases such as UDP-glucose, UDP-galactose, UDP-GlcNAc, UDP-GalNAc, UDP-xylose, UDP-glucuronic acid, GDP-mannose, GDP-fucose, and CMP-sialic acid. Other organisms have an extensive range of nucleotide sugar donors (Varki et al., 2008). Based on the structural studies, we have designed an intelligent platform for the LARGE protein, a golgi glycosyltransferase. The LARGE is a member of glycosyltransferase which has been studied in protein glycosylation (Fukuda & Hindsgaul, 2000). It was originally isolated from a region in chromosome 22 of the human genome which was frequently deleted in human meningiomas with alteration in glycosphingolipid composition. This led to a suggestion that the LARGE may have possible role in complex lipid glycosylation (Dumanski et al., 1987; Peyrard et al., 1999).

2. LARGE

LARGE is one of the largest genes present in the human genome and it is comprised of 660 kb of genomic DNA and contains 16 exons encoding a 756-amino-acid protein. It showed 98% amino acid identity to the mouse homologue and similar genomic organization. The expression of LARGE is ubiquitous but the highest levels of LARGE mRNA are present in heart, brain and skeletal muscle (Peyrard et al., 1999).
LARGE encodes a protein which has an N-terminal transmembrane anchor, coiled coil motif and two putative catalytic domains with a conserved DXD (Asp-any-Asp) motif typical of many glycosyltransferases that uses nucleoside diphosphate sugars as donors (Longman et al., 2003 & Peyrard et al., 1999). The proximal catalytic domain in the LARGE was most homologous to the bacterial glycosyltransferase family 8 (GT8 in CAZy database) members (Coutinho et al., 2003). The members of this family are mainly involved in the synthesis of bacterial outer membrane lipopolysaccharide. The distal domain resembled the human β1,3-N-acetytglucosaminyltransferase (iGnT), a member of GT49 family. The iGnT enzyme is required for the synthesis of the poly-N-acetyllactosamine backbone which is part of the erythrocyte *i* antigen (Sasaki et al., 1997). The presence of two catalytic domains in the LARGE is extremely unusual among the glycosyltransferase enzymes.

2.1 Functions of LARGE

2.1.1 Dystroglycan glycosylation

The Dystroglycan (DG) is an important constituent of the dystrophin-glycoprotein complex (DGC). This complex plays an essential role in the maintaining the stability of the muscle membrane and for the correct localization and/or ligand-binding activity, the glycosylation of some of these components are required (Durbeej et al., 1998). The DG comprises of two subunits, the extracellular α-DG and the transmembrane β-DG (Barresi, 2004). Various components present in the extracellular matrix including laminin (Smalheiser & Schwartz 1987), agrin (Gee et al., 1994), neurexin, (Sugita et al., 2001), and perlecan (Peng et al., 1998) interacts with α-DG. The carbohydrate moieties present in the α-DG are essential to bind with laminin and other ligands. The α-DG is modified by three different types of glycans such as: mucin type *O*-glycosylation, *O*-mannosylation, and *N*-glycosylation. The glycosylated α-DG is essential for the protein's ability to bind the laminin globular domain-containing proteins of the Extracellular Matrix (Kanagawa, 2005). LARGE is required for the generation of functional, properly glycosylated forms of α-DG (Barresi, 2004).

2.1.2 Human LARGE and α-Dystroglycan

The α-DG functional glycosylation by LARGE is likely to be involved in the generation of a glycan polymer which gives rise to the broad molecular weight range observed for α-DG detected by VIA4-1 and IIH6 antibodies. Both the human and mouse LARGE C-terminal glycosyltransferase domain is similar to β3GnT6, which adds GlcNAc to Gal to generate linear polylactosamine chains (Sasaki et al., 1997), the chain formed by LARGE might also be composed of GlcNAc and Glc.

In 1963, Myodystrophy, *myd*, was first described (Lane et al., 1976) as a recessive myopathy mapping to chromosome (Chr) 8, was identified as an intragenic deletion within the glycosyltransferase gene, LARGE. In $Large^{myd}$ and enr mice, the hypoglycosylation of α-DG in DGC was due to the mutation in LARGE (Grewal et al., 2001). The α-DG function was restored in $Large^{myd}$ skeletal muscle and ameliorates muscular dystrophy when LARGE gene was transferred, which indicated that adjustment in the glycosylation status of α-DG can improve the muscle phenotype.

The patients with clinical spectrum ranging from severe congenital muscular dystrophy (CMD), structural brain and eye abnormalities [Walker-Warburg syndrome (WWS), MIM 236670] to a relative mild form of limb-girdle muscular dystrophy (LGMD2I, MIM 607155) are linked to the abnormal O-linked glycosylation of α-DG (van Reeuwijk et al., 2005). A study made by Barresi R. *et al.* (2004) revealed the existence of dual and concentration dependent functions of LARGE. In physiological concentration, LARGE may be involved in regulating the α-DG O-mannosylation pathway. But when the LARGE is expressed by force, it may trigger some other alternative pathways for the *O*-glycosylation of α-DG which can generate a type of repeating polymer of variable lengths, such as glycosaminoglycan-like or core 1 or core 2 structures. This alternative glycan mimics the *O*-mannose glycan in its ability to bind α-DG ligands and can compensate for the defective tetrasaccharide. The functional LARGE protein is also required for neuronal migration during CNS development and it rescues α-DG in MEB fibroblasts and WWS cells (Barresi R. *et al.*, 2004).

2.1.3 LARGE in visual signal processing

The role of LARGE in proper visual signal processing was studied from the retina retinal pathology in $Large^{myd}$ mice. The functional abnormalities of the retina was investigated by a sensitive tool called Electroretinogram (ERG). In $Large^{myd}$ mice, the normal a-wave indicated that the mutant glycosyltransferase does not have any effect on its photoreceptor function.

But the alteration in b-wave may have resulted in downstream retinal circuitry with altered signal processing (Newman & Frishman, 1991). The DGC may also have a possible role in this aspect of the phenotype. The abnormal b-wave was responsible for the loss of retinal isoforms of dystrophin in humans and mice similar to the $Large^{myd}$ mice.

2.2 LARGE homologues

A homologous gene to LARGE was identified and named as LARGE2. It is found to be involved in α-DG maturation as like LARGE, according to Fujimura et al., (2005). It is still not well understood whether these two proteins are compensatory or cooperative. The co-expression of LARGE and LARGE2 did not increase the maturation of α-DG in comparison with either one of them alone and it proved that for the maturation of α-DG, the function of LARGE2 is compensatory and not cooperative. Gene therapy for muscular dystrophy using the LARGE gene is a current topic of research (Barresi R. *et al.*, 2004; Braun, 2004). When compared to LARGE, LARGE2 gene may be more effective because it can glycosylate heavily than LARGE and it also prevents the harmful and immature α-DG production.

The closely related homologues of LARGE are found in the human genome, (glycosyltransferase-like 1B; GYLTL1B), mouse genome (Glylt1b; also called LARGE-Like or LargeL) and in some other vertebrate species (Grewal & Hewitt, 2002). The homologue gene is positioned on the chromosome 11p11.2 of the human genome and it encodes 721 amino acid protein which has 67% identity with LARGE, suggests that the two genes may have risen by gene duplication. Like LARGE, it is also predicted to have two catalytic domains, though it lacks the coiled-coiled motif present in the former protein. The hyperglycosylation of α-dystroglycan by the overexpression of GYLTL1B increased its ability to bind laminin and both the genes showed the same level of increase in laminin binding ability (Brockington, et al., 2005).

3. Bioinformatics workflow and platform design

Many public databases and bioinformatics tools have been developed and are currently available for use (Ding & Berleant, 2002). The primary goal of bioinformaticians is to develop reliable databases and effective analysis tools capable of handling bulk amount of biological data. But the objective of laboratory researchers is to study specific areas within the life sciences, which requires only a limited set of databases and analysis tools. Thus the existing free bioinformatics tools are sometimes too complicated for the biologists to choose. One solution is to have an expert team who are familiar with both bioinformatics databases and to know the needs of a research group in a particular field. The expert team will recommend a workflow by using selected bioinformatics tools and databanks and also helps the scientists with the complicated issue of tools and databases. Moreover, such a team could organize large number of heterogeneous sources of biological information into a specific, expertly annotated databank.

The team can also regularly and systematically update the information essential to help biologists overcome the problems of integrating and keeping up-to-date with heterogeneous biological information (Gerstein, 2000).

We have built a novel information management platform, LGTBase (Hyperlink).This composite knowledge management platform includes the "LARGE-like GlcNAc Transferase Database" by integrating specific public databases like CAZy database, and the workflow analysis combined the usage of specific, public & designed bioinformatics tools to identify the members of the LARGE-like protein family.

4. Tools and database selection

To analyze a novel protein family, biologists need to understand many different types of information. Moreover, the speed of discovery in biology has been expanding exponentially in recent years. So the biologists have to pick the right information available from the vast resources available. To overcome these obstacles, a bioinformatics workflow can be designed for analysing a specific protein family. In our study, a workflow was designed based on the structure and characteristics of LARGE protein as shown in Figure 1 (Hwa et al., 2007). The unknown DNA/protein sequences will be first identified as members of the known gene families by using the Basic Local Alignment Search Tool (BLAST). The *blastp* search tool is used to look for new LARGE-like proteins present in different organisms. The researchers who wish to use our platform can obtain the protein sequences either from the experimental data or through the *blastp* results. The search results were then analyzed with the following tools. To begin with, the sequences are searched for the aspartate-any residue-aspartate (DXD) motif. The DXD motifs present in some glycosyltransferase families are essential for its enzyme activity.

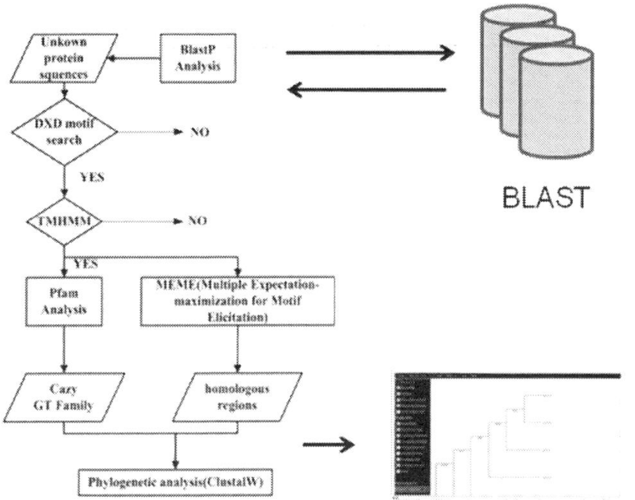

Fig. 1. Bioinformatics workflow of LGTBase.

The DXD motif prediction was then followed by the transmembrane domain prediction by using the TMHMM program (version 2.0; Center for Biological Sequence Analysis, Technical University of Denmark [http://www.cbs.dtu.dk/services/TMHMM-2.0/]). The transmembrane domain is a characteristic feature of the Golgi enzymes.

The sequence motifs are then identified by MEME (Multiple Expectation-maximization for Motif Elicitation) program (version 3.5.4; San Diego Supercomputer Center, UCSD [http://meme.sdsc.edu/meme/]).

This program finds the motif-homology between the target sequence and other known glycosyltransferases. In addition to all the above mentioned tools, the Pfam search (Sanger Institute [http://www.sanger.ac.uk/Software/Pfam/search.shtml]) can also be used to find the multiple sequence alignments and hidden Markov models in many existing protein domains and families. The Pfam results will indicate what kind of protein family the peptide belongs to. If it is a desired protein, investigators can then identify the evolutionary relationships by using phylogenetic analysis.

4.1 LARGE-like GlcNAc transferase database

The specific annotation entries used in the LGTBase are currently being used in a configuration that uses the information retrieved from several databases.

In CAZy database (Carbohydrate- Active enZymes) database ([http://afmb.cnrs-mrs.fr/CAZY/]), the glycosyltransferases are classified as families, clans, and folds based on their structural and sequence similarities, and also on their mechanistic investigation. The other databases used in this platform were listed in Table 1.

Database	Description	Website
EntrezGene	NCBI's repository for gene-specific information	http://www.ncbi.nlm.nih.gov/sites/entrez?db=gene
GenBank	NIH genetic sequence database, an annotated collection of all publicly available DNA sequences	http://www.ncbi.nlm.nih.gov/sites/entrez?db=nucleotide
Dictybase	Database for model organism *Dictyostelium discoideum*	http://dictybase.org/
UniProtKB/Swiss-Prot	High-quality, manually annotated, non-redundant protein sequence database	http://www.uniprot.org/
InterPro	Database of protein families, domains and functional sites	http://www.ebi.ac.uk/interpro/
MGI	Database provides integrated genetic, genomic, and biological data of the laboratory mouse	http://www.informatics.jax.org/
Ensembl	It provides genome- annotation information	http://www.ensembl.org/index.html
HGMD	Human Gene Mutation Database (HGMD) provides comprehensive data on human inherited disease mutations	http://www.hgmd.cf.ac.uk/ac/index.php
UniGene	NCBI database of the transcriptome	http://www.ncbi.nlm.nih.gov/unigene
GeneWiki	The database transfers information on human genes to Wikipedia article	http://en.wikipedia.org/wiki/Gene_Wiki
TGDB	Database with information about the genes involved in cancers	http://www.tumor-gene.org/TGDB/tgdb.html
HUGE	The database provides the results of the Human cDNA project at the Kazusa DNA Research Institute	http://zearth.kazusa.or.jp/huge/
RGD	Database with collection of genetic and genomic information on the rat	http://rgd.mcw.edu/
OMIM	Database provides information on human genes and genetic disorders.	http://www.ncbi.nlm.nih.gov/sites/entrez?db=omim
CGAP	Information of gene expression profiles of normal, precancer, and cancer cells.	http://cgap.nci.nih.gov/
PubMed	Database with 20 million citations for biomedical literature from medical journals, life science journals, related books.	http://www.ncbi.nlm.nih.gov/PubMed/
GO	Representation of gene and gene product attributes across all species	http://www.geneontology.org/

Table 1. The information sources of LARGE-like GlcNAc Transferase Database

All the information related to the LARGE-like protein family was retrieved from the different biological databases. In order to confirm that the information obtained was reliable, the data was scrutinized at two levels. First the information was selected from the above mentioned biological databases with customized programs (using the *perl* compatible regular expressions). Then the obtained information was annotated and validated by experts in glycobiology and bioinformatics.

The annotated data in the LGTBase database was divided into nine categories (Figure 2). The first category is related to genomic location, displays the chromosome, the cytogenetic band and the map location of the gene. The second is related to aliases and descriptions, displays synonyms and aliases for the relevant gene, and descriptions of its function, cellular localization and effect on phenotype. The third category on proteins provides annotated information about the proteins encoded by the relevant genes. The fourth is about protein domains and families, provides annotated information about protein domains and families and the fifth on protein function which provides annotated information about gene function. The sixth category is related to pathways and interactions, provides links to pathways and interactions followed by the seventh on disorders and mutations which draws its information from OMIM and UniProt. The eighth category is on expression in specific tissues, shows that the tissue expression values are available for a given gene. The last category is about research articles, lists the references related to the proteins which are studied. In addition, the investigator can also use DNA or protein sequences to assemble the dataset for the analysis using this workflow.

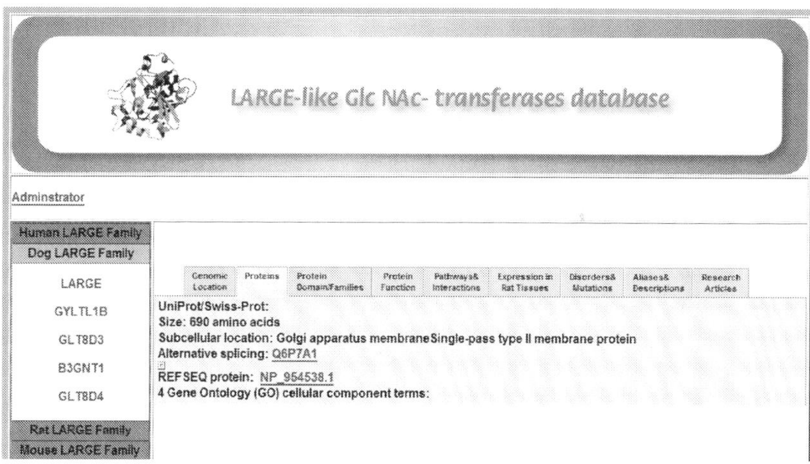

Fig. 2. The contents of LGTBase database

4.2 LARGE-like GlcNAc transferase workflow

4.2.1 Reference sequences search

The unknown DNA/protein sequences are identified as members of the known gene families using the Basic Local Alignment Search Tool (BLAST). BlastP is one of the BLAST programs and it searches protein databases using a protein query. We used BlastP to look for new LARGE-like proteins from different species and gathered the protein sequences of

LARGE like GlcNAc Transferases and built a protein database of 'LARGE-like protein'. This database would assist in search for more reference sequences of LARGE-like protein.

4.2.2 DXD motif search

In several glycosyltransferase families, the DXD motif is essential for the enzymatic activity (Busch et al. 1998). So we first searched for aspartate-any residue-aspartate (DXD) motif, commonly found in glycosyltransferase. Therefore, the 'DXD Motif Search' tool was designed. The input protein sequences are loaded or pasted in this tool and the results indicate the presence or absence of DXD motif.

4.2.3 Transmembrane helices search

The LARGE protein is a member of the N-acetylglucosaminyltransferase family. The presence of transmembrane domain is a characteristics feature of this family. TMHMM program is used to predict the transmembrane helices based on the hidden Markov model. The prediction gives the most probable location and orientation of transmembrane helices in the sequence. TMHMM can predict the location of transmembrane alpha helices and the location of intervening loop regions. This program also predicts the location of the loops that are present between the helices either inside or outside of the cell or organelle. The program is designed based on a 20 amino acids long alpha helix which contains hydrophobic amino acids that can span through a cell membrane.

4.2.4 MEME analysis

A motif is a sequence pattern that occurs repeatedly in a group of related protein or DNA sequences. MEME (Multiple Expectation-maximization for Motif Elicitation) represents motifs as position-dependent letter-probability matrices which describe the probability of each possible letter at each position in the pattern. The program can search for homologous sequences among the input protein sequences.

4.2.5 Protein families search

The Pfam HMM search was used to identify the protein family to which the input protein sequences belong. The Pfam database contains the information about most of the protein domains and families. The results from the Pfam HMM search will show the relation of input protein sequences with the existing protein families and domains.

4.2.6 Phylogenetic analysis

The phylogenetic analysis was performed to find any significant evolutionary relationship between the new protein sequences and the LARGE protein family and to support our previous findings. ClustalW, a multiple alignment program which aligns two or more sequences to determine any significant consensus sequences between them (Thompson et al., 1994). This approach can also be used for searching patterns in the sequence. The phylogenetic tree was constructed by using PHYLIP program (v.3.6.9) and viewed by Treeview software (v.1.6.6). In GlcNAc-transferase phylogenetic analysis, once the multiple alignment of all GlcNAc-transferase has been done, it can be used to construct the phylogenetic tree. About 25 protein sequences were identified as the LARGE-like protein family. By using the neighbor joining distance method, the phylogenetic tree showed that these proteins can be divided into 6 groups (Figure 3). The evolutionary history inferred from phylogenetic analysis is usually depicted as branching, tree-like diagrams which represents an estimated pedigree of the inherited relationships among the protein sequences from different species. These evolutionary relationships can be viewed either as Cladograms (Chenna et al., 2003) or Phylograms (Radomski & Slonimski, 2001).

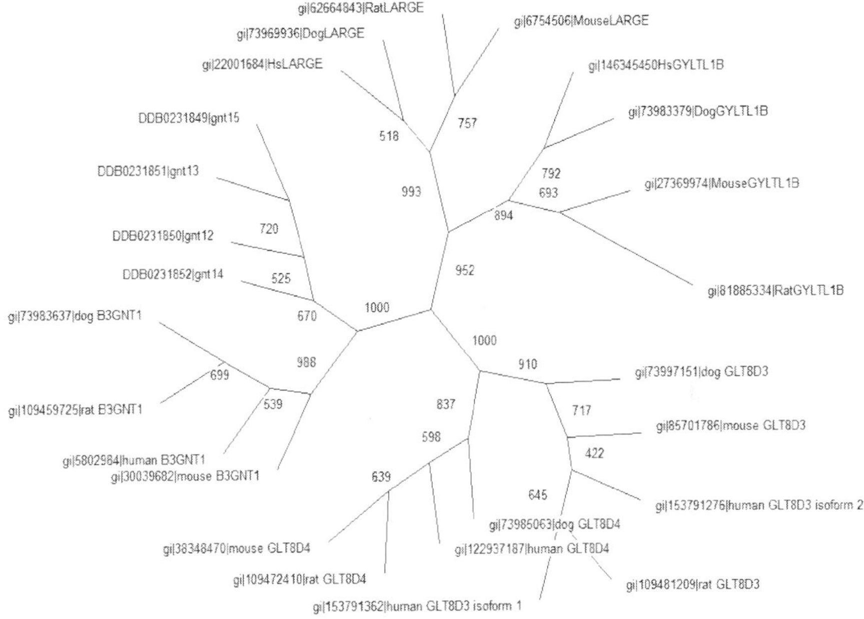

Fig. 3. Phylogenetic tree of LARGE-like Protein Family

4.3 Organization of the LGTBase platform

The data obtained from the analyses were stored in a MySQL relational database and the web interface was built by using PHP and CGI/Java scripts. According to the characteristics of LARGE-like GlcNAc transferase proteins, the workflow was designed and developed by using Java language and several open source bioinformatics programs. Tools with different languages, C, *perl*, java were integrated by using Java language (Figure 4). Adjustable parameters of the tools were reserved to fulfill the needs in future.

5. Application with LARGE protein family

A protein sequence (fasta format) can be entered into the BlastP assistant interface, enabling the other known proteins with similar sequences to be identified (Figure 5). The investigator can select all the resulting sequences or use only some of them. The data can then be transferred to the DXD analysis page (Figure 6). The rationale behind choosing the DXD analysis was since they are represented in many families of glycosyltransferases and it will be easy to narrow down the analysis of putative protein sequences to particular protein families or domains. There were many online tools available for the identification and characterization of unknown protein sequences. So depending upon the target protein of study, one can pick the tools to characterize it.

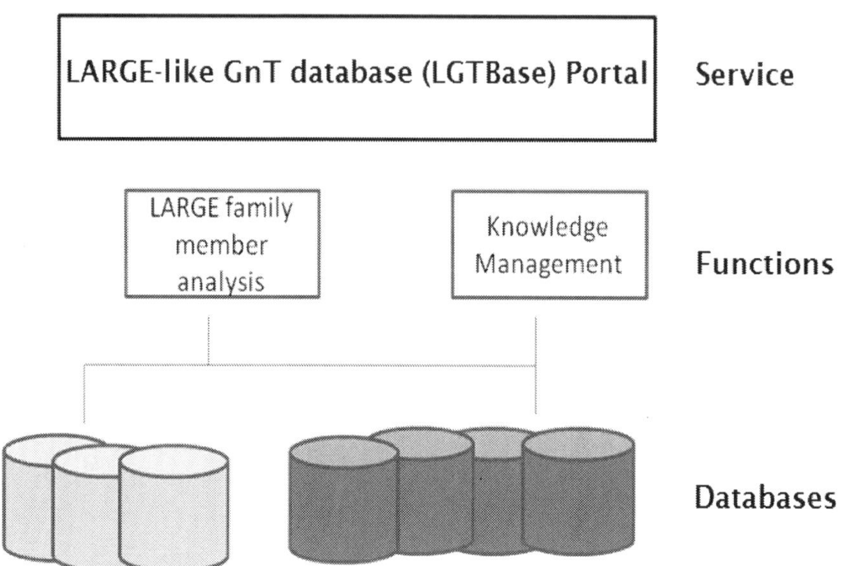

Fig. 4. Database selected for construction of the knowledge management platform

The sequences are analyzed with the DXD motif search tool (Figure 6), which selects those sequences containing the DXD motif for the TMHMM analysis. The transmembrane helices can be predicted with TMHMM analysis (Figure 7). The transmembrane domains are predicted by the hydrophobic nature of the proteins and mainly used to identify the cellular location of the proteins. Similar to transmembrane domain prediction, there were several other domains that can be predicted based on the protein's characters like hydrophobic, hydrophilic etc., The dataset containing DXD motifs and transmembrane helices are then selected for MEME (Figure 8) and Pfam analysis (Figure 9). Some sequence motifs occur repeatedly in the data set and are conjectured to have a biological significance are predicted by MEME analysis. This application plays a significant role in characterization of the putative protein sequences after the initial studies with the DXD motif, transmembrane domain, and other tools. This tool can be used for all kind of protein sequences since its prediction is based on the pattern of sequences present in the study. The protein sequences in the dataset can be identified to the known protein families by Pfam analysis. The pfam classification can also be used for almost all the putative protein sequences because of its large collection of protein domain families represented by multiple sequence alignments and Hidden Markov Models. After the MEME and Pfam analysis were done, ClustalW and Phylip programs were used for Phylogenetic Analysis (Figure 9) to see the evolutionary relationship among the data sets (Figure 10). Finally, these results can be used to design experiments to be performed in the laboratory.

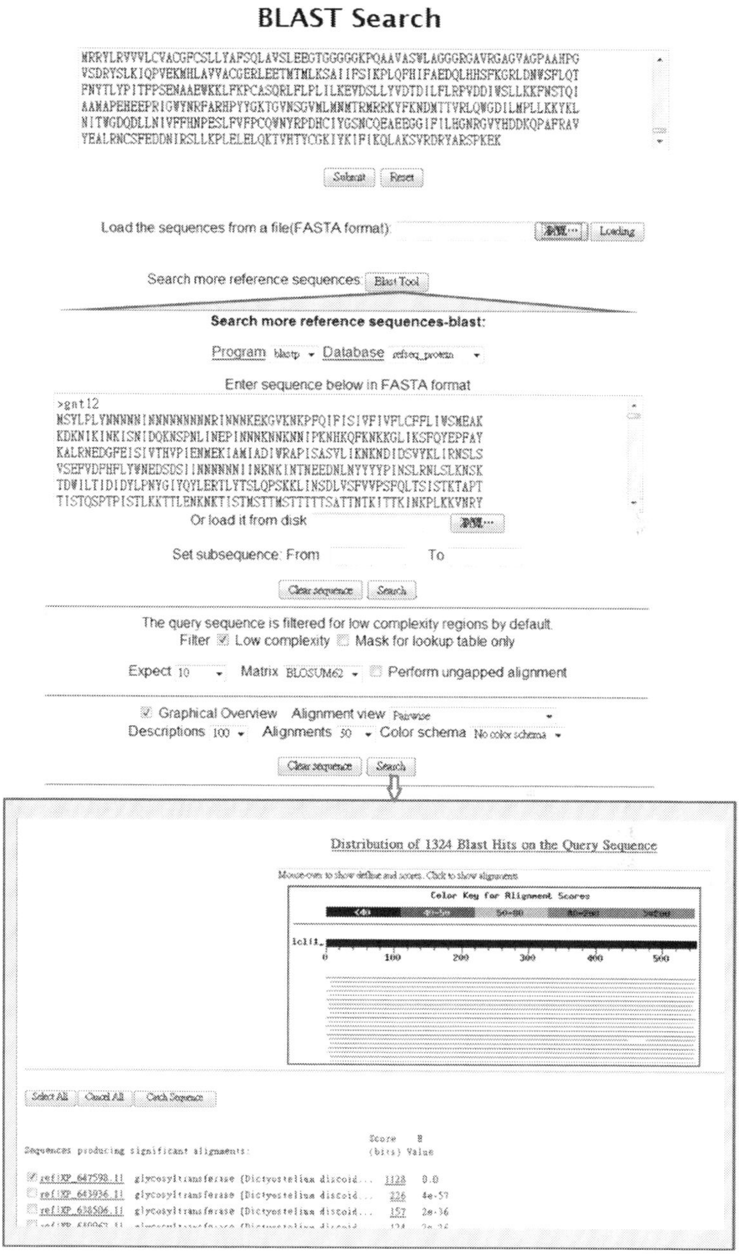

BLAST Search

Submit Reset

Load the sequences from a file(FASTA format):

Search more reference sequences: Blast Tool

Search more reference sequences-blast:

Program blastp Database refseq_protein

Enter sequence below in FASTA format

>gnt12

Or load it from disk

Set subsequence: From To

Clear sequence Search

The query sequence is filtered for low complexity regions by default.
Filter Low complexity Mask for lookup table only

Expect 10 Matrix BLOSUM62 Perform ungapped alignment

Graphical Overview Alignment view Pairwise
Descriptions 100 Alignments 50 Color schema No color schema

Clear sequence Search

Distribution of 1324 Blast Hits on the Query Sequence

Color Key for Alignment Scores

Select All Cancel All Catch Sequence

Sequences producing significant alignments: Score (bits) E Value

ref|XP_647598.1| glycosyltransferase (Dictyostelium discoid... 1128 0.0
ref|XP_643936.1| glycosyltransferase (Dictyostelium discoid... 226 4e-57
ref|XP_638506.1| glycosyltransferase (Dictyostelium discoid... 157 2e-36

Fig. 5. The BlastP tool of the LGTBase platform to find similar sequences to LARGE

DXD Motif Search

Results from DXD motif search to Transmembrane Helices search

Sequences with DXD motif

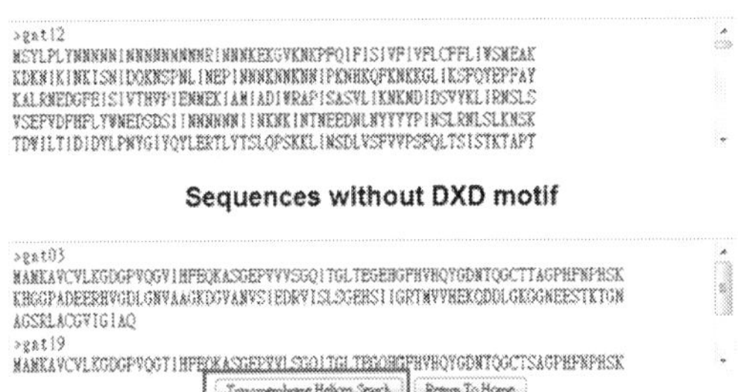

Fig. 6. DXD motif search tool of the LGTBase platform for DXD motif prediction

Results from Transmembrane Helices to MEME analysis

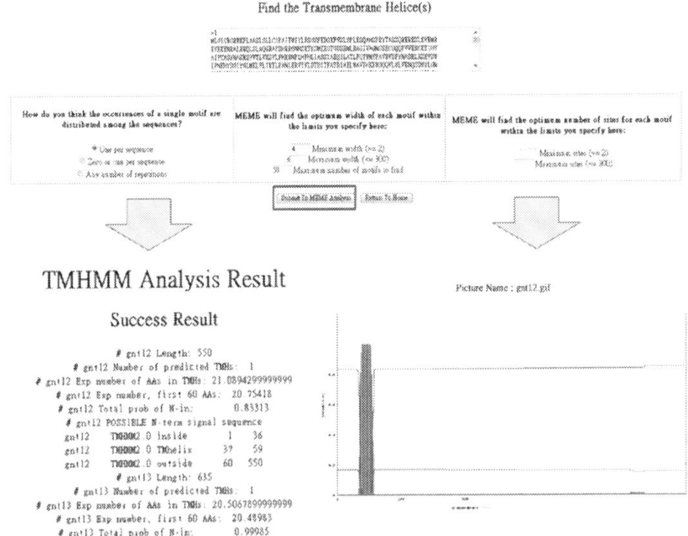

Fig. 7. TMHMM analysis tool of the LGTBase platform for Transmembrane domain prediction

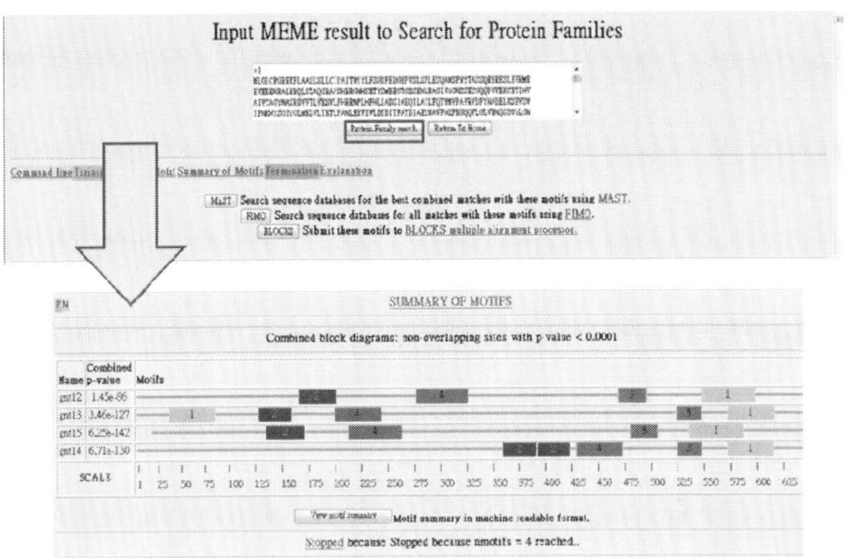

Fig. 8. MEME analysis tool of the LGTBase platform to predict the sequence motifs

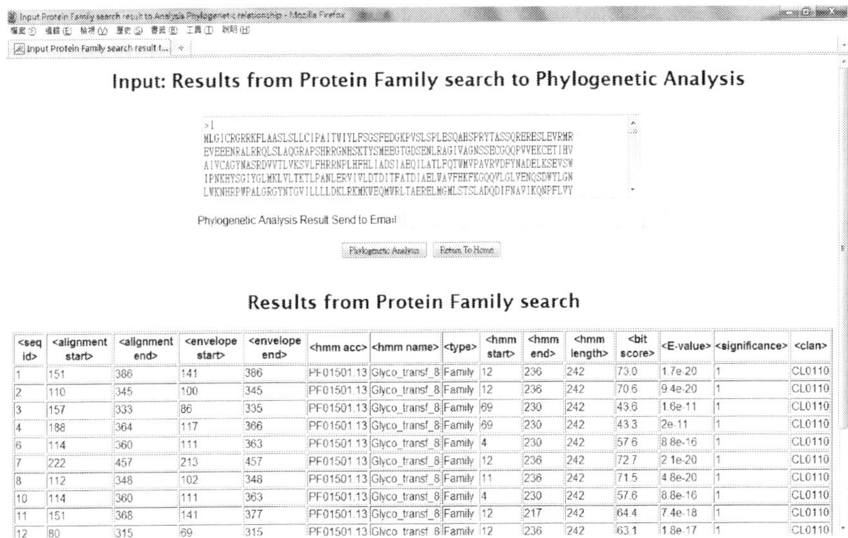

<seq id>	<alignment start>	<alignment end>	<envelope start>	<envelope end>	<hmm acc>	<hmm name>	<type>	<hmm start>	<hmm end>	<hmm length>	<bit score>	<E-value>	<significance>	<clan>
1	151	386	141	386	PF01501.13	Glyco_transf_8	Family	12	236	242	73.0	1.7e-20	1	CL0110
2	110	345	100	345	PF01501.13	Glyco_transf_8	Family	12	236	242	70.6	9.4e-20	1	CL0110
3	157	333	86	335	PF01501.13	Glyco_transf_8	Family	69	230	242	43.6	1.6e-11	1	CL0110
4	188	364	117	366	PF01501.13	Glyco_transf_8	Family	69	230	242	43.3	2e-11	1	CL0110
6	114	360	111	363	PF01501.13	Glyco_transf_8	Family	4	230	242	57.6	8.8e-16	1	CL0110
7	222	457	213	457	PF01501.13	Glyco_transf_8	Family	12	236	242	72.7	2.1e-20	1	CL0110
8	112	348	102	348	PF01501.13	Glyco_transf_8	Family	11	236	242	71.5	4.8e-20	1	CL0110
10	114	360	111	363	PF01501.13	Glyco_transf_8	Family	4	230	242	57.6	8.8e-16	1	CL0110
11	151	368	141	377	PF01501.13	Glyco_transf_8	Family	12	217	242	64.4	7.4e-18	1	CL0110
12	80	315	69	315	PF01501.13	Glyco_transf_8	Family	12	236	242	63.1	1.8e-17	1	CL0110

Fig. 9. Pfam analysis tool of the LGTBase platform to identify the known protein family of the target protein which is studied

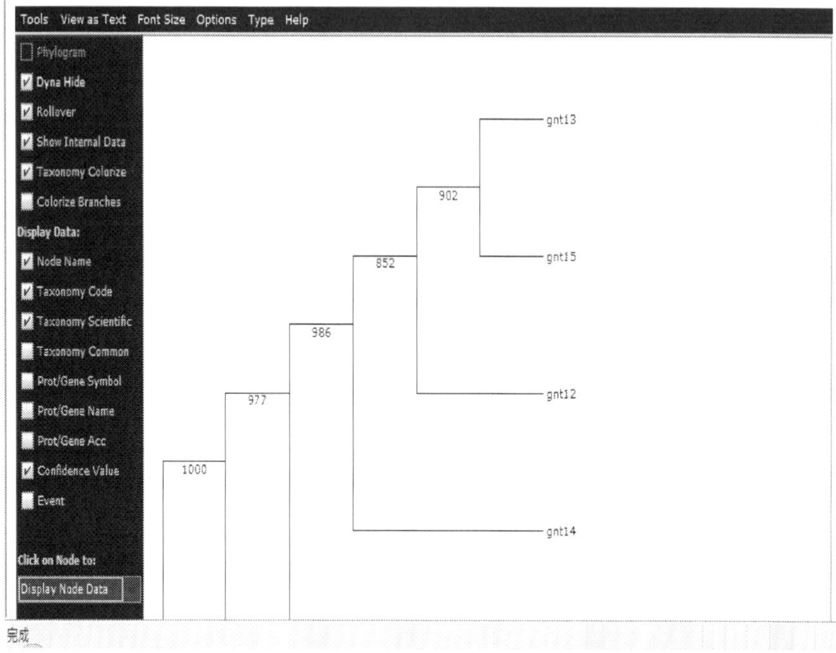

Fig. 10. Phylogenetic analysis tool of the LGTBase platform to study the evolutionary relationship of the target protein

6. Future direction

We have described how to construct a computational platform to analyze the LARGE protein family. Since the platform was built based on several commonly shared protein domains and motifs, it can also be modified for analyzing other golgi glycosyltransferases. Furthermore, the phylogenetic analysis (Figure 3) revealed that LARGE protein family is related to β-1,3-*N*-acetylglucosaminyltransferase 1 (β3GnT). β3GnT (EC 2.4.1.149) is a group of enzymes belong to glycosyltransferases family. Some β3GnT enzymes catalyze the transfer of GlcNAc from UDP-GlcNAc to Gal in the Galβ1-4 Glc(NAc) structure with β-1,3 linkage. These enzymes were grouped into GT family 31, 49 in the CAZy database. The enzyme uses 2 substrates namely, UDP-N-acetyl-D-glucosamine and D-galactosyl-β-1,4-N-acetyl-D-glucosaminyl-R and the products are formed as UDP, N-acetyl-β-D- glucosaminylβ-1,3 -D-galactosamine. These enzymes participate in the formation of keratan sulfate, glycosphingolipid biosynthesis, neo-lacto series and N-linked glycans. There are currently 9 members known from the β3GnT family.

The β3GnT1 (iGnT) was the first enzyme to be isolated when cDNA of a human β-1,3-N-acetylglucosaminyltransferase essential for poly-N-acetyllactosamine synthesis was studied (Zhou et al., 1999). The poly-N-acetyllactosamine synthesized by iGnT provides critical backbone structure for the addition of functional oligosaccharides such as Sialyl Lewis X. It has been reported recently that β3GnT1 is involved in attenuating prostate cancer cell locomotion by regulating the synthesis of laminin-binding glycans on α-DG (Bao et al., 2009). Since there are several common shared domains similar to the LARGE protein, the new platform for β3GnT protein family can be constructed based on the original platform. Apart from β3GnT1, β3GnT2 enzyme is responsible for elongation of poly-lactosamine chains. This enzyme was isolated based on structural similarity with the β3GalT family. Studies showed that on a panel of invasive and noninvasive fresh transitional cell carcinomas (TCCs) showed strong down regulation of β3GnT2 in the invasive lesions, suggesting that a decline in the expression levels of some members of the glycosyltransferase (Gromova et al., 2001).

The β3GnT3 and β3GnT4 enzymes were subsequently isolated based on the structural similarity with β3GalT family. β3GnT3 is a type II transmembrane protein and contains a signal anchor that is not cleaved. It prefers the substrates of lacto-N-tetraose and lacto-N-neotetraose, and it is also involved in the biosynthesis of poly-N-acetyllactosamine chains and the biosynthesis of the backbone structure of dimeric sialyl Lewis A. It plays dominant role in the L-selectin ligand biosynthesis, lymphocyte homing and lymphocyte trafficking. The β3GnT3 enzyme is highly expressed in the non-invasive colon cancer cells. β3GnT4 is involved in the biosynthesis of poly-N-acetyllactosamine chains and prefers lacto-N-neotetraose as the substrate. It is a type II transmembrane protein and it is expressed more in bladder cancer cells (Shiraishi et al., 2001). β3GnT5 is responsible for lactosyltriaosylceramide synthesis, an essential component of lacto/neolacto series glycolipids (Togayachi et al., 2001). The expression of the HNK-1 and Lewis x antigens on the lacto/neo-lacto-series of glycolipids is developmentally and tissue-specifically regulated by β3GnT5. The overexpression of β3GnT5 in human gastric carcinoma cell lines led to increased sialyl-Lewis X expression and increased *H.pylori* adhesion (Marcos et al., 2008).

The β3GnT6 synthesizes the core 3 O-glycan structure and speculates that this enzyme plays an important role in the synthesis and function of mucin O-glycan in the digestive organs. In addition, the expression of β3GnT6 was markedly down regulated in gastric and colorectal carcinomas (Iwai et al., 2005). Expression of β3GnT7 has been reported to be down-regulated upon malignant transformation (Kataoka et al., 2002). Elongation of the carbohydrate backbone of keratan sulfate proteoglycan is catalyzed by β3GnT7 and β1,4-galactosyltransferase 4 (Hayatsu et al., 2008). β3GnT7 can transfer GlcNAc to Gal to synthesize a polylactosamine chain with each enzyme differing in its acceptor molecule preference. The polylactosamine and related structures plays crucial role in cell-cell interaction, cell-extracellular matrix interaction, immune response and determining metastatic capacity. The β3GnT8 enzyme extends a polylactosamine chain specifically on a tetraantennary N-glycans. β3GnT8 transfers GlcNAc to the non-reducing terminus of the

Galβ1-4GlcNAc of tetra antennary *N*-glycan *in vitro*. Intriguingly, β3GnT8 is significantly upregulated in colon cancer tissues than in normal tissue (Ishida et al., 2005). The co-transfection of β3GnT8 and β3GnT2 resulted in synergistic enhancement of the activity of the polylactosamine synthesis. This indicates that these two enzymes interact and complement each other's function in the cell. As a summary, the members of the β3GnT protein family are important in human cancer biology.

Our initial motif analysis showed that there are 3 important functional domains predicted are commonly found among the β3GnT enzymes. The first motif is a structural motif necessary for maintaining the protein fold. The second, DXD motif represented in many glycosyltransferases is involved in the binding of the nucleotide-sugar donor substrate, both directly and indirectly through coordination of metal ions such as magnesium or manganese in the active site. A glycine-rich loop is the third motif found at the bottom of the active site cleft. This loop is likely to play a role in the recognition of both the GlcNAc portion of the donor and the substrate. Since the three common domains of β3GnT are similar to the LARGE protein family, it is feasible to modify the current LARGE platform to analyze other golgi glycosyltransferases such as β3GnT.

7. References

Bao, X., Kobayashi, M., Hatakeyama, S., Angata, K., Gullberg, D., Nakayama, J., Fukuda, M.N. & Fukuda, M. (2009). Tumor suppressor function of laminin-binding α-dystroglycan requires a distinct β-3-N-acetylglucosaminyltransferase. *Proceedings of the National Academy of Sciences USA,* Vol.106, No.29, (July 2009), pp. 12109-12114

Barresi, R., Michele, D.E., Kanagawa, M., Harper, H.A., Dovico, S.A., Satz, J.S., Moore, S.A., Zhang, W., Schachter, H., Dumanski, J.P., Cohn, R.D., Nishino, I. & Campbell, K.P. (2004). LARGE can functionally bypass alpha-dystroglycan glycosylation defects in distinct congenital muscular dystrophies. *Nature Medicine,* Vol.10, No.7, (July 2004), pp. 696-703.

Braun, S. (2004). Naked plasmid DNA for the treatment of muscular dystrophy. *Current Opinion in Molecular Therapeutics,* Vol.6, (October 2004), pp. 499-505.

Brockington, M., Torelli, S., Prandini, P., Boito, C., Dolatshad, N.F., Longman, C., Brown, S.C., Muntoni, F. (2005). Localization and functional analysis of the LARGE family of glycosyltransferases: significance for muscular dystrophy. *Human Molecular Genetics,* Vol.14, No.5, (March 2005), pp. 657-665.

Busch, C., Hofmann, F., Selzer, J., Munro, S., Jeckel, D. & Aktories, K. (1998). A common motif of eukaryotic glycosyltransferases is essential for the enzyme activity of large clostridial cytotoxins. *Journal of Biological Chemistry,* Vol.273, No.31, (July 1998), pp.19566-19572.

Cantarel, B.L., Coutinho, P.M., Rancurel, C., Bernard, T., Lombard, V. & Henrissat, B. (2009) The Carbohydrate-Active EnZymes database (CAZy): an expert resource for Glycogenomics. *Nucleic Acids Research, Vol.* 37, (January 2009), pp. D233-238

Chenna, R., Sugawara, H., Koike, T., Lopez, R., Gibson, T.J. & Higgins, D.G. Thompson JD (2003). Multiple sequence alignment with the Clustal series of programs. *Nucleic Acids Res.* Vol.31, No.13, (July 2003), pp. 3497-3500.

Coutinho, P.M., Deleury, E., Davies, G.J. & Henrissat, B. (2003). An evolving hierarchical family classification for glycosyltransferases. *Journal of Molecular Biology,* Vol.328, (April 2003), pp. 307-317.

Ding, J., Berleant, D., Nettleton, D. & Wurtele, E. (2002). Mining MEDLINE: abstracts, sentences, or phrases? *Pacific Symposium on Biocomputing,* Vol.7, pp. 326-337.

Dumanski, J.P., Carlbom, E., Collins, V.P., Nordenskjold, M. (1987). Deletion mapping of a locus on human chromosome 22 involved in the oncogenesis of meningioma. *Proceedings of the National Academy of Sciences USA,* Vol.84, (December 1987), pp. 9275-9279.

Durbeej, M., Henry, M.D. & Campbell, K.P. (1998). Dystroglycan in development and disease. *Current Opinions in Cell Biology* Vol. 10, (October 1998), pp. 594-601.

Fujimura, K., Sawaki, H., Sakai, T., Hiruma, T., Nakanishi, N., Sato, T., Ohkura, T., Narimatsu, H. (2005). LARGE2 facilitates the maturation of *α*-dystroglycan more effectively than LARGE. *Biochemical and Biophysical Research Communications,* Vol.329, No.3, (April 2005), pp. 1162-1171

Fukuda, M., Hindsgaul, O., Hames, B.D. & Glover, D.M. (1994). In *Molecular Glycobiology,* Oxford Univ. Press, Oxford.

Fukuda, M., & Hindsgaul, O. (2000). *Molecular and Cellular Glycobiology* (2nd ed.), Oxford Univ. Press, Oxford.

Gee, S.H., Montanaro, F., Lindenbaum, M.H., and Carbonetto, S. (1994). Dystroglycan-α, a dystrophin-associated glycoprotein, is a functional agrin receptor. *Cell,* Vol.77, (June 1994), pp. 675-686.

Gerstein, M. (2000). Integrative database analysis in structural genomics. *Nature Structural Biology,* Vol.7, (November 2000), Suppl: 960-963.

Grewal, K., Holzfeind, P.J., Bittner, R.E. & Hewitt, J.E., (2001). Mutant glycosyltransferase and altered glycosylation of alpha-dystroglycan in the myodystrophy mouse. *Nature Genetics,* Vol.28, (June 2001), pp.151-154.

Grewal, P.K. & Hewitt, J.E. (2002). Mutation of Large, which encodes a putative glycosyltransferase, in an animal model of muscular dystrophy. *Biochimica et Biophysica Acta,* Vol.1573, (December 2002), pp. 216-224.

Gromova, I., Gromov, P. & Celis J.E. (2001). A Novel Member of the Glycosyltransferase Family, β3GnT2, highly down regulated in invasive human bladder Transitional Cell Carcinomas. *Molecular Carcinogenesis,* Vol. 32, No. 2, (October 2001), pp. 61-72

Hayatsu, N., Ogasawara, S., Kaneko, M.K., Kato, Y. & Narimatsu, H. (2008). Expression of highly sulfated keratan sulfate synthesized in human glioblastoma cells. *Biochemical and Biophysical Research Communications,* Vol. 368, No. 2, (April 2008), pp. 217-222

Hwa, K.Y., Pang, T.L. & Chen, M.Y. (2007). Classification of LARGE-like GlcNAc-Transferases of *Dictyostelium discoideum* by Phylogenetic Analysis. *Frontiers in the Convergence of Bioscience and Information Technologies,* pp. 289-293.

Ishida, H., Togayachi, A., Sakai, T., Iwai, T., Hiruma, T., Sato, T., Okubo, R., Inaba, N., Kudo, T., Gotoh, M., Shoda, J., Tanaka, N., & Narimatsu, H. A novel beta1,3-N-acetylglucosaminyltransferase (beta3Gn-T8), which synthesizes poly-N-acetyllactosamine, is dramatically upregulated in colon cancer. *FEBS Letters.* (January 2005), Vol. 579, No.1, pp. 71-78.

Ishida, H., Togayachi, A., Sakai, T., Iwai, T., Hiruma, T., Sato, T., Okubo, R., Inaba, N., Kudo, T., Gotoh, M., Shoda, J., Tanaka, N. & Narimatsu, H. (2005). A novel beta1,3-N-acetylglucosaminyltransferase (beta3Gn-T8), which synthesizes poly-N-acetyllactosamine, is dramatically upregulated in colon cancer. *FEBS Letters,* Vol.579, No.1, (January 2005), pp. 71-8.

Iwai, T., Kudo, T., Kawamoto, R., Kubota, T., Togayachi, A., Hiruma, T., Okada, T., Kawamoto, T., Morozumi, K. & Narimatsu, H. (2005). Core 3 synthase is down-regulated in colon carcinoma and profoundly suppresses the metastatic potential of carcinoma cells. *Proceedings of the National Academy of Sciences USA,* Vol.102, No.12, (March 2005), pp. 4572-4577

Kanagawa, M., Michele, D.E., Satz, J.S., Barresi, R., Kusano, H., Sasaki, T., Timpl, R., Henry, M. D., and Campbell, K.P. (2005). Disruption of Perlecan Binding and Matrix Assembly by Post-Translational or Genetic Disruption of Dystroglycan Function. *FEBS Letters,* Vol.579, No.21, (August 2005), pp. 4792-4796.

Kataoka, K. & Huh, N.H. (2002). A novel β1,3-N-acetylglucosaminyltransferase involved in invasion of cancer cells as assayed *in vitro. Biochemical and Biophysical Research Communications,* Vol. 294, No.4, (June 2002), pp. 843-848

Lane, P.W., Beamer, T.C. & Myers, D.D. (1976). Myodystrophy, a new myopathy on chromosome 8 of the mouse. *Journal of Heredity,* Vol. 67, No.3 (May-June 1976), pp. 135-138.

Longman, C., Brockington, M., Torelli, S., Jimenez-Mallebrera, C., Kennedy, C., Khalil, N., Feng, L., Saran, R.K., Voit, T., Merlini, L., Sewry, C.A., Brown, S.C. & Muntoni F. (2003). Mutations in the human LARGE gene cause MDC1D, a novel form of congenital muscular dystrophy with severe mental retardation and abnormal glycosylation of alpha dystroglycan. *Human Molecular Genetics,* Vol.12, No.21, (November 2003), pp. 2853-2861.

Marcos, N.T., Magalhães, A., Ferreira, B., Oliveira, M.J., Carvalho, A.S., Mendes, N., Gilmartin, T., Head, S.R., Figueiredo, C., David, L., Santos-Silva, F. & Reis, C.A. (2008). *Helicobacter pylori* induces β3GnT5 in human gastric cell lines, modulating expression of the SabA ligand Sialyl-Lewis X. *Journal of Clinical Investigation,* Vol. 118, No. 6, (June 2008), pp.2325-2336

Narimatsu, H. (2006). Human glycogene cloning: focus on beta 3-glycosyltransferase and beta 4-glycosyltransferase families. *Current Opinions in Structural Biology.* Vol.16, No.5, (October 2006), pp. 567-575.

Newman, E.A. & Frishman, L.J. (1991). The b-wave. In Arden, G.B. (ed.), *Principles and Practice of Clinical Electrophysiology of Vision,* Mosby-Year Book, St Louis, MO.

Peng, H.B., Ali, A.A., Daggett, D.F., Rauvala, H., Hassell, J.R., & Smalheiser, N.R. (1998). The relationship between perlecan and dystroglycan and its implication in the formation of the neuromuscular junction. *Cell Adhesion and Communication,* Vol.5, No.6, (September 1998), pp. 475-489

Peyrard, M., Seroussi, E., Sandberg-Nordqvist, A.C., Xie, Y.G., Han, F.Y.,Fransson, I., Collins, J., Dunham, I., Kost-Alimova, M., Imreh, S.,Dumanski, J.P., (1999). The human LARGE gene from 22q12.3-q13.1 is a new, distinct member of the glycosyltransferase gene family. *Proceedings of the National Academy of Sciences USA,* Vol.96, No.2, (January 1999), pp. 589-603.

Radomski, J.P. & Slonimski, P.P. (2001). Genomic style of proteins: concepts, methods and analyses of ribosomal proteins from 16 microbial species. *FEMS Microbiol Reviews,* Vol.25, No.4, (August 2001), pp. 425-435.

Sasaki, K., Kurata-Miura, K., Ujita, M., Angata, K., Nakagawa, S., Sekine, S., Nishi, T. & Fukuda, M. (1997). Expression cloning of cDNA encoding a human beta-1,3-N-acetylglucosaminyl transferase that is essential for poly-N-acetyllactosamine synthesis. *Proceedings of the National Academy of Sciences USA,* Vol.94, No.26, (December 1997), pp. 14294-14299.

Shiraishi, N., Natsume, A., Togayachi, A., Endo, T., Akashima, T., Yamada, Y., Imai, N., Nakagawa, S., Koizumi, S., Sekine, S., Narimatsu, H. & Sasaki K. (2001). Identification and characterization of 3 novel β1,3-N-Acetylglucosaminyltransferases. Structurally Related to the β1,3-Galactosyltransferase family. *The Journal of Biological Chemistry,* Vol. 276, No.5, (February 2001), pp. 3498-3507

Smalheiser, N. R., and Schwartz, N. B. (1987) Cranin: a laminin-binding protein of cell membranes. *Proceedings of the National Academy of Sciences USA,* Vol.84, No.18, (September 1987), pp. 6457-6461.

Sugita, S., Saito, F., Tang, J., Satz, J., Campbell, K., & Sudhof, T.C. (2001). A stoichiometric complex of neurexins and dystroglycan in brain. *Journal of Cell Biology,* Vol.154, No.2, (July 2001), pp. 435-445

Thompson, J.D., Higgins, D.G. & Gibson, T.J. (1994). CLUSTAL W: improving the sensitivity of progressive multiple sequence alignment through sequence weighting, position-specific gap penalties and weight matrix choice. *Nucleic Acids Research,* Vol.22, No.22, (November 1994), pp. 4673-4680.

Togayachi, A., Akashima, T., Ookubo, R., Kudo, T., Nishihara, S., Iwasaki, H., Natsume, A., Mio, H. Inokuchi J. and T. Irimura *et al.,* Molecular cloning and characterization of UDP-GlcNAc: Lactosylceramide β1,3-*N*-acetylglucosaminyltransferase (β3Gn-T5), an essential enzyme for the expression of HNK-1 and Lewis X epitopes on glycolipids, *Journal of Biological Chemistry,* Vol. 276, No.5, (March 2001), pp. 22032-22040

van Reeuwijk, J., Brunner, H.G., van Bokhoven, H. (2005). Glyc-O-genetics of Walker-Warburg syndrome. *Clinical Genetics,* Vol. 67, No.4, (April 2005), pp. 281-289.

Varki, A., Cummings, R.D., Esko, J.D., Freeze, H.H., Stanley, P., Bertozzi, C.R., Hart, G.W. & Etzler, M.E. (2008). *Essentials of Glycobiology,* (2nd ed.) Plainview (NY): Cold Spring Harbor Laboratory Press

Zhou, D., Dinter, A., Gutiérrez Gallego, R., Kamerling, J.P., Vliegenthart, J.F., Berger, E.G. & Hennet, T. (1999). A β-1,3-N-acetylglucosaminyltransferase with poly-N-acetyllactosamine synthase activity is structurally related to β-1,3-galactosyltransferases. *Proceedings of the National Academy of Sciences USA*, Vol. 96, No. 2, (January 1999), pp. 406-411

5

MicroArray Technology - Expression Profiling of MRNA and MicroRNA in Breast Cancer

Aoife Lowery, Christophe Lemetre, Graham Ball and Michael Kerin
[1]*Department of Surgery, National University of Ireland Galway,*
[2]*John Van Geest Cancer Research Centre, School of Science & Technology, Nottingham Trent University, Nottingham*
[1]*Ireland*
[2]*UK*

1. Introduction

Breast cancer is the most common form of cancer among women. In 2009, an estimated 194,280 new cases of breast cancer were diagnosed in the United States; breast cancer was estimated to account for 27% of all new cancer cases and 15% of cancer-related mortality in women (Jemal et al, 2009). Similarly, in Europe in 2008, the disease accounted for some 28% and 17% of new cancer cases and cancer-related mortality in women respectively (Ferlay et al, 2008). The increasing incidence of breast cancer worldwide will result in an increased social and economic burden; for this reason there is a pressing need from a health and economics perspective to develop and provide appropriate, patient specific treatment to reduce the morbidity and mortality of the disease. Understanding the aetiology, biology and pathology of breast cancer is hugely important in diagnosis, prognostication and selection of primary and adjuvant therapy. Breast tumour behaviour and outcome can vary considerably according to factors such as age of onset, clinical features, histological characteristics, stage of disease, degree of differentiation, genetic content and molecular aberrations. It is increasingly recognised that breast cancer is not a single disease but a continuum of several biologically distinct diseases that differ in their prognosis and response to therapy (Marchionni et al, 2008; Sorlie et al, 2001). The past twenty years has seen significant advances in breast cancer management. Targeted therapies such as hormonal therapy for estrogen receptor (ER) positive breast tumours and trastuzumab for inhibition of HER2/neu signalling have become an important component of adjuvant therapy and contributed to improved outcomes (Fisher et al, 2004; Goldhirsch et al, 2007; Smith et al, 2007). However, our understanding of the molecular basis underlying breast cancer heterogeneity remains incomplete. It is likely that there are significant differences between breast cancers that reach far beyond the presence or absence of ER or HER2/neu amplification. Patients with similar morphology and molecular phenotype based on ER, PR and HER2/neu receptor status can have different clinical courses and responses to therapy. There are small ER positive tumours that behave aggressively while some large high grade ER negative, HER2/neu receptor positive tumours have an indolent course. ER-positive tumours are typically associated with better clinical outcomes and a good response to hormonal therapies such as tamoxifen (Osborne et al, 1998). However, a subset of these patients recur and up to 40% develop resistance to hormonal therapy (Clarke et al, 2003). Furthermore, clinical studies have shown that adding adjuvant chemotherapy to tamoxifen in the treatment of node negative, ER positive breast cancer improves disease outcome (Fisher et al, 2004). Indeed, treatment with tamoxifen alone is only associated with a 15% risk of distant recurrence, indicating that 85% of these patients would do well without, and could be spared the cytotoxic side-effects of adjuvant chemotherapy.

The heterogeneity of outcome and response to adjuvant therapy has driven the discovery of further molecular predictors. Particular attention has focused on those with prognostic significance which may help target cancer treatment to the group of patients who are likely to derive benefit from a particular therapy. There has been a huge interest in defining the *gene expression profiles* of breast tumours to further understand the aetiology and progression of the disease in order to identify novel prognostic and therapeutic markers. The sequencing of the human genome and the advent of high throughput molecular profiling has facilitated comprehensive analysis of transcriptional variation at the genomic level. This has resulted in an exponential increase in our understanding of breast cancer molecular biology. Gene expression profiling using microarray technology was first introduced in 1995 (Schena et al, 1995). This technology enables the measurement of expression of tens of thousands of mRNA sequences simultaneously and can be used to compare gene expression within a sample or across a number of samples. Microarray technology has been productively applied to breast cancer research, contributing enormously to our understanding of the molecular basis of breast cancer and helping to achieve the goal of individualised breast cancer treatment. However as the use of this technology becomes more widespread, our understanding of the inherent limitations and sources of error increases. The large amount of data produced from such high throughput systems has necessitated the use of complex computational tools for management and analysis of this data; leading to rapid developments in bioinformatics.

This chapter provides an overview of current gene expression profiling techniques, their application to breast cancer prognostics and the bioinformatic challenges that must be overcome to generate meaningful results that will be translatable to the clinical setting. A literature search was performed using the PubMed database to identify publications relevant to this review. Citations from these articles were also examined to yield further relevant publications.

2. Microarray technology – principles & technical considerations

2.1 High throughput genomic technology

There are a multitude of high throughput genomic approaches which have been developed to simultaneously measure variation in thousands of DNA sequences, mRNA transcripts, peptides or metabolites:

- DNA microarray measures gene expression
- Microarray comparative genomic hybridisation (CGH) measures genomic gains and losses or identifies differences in copy number for genes involved in pathological states (Oosterlander et al, 2004)
- Single nucleotide polymorphism (SNP) microarray technology (Huang et al, 2001) has been developed to test for genetic aberrations that may predispose an individual to disease development.

- CpG arrays (Yan et al , 2000) can be used to determine whether patterns of specific epigenetic alterations correlate with pathological parameters.
- Protein microarrays (Stoll et al, 2005) consisting of antibodies, proteins, protein fragments, peptides or carbohydrate elements, are used to detect patterns of protein expression in diseased states.
- ChIP-on-chip (Oberley et al, 2004) combines chromatin immunoprecipitation (ChIP) with glass slide microarrays (chip) to detect how regulatory proteins interact with the genome.

All of these approaches offer unique insights into the genetic and molecular basis of disease development and progression.

This chapter focuses primarily on gene expression profiling and cDNA microarrays, however many of the issues raised, particularly in relation to bioinformatics are also applicable to the other "-omic" technologies.

Gene expression which is a measurement of gene "activity" can be determined by the abundance of its messenger RNA (mRNA) transcripts or by the expression of the protein which it encodes. ER, PR and HER2/neu receptor status are determined in clinical practice using immunohistochemistry (IHC) to quantitate protein expression or fluorescence in situ hybridisation (FISH) to determine copy number. These techniques are semi-quantitative and are optimal when determining the expression of individual or a small number of genes.

Microarray technology is capable of simultaneously measuring the expression levels of thousands of genes in a biological sample at the mRNA level. The abundance of individual mRNA transcripts in a sample is a reflection of the expression levels of corresponding genes. When a complementary DNA (cDNA) mixture reverse transcribed from the mRNA is labelled and hybridised to a microarray, the strength of the signal produced at each address shows the relative expression levels of the corresponding gene.

cDNA microarrays are miniature platforms containing thousands of DNA sequences which act as gene specific probes, immobilised on a solid support (nylon, glass, silicon) in a parallel format. They are reliant on the complementarity of the DNA duplex i.e. reassembly of strands with base pairing A to T and C to G which occurs with high specificity. There are microarray platforms available containing bound librarys of oligonucleotides representing literally all known human genes e.g. Affymetrix GeneChip (Santa Clara, CA), Agilent array (Santa Clara, CA), Illumina bead array (San Diego, CA). When fluorescence-labelled cDNA is hybridised to these arrays, expression levels of each gene in the human genome can be quantified using laser scanning microscopes. These microscopes measure the intensity of the signal generated by each bound probe; abundant sequences generate strong signals and rare sequences generate weaker signals. Despite differences in microarray construction and hybridization methodologies according to manufacturing, microarray-based measurements of gene expression appear to be reproducible across a range of different platforms when the same starting material is used, as demonstrated by the MicroArray Quality Control project (Shi et al, 2006).

2.2 Experimental approach

There are experimental design and quality control issues that must be considered when undertaking a microarray experiment. The experiment should be designed appropriately to answer a specific question and samples must be acquired from either patients or cultured cells which are appropriate to the experimental setup. If the aim of a microarray experiment is to identify differentially expressed genes between two groups of samples i.e. "experiment" and "control", it is critical that the largest source of variation results from the phenotype under investigation (e.g. patient characteristic or treatment). The risk of confounding factors influencing the results can be minimised by ensuring that the groups of samples being compared are matched in every respect other than the phenotype under investigation. Alternatively, large sample numbers can be used to increase the likelihood that the experimental variable is the only consistent difference between the groups.

For a microarray experiment, fresh frozen tissue samples are required which have been snap-frozen in liquid nitrogen or collected in an RNARetain™ or RNA Later™ solution to preserve the quality of the RNA. Formalin-fixed and paraffin embedded tissue samples are generally unsuitable for microarray studies as the RNA in the sample suffers degradation during tissue processing (Cronin et al, 2004; Masuda et al, 1999, Paik et al, 2005).

Due to the omnipresence of ribonucleases and the inherent instability of RNA, it is essential to measure the integrity of RNA after extraction. Only samples of the highest integrity should be considered for reverse transcription to cDNA and hybridisation to the microarray platform (figure 1). Once obtained, intensity readings must be background adjusted and transformed; this data is then normalised and analysed and results are generally interpreted according to biological knowledge. The success of microarray experiments is highly dependent on replication. Technical replication refers to the repeated assaying of the same biological sample to facilitate quality assessment. Even more important is biological replication on larger sample sets. The accuracy of microarray expression measurements must be confirmed using a reliable independent technology, such as real-time quantitative PCR, and validated on a larger set of independent biological samples. It is independent validation studies that determine the strength or clinical relevance of a gene expression profile.

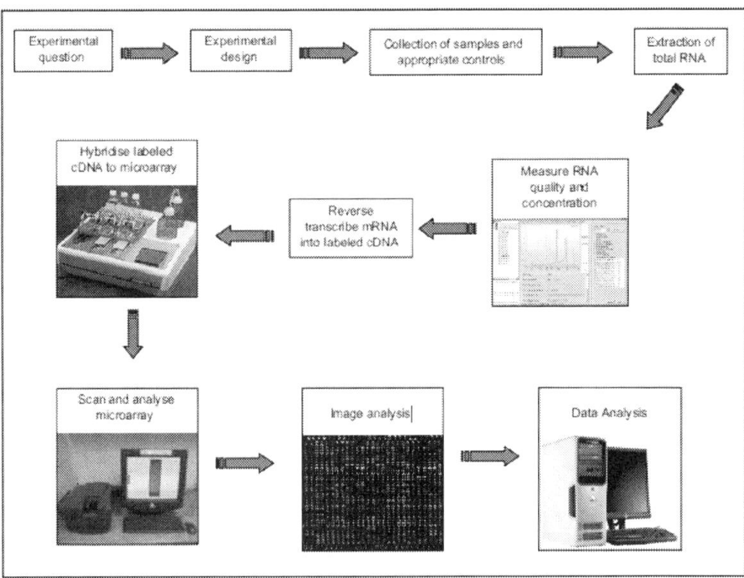

Fig. 1. The steps involved in a cDNA microarray experiment

3. Molecular profiling – unlocking the heterogeneity of breast cancer

Breast cancer researchers were quick to adopt high throughput microarray technology, which is unsurprising considering the opportunity it provides to analyse thousands of genes simultaneously.

3.1 Class discovery

Microarray studies can be used in three different manners;

- class comparison
- class prediction
- class discovery (Simon et al, 2003)

All of these approaches have been applied to the study of breast cancer.

Class discovery involves analyzing a given set of gene expression profiles with the goal of discovering subgroups that share common features. The early gene expression profiling studies of breast cancer (Perou et al, 2000; Sorlie et al, 2001) were class discovery studies. Researchers used an unsupervised method of analysis, in which tumours were clustered into subgroups by a 496-gene "intrinsic" gene set that reflects differences in gene expression between tumours without using selection criteria. The tumour subtype groupings consist of luminal like subtypes which are predominantly ER and PR positive, basal-like subtypes which are predominantly triple negative for ER, PR and HER2/neu, HER2/neu-like subtypes which have increased expression of the HER2/neu amplicon and a normal-like subtype (Perou et al, 2000). Subsequent studies from the same authors, on a larger cohort of patients with follow-up data showed that the luminal subgroup could be further subdivided into at least two groups, and that these molecular subtypes were actually associated with distinct clinical outcomes (Sorlie et al 2001). These molecular subtypes of breast cancer have been confirmed and added to in subsequent microarray datasets (Hu et al, 2006; Sorlie et al, 2003; Sotiriou et al, 2003). Given the importance of the ER in breast cancer biology, it is not surprising that the most striking molecular differences were identified between the ER-positive (luminal) and ER-negative subtypes. These differences have been repeatedly identified and validated with different technologies and across different platforms (Fan et al, 2006; Farmer et al, 2005; Sorlie et al, 2006). The luminal subgroup has been subdivided into two subgroups of prognostic significance:

- *luminal A tumours* which have high expression of ER -activated genes, and low expression of proliferation related genes
- *luminal B tumours* which have higher expression of proliferation related genes and a poorer prognosis than luminal A tumours (Geyer et al, 2009; Paik et al, 2000; Parker et al, 2009; Sorlie et al, 2001, 2003).

The ER negative tumours are even more heterogeneous and comprise the:

- *basal-like subgroup* which lack ER and HER2/neu expression and feature more frequent overexpression of basal cytokeratins, epidermal growth factor receptor and c-Kit (Nielsen et al, 2004)
- *HER2/neu subgroup* which overexpress HER2/neu and genes associated with the HER2/neu pathway and/or the HER2/neu amplicon on chromosome 17.

The HER2/neu and basal-like subtypes have in common an aggressive clinical behaviour but appear to be more responsive to neoadjuvant chemotherapy than the luminal subtypes (Carey et al, 2007; Rouzier et al, 2005). Also clustering with the ER negative tumours are the normal-like breast cancers; these are as yet poorly characterised and have been shown to cluster with fibroadenoma and normal breast tissue samples (Peppercorn et al, 2008). It is important at this point to acknowledge the limitations of this molecular taxonomy; intrasubtype heterogeneity has been noted despite the broad similarities defined by these large subtypes (Parker et al, 2009). In particular the basal-like subgroup can be divided into multiple additional subgroups (Kreike et al, 2007; Nielsen et al, 2004). Additionally, although the luminal tumours have been separated into subgroups of prognostic significance, meta-analysis of published expression data has suggested that these luminal tumours actually form a continuum and their separation based on expression of proliferation genes may be subjective (Shak et al, 2006; Wirapati et al, 2008). Furthermore, the clinical significance of the normal-like subtype is yet to be determined; it has been proposed that this subgroup may in fact represent an artefact of sample contamination with a high content of normal breast tissue (Parker et al, 2009; Peppercorn et al, 2008). Due to these limitations and the subjective nature of how the molecular subtypes were identified, the translation of this taxonomy to the clinical setting as a definitive classification has been difficult (Pustzai et al, 2006). The development of a prognostic test based on the intrinsic subtypes has not been feasible to date. However, the seminal work by Sorlie and Perou (Perou et al, 2000; Sorlie et al, 2001) recognized for the first time the scale of biological heterogeneity within breast cancer and led to a paradigm shift in the way breast cancer is perceived.

3.2 Class comparison

A number of investigators undertaking microarray expression profiling studies in breast cancer have since adopted class comparison studies. These studies employ supervised analysis approaches to determine gene expression differences between samples which already have a predefined classification. The "null hypothesis" is that a given gene on the array is not differentially expressed between the two conditions or classes under study. The alternative hypothesis is that the expression level of that gene *is* different between the two conditions. An example of this approach is the microarray experiments that have been undertaken to define differences between invasive ductal and invasive lobular carcinomas (Korkola, 2003; Weigelt, 2009; Zhao, 2004), between hereditary and sporadic breast cancer (Berns, 2001; Hedenfalk, 2001) and between different disease stages of breast cancer (Pedraza, 2010).

3.3 Class prediction

Perhaps the most clinically relevant use of this technology, however, are the microarray class prediction studies which have been designed to answer specific questions regarding gene expression in relation to clinical outcome and response to treatment. The latter approach attempts to identify predictive markers, as opposed to the prognostic markers which were identified in the "intrinsic gene-set". There is frequently some degree of confusion regarding the terms of "prognostic" and "predictive biomarkers". This is partially due to the fact that many prognostic markers also predict response to adjuvant therapy. This is particularly true in breast cancer where, for example, the ER is prognostic, and predictive of response to hormonal therapy, but also predictive of a poorer response to chemotherapy (Carey 2007; Kim, 2009; Rouzier 2005,).

One of the first microarray studies designed to identify a gene-set predictive of prognosis in breast cancer was that undertaken by van't Veer and colleagues (van't Veer et al, 2002). They developed a 70-gene set capable of predicting the development of metastatic disease in a group of 98 patients made up of 34 who had developed metastasis within 5-years of follow-up, 40 patients who remained disease-free at 5-years, 18 patients with a BRCA-1 mutation, and 2 patients with a BRCA-2 mutation. The 70-gene signature was subsequently validated in a set of 295 breast cancers, including the group used to train the model, and shown to be more accurate than standard histopathological parameters at predicting outcome in these breast cancer patients (van de Vijver et al, 2002). The signature includes many genes involved in proliferation, and genes associated with invasion, metastasis, stromal integrity and angiogenesis are also represented. This 70-gene prognostic signature classifies patients based on correlation with a "good-prognosis" gene expression profile; a coefficient of greater than 0.4 is classified as good prognosis. The signature was initially criticised for the inclusion of some patients in both the discovery and validation stages (van de Vijver et al, 2002). However, it has been subsequently validated in multiple cohorts of node-positive and node-negative patients and has been shown to outperform traditional clinical and histological parameters at predicting prognosis (Buyse et al, 2006; Mook et al, 2009).

3.3.1 Mammaprint assay

The 70-gene signature was approved by the FDA to become the MammaPrint Assay (Agendia BV, Amsterdam, The Netherlands); the first fully commercialized microarray based multigene assay for breast cancer. This prognostic tool is now available and can be offered to women under the age of 61 years with lymph node negative breast cancer. The MammaPrint test results are dichotomous, indicating either a high or low risk of disease recurrence, and the test performs best at the extremes of the spectrum of disease outcome i.e. identifying patients with a very good or a very poor prognosis.

The MammaPrint signature is a purely prognostic tool, and its role as a predictive marker for response to therapy was not examined at the time it was developed. Its' clinical utility is currently being assessed, however, in a prospective clinical trial called microarray in node negative and 1 to 3 positive lymph node disease may avoid chemotherapy (MINDACT) trial (Cardoso et al, 2008). The trial aims to recruit 6000 patients, all of whom will be assessed by standard clinicopathologic prognostic factors and by the MammaPrint assay. In cases where there is concordance between the standard prognostic factors and the molecular assay, patients will be treated accordingly with adjuvant chemotherapy with or without endocrine therapy for poor prognosis patients. If both assays predict a good prognosis, no adjuvant chemotherapy is given, and adjuvant hormonal therapy is given alone where indicated. In cases where there is disconcordance between the standard clinicopathological prognostic factors and the MammaPrint assays' prediction of prognosis the patients are randomised to receive adjuvant systemic therapy based on either the clinicopathological or the MammaPrint prognostic prediction results. The expected outcome is that there will be a reduction of 10-15% in the number of patients requiring adjuvant chemotherapy based on the MammaPrint assay prediction. It is envisaged that this trial will answer the questions of what patients can be spared chemotherapy and still have a good prognosis, thus accelerating progress towards the goal of more tailored therapy for breast cancer patients.

3.3.2 Oncotype Dx assay

While MammaPrint was developed as a prognostic assay, the other most widely established commercialized multigene assay Oncotype Dx was developed in a more context specific manner as a prognostic *and* predictive test to determine the benefit of chemotherapy in women with node-negative, ER-positive breast cancer treated with tamoxifen (Paik et al, 2004). The authors used published microarray datasets, including those that identified the intrinsic breast cancer subtypes and the 70-gene prognostic signature identified by the Netherlands group to develop real time quantitative polymerase chain reaction (RQ-PCR) tests for 250 genes. Research undertaken by the National Surgical Adjuvant Breast and Bowel Project (NSABP) B14 protocol using three independent clinical series, resulted in the development of an optimised 21-gene predictive assay (Paik et al, 2004). The assay has been commercialised as Onco*type*® DX by Genomic Health Inc[1] and consists of a panel of 16 discriminator genes and 5 endogenous control genes which are detected by RQ-PCR using formalin-fixed paraffin embedded (FFPE) sections from standard histopathology blocks. The ability to use FFPE tissue facilitates clinical translation and has allowed retrospective analysis of archived tissue in large cohorts with appropriate follow up data. The assay has been used to generate Recurrence Scores (RS) by differentially weighting the constituent genes which are involved in:

- proliferation (MKI67, STK15, BIRC5/Survivin, CCNB1, MYBL2)
- estrogen response (*ER, PGR, SCUBE2*)
- HER2/neu amplicon (HER2/neu/ERBB2, GRB7),
- invasion (MMP11, CTSL2)
- apoptosis (*BCL2, BAG1*)
- drug metabolism (*GSTM1*)
- macrophage response (*CD68*).

The assay was evaluated in 651 ER positive lymph node negative breast cancer patients who were treated with either tamoxifen or tamoxifen and chemotherapy as part of the NSABP B20 protocol (Paik et al, 2006). It was found that patients with high recurrence scores had a large benefit from chemotherapy, with a 27.6% mean decreased in 10 year distance recurrence rates, while those with a low recurrence score derived virtually no benefit from chemotherapy. The RS generated by the expression of the 21 genes is a continuous variable ranging from 1-100, but has been divided into three groups for clinical decision making; low (<18), intermediate (18-31) and high (>31). It has been shown in a number of independent datasets that ER positive breast cancer patients with a low RS have a low risk of recurrence and derive little benefit from chemotherapy. Conversely, ER positive patients with high RS have a high risk of recurrence but do benefit from chemotherapy (Goldstein, 2006; Habel, 2006; Mina, 2007; Paik, 2006). The ability of the 21-gene signature to so accurately predict prognosis has led to the inclusion of the Oncotype Dx assay in American Society of Clinical Oncology (ASCO) guidelines on the use of tumour markers in breast cancer as a predictor of recurrence in ER-positive, node-negative patients. However, despite the accurate performance of the assay for high and low risk patients, there remains uncertainty regarding the management of patients with intermediate RS (18-31). This issue is being addressed in a prospective randomized trial assigning individual options for treatment (TAILORx) sponsored by the National Cancer Institute (Lo et al, 2007). This multicentre trial aims to recruit 10,000 patients with ER -positive, lymph node negative breast cancer who are assigned to one of three groups based on their RS; low<11, intermediate 11-25 and high >25. Notably, the RS criteria have been changed for the TAILORx trial, with the intermediate

[1]http://www.genomichealth.com/OncotypeDX

range being changed from RS 18-30 to RS 11-25 to avoid exluding patients who may derive a small benefit from chemotherapy (Sparano et al, 2006). Patients in the intermediate RS group are randomly assigned to receive either adjuvant chemotherapy and hormonal therapy, or hormonal therapy alone. The primary aim of the trial is to determine if ER positive patients with an intermediate RS benefit from adjuvant chemotherapy or not.

The MammaPrint and Oncotype Dx gene signatures both predict breast cancer behaviour, however there are fundamental differences between them (outlined in table 1). This chapter has focused on these signatures as they were the first to be developed, have been extensively validated, and are commercially available. However it is important to note that there are other multi-gene based assays that have been developed and commercialized but are not discussed in detail as they are not yet as widely utilized (Loi et al, 2007; Ma et al, 2008; Ross et al, 2008; Wang et al, 2005).

Assay	MammaPrint	Oncotype Dx
Manufacturer	Agendia BV	Genomic Health, Inc.
Development of Signature	From candidate set of 25,000 genes in 98 patients	From candidate set of 250 genes in 447 patients
Gene signature	70 genes	21 genes
Patient cohort	Stage I & II breast cancer Lymph node negative <55yrs	Stage I & II breast cancer Lymph node negative ER positive Receiving Tamoxifen
Platform	cDNA Microarray	RQ-PCR
Sample requirements	Fresh frozen tissue or collected in RNA preservative	FFPE tissue
Outcome	5-year distant relapse free survival	10-year distant relapse free survival
Test Results	Dichotomous correlation coefficient >4.0 = good prognosis <4.0 = poor prognosis	Continuous recurrence score <18 = low risk 18-31= intermediate risk >31 = high risk
Predictive	No; purely prognostic	Yes
Prospective Trial	MINDACT	TAILORx
FDA approved	Yes	No
ASCO Guidelines	No	Yes

Table 1. Comparison of commercially available prognostic assays MammaPrint and Oncotype Dx

4. Microarray data integration

4.1 Setting standards for microarray experiments

It must be acknowledged that despite the multitude of breast cancer prognostic signatures available, the overlap between the gene lists is minimal (Ahmed, 2005; Brenton, 2005; Fan et al, 2006; Michiels et al, 2005). This lack of concordance has called into question the applicability of microarray analysis across the entire breast cancer population. In order to facilitate external validation of signatures and meta-analysis in an attempt to devise more robust signatures, it is important that published microarray data be publicly accessible to the scientific community. In 2001 the Microarray Gene Expression Data Society proposed experimental annotation standards known as minimum information about a microarray experiment (MIAME), stating that raw data supporting published studies should be made publicly available in one of a number of online repositories (table 2), these standards are now upheld by leading scientific journals and facilitating in depth interrogation of multiple datasets simultaneously.

Public Database for Microarray Data	URL	Organization	Description
Array Express	http://www.ebi.ac.uk/arrayexpress/	European Bioinformatics Institute (EBI)	Public data deposition and queries
GEO Gene Expression Omnibus	http://www.ncbi.nlm.nih.gov/geo/	National Centre for Biotechnology Information (NCBI)	Public data deposition and queries
CIBEX Center for Information Biology Gene Expression Database	http://cibex.nig.ac.jp/index.jsp	National Institute of Genetics	Public data deposition and queries
ONCOMINE Cancer Profiling Database	http://www.oncomine.org/main/index.jsp	University of Michigan	Public queries
PUMAdb Princeton University MicroArray database	http://puma.princeton.edu/	Princeton University	Public queries
SMD Stanford Microarray Database	http://genome-www5.stanford.edu/	Stanford Univeristy	Public queries
UNC Chapel Hill Microarray database	https://genome.unc.edu/	University of North Carolina at Chapel Hill	Public queries

Table 2. List of Databases with Publicly Available Microarray Data

4.2 Gene ontology

The volume of data generated by high throughput techniques such as microarray poses the challenge of how to integrate the genetic information obtained from large scale experiments with information about specific biological processes, and how genetic profiles relate to functional pathways. The development of the Gene Ontology (GO) as a resource for experimentalists and bioinformaticians has contributed significantly to overcoming this challenge (Ashburner et al, 2000). The GO Consortium was established with the aim of producing a structured, precisely defined, common, controlled vocabulary for describing the roles of genes and gene products in any organism. Initially a collaboration between three organism databases: Flybase (The Flybase Consortium, 1999), Mouse Genome Informatics (Blake et al, 2000) and the Saccharomyces Genome Database (Ball et al, 2000), the GO Consortium has grown to include several of the world's major repositories for plant, animal and microbial genomes.

The Gene Ontology provides a structure that organizes genes into biologically related groups according to three criteria. Genes and gene products are classified according to:

- Molecular Function: biochemical activity of gene products at the molecular level
- Biological Process: biological function of a gene product
- Cellular Component: location in the cell or extracellular environment where molecular events occur

Every gene is described by a finite, uniform vocabulary. Each GO entry is defined by a numeric ID in the format GO#######. These GO identifiers are fixed to the textual definition of the term, which remains constant. A GO annotation is the specific association between a GO identifier and a gene or protein and has a distinct evidence source that supports the association. A gene product can take part in one or more biological process and perform one or more molecular functions. Thus, a well characterized gene product can be annotated to multiple GO terms in the three GO categories outlined above. GO terms are related to each-other such that each term is placed in the context of all of the other terms in a node-directed acyclic graph (DAC). The relationships used by the GO are: "is_a", "part_of", "regulates", "positively_regulates", "negatively_regulates" and "disjoint_from". Each term in the DAC may have one or more parent terms and possibly one or more child nodes, and the DAC gives a graphical representation of how GO terms relate to each other in a hierarchical manner.

The development of Gene Ontology has facilitated analysis of microarray gene sets in the context of the molecular functions and pathways in which they are involved (Blake & Harris, 2002). GO-term analysis can be used to determine whether genetic "hits" show enrichment for a particular group of biological processes, functions or cellular compartments. One approach uses statistical analysis to determine whether a particular GO is over or under-represented in the list of differentially expressed genes from a microarray experiment. The statistical tests used for such analysis include hypergeometric, binomial or Chi-square tests (Khatri et al, 2005).

An alternative approach known as "gene-set testing" has been described which involves beginning with a known set of genes and testing whether this set as a whole is differentially expressed in a microarray experiment (Lamb et al, 2003; Mootha et al, 2003). The results of such analyses inform hypotheses regarding the biological significance of microarray analyses.

Several tools have been developed to facilitate analysis of microarray data using GO, and a list of these can be found at: http://www.geneontology.org/GO.tools.microarray.shtml

Analysing microarray datasets in combination with biological knowledge provided by GO makes microarray data more accessible to the molecular biologist and can be a valuable strategy for the selection of biomarkers and the determination of drug treatment effect in breast cancer (Arciero et al, 2003; Cunliffe et al, 2003).

4.3 Microarray meta-analysis – combining datasets

Meta-analyses have confirmed that different prognostic signatures identify similar biological subgroups of breast cancer patients (Fan et al, 2006) and have also shown that the designation of tumours to a "good prognosis"/"low risk" group or a "poor prognosis"/"high risk" group is largely dependent on the expression patterns of proliferative genes. In fact, some of these signatures have been shown to have improved performance when only the proliferative genes are used (Wirapati, 2008). Metanalyses of the signatures have also proposed that the prognostic ability of the signatures is optimal in the ER positive and HER2-negative subset of breast tumours (Desmedt, 2008; Wirapati, 2008), the prognosis of this group of tumours being governed by proliferative activity.

Despite obvious clinical application, none of these prognostic assays are perfect, and they all carry a false classification rate. The precise clinical value for these gene expression profiles remains to be established by the MINDACT and TAILORx trials. In the interim the performance of these assays is likely to be optimised by combining them with data from traditional clinicopathological features, an approach which has been shown to increase prognostic power (Sun et al, 2007).

Microarray technology has undoubtedly enhanced our understanding of the molecular mechanisms underlying breast carcinogenesis; profiling studies have provided a myriad of candidate genes that may be implicated in the cancer process and are potentially useful as prognostic and predictive biomarkers or as therapeutic targets. However, as yet there is little knowledge regarding the precise regulation of these genes and receptors, and further molecular categories are likely to exist in addition to and within the molecular subtypes already delineated. Accumulating data reveal the incredible and somewhat foreboding complexity and variety of breast cancers and while mRNA expression profiling studies are ongoing, a new player in breast cancer biology has come to the fore in recent years; a recently discovered RNA species termed MiRNA (miRNA) which many scientists believe may represent a crucial link in the cancer biology picture.

5. MicroRNA - a recently discovered layer of molecular complexity

It has been proposed that the discovery of miRNAs as regulators of gene expression represents a paradigm changing event in biology and medicine. This discovery was made in 1993 by researchers at the Ambros laboratory in Dartmouth Medical School, USA at which time it was thought to be a biological entity specific to the nematode *C. Elegans* (Lee et al, 1993). In the years following this discovery, hundreds of miRNAs were identified in animals and plants. However it is only in the past 5 years that the field of miRNA research has really exploded with the realisation that miRNAs are critical to the development of multicellular organisms and the basic functions of cells (Bartel, 2004). MiRNAs are fundamental to genetic regulation, and their aberrant expression and function have been linked to numerous diseases and disorders (Bartel, 2004; Esquela-Kerscher & Slack, 2006). Importantly, miRNA have been critically implicated in the pathogenesis of most human cancers, thus uncovering an entirely new repertoire of molecular factors upstream of gene expression.

5.1 MicroRNA - novel cancer biomarkers

The first discovery of a link between miRNAs and malignancy was the identification of a translocation-induced deletion at chromosome 13q14.3 in B-cell Chronic Lymphocytic

Leukaemia (Calin et al, 2002). Loss of *miR-15a* and *miR-16-1* from this locus results in increased expression of the anti-apoptotic gene *BCL2*. Intensifying research in this field, using a range of techniques including miRNA cloning, quantitative PCR, microarrays and bead-based flow cytometric miRNA expression profiling has resulted in the identification and confirmation of abnormal miRNA expression in a number of human malignancies including breast cancer (Heneghan et al, 2010; Lowery et al, 2007). MiRNA expression has been observed to be upregulated or downregulated in tumours compared with normal tissue, supporting their dual role in carcinogenesis as either oncogenic miRNAs or tumour suppressors respectively (Lu et al, 2005). The ability to profile miRNA expression in human tumours has led to remarkable insight and knowledge regarding the developmental lineage and differentiation states of tumours. It has been shown that distinct patterns of miRNA expression are observed within a single developmental lineage, which reflect mechanisms of transformation, and support the idea that miRNA expression patterns encode the developmental history of human cancers. In contrast to mRNA profiles it is possible also to successfully classify poorly differentiated tumours using miRNA expression profiles (Volinia et al, 2006). In this manner, miRNA expression could potentially be used to accurately diagnose poorly differentiated tissue samples of uncertain histological origin, e.g. metastasis with an unknown primary tumour, thus facilitating treatment planning.

MicroRNAs exhibit unique, inherent characteristics which make them particularly attractive for biomarker development. They are known to be dysregulated in cancer, with pathognomonic or tissue specific expression profiles and even a modest number of miRNAs is sufficient to classify human tumours, which is in contrast to the relatively large mRNA signatures generated by microarray studies (Lu et al, 2005). Importantly, miRNA are remarkably stable molecules. They undergo very little degradation even after processing such as formalin fixation and remain largely intact in FFPE clinical tissues, lending themselves well to the study of large archival cohorts with appropriate follow-up data (Li et al, 2007; Xi et al, 2007). The exceptional stability of miRNAs in visceral tissue has stimulated investigation into their possible preservation in the circulation and other bodily fluids (urine, saliva etc.). The hypothesis is that circulating miRNAs, if detectable and quantifiable would be the ideal biomarker accessible by minimally invasive approaches such as simple phlebotomy (Cortez et al, 2009; Gilad et al, 2008; Mitchell et al, 2008).

5.2 MicroRNA microarray

The unique size and structure of miRNAs has necessitated the modification of existing laboratory techniques, to facilitate their analysis. Due to the requirement for high quality large RNA molecules, primarily for gene expression profiling, many laboratories adopted column-based approaches to selectively isolate large RNA molecules, discarding small RNA fractions which were believed to contain degradation products. Modifications to capture miRNA have been made to existing protocols to facilitate analysis of the miRNA fraction. Microarray technology has also been modified to facilitate miRNA expression profiling. Labelling and probe design were initially problematic due to the small size of miRNA molecules. Reduced specificity was also an issue due to the potential of pre-miRNA and pri-miRNAs to produce signals in addition to active mature miRNA. Castoldi *et al* described a novel miRNA microarray platform using locked nucleic acid (LNA)-modified capture probes (Castoldi et al, 2006). LNA modification improved probe thermostability and increased specificity, enabling miRNAs with single nucleotide differences to be discriminated - an important consideration as sequence-related family members may be involved in different physiological functions (Abbott et al, 2005). An alternative high throughput miRNA profiling technique is the bead-based flow cytometric approach developed by Lu *et al.*; individual polystyrene beads coupled to miRNA complementary probes are marked with fluorescent tags (Lu et al, 2005). After hybridization with size-fractioned RNAs and streptavidin-phycoerythrin staining, the beads are analysed using a flow-cytometer to measure bead colour and pycoerythrin, denoting miRNA identity and abundance respectively. This method offered high specificity for closely related miRNAs because hybridization occurs in solution. The high-throughput capability of array-based platforms make them an attractive option for miRNA studies compared to lower throughput techniques such as northern blotting and cloning; which remain essential for the validation of microarray data.

5.2.1 MicroRNA microarray - application to breast cancer

Microarray analysis of miRNA expression in breast cancer is in its' infancy relative to expression profiling of mRNA. However, there is increasing evidence to support the potential for miRNAs as class predictors in breast cancer. The seminal report of aberrant miRNA expression in breast cancer by Iorio et al. in 2005 identified 29 miRNAs that were differentially expressed in breast cancer tissue compared to normal, a subset of which could correctly discriminate between tumour and normal with 100% accuracy (Iorio et al, 2005). Among the leading miRNAs differentially expressed; *miR-10b*, *miR-125b* and *mR-145* were downregulated whilst *miR-21* and *miR-155* were consistently over-expressed in breast tumours. In addition, miRNA expression correlated with biopathological features such as ER and PR expression (*miR-30*) and tumour stage (*miR-213* and *miR-203*). Mattie et al. subsequently identified unique sets of miRNAs associated with breast tumors defined by their HER2/neu or ER/PR status (Mattie et al, 2006). We have described 3 miRNA signatures predictive of ER, PR and Her2/neu receptor status, respectively, which were identified by applying artificial neural network analysis to miRNA microarray expression data (Lowery et al, 2009). Blenkiron et al used an integrated approach of both miRNA and mRNA microarray expression profiling to classify tumours according to "intrinsic subtype". This approach identified a number of miRNAs that are differentially expressed according to intrinsic breast cancer subtype and associated with clinicopathological factors including ER status and tumour grade. Importantly, there was overlap between the differentially expressed miRNAs identified in these studies.

There has been interest in assessing the prognostic value of miRNAs, and expression studies in this regard have focused on detecting differences in miRNA expression between primary breast tumours and metastatic lymph nodes. This approach has identified numerous miRNA that are dysregulated in primary breast tumours compared to metastatic lymph nodes (Baffa et al 2009; Huang et al, 2008). MiRNA have also been identified that are differentially expressed in patients who had a "poor prognosis" or a short time to development of distant metastasis (Foekens et al, 2008); miR-516-3p, miR-128a, miR-210, and miR-7 were linked to aggressiveness of lymph node-negative, ER-positive human breast cancer.

The potential predictive value of miRNA is also under investigation. Preclinical studies have reported associations between miRNA expression and sensitivity to adjuvant breast cancer therapy including chemotherapy, hormonal therapy and HER2/neu targeted therapy (Ma et al, 2010; Tessel et al, 2010; Wang et al, 2010), prompting analysis of tumour response in clinical samples. Rodriguez-Gonzalez et al attempted to identify miRNAs related to response to tamoxifen therapy by exploiting the Foekens dataset (Foekens, 2008) which comprised miRNA expression levels of 249 miRNAs in 38 ER positive breast cancer patients. Fifteen of these patients were hormone naive and experienced relapse, which was treated with tamoxifen. Ten patients responded and five did not, progressing within 6 months. Five miRNAs (miR-4221, miR-30a-3p, miR-187, miR-30c and miR-182) were the most differentially expressed between patients who benefitted from tamoxifen and those who failed therapy. The predictive value for these miRNAs was further assessed in 246 ER positive primary tumours of hormone naive breast cancer patients who received tamoxifen as monotherapy for metastatic disease. MiR-30a-3p, miR-30c and miR-182 were significantly associated with response to tamoxifen, but only miR-30c remained an independent predictor on multivariate analysis (Rodriguez-Gonzalez, 2010).

Microarray-based expression profiling has also been used to identify circulating miRNAs which are differentially expressed in breast cancer patients and matched healthy controls. Zhao et al profiled 1145 miRNAs in the plasma of 20 breast cancer patients and 20 controls, identifying 26 miRNAs with at least two-fold differential expression which reasonably separated the 20 cases from the 20 controls (Zhao et al, 2010). This is the first example of genome-wide miRNA expression profiling in the circulation of breast cancer patients and indicates potential for development of a signature of circulating miRNAs that may function as a diagnostic biomarker of breast cancer.

At present diagnostic, prognostic and predictive miRNA signatures and markers remain hypothesis generating. They require validation in larger, independent clinical cohorts prior to any consideration for clinical application. Furthermore as additional short non-coding RNAs are continuously identified through biomarker discovery programmes, the available profiling technologies must adapt their platforms to incorporate newer potentially relevant targets. MicroRNAs possess the additional attraction of potential for development as therapeutic targets due to their ability to regulate gene expression. It is likely that future microarray studies will adopt and integrated approach of miRNA and mRNA expression analysis in an attempt to decipher regulatory pathways in addition to expression patterns.

6. Limitations of microarray technology & bioinformatic challenges

In addition to the great promises and opportunities held by microarray technologies, several issues need to be borne in mind and appropriately addressed in order to perform reliable and non-questionable experiments. As a result, several steps need to be addressed in order to identify and validate reliable biomarkers in the scope of potential future clinical application. This is one of the reasons why, despite the promises of using powerful high-throughput technologies as such as microarray, only very few useful biomarkers have been identified so far and/or have been translated to useful clinical assay or companion diagnostics (Mammaprint®, Oncotype DX®). There still remains a lack of clinically relevant biomarkers (Rifai et al, 2006). Amongst the limitations and pitfalls around the technology and the use of microarrays, some of the most important are the reported lack of reproducibility, as well as the massive amount of data generated, often extremely noisy and with an increasing complexity. As for example, in the recent Affymetrix GeneChip 1.0 ST microarray platform (designed to target all known and predicted exons in human, mouse and rat genomes), where there is approximately 1.2 million exon clusters corresponding to

over 1.4 million probesets (Lancashire et al, 2009). As a result, it appears clearly that extracting any relevant key component from such datasets requires robust mathematical and/or statistical models running on efficient hardware to perform the appropriate analyses.
With this in mind, it is clear that the identification of new biomarkers still requires a concerted, multidisciplinary effort. It requires the expertise of the biologist or pathologist, to extract the samples, the scientist to perform the analysis on the platform and then the bioinformatician/biostatistician to analyse and interpret the output. The data-mining required to cope with these types of data needs careful consideration and specific computational tools, and as such remains a major challenge in bioinformatics.

6.1 Problems with the analysis of microarray data

6.1.1 Dimensionality and false discovery

The statistical analysis of mRNA or miRNA array data poses a number of challenges. This type of data is of extremely high dimensionality *i.e.* has a large number of variables. Each of these variables represents the relative expression of a mRNA or miRNA in a sample. Each of these components contain noise, are non-linear may not follow a normal distribution through a population and may be strongly correlated with other probes in the profile. These characteristics mean that the data may violate many of the assumptions of conventional statistical techniques, particularly with parametric tests.
The dimensionality of the data poses a significant problem, and remains as one of the most critical when analysing microarray data. When one analyses this type of data, one has to consider what is referred to as *the curse of dimensionality*, firstly described by Bellman in 1961 as the *"exponential growth of the search space as a function of dimensionality"* (Bellman, 1961; Bishop, 1995). This occurs in highly dimensional systems where the number of dimensions masks the true importance of an individual single dimension (variable). It is particularly true in a microarray experiment when the number of probes representing the number of miRNA/mRNA studied far exceeds the number of available samples. So there is the potential for a probe that is in reality of high importance to be missed when considered with a large number of other probes. This problem is overcome by breaking down the analysis into single or small groups of variables and repeating the analysis rather than considering the whole profile in one single analysis. Other methods consists of using pre-processing methods and feature extraction algorithms in order to only analyse a subset of the data supposed to hold the most relevant features (Bishop, 1995), as determined by the pre-processing steps.
High dimensionality also creates problems due to false discovery. The false discovery rate (FDR) introduced by Benjamini and Hochberg (Benjamini and Hochberg, 1995) is a measure of the number of features incorrectly identified as "differential" and various approaches have been suggested to accurately control the FDR. In this case if one has a high number of dimensions and analyses each singly (as above) a proportion can appear to be of high importance due to random chance considering the distribution, even when they are not. To overcome this one has to examine a rank order of importance and when testing for significance one has to correct the threshold for significance by dividing it by the number of dimensions. So for example when analysing the significance of single probes from a profile with 4,000 probes in it the threshold becomes $P < 0.05$ divided by 4,000 i.e. $P < 0.0000125$.

6.1.2 Quality and noise

Noise also poses a problem in the analysis of mRNA or miRNA data. The inherent technical and biological variability necessarily induces noise within the data, eventually leading to biased results. The noise may lead to misinterpretation of sample groups that may actually have no biological relevance. As a consequence extreme care needs to be taken to address the problem of noise.

Noise may be random where it is applied to all parts of the miRNA equally or systematic where particular probes inherently have more noise than others because of the nature of the component miRNA or genomic code that they represent.

It is now widely acknowledged that the reported high level of noise found in microarray data is the most critical pull-back of microarray-based studies, as it is pointed by the MAQC Consortium (Shi et al, 2006; Klebanov and Yakovlev, 2007).

6.1.3 Complexity and non-normality

Because of the complex nature of the profile a particular mRNA or miRNA may be non-normally distributed through a population. Such non-normality will immediately invalidate any statistical test that uses parametric statistics i.e. depends on the assumption of a normal distribution. Invalidated tests would include ANOVA and t-test. To overcome this, the data would have to be transformed mathematically to follow a normal distribution or an alternative non parametric test would have to be employed. Examples of non-parametric tests include Kruskal-Wallis and Mann Whitney U which are ANOVA and unpaired T-Test alternatives respectively. Generally non-parametric tests lack power compared to their parametric alternatives and this may prove to be a problem in high dimensional space due to the reasons described previously.

6.1.4 Reproducibility

Reproducibility has a marked effect on the accuracy of any analysis conducted. Furthermore reproducibility has a profound effect on the impact of other issues such as dimensionality and false detection. Robust scientific procedures requires that the results have to be reproducible in order to reduce the within sample variability, the variability between sample runs and the variability across multiple reading instruments. Aspects of variability can be addressed using technical and experimental replicates. The averaging of samples profiles can be used to increase the confidence in the profiles for comparison (Lancashire et al., 2009). Technical replicates provide information on the variability associated with instrumental variability whilst experimental (or biological) replicates give a measure of the natural sample to sample variation. Problems in data analysis occur when the technical variability is high. In this situation the problem in part can be resolved by increasing the number of replicates. If however the technical variation is higher than the biological variation then the sample cannot be analysed.

6.1.5 Auto-correlation or co-correlation

Auto correlation exists when two components within a system are strongly linearly correlated with one another. In any complicated system there are likely to be a number of components that are auto correlated. This is especially true in array profiling of biological samples. Firstly due to biological processes one protein in a set of samples is likely to interact or correlate with another through a population.

Auto correlation becomes a problem when using linear based regression approaches. This is because one of the assumptions of regression using multiple components is that the components are not auto correlated. If intensity for multiple miRNA probes are to be added into a regression to develop a classifier these components should not be auto correlated. Auto correlation can be tested for using the Durbin Watson test.

6.1.6 Generality

The whole purpose of biomarker (or set of biomarkers) identification, using high-throughput technologies or any other, is to provide the clinicians with an accurate model in order to assess a particular aspect. However, a model is only as good as its ability to generalize to unseen real world data. A model only able to explain the population on which it was developed would be purely useless for any application.

As a result, if one is to develop classifiers from mRNA or miRNA array data the features identified should be generalised. That is they will predict for new cases in the general population of cases. When analysing high dimensional data there is an increased risk of over fitting, particularly when the analysis methods imply supervised training on a subset of the population. So for example, when a large number of mRNA or miRNA are analysed there is the potential for false detection to arise. If a random element identified through false detection is included as a component of a classifier (model) then the generality of that classifier will be reduce; i.e. it is not a feature that relates to the broader population but is a feature specific to the primary set of data used to develop the classifier. Standards of validation required to determine generality have been defined by Michiels et al, 2007.

Generality of classifiers can be increased by the application of bootstrapping or cross validation approaches.

Some algorithms and approaches, that usually involve supervised training, suffer from over-fitting (sometimes called memorisation). This is a process where a classifier is developed for a primary dataset but models the noise within the data as well as the relevant features. This means that the classifier will not accurately classify for new cases i.e. it does not represent a general solution to the problem which is applicable to all cases. This is analogous, for example, to one developing a classifier that predicts well the risk of metastasis for breast cancer patients from Nottingham but will not predict well for a set of cases from Denmark. Over fitted classifiers seldom represent the biology of the system being investigated and the features identified are often falsely detected.

One of the most common solutions to avoid over-fitting is to apply a Cross Validation technique in combination with the supervised training. Random sample cross validation is a process of mixing data. Firstly the data are divided into two or three parts (figure 2); the first part is used to develop the classifier and the second or second and third parts are used to test the classifier. These parts are sometimes termed training, test and validation data sets respectively. In certain classifiers such as Artificial Neural Network based classifiers the second blind set is used for optimisation and to prevent over fitting. In random sample cross validation the random selection and training process is repeated a number of times to create a number of models each looking at the global dataset in a number of different ways (figure 2). Often the mean performance of these models is considered.

Leave one out cross validation is an approach also used to validate findings. In this case one sample is left out of the analysis. Once training is complete the sample left out is tested. This process is repeated a number of times to determine the ability of a classifier to predict unseen cases. This approach of random sample cross validation drives the classifier solution to a generalised one by stopping the classifier from training too much on a seen dataset and stopping the training earlier based on a blind dataset.

7. Methods used to analyse microarray data and their limitations

With the advent of cutting edge new technologies such as microarrays, the analysis tools for the data produced need to be appropriately applied. Although expression arrays have brought high hopes and expectations, they have brought tremendous challenges with them. They have been proven to suffer from different limitations as previously discussed. However, innovative computational analysis solutions have been developed and have been proven efficient and successful at identifying markers of interest regarding particular questions. This section presents some of the most common methods employed to overcome the limitations discuss above, and to analyse expression array data.

7.1 Application of ordination techniques

If we are to utilise the mRNA or miRNA profile we have to identify robust features despite its high dimensionality that are statistically valid for the general population not just for a subset. Ordination techniques are used to map the variation in data. They are not directly predictive and cannot classify directly unless combined with another classification technique.

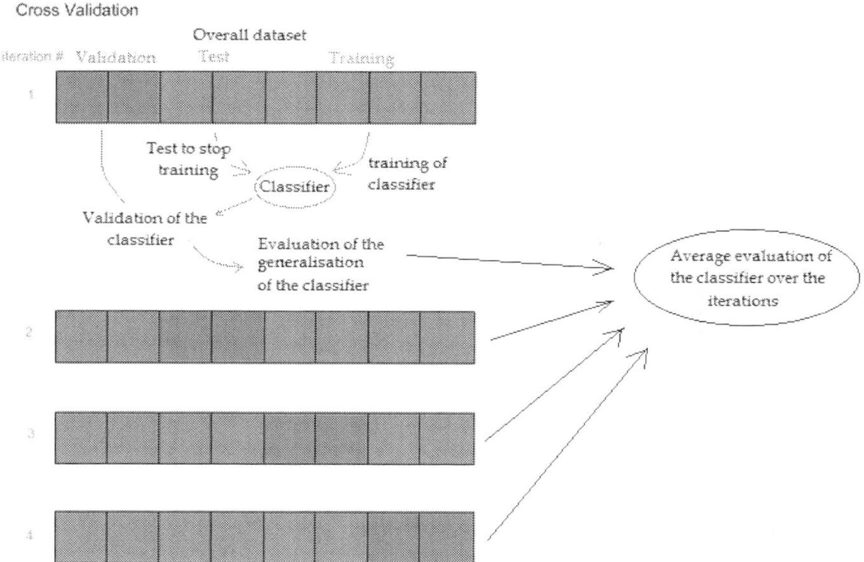

Fig. 2. Illustration of Cross Validation technique, here with three subsets: the training subset used to train the classifier, the test subset used to stop the training when it has reached an optimal performance on this subset, and a validation subset to evaluate the performance (generalization ability) of the trained classifier.

7.1.1 Principal components analysis

PCA is usually a method of choice for dimensionality reduction. It is a multivariate exploratory technique used to simplify complex data space (Raychaudhuri et al, 2000) by translating the data space into a new space defined by the principal components. It works by identifying the main (principal) components that explain best the shape (variance) of a data set. Each principal component is a vector (line) through the data set that explains a proportion of the variance, it is the expression of a linear combination of the data. In PCA the first component that is added is the one that explains the most variance the second component added is then orthogonal to the first. Subsequent orthogonal components are added until all of the variation is explained. The addition of vectors through a multidimensional data set is difficult to visualise in print, we have tried to illustrate it with 3 dimensions in figure 3. In mRNA/miRNA profile data where thousands of dimensions exist, PCA is a useful technique as it reduces the dimensionality to a manageable number of principal components. If the majority of the variance is explained in 2 or 3 principal components these can be used to visualise the structure of the population using 2 or 3 dimensional plots. A limited parameterisation can also be conducted to determine the contribution of each parameter (miRNA) to each of the principal components. This however suffers from the curse of dimensionality in high dimensional systems. Thus the main limitation of using PCA for gene expression data is the inability to verify the association of a principal component vector with the known experimental variables (Marengo et al, 2004). This often makes it difficult to accurately identify the importance of the mRNA or miRNA in the system, and make it a valuable tool only for data reduction.

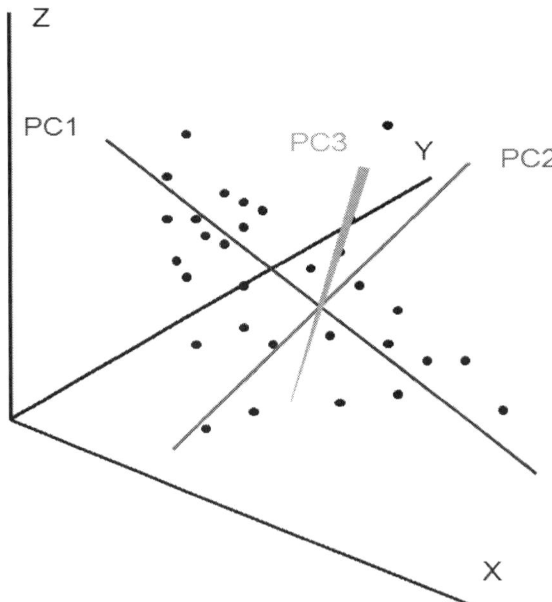

Fig. 3. Example of a 3 dimension PCA with the 3 orthogonal PCs.

7.1.2 Hierarchical clustering

Although several clustering techniques exist, the most used in the context of microarray data analysis is hierarchical clustering. Hierarchical clustering is used to identify the structure of a given population of cases or a given set of markers such as proteins. Every case is considered to have a given position in multidimensional space. Hierarchical clustering determines the similarity of cases in this space based on the distance between points. There are various linkage methods used for calculating distance, such as single linkage, complete linkage and average linkage. Single linkage computes the distance as the distance between the two nearest points in the clusters being compared. Complete linkage computes the distance between the two farthest points, whilst average linkage averages all distances across all the points in the clusters being compared. One commonly used distance measure is Euclidian distance which is the direct angular distance between two points. In fact it considers the distance in multidimensional space between each point and every other point. In this way a hierarchy of distances is determined. This hierarchy is plotted in the form of a dendrogram (figure 4). From this dendrogram we can identify clusters of cases or markers that are similar at a given distance.
The one major problem concerning clustering is that it suffers from the curse of dimensionality when analysing complex datasets. In a high dimensional space, it is likely that for any given pair of points within a cluster there will exist dimensions on which these points are far apart from one another. Therefore distance functions using all input features equally may not be truly effective (Domeniconi et al, 2004). Furthermore, clustering methods will often fail to identify coherent clusters due to the presence of many irrelevant and redundant features (Greene et al, 2005). Additionally, the important number of different distance measure may add an additional bias: it has been reported that the choice of a distance measure can greatly affect the results and produce different outcomes after the analysis (Quackenbush, 2001). Dimensionality is also of importance when one is examining the structure of a population through ordination techniques. This is particularly the case when utilising hierarchical cluster analysis. This approach is of limited suitability for high dimensional data as in a high dimensional space the distance between individual cases reaches convergence making all cases appear the same (Domeniconi et al, 2004). This makes it difficult to identify the real structure in the data or clusters of similar cases.

7.2 Application of modelling techniques

This second part of the section focusing on analysis tools considers more evolved techniques with what is known as *machine learning*. There are however a number of other techniques that can be employed in a predictive or classification capacity. Others include hidden Markov and Bayesian methods. These are widely described in the literature.

7.2.1 Decision tree based methodologies

Decision tree methodologies include, boosted decision trees, classification and regression trees, random forest methodologies. This approach is based on splitting a population into groups based on a hierarchy of rules (figure 5). Thus a given case is split into a given class based on a series of rules. This approach has been modified in a number of ways. Generally, a decision is made based on a feature that separates classes (one branch of the cluster dendrogram from another) within the population. This decision is based on a logical or numerical rule. Although their use in the analysis of miRNA data has been limited, decision trees have been used in the analysis of miRNA data derived to classify cancer patients (Xu, et al, 2009).

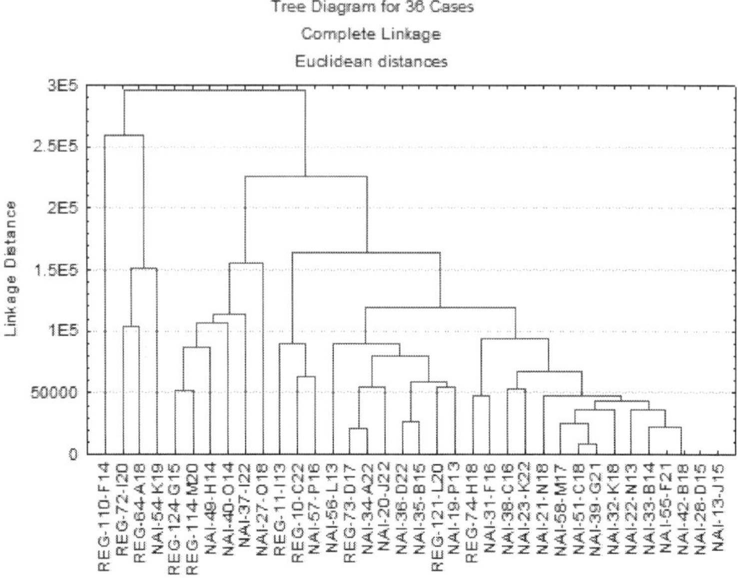

Fig. 4. Example of a hierarchical clustering analysis result aiming to find clusters of similar cases.

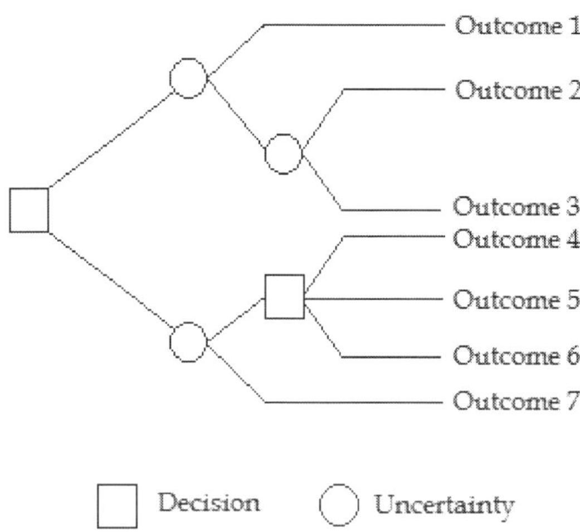

Fig. 5. Schematic example of the basic principle of Decision Trees

Boosted decision trees take the primary decision tree algorithm and boost it. Boosting is a process where classifiers are derived to allow prediction of those not correctly predicted by earlier steps. This means that a supervised classification is run where the actual class is known. A decision tree is created that classifies correctly as many cases as possible. Those cases that are incorrectly classified are given more weighting. A new tree is then created with these boosted weights. This process is similar to the iterative leaning that is conducted with the Artificial Neural Network back propagation algorithm.

Random forest approaches take the basic decision tree algorithm and couple it with random sample cross validation. In this way a forest of trees is created. Integration of a number of decision trees identifies a combined decision tree which, as it is developed on blind cases, represents what approaches a generalised solution for the problem being modelled (Breiman et al, 2001). This approach has been shown to be very good at making generalised classifications. The approach essentially derives each tree from a random vector with equivalent distribution from within the data set, essentially an extensive form of cross validation. Yousef et al, (2010) have used random forest as one method for the identification of gene targets for miRNAs. Segura et al (2010) have used random forests as a part of an analysis to define post recurrence survival in melanoma patients.

7.2.2 Artificial Neural Networks

Artificial Neural Networks are a non linear predictive system that may be used as a classifier. A popular form of ANN is the multi-layer perceptron (MLP) and is used to solve many types of problems such as pattern recognition and classification, function approximation, and prediction. The approach is a form of artificial intelligence in that it "learns" a solution to a problem from a preliminary set of samples. This is achieved by comparing predicted versus actual values for a seen data set (the training data set described earlier) and using the error of the predicted values from the ANN to iteratively develop a solution that is better able to classify. In MLP ANNs, learning is achieved by updating the weights that exist between the processing elements that constitute the network topology (figure 6). The algorithm fits multiple activation functions to the data to define a given class in an iterative fashion, essentially an extension of logistic regression. Once trained, ANNs can be used to predict the class of an unknown sample of interest. Additionally, the variables of the trained ANN model may be extracted to assess their importance in the system of interest. ANNs can be coupled with Random sample cross validation or any other cross validation method (LOO or MCCV) in order to ensure that the mode developed is not over fitted. One of the advantages of ANNs is that the process generates a mathematical model that can be interrogated and explored in order to elucidate further biological details and validate the model developed on a wide range of cases. A review of their use is in a clinical setting presented in Lisboa and Taktak (2006). Back propagation MLP ANNs have been proposed for use in the identification of biomarkers from miRNA data by Lowery et al, 2009.

7.2.3 Linear Discriminant Analysis (LDA)

Linear discriminant analysis attempts to separate the data into two subgroups by calculating the optimal linear line that best splits the population. Calculation of this discriminating line is conducted by taking into account sample variation within similar classes, and minimizing it between classes. As a result, any additional sample has its class determined by the side of the discriminating line it falls.

LDA can outperform other linear classification methods as LDA tries to consider the variation within the sample population. Nevertheless, LDA still suffers from its linear characteristic, and often fails to accurately classify non-linear problems, which is mostly the case in biomedical sciences (Stekel et al, 2003). This is the reason why non-linear classifiers are recommended.

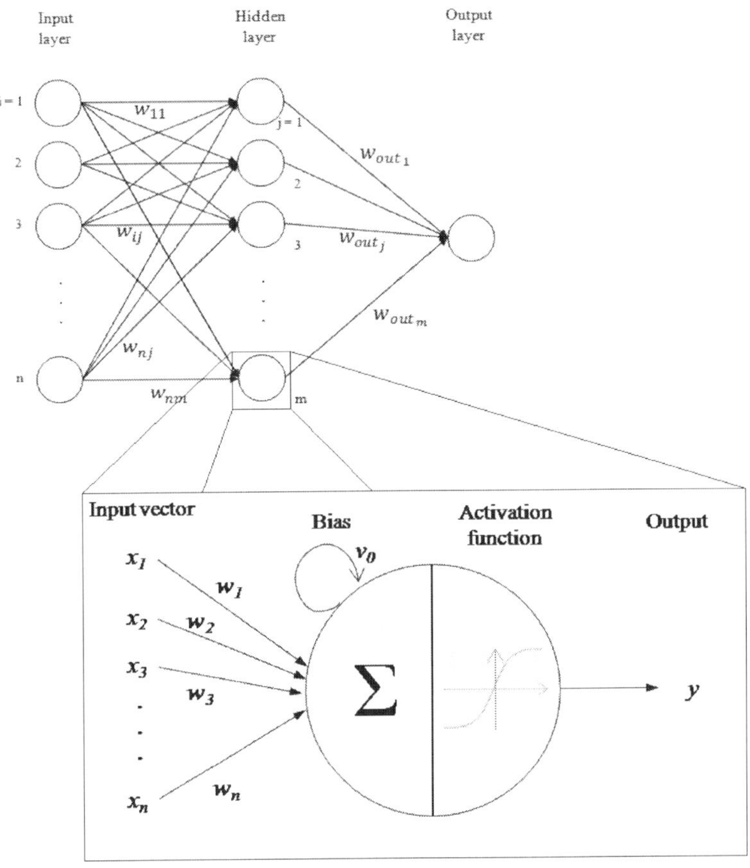

Fig. 6. Example of a classical MLP ANN topology with the details of a node (or neurone)

7.2.4 Support Vector Machines

Support Vector Machines (SVMs) are another popular form of machine learning algorithms in the field of analyzing MA data for non-linear modeling (Vapnik and Lerner, 1963). They are an evolution of LDA in the sense that they work by separating the data into 2 sub-groups. They work by separating the data into two regions by constructing a straight line or hyper plane that best separates between classes (figure 7). In the common example of a two-class classification problem, SVMs attempt to find a linear "maximal margin hyperplane" able to accurately discriminate the classes (Dreiseitlet al, 2001), similarly to what does Linear Discriminant Analysis. If no such linear hyperplane can be found, usually due to the inherent non-linearity of the dataset, the data are mapped into a high-dimensional feature space using a kernel function (for example polynomial or radial basis functions) in which the two classes can now be separated by a hyperplane which corresponds to a non-linear classifier (Furey et al, 2000). The class of the unknown sample is then determined by the side of the "maximal marginal hyper plane" on which it lies. SVMs have been used to analyse miRNA data by Xue et al, 2005.

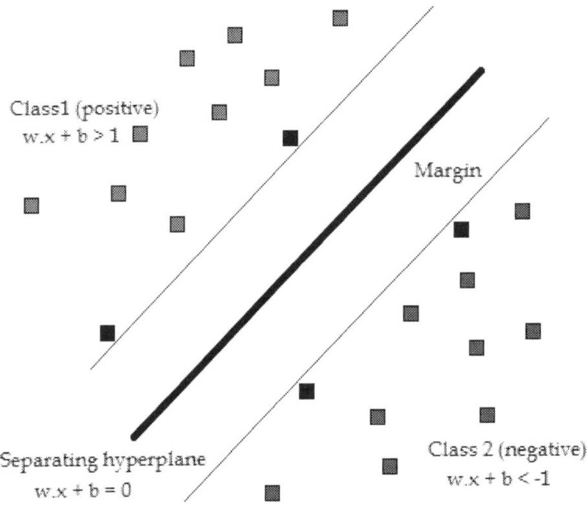

Fig. 7. Schematic representation of the principle of SVM. SVM tries to maximise the margin from the hyperplane in order to best separate the two classes (red positives from blue negatives).

8. Conclusion

The capability of microarray to simultaneously analyse expression patterns of thousands of DNA sequences, mRNA or miRNA transcripts has the potential to provide a unique insight into the molecular biology of malignancy. However, the clinical relevance and value of microarray data is highly dependent on a number of crucial factors including appropriate experimental design and suitable bioinformatic analysis. Breast cancer is a heterogeneous disease with many biological variables which need to be considered to generate meaningful results. Cohort selection is critical and sufficient biological and technical replicates must be included as part of microarray study design. Experimental protocols should be appropriate to the research question. The research community have enthusiastically applied high throughput technologies to the study of breast cancer. Class prediction, class comparison and class discovery studies have been undertaken in an attempt to unlock the heterogeneity of breast cancer and identify novel biomarkers. Molecular signatures have been generated which attempt to outperform current histopathological parameters at prognostication and prediction of response to therapy. Two clinical tests based on gene expression profiling (Oncotype DX and Mammaprint) are already in clinical use and being evaluated in multicentre international trials. It is essential that the potential of microarray signatures is carefully validated before they are adopted as prognostic tools in the clinical setting. Standards have been set for the reporting of microarray data (MIAME) and such data is publically available to facilitate external validation and meta-analysis. It is imperative that the data is integrated with knowledge normally processed in the clinical setting if we are to overcome the difficulties in reproducibility, standardization and lack of proof of significance beyond traditional clinicopathological tools that are limiting the incorporation of microarray based tools into today's standard of care.

Deriving biologically and clinically relevant results from microarray data is highly dependent on bioinformatic analysis. Microarray data is limited by inherent characteristics that render traditional statistical approaches less effective. These include high dimensionality, false discovery rates, noise, complexity, non-normality and limited reproducibility. High dimensionality remains one of the most critical challenges in the analysis of microarray data. Hierarchical clustering approaches, which have been widely used in the analysis of breast cancer microarray data, do not cope well with dimensionality. In overcoming this challenge supervised machine learning techniques have been adapted to the clinical setting to complement the existing statistical methods. The majority of machine learning techniques originated in weak-theory domains such as business and marketing. However, these approaches including Artificial Neural Networks and Support Vector Machines have been successfully applied to the analysis of miRNA microarray data in the context of clinical prognostication and prediction.

It is clear that the goal of translating microarray technology to the clinical setting requires close collaboration between the involved scientific disciplines.If the current momentum in microarray-based miRNA and mRNA translational research can be maintained this will add an exciting new dimension to the field of diagnostics and prognostics and will bring us closer to the ideal of individualized care for breast cancer patients.

9. References

Abbott AL, Alvarez-Saavedra E, Miska EA et al(2005) The let-7 MiRNA family members mir-48, mir-84, and mir-241 function together to regulate developmental timing in Caenorhabditis elegans. Dev Cell. 9(3):403-14.

Adam BL, Qu Y, Davis JW, Ward MD et al (2002) Serum protein fingerprinting coupled with a pattern-matching algorithm distinguishes prostate cancer from benign prostate hyperplasia and healthy men. Cancer Research 62:3609-3614.

Ahmed AA, Brenton JD (2005) Microarrays and breast cancer clinical studies: forgetting what we have not yet learnt. Breast Cancer Res 7:96–99.

Arciero C, Somiari SB, Shriver CD, et al. (2003). Functional relationship and gene ontology classification of breast cancer biomarkers. *Int. J. Biol. Markers* 18: 241-272.

Ashburner M, Ball CA, Blake JA et al (2000). Gene Ontology: tool for the unification of biology. Nat Genet. 25(1): 25–29.

Baffa R, Fassan M, Volinia S et al.(2009) MicroRNA expression profiling of human metastatic cancers identifies cancer gene targets. J Pathol, 219(2), 214-221

Ball CA, Dolinski K, Dwight SS, et al (2000). Integrating functional genomic information into the *Saccharomyces* Genome Database. Nucleic Acids Res;28:77–80

Ball G, Mian S, Holding F, et al (2002) An integrated approach utilizing artificial neural networks and SELDI mass spectrometry for the classification of human tumours and rapid identification of potential biomarkers. Bioinformatics 18:3395-3404.

Bartel DP. (2004) MiRNAs: genomics, biogenesis, mechanism and function. Cell; 116:281-97.

Bellman RE (1961) Adaptive Control Processes. Princeton University Press, Princeton, NJ

Benjamini Y, Hochberg Y (1995) Controlling the False Discovery Rate: A Practical and Powerful Approach to Multiple Testing. Journal of the Royal Statistical Society. Series B (Methodological);57:289-300.

Berns EM, van Staveren IL, Verhoog L et al (2001) Molecular profiles of BRCA1-mutated and matched sporadic breast tumours: relation with clinico-pathological features.British journal of cancer;85(4):538-45.

Bishop C (1995) Neural networks for pattern recognition. Oxford University Press.

Blake JA, Eppig JT, Richardson JE, Davisson MT (2000). The Mouse Genome Database (MGD): expanding genetic and genomic resources for the laboratory mouse. Nucleic Acids Res.;28:108–111

Blake JA, Harris MA (2002) The Gene Ontology (GO) project: structured vocabularies for molecular biology and their application to genome and expression analysis. Curr Protoc Bioinformatics. Chapter 7:Unit 7.2.

Blenkiron C, Goldstein LD, Thorne NP, et al (2007) MicroRNA expression profiling of human breast cancer identifies new markers of tumor subtype. Genome Biol 8(10):R214

Brenton JD, Carey LA, Ahmed AA, Caldas C (2005). Molecular classification and molecular forecasting of breast cancer: ready for clinical application? J Clin Oncol 23:7350–7360.

Breiman L. Random Forests (2001) Machine Learning 45:5-32.

Buyse M, Loi S, van't Veer L, et al. (2006) Validation and clinical utility of a 70-gene prognostic signature for women with node-negative breast cancer. J Natl Cancer Inst. 98(17):1183-1192.

Calin GA, Dumitru CD, Shimizu M, et al (2002) Frequent deletions and down-regulation of micro- RNA genes miR15 and miR16 at 13q14 in chronic lymphocytic leukemia PNAS.;99(24):15524-9.

Cardoso F, Van't Veer L, Rutgers E, et al. (2008) Clinical application of the 70-gene profile: the MINDACT trial. J Clin Oncol; 26:729–735.

Carey LA, Dees EC, Sawyer L et al (2007). The triple negative paradox: Primary tumor chemosensitivity of breast cancer subtypes. Clin Cancer Res 2007; 13:2329 -2334.

Castoldi M, Schmidt S, Benes V, et al (2006) A sensitive array for MiRNA expression profiling (miChip) based on locked nucleic acids (LNA). RNA;12(5):913-20.

Clarke R, Liu MC, Bouker KB, et al (2003). Antiestrogen resistance in breast cancer and the role of estrogen receptor signaling. Oncogene;22(47):7316-39.

Cortez MA, Calin GA (2009). MiRNA identification in plasma and serum: a new tool to diagnose and monitor diseases. Expert Opin Biol Ther;9(6):703-711.

Cronin, M, Pho M, Dutta D et al (2004). Measurement of gene expression in archival paraffin-embedded tissues: development and performance of a 92-gene reverse transcriptase-polymerase chain reaction assay. Am J Pathol. 164:35–42

Cunliffe HE, Ringner M, Bilke S, et al. (2003). The gene expression response of breast cancer to growth regulators: patterns and correlation with tumor expression profiles. *Cancer Res*. 63:7158-7166.

Desmedt C, Haibe-Kains B, Wirapati P, et al (2008). Biological processes associated with breast cancer clinical outcome depend on the molecular subtypes. Clin Cancer Res ;14:5158-65.

Domeniconi C, Papadopoulos D, Gunopulos D, et al (2004). Subspace clustering of high dimensional data. Proceedings 4th SIAM International Conference on Data Mining, pp. 517-521. Lake Buena Vista, FL, SIAM, 3600 UNIV CITY SCIENCE CENTER, PHILADELPHIA, PA 19104-2688 USA.

Dreiseitl S, Ohno-Machado L, Kittler H, et al (2001). A Comparison of Machine Learning Methods for the Diagnosis of Pigmented Skin Lesions. Journal of Biomedical Informatics;34:28-36.

Esquela-Kerscher A, Slack FJ.(2006) Oncomirs - MiRNAs with a role in cancer. Nature reviews;6(4):259-69.

Fan C, Oh DS, Wessels L, et al (2006) Concordance among gene-expression-based predictors for breast cancer. N Engl J Med;355:560–9.

Farmer P, Bonnefoi H, Becette V, et al (2005) Identification of molecular apocrine breast tumours by microarray analysis.Oncogene;24:4660–71.

Ferlay J, Parkin DM, Steliarova-Foucher E (2010) Estimates of cancer incidence and mortality in Europe in 2008. Eur J Cancer; 46:765-781.

Fisher B, Jeong JH, Bryant J, et al (2004). Treatment of lymph-node-negative, oestrogen receptor-positive breast cancer: Long-term findings from National Surgical
Adjuvant Breast and Bowel Project randomised clinical trials. Lancet;364:858–868

Foekens JA, Sieuwerts AM, Smid M et al (2008) Four miRNAs associated with aggressiveness of lymph node negative, estrogen receptor-positive human breast cancer.PNAS;105(35):13021-6.

Furey T S, Cristianini N, Duffy N, et al (2000). Support vector machine classification and validation of cancer tissue samples using microarray expression data. Bioinformatics16:906-914.

Geyer FC, Lopez-Garcia MA, Lambros MB, Reis-Filho JS (2009) Genetic Characterisation of Breast Cancer and Implications for Clinical Management. J Cell Mol Med (10):4090 103.

Gilad S, Meiri E, Yogev Y, et al (2008). Serum MiRNAs are promising novel biomarkers. PLoS ONE. ;3(9):e3148.

Goldhirsch A, Wood WC, Gelber RD,et al (2007). Progress and promise: highlights of the international expert consensus on the primary therapy of early breast cancer 2007. Ann Oncol;18(7):1133-44.

Goldstein LJ, Gray R, Badve S, et al (2008) Prognostic utility of the 21-gene assay in hormone receptor-positive operable breast cancer compared with classical clinicopathologic features. J Clin Oncol;26:4063–4071

Greene D, Cunningham P, Jorge A, et al. (2005). Producing accurate interpretable clusters from high-dimensional data, Proceedings 9th European Conference on Principles and Practice of Knowledge Discovery in Databases (PKDD), pp. 486-494, Porto Portugal.

Habel LA, Shak S, Jacobs MK, et al (2006). A population-based study of tumor gene expression and risk of breast cancer death among lymph node-negative patients. Breast Cancer Res;8:R25.

Hedenfalk I, Duggan D, Chen Y, et al (2001) Gene expression profiles in hereditary breast cancer. N Engl J Med.;344(8):539-48.

Heneghan HM, Miller N, Kerin MJ. (2010)MiRNAs as biomarkers and therapeutic targets in cancer. Curr Opin Pharmacol;10(5):543-50.

Hu Z, Fan C, Oh DS,, et al. (2006) The molecularportraits of breast tumors are conserved across microarray platforms. BMC Genom ;7:96.

Huang JX, Mehrens D, Wiese R,, et al. 2001. High-throughput genomic and Proteomic analysis using microarray technology. Clinical Chem, 47: 1912-16.

Huang Q, Gumireddy K, Schrier M et al.(2008) The microRNAs miR-373 and miR-520c promote tumour invasion and metastasis. Nat Cell Biol;10(2):202-210

Izmirlian G (2004) Application of the random forest classification algorithm to a SELDI-TOF proteomics study in the setting of a cancer prevention trial. Annals of the New York Academy of Sciences; 1020:154-174

Iorio MV, Ferracin M, Liu CG, et al (2005). MicroRNA gene expression deregulation in human breast cancer. Cancer research;65: 7065-70.

Jemal A, Siegel R, Ward E, et al. (2009) Cancer statistics, 2009. *CA Cancer J Clin*;59:225-249.

Khatri, P., Draghici, S. (2005), Ontological analysis of gene expression data: current tools, limitations, and open problems. Bioinformatics; 21: 3587–3595.

Kim C, Taniyama Y, Paik S (2009). Gene-expression-based prognostic and predictive markers for breast cancer- A primer for practicing pathologists Crit Rev Oncol Hematol.;70(1):1-11.

Klebanov L, Yakovlev A (2007) How high is the level of technical noise in microarray data? Biology Direct;2:9.

Kreike B, van Kouwenhove M, Horlings H et al (2007). Gene expression profiling and histopathological characterization of triple-negative/basal-like breast carcinomas. Breast Cancer Res;9:R65.

Korkola JE, DeVries S, Fridlyand J, et al (2003). Differentiation of lobular versus ductal breast carcinomas by expression microarray analysis. Cancer Res;63:7167–7175.

Lamb J, Ramaswamy S, Ford HL, et al (2003). A mechanism of cyclin D1 action encoded in the patterns of gene expression in human cancer. Cell; 114(3):323-34.

Lancashire LJ, Lemetre C, Ball GR (2009). An introduction to artificial neural networks in bioinformatics--application to complex microarray and mass spectrometry datasets in cancer studies. Briefings in Bioinformatics;10:315-329.

Lee RC, Feinbaum RL, Ambros V. (1993) The C. elegans heterochronic gene lin-4 encodes small RNAs with antisense complementarity to lin-14. Cell.;75(5):843-54.

Leopold E, Kindermann J (2006). Content Classification of Multimedia Documents using Partitions of Low-Level Features. Journal of Virtual Reality and Broadcasting 3(6).

Li J, Smyth P, Flavin R, et al. (2007) Comparison of miRNA expression patterns using total RNA extracted from matched samples of formalin- fixed paraffin-embedded (FFPE) cells and snap frozen cells. BMC biotechnology;7:36

Lisboa PJ, Taktak AF(2006). The use of artificial neural networks in decision support in cancer: A systematic review. *Neural Networks*;19:408-415.

Lo SS, Norton J, Mumby PB et al(2007). Prospective multicenter study of the impact of the 21-gene recurrence score (RS) assay on medical oncologist (MO) and patient (pt) adjuvant breast cancer (BC) treatment selection. J Clin Oncol;25(18 suppl):577

Loi S, Haibe-Kains B, Desmedt C, *et al* (2007). Definition of clinically distinct molecular subtypes in estrogen receptor-positive breast carcinomas through genomic grade. J. Clin. Oncol. 25, 1239–1246

Lowery AJ, Miller N, McNeill RE, Kerin MJ (2008). MicroRNAs as prognostic indicators and therapeutic targets: potential effect on breast cancer management. Clin Cancer Res. ;14(2):360-5.

Lowery AJ, Miller N, Devaney A, et al (2009) . MicroRNA signatures predict estrogen receptor, progesterone receptor and Her2/neu receptor status in breast cancer. Breast Cancer Res.;11(3):R27.

Lu J, Getz G, Miska EA, et al.(2005) MiRNA expression profiles classify human cancers. Nature. 2005;435(7043):834-8

Ma XJ, Hilsenbeck SG, Wang W et al (2006). The HOXB13:IL17BR expression index is a prognostic factor in early-stage breast cancer. J Clin Oncol; 24:4611- 4619.

Ma J, Dong C, Ji C (2010). MicroRNA and drug resistance. Cancer Gene Ther, 17(8), 523-531

Manning AT, Garvin JT, Shahbazi RI, et al (2007). Molecular profiling techniques and bioinformatics in cancer research Eur J Surg Oncol;33(3):255-65.

Marchionni L, Wilson RF, Wolff AC, et al (2008). Systematic review: gene expression profiling assays in early-stage breast cancer. Ann Intern Med.;148(5):358-369.

Marengo E, Robotti E, Righetti PG, et al (2004). Study of proteomic changes associated with healthy and tumoral murine samples in neuroblastoma by principal component analysis and classification methods. Clinica Chimica Acta;345:55-67.

Masuda N, Ohnishi T, KawamotoS , et al (1999) Analysis of chemical modification of RNA from formalin-fixed samples and optimization of molecular biology applications for such samples. Nucleic Acids Res. 27, 4436–4443

Matharoo-Ball B, Ratcliffe L, Lancashire L, et al (2007). Diagnostic biomarkers differentiating metastatic melanoma patients from healthy controls identified by an integrated MALDI-TOF mass spectrometry/bioinformatic approach. Proteomics Clinical Applications; 1:605-620

Mattie MD, Benz CC, Bowers J, et al (2006). Optimized high-throughput MiRNA expression profiling provides novel biomarker assessment of clinical prostate and breast cancer biopsies. Molecular cancer;5:24

Michiels S, Koscielny S, Hill C (2005). Prediction of cancer outcome with microarrays: a multiple random validation strategy. Lancet;365: 488-92.

Michiels S, Koscielny S, Hill C (2007). Interpretation of microarray data in cancer. British Journal of Cancer;96:1155–1158.

Mina L, Soule SE, Badve S, et al. (2007) Predicting response to primary chemotherapy: gene expression profiling of paraffin-embedded core biopsy tissue. Breast Cancer Res Treat ;103:197–208.

Mitchell PS, Parkin RK, Kroh EM, et al (2008).Circulating MiRNAs as stable blood- based markers for cancer detection.PNAS;105(30):10513-8

Mook S, Schmidt MK, Viale G, et al (2009). The 70-gene prognosis signature predicts disease outcome in breast cancer patients with 1-3 positive lymph nodes in an independent validation study. Breast Cancer Res Treat;116:295–302.

Mootha VK, Lindgren CM, Eriksson KF, et al (2003). PGC-1alpha Responsive Genes Involved in Oxidative Phosphorylation are Coordinately Downregulated in Human Diabetes, Nature Genetics 34(3):267-73

Nielsen TO, Hsu FD, Jensen K et al (2004). Immunohistochemical and clinical characterization of the basal-like subtype of invasive breast carcinoma. Clin Cancer Res ;10:5367-74.

Oberley MJ, Tsao J, Yau P, Farnham PJ (2004). Highthroughput screening of chromatin immunoprecipitates using CpG-island microarrays. Methods Enzymol;376: 315-34.

Oostlander AE, Meijer GA, Ylstra B (2004). Microarraybased comparative genomic hybridization and its applications in human genetics. Clin Genet, 66: 488-495.

Osborne CK(1998) Tamoxifen in the treatment of breast cancer. N Engl J Med;339(22):1609-18.

Paik S, Shak S, Tang G, et al (2004). A multigene assay to predict recurrence of tamoxifen-treated, node-negative breast cancer. N Engl J Med;351(27):2817-26.

Paik, S. Kim, C. Y, Song, Y. K. & Kim, W. S. (2005) Technology insight: application of molecular techniques to formalin-fixed paraffin-embedded tissues from breast cancer. Nat. Clin. Pract. Oncol;2:246–254

Paik S, Tang G, Shak S, , et al(2006). Gene expression and benefit of chemotherapy in women with node-negative, estrogen receptor-positive breastcancer. J Clin Oncol; 24 (23) : 3726-34.

Parker JS, Mullins M, Cheang MC, et al (2009). Supervised risk predictor of breast cancer based on intrinsic subtypes. J Clin Oncol;27:1160–1167.

Pedraza V, Gomez-Capilla JA, Escaramis G, et al (2010) Gene expression signatures in breast cancer distinguish phenotype characteristics, histologic subtypes, and tumor invasiveness. Cancer.;116(2):486-96.

Peppercorn J, Perou CM, Carey LA. (2008) Molecular subtypes in breast cancer evaluation and management: divide and conquer. Cancer Invest;26:1–10.

Perou CM, Sorlie T, Eisen MB, et al (2000). Molecular portraits of human breast tumours. Nature;406: 747-52.

Pusztai L, Mazouni C, Anderson K, et al (2006). Molecular classification of breast cancer: limitations and potential. Oncologist;11:868–877.

Quackenbush J (2001). Computational analysis of microarray data. Nature Reviews Genetics ;2:418-27.

Raychaudhuri S, Stuart JM, Altman RB (2000). Principal components analysis to summarize microarray experiments: application to sporulation time series. In Pacific Symposium on Biocomputing, pp. 455–466.

Rifai N, Gillette MA, Carr SA (2006). Protein biomarker discovery and validation: the long and uncertain path to clinical utility. Nature biotechnology;24:971-983

Rouzier R, Perou CM, Symmans WF et al (2005). Breast cancer molecular subtypes respond differently to preoperative chemotherapy. Clin Cancer Res;11:5678 - 5685.

Schena M, Shalon D, Davis RW, Brown PO (1995). Quantitative monitoring of gene expression patterns with a complementary DNA microarray. Science;270:467-70.

Segura MF, Belitskaya-Lévy I, Rose A, et al (2010) Melanoma MicroRNA Signature Predicts Post-Recurrence Survival. Clinical Cancer Research;16:1577.

Shak S, Baehner FL, Palmer G, et al (2006) Subtypes of breast cancer defined by standardized quantitative RT-PCR analysis of 10 618 tumors. Breast Cancer Res Treat 2006;100:S295–295.

Shi L, Reid LH, Jones WD, et al (2006)The microarray quality control (MAQC) project shows inter- and intraplatform reproducibility of gene expression measurements. Nature Biotechnologie;24:1151–1161.

Simon RM, Korn EL, McShane LM, et al (2003). Design and analysis of DNA microarray investigations. Springer New York

Smith I, Procter M, Gelber RD, et al (2007). 2-year follow up of trastuzumab after adjuvant chemotherapy in HER2-positive breast cancer: a randomised controlled trial. Lancet. ;369(9555):29-36.

Sorlie T, Perou CM, Tibshirani R, et al (2001) Gene expression patterns of breast carcinomas distinguish tumor subclasses with clinical implications.PNAS;98: 10869-74.

Sorlie T, Tibshirani R, Parker J, et al (2003). Repeated observation of breast tumor subtypes in independent gene expression data sets. PNAS;100:8418–8423.

Sorlie T, Perou CM, Fan C, et al (2006) Gene expression profiles do not consistently predict the clinical treatment response in locally advanced breast cancer. Mol Cancer Ther;5:2914–8.

Sotiriou C, Neo SY, McShane LM, et al (2003) Breast cancer classification and prognosis based on gene expression profiles from a population based study.PNAS;100:10393–10398

Sparano JA. (2006). TAILORx: Trial assigning individualized options for treatment (Rx). Clin Breast Cancer;7:347–350.

Stekel D. (2003). Microarray bioinformatics. Cambrigde University Press,

Stoll D, Templin MF, Bachmann J, Joos TO (2005). Protein microarrays: applications and future challenges. Curr Opin Drug Discov Devel, 8: 239-252.

Sun Y, Goodison S. Li J, Liu L., Farmerie W (2007). Improved breast cancer prognosis through the combination of clinical and genetic markers, Bioinformatics;23:30–37

Tessel MA, Krett NL, Rosen ST (2010).Steroid receptor and microRNA regulation in cancer. Curr Opin Oncol;22(6):592-597

The FlyBase Consortium (1999). The FlyBase database of the Drosophila Genome Projects and community literature. Nucleic Acids Res;27:85–88.

van de Vijver M, He Y, van't Veer L, et al (2002). A gene-expression signature as a predictor of survival in breast cancer. N Engl J Med;347:1999–2009.

van't Veer L, Dai H, van de Vijver M, et al (2002). Gene expression profiling predicts clinical outcome of breast cancer. Nature;415:530–6.

Vapnik V, Lerner A (1963). Pattern recognition using generalized portrait method. Automation and Remote Control 1963;24:774-780.

Volinia S, Calin GA, Liu CG, et al (2006). A MiRNA expression signature of human solid tumors defines cancer gene targets. PNAS.;103(7):2257-61.

Wadsworth JT, Somers KD, Cazares LH, et al (2004) Serum protein profiles to identify head and neck cancer. Clinical Cancer Research;10:1625-1632.

Wang Y, Klijn JG, Zhang Y et al (2005). Gene-expression profiles to predict distant metastasis of lymph-node-negative primary breast cancer. Lancet;365:671–679.

Warnat P, Eils R, Brors B (2005). Cross-platform analysis of cancer microarray data improves gene expression based classification of phenotypes. BMC bioinformatics 2;6: 265.

Weigelt B, Geyer FC, Natrajan R, *et al* (2010) The molecular underpinning of lobular histological growth pattern: a genome-wide transcriptomic analysis of invasive lobular carcinomas and grade- and molecular subtype-matched invasive ductal carcinomas of no special type. J Pathol;220(1):45-57

Wirapati P, Sotiriou C, Kunkel S, *et al (2008)*. Meta-analysis of gene expression profiles in breast cancer: toward a unified understanding of breast cancer subtyping and prognosis signatures. *Breast Cancer Res*;10:R65.

Wong JWH, Cagney G, Cartwright HM (2005). SpecAlign—processing and alignment of mass spectra datasets. Bioinformatics;21:2088-2090

Xi Y, Nakajima G, Gavin E, et al (2007). Systematic analysis of MiRNA expression of RNA extracted from fresh frozen and formalin-fixed paraffin-embedded samples. RNA;13(10):1668-74.

Xue C, Li F, He T, Liu GP, Li Y, Xuegong Z (2005). Classification of real and pseudo microRNA precursors using local structure-sequence features and support vector machine. BMC Bioinformatics;6:310.

Xu R, Xu J, Wunsch DC (2009). Using default ARTMAP for cancer classification with MicroRNA expression signatures, *International Joint Conference on Neural Networks*, pp.3398-3404,

Yan PS, Perry MR, Laux DE, et al (2000). CpG island arrays: an application toward deciphering epigenetic signatures of breast cancer. Clinical Cancer Research; 6: 1432-38.

Yousef M, Najami N, Khalifa W (2010). A comparison study between one-class and two-class machine learning for MicroRNA target detection. Journal of Biomedical Science and Engineering ;3:247-252.

Zhao H, Langerod A, Ji Y, et al (2004) Different gene expression patterns in invasive lobular and ductal carcinomas of the breast. Mol Biol Cell;15:2523-2536.

Zhao H, Shen J, Medico L, et al (2010). A Pilot Study of Circulating miRNAs as Potential Biomarkers of Early Stage Breast Cancer. PLoS One;5(10),e137:5 , 2010

Zheng T, Wang J, Chen X, Liu L (2010) Role of microRNA in anticancer drug resistance. Int J Cancer;126(1):2-10.

6

Computational Tools for Identification of microRNAs in Deep Sequencing Data Sets

Manuel A. S. Santos and Ana Raquel Soares
University of Aveiro
Portugal

1. Introduction

MicroRNAs (miRNAs) are a class of small RNAs of approximately 22 nucleotides in length that regulate eukaryotic gene expression at the post-transcriptional level (Ambros 2004; Bartel 2004; Filipowicz *et al.* 2008). They are transcribed as long precursor RNA molecules (pri-miRNAs) and are successively processed by two key RNAses, namely Drosha and Dicer, into their mature forms of ~22 nucleotides (Kim 2005; Kim *et al.* 2009). These small RNAs regulate gene expression by binding to target sites in the 3′ untranslated region of mRNAs (3′UTR). Recognition of the 3′UTR by miRNAs is mediated through complementary hybridization at least between nucleotides 2-8, numbered from the 5′ end (seed sequences) of the small RNAs, and complementary sequences present in the 3′UTRs of mRNAs (Ambros 2004; Bartel 2004; Zamore and Haley 2005). Perfect or nearly perfect complementarities between miRNAs and their 3′UTRs induce mRNA cleavage by the RNA-induced silencing complex (RISC), whereas imperfect base pair matching may induce translational silencing through various molecular mechanisms, namely inhibition of translation initiation and activation of mRNA storage in P-bodies and/or stress granules (Pillai *et al.* 2007).

This class of small RNAs is well conserved between eukaryotic organisms, suggesting that they appeared early in eukaryotic evolution and play fundamental roles in gene expression control. Each miRNA may repress hundreds of mRNAs and regulate a wide variety of biological processes, namely developmental timing (Feinbaum and Ambros 1999; Lau *et al.* 2001), cell differentiation (Tay *et al.* 2008), immune response (Ceppi *et al.* 2009) and infection (Chang *et al.* 2008). For this reason, their identification is essential to understand eukaryotic biology. Their small size, low abundance and high instability complicated early identification, but these obstacles have been overcome by next generation sequencing approaches, namely the Genome Sequencer™ FLX from Roche, the Solexa/Illumina Genome Analyzer and the Applied Biosystems SOLiD™ Sequencer which are currently being routinely used for rapid miRNA identification and quantification in many eukaryotes (Burnside *et al.* 2008; Morin *et al.* 2008; Schulte *et al.* 2010).

As in other vertebrates, miRNAs control gene expression in zebrafish, since defective miRNA processing arrest development (Wienholds *et al.* 2003). Also, a specific subset of miRNAs is required for brain morphogenesis in zebrafish embryos, but not for cell fate determination or axis formation (Giraldez *et al.* 2005). In other words, miRNAs play an important role in zebrafish organogenesis and their expression at specific time points is relevant to organ formation and differentiation. Since identification of the complete set of miRNAs is fundamental to fully understand biological processes, we have used high throughput 454 DNA pyrosequencing technologies to fully characterize the zebrafish miRNA population (Soares *et al.* 2009). For this, a series of cDNA libraries were prepared from miRNAs isolated at different embryonic time points and from fully developed organs sequenced using the Genome Sequencer™ FLX. This platform yields reads of up to 200 bases each and can generate up to 1 million high quality reads per run, which provides sufficient sequencing coverage for miRNA identification and quantification in most organisms. However, deep sequencing of small RNAs may pose some problems that need to be taken into consideration to avoid sequencing biases. For example, library preparation and computational methodologies for miRNA identification from large pool of reads need to be optimized. There are many variables to consider, namely biases in handling large sets of data, sequencing errors and RNA editing or splicing. If used properly, deep sequencing technologies have enormous analytical power and have been proven to be very robust in retrieving novel small RNA molecules. One of the major challenges when analyzing deep sequencing data is to differentiate miRNAs from other small RNAs and RNA degradation products.

Different research groups are developing dedicated computational methods for the identification of miRNAs from large sets of sequencing data generated by next generation sequencing experiments. miRDeep (http://www.mdc-berlin.de/en/research/research_teams/systems_biology_of_gene_regulatory_elements/projects/miRDeep/index.html) (Friedlander *et al.* 2008) and miRanalizer (http://web.bioinformatics.cicbiogune.es/microRNA/miRanalyser.php) (Hackenberg *et al.* 2009) can both detect known miRNAs annotated in miRBase and predict new miRNAs (although using different prediction algorithms) from small RNA datasets generated by deep sequencing. Although these online algorithms are extremely useful for miRNA identification, custom-made pipeline analysis of deep sequencing data may be performed in parallel to uncover the maximum number of small non-coding RNA molecules present in the RNA datasets.

In this chapter, we discuss the tools and computational pipelines used for miRNA identification, discovery and expression from sequencing data, based on our own experience of deep sequencing of zebrafish miRNAs, using the Genome Sequencer™ FLX from Roche. We show how a combination of a public available, user-friendly algorithm, such as miRDeep, with custom-built analysis pipelines can be used to identify non-coding RNAs and uncover novel miRNAs. We also demonstrate that population statistics can be applied to statistical analysis of miRNA populations identified during sequencing and we demonstrate that robust computational analysis of the data is crucial for extracting the maximum information from sequencing datasets.

2. miRNA identification by next-generation sequencing

2.1 Extraction of next-generation sequencing data

Next generation sequencing methods have been successfully applied in the last years to miRNA identification in a variety of organisms. However, the enormous amount of data generated represents bioinformatics challenges that researchers have to overcome in order to extract relevant data from the datasets.

We have used the Genome Sequencer™ FLX system (454 sequencing) to identify zebrafish miRNAs from different developmental stages and from different tissues. For this, cDNA libraries are prepared following commonly used protocols (Droege M and Hill B. 2008; Soares *et al.* 2009). These libraries contain specific adaptors for the small RNA molecules containing specific priming sites for sequencing. After sequencing, raw data filtration and extraction is performed using specialist software incorporated into the Genome Sequencer™ FLX system (Droege M and Hill B. 2008). Raw images are processed to remove background noise and the data is normalized. Quality of raw sequencing reads is based on complete read through of the adaptors incorporated into the cDNA libraries. The 200 base pair of 454 sequencing reads provide enough sequencing data for complete read through of the adaptors and miRNAs. During quality control, the adaptors are trimmed and the resulting sequences are used for further analysis. Sequences ≥ 15 nucleotides are kept for miRNA identification, and constitute the small RNA sequencing data.

Other sequencing platforms, such as Illumina/Solexa and SOLiD™, also have specialist software for raw data filtration. DSAP, for example, is an automated multiple-task web service designed to analyze small RNA datasets generated by the Solexa platform (Huang *et al.* 2010). This software filters raw data by removing sequencing adaptors and poly-A/T/C/G/N nucleotides. In addition, it performs non-coding RNA matching by sequence homology mapping against the non-coding RNA database Rfam (rfam.sanger.ac.uk/) and detects known miRNAs in miRBase (Griffiths-Jones *et al.* 2008), based on sequence homology.

The SOLiD™ platform has its own SOLiD™ System Small RNA Analysis Pipeline Tool (RNA2MAP), which is available online (http://solidsoftwaretools.com/gf/project/rna2map). This software is similar to DSAP, as it filters raw data and identifies known miRNAs in the sequencing dataset by matching reads against miRBase sequences and against a reference genome. Although these specialist software packages are oriented for miRNA identification in sequencing datasets they are not able to identify novel miRNAs. For this, datasets generated from any of the sequencing platforms available have to be analyzed using tools that include algorithms to identify novel miRNAs.

2.2 miRNA identification from next generation sequencing databases

miRNA identification (of both known and novel molecules) from datasets generated by deep-sequencing has been facilitated by the development of public user friendly algorithms, such as miRDeep (Friedlander *et al.* 2008), miRanalyzer (Hackenberg *et al.* 2009) and miRTools (Zhu *et al.* 2010).

We used miRDeep to identify miRNAs in our sequencing datasets (Figure 1). miRDeep was the first public tool available for the analysis of deep-sequencing miRNA data. This software was developed to extract putative precursor structures and predict secondary structures using RNAfold (Hofacker 2003) after genome alignment of the sequences retrieved by next-generation sequencing. This algorithm relies on the miRNA biogenesis model. Pre-miRNAs are processed by DICER, which originates three different fragments, namely the mature miRNA, the star and the hairpin loop sequences (Kim *et al.* 2009). miRDeep scores the compatibility of the position and frequency of the sequenced RNA with the secondary structures of the miRNA precursors and identifies new, conserved and non-conserved miRNAs with high confidence. It distinguishes between novel and known miRNAs, by evaluating the presence or absence of alignments of a given sequence with the stem loop sequences deposited in miRBase. The sequence with the highest expression is always considered as the mature miRNA sequence by the miRDeep algorithm. All hairpins that are not processed by DICER will not match a typical secondary miRNA structure and are filtered out.

After aligning the sequences against the desired genome using megaBlast, the blast output is parsed for miRDeep uploading. As sequencing errors, RNA editing and RNA splicing may alter the original miRNA sequence, one can re-align reads that do not match the genome using SHRiMP (http://compbio.cs.toronto.edu/shrimp/). The retrieved alignments are also parsed for miRDeep for miRNA prediction. miRDeep itself allows up to 2 mismatches in the 3′ end of each sequence, which already accounts with some degree of sequencing errors that might have occurred.

Reads matching more than 10 different genome loci are generally discarded, as they likely constitute false positives. The remaining alignments are used as guidelines for excision of the potential precursors from the genome. After secondary structure prediction of putative precursors, signatures are created by retaining reads that align perfectly with those putative precursors to generate the signature format. miRNAs are predicted by discarding non-plausible DICER products and scoring plausible ones. The latter are blasted against mature miRNAs deposited in miRBase, to extract known and conserved miRNAs. The remaining reads are considered novel miRNAs.

In order to evaluate the sensitivity of the prediction and data quality, miRDeep calculates the false positive rate, which should be below 10%. For this, the signature and the structure-pairings in the input dataset are randomly permutated, to test the hypothesis that the structure (hairpin) of true miRNAs is recognized by DICER and causes the signature.

miRanalizer (Hackenberg *et al.* 2009) is a recently developed web server tool that detects both known miRNAs annotated in miRBase and other non-coding RNAs by mapping sequences to non-coding RNA libraries, such as Rfam. This feature is important, as more classes of small non coding RNAs are being unravelled and their identification can provide clues about their functions. At the same time, by removing reads that match other non coding RNA classes, it reduces the false positive rate in the prediction of novel miRNAs, as these small non coding RNAs can be confused with miRNAs. For novel miRNA prediction, miRanalizer implements a machine learning approach based on the random forest method, with the number of trees set to 100 (Breiman 2001). miRanalyzer can be applied to miRNA discovery in different models, namely human, mouse, rat, fruit-fly, round-worm, zebrafish and dog, and uses datasets from different models to build the final prediction model. In comparison to miRDeep, this is disadvantageous as the latter can predict novel miRNAs from any model. All pre-miRNAs candidates that match known miRNAs are extracted from the experimental dataset and labelled as positive instances. Next, an equal amount of pre-miRNA candidates from the same dataset are selected by random selection with the known miRNAs removed and labelled as negative. Pre-processing of reads corresponding to putative new miRNAs includes clustering of all reads that overlap with the genome, testing whether the start of the current read overlaps less than 3 nucleotides with the end position of previous reads. This avoids DICER products grouping together and be considered non-miRNAs products, which would increase false negatives. Besides, clusters of more than 25 base pairs in length are discarded and the secondary structure of the miRNA is predicted via RNAfold (Hofacker 2003). Structures where the cluster sequence is not fully included and where part of the stem cannot be identified as a DICER product are discarded.

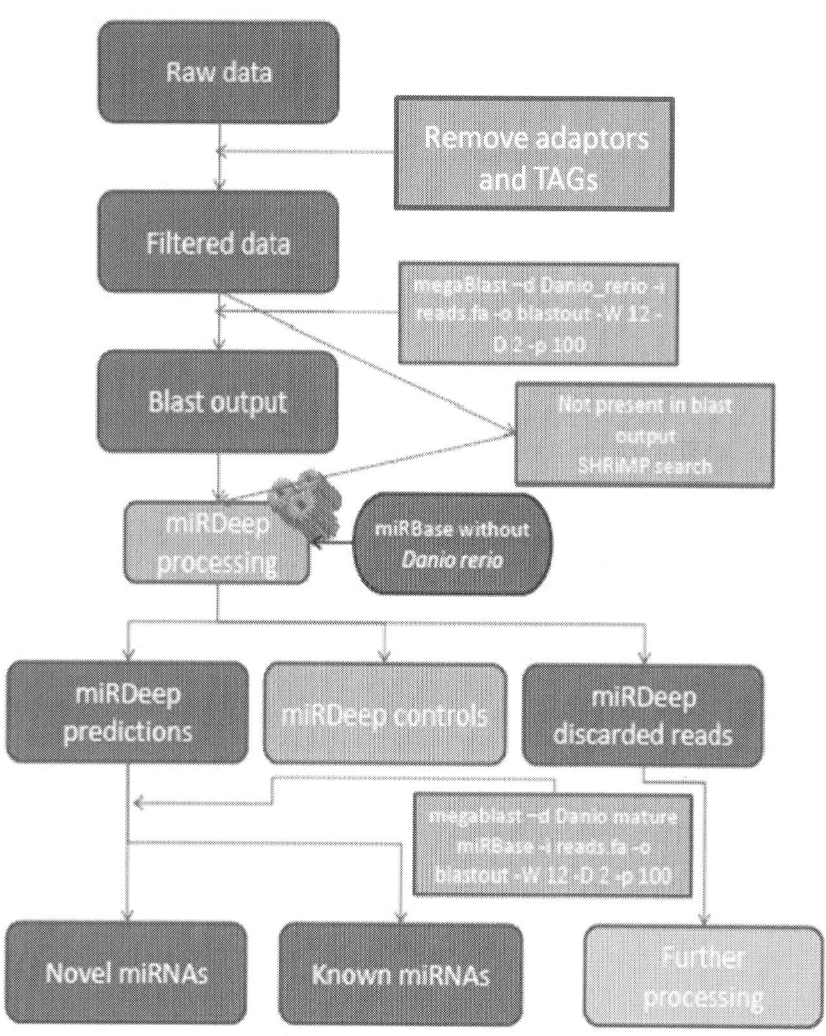

Fig. 1. Data pipeline analysis using miRDeep.

miRTools is a comprehensive web server that can be used for characterization of the small RNA transcriptome (Zhu *et al.* 2010). It offers some advantages relative to miRDeep and miRanalyzer, since it integrates multiple computational approaches including tools for raw data filtration, identification of novel miRNAs and miRNA expression profile generation. In order to detect novel miRNAs, miRTools analyze all sequences that are not annotated to known miRNAs, other small non-coding RNAs and genomic repeats or mRNA that match the reference genome. These sequences are extracted and their RNA secondary structures are predicted using RNAfold (Hofacker 2003) and novel miRNAs are identified using miRDeep.

2.3 Analysis of discarded reads by miRNA identification algorithms can identify new miRNAs

Since miRDeep and miRanalyzer are highly stringent algorithms, some miRNAs may escape detection. The false negative discovery rate can, however be calculated by simply performing a megaBlast search of the sequencing data against the miRNAs deposited in miRBase. Perfect alignments are considered true positives. The list of known miRNAs identified by this method is compared to the list of known miRNAs identified by miRDeep or miRanalyzer. False negatives are those miRNAs present in the blast analysis, but which were missed by the miRNA prediction algorithms. This is, in our opinion, an essential control, as it gives information about the percentage of miRNAs that may have escaped miRDeep or miRanalyzer analysis. We have detected ~19% of false negatives, which prompted us to develop a parallel pipeline to analyze reads that may have been incorrectly discarded by the original algorithm (Figure 2). This analysis can and should be performed independently of the algorithm used to retrieve miRNAs from deep sequencing data.

To overcome the lack of sensitivity of miRDeep, our parallel bioinformatics pipeline includes a megaBlast alignment between the dataset of discarded reads by miRDeep and mature sequences deposited in miRBase. Besides, novel transcripts encoding miRNAs predicted by computational tools can be retrieved from the latest Ensembl version using BioMart and also from literature predictions. These sequences are then used to perform a megaBlast search against the sequencing data. The transcripts with perfect matches and alignment length > 18 nucleotides are kept for further processing. These transcripts are then compared with the mature miRNAs deposited in miRBase and those that produce imperfect alignments or do not produce alignments are considered novel miRNAs. Imperfect alignments may identify conserved miRNAs if there is a perfect alignment in the seed region.

Complementary alignments of our dataset reads against the zebrafish genome with SHRiMP alignments and complementary miRDeep analysis with an analysis of the reads discarded by this algorithm, allowed us to identify 90% of the 192 zebrafish miRNAs previously identified, plus 107 miRNA star sequences and 25 novel miRNAs.

2.4 Generation of miRNA profiles from deep sequencing data

Deep sequencing of miRNAs can also be used to generate miRNA expression profiles as the absolute number of sequencing reads of each miRNA is directly proportional to their relative abundance. miRNA profiles can be generated based on the number of reads of each particular miRNA. However, a normalization step is essential to compare miRNA expression levels between different samples. The variation in the total number of reads between samples leads to erroneous interpretation of miRNA expression patterns by direct comparison of read numbers (Chen *et al.* 2005). Normalization assumes that the small RNA population is constant and is represented by an arbitrary value (e.g. 1000), and can be calculated as indicated below:

$$\text{miRNA relative expression} = \frac{1000 \times (NRmiRNA_X{}^Y)}{TNRmiRNAs^Y}$$

where $NRmiRNA_X{}^Y$ is the number of reads of $miRNA_X$ (X = any miRNA) in sample Y, and $TNRmiRNAs^Y$ is the total number of miRNAs in sample Y. 1000 is an arbitrary number of reads that allows for data normalization across different samples. This calculates the relative expression of a specific miRNA in a given sample, relative to all miRNAs expressed.

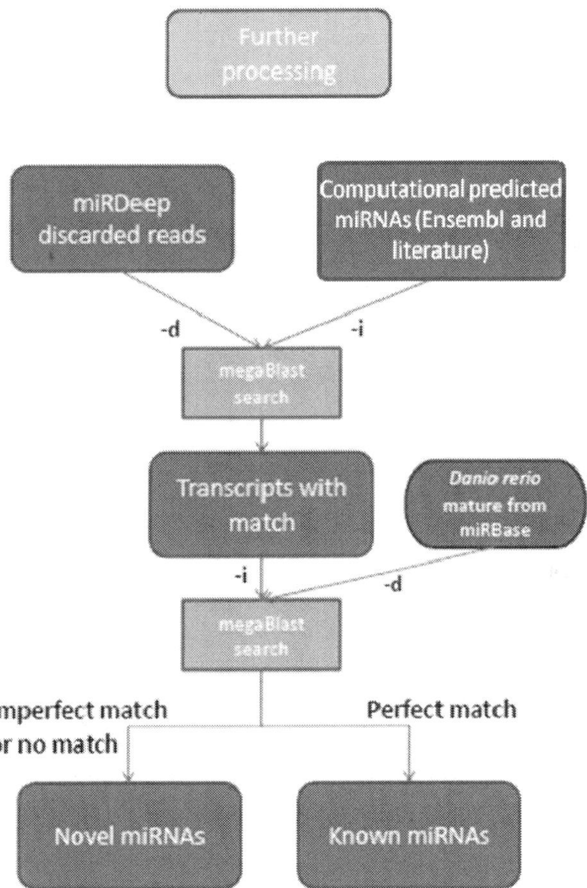

Fig. 2. Bioinformatics pipeline of reads discarded by miRDeep (-i and –d stand for query file and database respectively).

Using this formula it is possible to generate miRNA profiles for each sample sequenced. These profiles provide valuable information about relative miRNA expression, which is essential to understand miRNA function in different tissues. In order to compare miRNA profiles of two deep sequencing samples (e.g. condition vs control), a two-side t-test can be applied to determine miRNA levels. Sequence count values should be log-transformed to stabilize variance (Creighton *et al.* 2009). miRTools already include a computational approach to identify significantly differentially expressed miRNAs (Zhu *et al.* 2010). It compares differentially expressed miRNAs in multiple samples after normalization of the read count of each miRNA with the total number of miRNA read counts which are matched to the reference genome. The algorithm calculates statistical significance (P-value) based on a Bayesian method (Audic and Claverie 1997), which accounts for sampling variability of tags with low counts. Significantly differentially expressed miRNAs are those that show P-values <0.01 and at least 2-fold change in normalized sequence counts.

2.5 Statistical analysis of miRNA population

The platforms available for miRNA sequencing offer different sequencing coverage, ranging from thousands to millions of reads. In principle, higher sequencing coverage will enable discovery of more miRNA molecules in a sequencing run. However, technical problems during sample preparation can interfere with good quality sequencing of small RNAs. One of the most common problems is the generation of primer dimmers during PCR amplification of cDNA libraries. This may indicate an excess of primers during amplification, when compared to the miRNA levels in a given cDNA library or low annealing temperature. This problem is often only detected after sequencing. When this happens, a large number of reads do not pass quality control filters and the number of reads corresponding to small RNAs is considerably lower than the initial sequencing coverage. Besides this, quality control filters do not consider reads with sequencing errors in the adaptors or without recognizable adaptors. For these reasons, a tool that verifies if the sequencing coverage is sufficient to retrieve most miRNAs in a given sample is important.

A useful approach to assess the representativeness of miRNA reads in a sequencing experiment is to apply population statistics to the overall miRNA population. We have developed a statistical tool to calculate how many miRNAs are expected in a given sequencing experiment and how many reads are needed to identify them. Rarefaction curves of the total number of reads obtained versus the total number of miRNA species identified are plotted and the total richness of the miRNA population is determined. Chao1, a non-parametric richness estimator (Chao 1987), can be used to determine the total richness of the miRNA population, as a function of the observed richness (S_{obs}), and the number of total sequences obtained by sequencing. The value obtained represents the number of different miRNAs that can be identified in a specific sequencing experiment. The rarefaction curve estimates the number of reads needed to identify the different miRNAs that may be present in a sequencing run. For example, 206 miRNAs are expected to be present in a sequencing experiment that retrieves approximately 40000 reads (Figure 3). The steep curve levels off towards an asymptote, indicating the point (~20000 reads) where additional sampling will not yield extra miRNAs. As that critical point is below the total number of reads obtained, we can conclude that the sequencing coverage is sufficient to identify all miRNAs predicted in the particular sample. Rarefaction curves and the Chao1 statistical estimator are computed using EstimateS8.0 (Colwell and Coddington 1994).

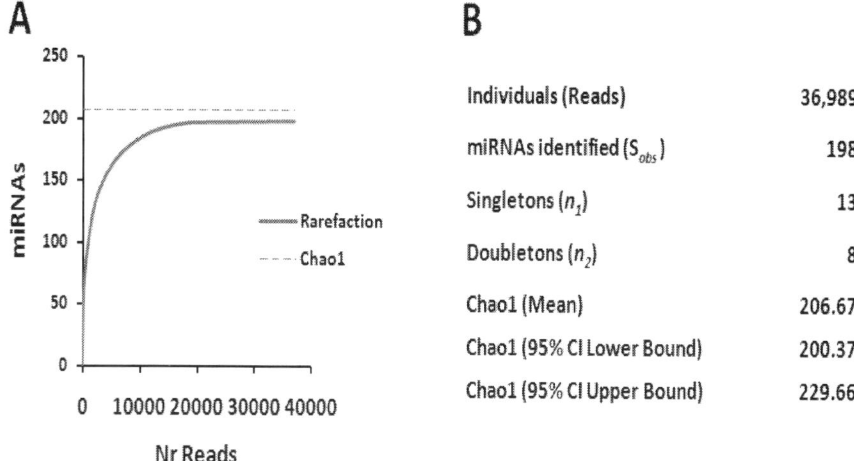

Fig. 3. Statistical analysis of miRNA population. **A)** A rarefaction curve of the total number of reads generated by deep sequencing versus the total number of miRNA species identified is shown. The steep curve levels off towards an asymptote, which indicates the point where additional sampling will not yield new miRNAs. **B)** Homogeneity of the miRNA population was assessed using population statistics and by determining the Chao1 diversity estimator. The Chao1 reached a mean stable value of 207, with lower and upper limits of 200.37 to 229.66, respectively, for a level of confidence of 95%.

3. Conclusion

Small non-coding RNAs are a class of molecules that regulate several biological processes. Identification of such molecules is crucial to understand the molecular mechanisms that they regulate. There are already several deep sequencing approaches to identify these molecules. However, correct interpretation of sequencing data depends largely on the bioinformatics and statistical tools available. There are online algorithms that facilitate identification of miRNAs and other small non-coding RNAs from large datasets. However, there are no tools to predict novel small non-coding RNAs beyond miRNAs. As those additional RNA classes, namely piRNAs, snRNAs and snoRNAs are processed differently, the development of algorithms based solely on their biogenesis is challenging. Moreover, the available algorithms have some limitations and additional data analysis should be performed with the discarded reads that can potentially hold non-conventional miRNA molecules. Analysis of deep sequencing data is a powerful methodology to identify novel miRNAs in any organism and determine their expression profiles. The challenge is to deal with increasing dataset size and to integrate the information generated by small RNA sequencing experiments. This will be essential to understand how different RNA classes are related. Computational tools to integrate small non-coding RNA data with gene expression data and target predictions are pivotal to understand the biological processes regulated by miRNAs and other small non-coding RNA classes.

4. References

Ambros, V. (2004). The functions of animal microRNAs. *Nature* 431(7006), 350-355.

Audic, S., and Claverie, J. M. (1997). The significance of digital gene expression profiles. *Genome Res.* 7(10), 986-995.

Bartel, D. P. (2004). MicroRNAs: genomics, biogenesis, mechanism, and function. *Cell* 116(2), 281-297.

Breiman, L. Random forests. Machine Learning (2001). 45:28.

Burnside, J., Ouyang, M., Anderson, A., Bernberg, E., Lu, C., Meyers, B. C., Green, P. J., Markis, M., Isaacs, G., Huang, E., and Morgan, R. W. (2008). Deep sequencing of chicken microRNAs. *Bmc Genomics* 9.

Ceppi, M., Pereira, P. M., Dunand-Sauthier, I., Barras, E., Reith, W., Santos, M. A., and Pierre, P. (2009). MicroRNA-155 modulates the interleukin-1 signaling pathway in activated human monocyte-derived dendritic cells. *Proc. Natl. Acad. Sci. U. S. A* 106(8), 2735-2740.

Chang, J. H., Cruo, J. T., Jiang, D., Guo, H. T., Taylor, J. M., and Block, T. M. (2008). Liver-specific MicroRNA miR-122 enhances the replication of hepatitis C virus in nonhepatic cells. *Journal of Virology* 82(16), 8215-8223.

Chao, A. (1987). Estimating the Population-Size for Capture Recapture Data with Unequal Catchability. *Biometrics* 43(4), 783-791.

Chen, P. Y., Manninga, H., Slanchev, K., Chien, M. C., Russo, J. J., Ju, J. Y., Sheridan, R., John, B., Marks, D. S., Gaidatzis, D., Sander, C., Zavolan, M., and Tuschl, T. (2005). The developmental miRNA profiles of zebrafish as determined by small RNA cloning. *Genes & Development* 19(11), 1288-1293.

Colwell, R. K., and Coddington, J. A. (1994). Estimating Terrestrial Biodiversity Through Extrapolation. *Philosophical Transactions of the Royal Society of London Series B-Biological Sciences* 345(1311), 101-118.

Creighton, C. J., Reid, J. G., and Gunaratne, P. H. (2009). Expression profiling of microRNAs by deep sequencing. *Brief. Bioinform.* 10(5), 490-497.

Droege M, and Hill B. The Genome Sequencer FLX trade mark System-Longer reads, more applications, straight forward bioinformatics and more complete data sets. J.Biotechnol. 136 (1-2): 3-10. 2008.

Feinbaum, R., and Ambros, V. (1999). The timing of lin-4 RNA accumulation controls the timing of postembryonic developmental events in Caenorhabditis elegans. *Dev. Biol.* 210(1), 87-95.

Filipowicz, W., Bhattacharyya, S. N., and Sonenberg, N. (2008). Mechanisms of post-transcriptional regulation by microRNAs: are the answers in sight? *Nat. Rev. Genet.* 9(2), 102-114.

Friedlander, M. R., Chen, W., Adamidi, C., Maaskola, J., Einspanier, R., Knespel, S., and Rajewsky, N. (2008). Discovering microRNAs from deep sequencing data using miRDeep. *Nature Biotechnology* 26(4), 407-415.

Giraldez, A. J., Cinalli, R. M., Glasner, M. E., Enright, A. J., Thomson, J. M., Baskerville, S., Hammond, S. M., Bartel, D. P., and Schier, A. F. (2005). MicroRNAs regulate brain morphogenesis in zebrafish. *Science* 308(5723), 833-838.

Griffiths-Jones, S., Saini, H. K., van, D. S., and Enright, A. J. (2008). miRBase: tools for microRNA genomics. *Nucleic Acids Res.* 36(Database issue), D154-D158.

Hackenberg, M., Sturm, M., Langenberger, D., Falcon-Perez, J. M., and Aransay, A. M. (2009). miRanalyzer: a microRNA detection and analysis tool for next-generation sequencing experiments. *Nucleic Acids Res.* 37(Web Server issue), W68-W76.

Hofacker, I. L. (2003). Vienna RNA secondary structure server. *Nucleic Acids Res.* 31(13), 3429-3431.

Huang, P. J., Liu, Y. C., Lee, C. C., Lin, W. C., Gan, R. R., Lyu, P. C., and Tang, P. (2010). DSAP: deep-sequencing small RNA analysis pipeline. *Nucleic Acids Res.* 38(Web Server issue), W385-W391.

Kim, V. N. (2005). MicroRNA biogenesis: coordinated cropping and dicing. *Nat. Rev. Mol. Cell Biol.* 6(5), 376-385.

Kim, V. N., Han, J., and Siomi, M. C. (2009). Biogenesis of small RNAs in animals. *Nat. Rev. Mol. Cell Biol.* 10(2), 126-139.

Lau, N. C., Lim, L. P., Weinstein, E. G., and Bartel, D. P. (2001). An abundant class of tiny RNAs with probable regulatory roles in Caenorhabditis elegans. *Science* 294(5543), 858-862.

Morin, R. D., O'Connor, M. D., Griffith, M., Kuchenbauer, F., Delaney, A., Prabhu, A. L., Zhao, Y., McDonald, H., Zeng, T., Hirst, M., Eaves, C. J., and Marra, M. A. (2008). Application of massively parallel sequencing to microRNA profiling and discovery in human embryonic stem cells. *Genome Research* 18(4), 610-621.

Pillai, R. S., Bhattacharyya, S. N., and Filipowicz, W. (2007). Repression of protein synthesis by miRNAs: how many mechanisms? *Trends Cell Biol.* 17(3), 118-126.

Schulte, J. H., Marschall, T., Martin, M., Rosenstiel, P., Mestdagh, P., Schlierf, S., Thor, T., Vandesompele, J., Eggert, A., Schreiber, S., Rahmann, S., and Schramm, A. (2010). Deep sequencing reveals differential expression of microRNAs in favorable versus unfavorable neuroblastoma. *Nucleic Acids Res.* 38(17), 5919-5928.

Soares, A. R., Pereira, P. M., Santos, B., Egas, C., Gomes, A. C., Arrais, J., Oliveira, J. L., Moura, G. R., and Santos, M. A. S. (2009). Parallel DNA pyrosequencing unveils new zebrafish microRNAs. *Bmc Genomics* 10.

Tay, Y. M. S., Tam, W. L., Ang, Y. S., Gaughwin, P. M., Yang, H., Wang, W. J., Liu, R. B., George, J., Ng, H. H., Perera, R. J., Lufkin, T., Rigoutsos, I., Thomson, A. M., and Lim, B. (2008). MicroRNA-134 modulates the differentiation of mouse embryonic stem cells, where it causes post-transcriptional attenuation of Nanog and LRH1. *Stem Cells* 26(1), 17-29.

Wienholds, E., Koudijs, M. J., van Eeden, F. J., Cuppen, E., and Plasterk, R. H. (2003). The microRNA-producing enzyme Dicer1 is essential for zebrafish development. *Nat. Genet.* 35(3), 217-218.

Zamore, P. D., and Haley, B. (2005). Ribo-gnome: the big world of small RNAs. *Science* 309(5740), 1519-1524.

Zhu, E., Zhao, F., Xu, G., Hou, H., Zhou, L., Li, X., Sun, Z., and Wu, J. (2010). mirTools: microRNA profiling and discovery based on high-throughput sequencing. *Nucleic Acids Res.* 38(Web Server issue), W392-W397.

7

Computational Methods in Mass Spectrometry-Based Protein 3D Studies

Rosa M. Vitale[1], Giovanni Renzone[2], Andrea Scaloni[2] and Pietro Amodeo[1]
[1]*Istituto di Chimica Biomolecolare, CNR, Pozzuoli*
[2]*Laboratorio di Proteomica e Spettrometria di Massa, ISPAAM, CNR, Naples*
Italy

1. Introduction

Mass Spectrometry (MS)-based strategies featuring chemical or biochemical probing represent powerful and versatile tools for studying structural and dynamic features of proteins and their complexes. In fact, they can be used both as an alternative for systems intractable by other established high-resolution techniques, and as a complementary approach to these latter, providing different information on poorly characterized or very critical regions of the systems under investigation (Russell et al., 2004). The versatility of these MS-based methods depends on the wide range of usable probing techniques and reagents, which makes them suitable for virtually any class of biomolecules and complexes (Aebersold et al., 2003). Furthermore, versatility is still increased by the possibility of operating at very different levels of accuracy, ranging from qualitative high-throughput fold recognition or complex identification (Young et al., 2000), to the fine detail of structural rearrangements in biomolecules after environmental changes, point mutations or complex formations (Nikolova et al.,1998; Millevoi et al., 2001; Zheng et al., 2007). However, these techniques heavily rely upon the availability of powerful computational approaches to achieve a full exploitation of the information content associated with the experimental data.
The determination of three-dimensional (3D) structures or models by MS-based techniques (MS3D) involves four main activity areas: 1) preparation of the sample and its derivatives labelled with chemical probes; 2) generation of derivatives/fragments of these molecules for further MS analysis; 3) interpretation of MS data to identify those residues that have reacted with probes; 4) derivation of 3D structures consistent with information from previous steps. Ideally, this procedure should be considered the core of an iterative process, where the final model possibly prompts for new validating experiments or helps the assignment of ambiguous information from the mass spectra interpretation step.
Both the overall MS3D procedure and its different steps have been the subject of several accurate review and perspective articles (Sinz, 2006; Back et al., 2003; Young et al., 2000; Friedhoff, 2005, Renzone, et al., 2007a). However, with the partial exception of a few recent papers (Van Dijk et al., 2005; Fabris et al., 2010; Leitner et al., 2010), the full computational detail behind 3D model building (step 4) has generally received less attention than the former three steps. Structural derivation in MS3D, in fact, is considered a special case of structural determination from sparse/indirect constraints (SD-SIC). Nevertheless, information for modelling derivable from MS-based experiments exhibits some peculiar

features that differentiate it from the data types associated with other experimental techniques involved in SD-SIC procedures, such as nuclear magnetic resonance (NMR), electron microscopy, small-angle X-ray scattering (SAXS), Förster resonance energy transfer (FRET) and other fluorescence spectroscopy techniques, for which most of the currently available SD-SIC methods have been developed and tailored (Förster et al., 2008; Lin et al., 2008; Nilges et al., 1988a; Aszodi et al., 1995).

In this view, this study will illustrate possible approaches to model building in MS3D, underlining the main issues related to this specific field and outlining some of the possible solutions to these problems. Whenever possible, alternative methods employing either different programs selected among most popular applications in homology modelling, threading, docking and molecular dynamics (MD), or different strategies to exploit the information contained in MS data will be described. Discussion will be limited to packages either freely available, or costing less than 1,000 US$ for academic users. For programs, the home web address has been reported, rather than references that are very often partial and/or outdated. Some examples, derived from the literature available in this field, or developed *ad hoc* to illustrate some critical features of the computational methods in MS3D, should clarify potentiality and current limitations of this approach.

2. General MS3D modelling procedures

2.1 Possible computational protocols for MS3D approaches

MS3D can be fruitfully applied to many structure-related problems; thus, it requires the (possibly combined) use of different modelling procedures. However, a very general scheme for a MS3D approach can still be sketched (Fig. 1). It includes:

- an initial generation of possible structures for the investigated system by some sampling algorithms (S1 or S2 stages);
- followed by classification, clustering and selection steps of the best sampled structures based on one or more criteria (F1 or F2a-F2b-F2c);
- an optional narrowing of the ensemble by a refinement of the selected models (R);
- followed by new classification, clustering and selection stages for the identification of the most representative models (FF).

Selection criteria are very often represented by more or less sophisticated combinations of different scoring (i.e. the higher, the better), penalty (i.e. the lower, the better) or target (i.e. the closer to its reference value, the better) functions. For the sake of brevity, from here onwards the term "scoring" will be indiscriminately used for either true scoring, or penalty, or target function, when their discrimination is not necessary.

The features characterizing a specific approach are: a) combination of sampling (and optimization) algorithms, b) scoring functions in sampling/optimization and classification/ clustering/selection stages, c) strategies to introduce MS-based experimental information.

A first major branching in this scheme already occurs in the earliest modelling stages (box A), depending if MS-based information is, at least in part, integrated in the structure generation stage (path S1-F1), or rather deferred to a subsequent model classification/ selection step (path S2-F2a-F2b-F2c).

Depending on information types, programs and strategies used in modelling (see next sections for theory and examples), MS-based data can be either all introduced during sampling (S1), or all used in the filtering stage (F2a), or subdivided between the two steps (S1+F1). The main advantage of the inclusion of MS-based information into sampling (path

S1-F1) is an increase in model generation efficiency by limitation of the conformational or configurational subspace to be explored. In several potentially problematic cases, i.e. large molecules with very limited additional information available, this reduction can transform a potentially insoluble problem into a reliable model generation, capable of correlating structural and functional features of the investigated system. However, for the very same reason, if information is introduced too abruptly or tightly during structural sampling, it can artificially freeze the models into a wrong, or at least incomplete, set of solutions (Latek et al., 2007; Bowers et al., 2000). Also the weight of erroneous restraints will be considerably amplified by the impossibility of a comparison with solutions characterized by some restraint violations, but considerably more favourable scoring function values, which are often diagnostic of inadequate sampling and/or errors in the experimental restraint set.

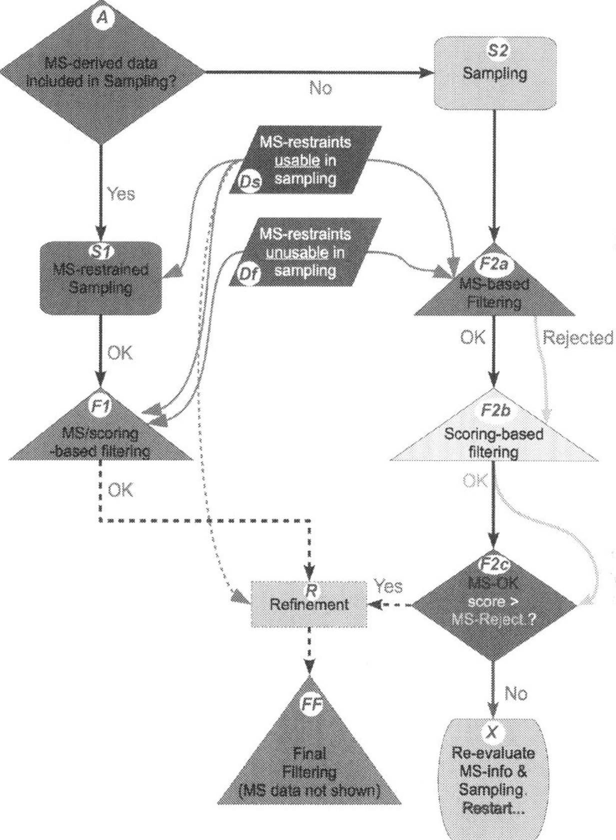

Fig. 1. Flowchart of a generic MS3D modelling approach. Magenta, violet and pink represent steps in which MS-based information is applied. Triangular arrows indicate use of MS-based data. Dotted lines and borders are used for optional refinement stages. Blue codes in white circles/ellipses label the corresponding stages within the text.

Accordingly, both the protocol used to implement MS-based information into modelling procedures and the MS-based data themselves generally represent very critical features, which require the maximum attention during computational setup and final analyses. In addition, implementation of restraints in the sampling procedure either requires some purposely programming activity, or severely limits the choice of modelling tools to programs already including suitable user-defined restraints.
Use of MS-based information in post-sampling analyses (path S2-F2a-F2b-F2c) to help classifying and selecting the final models exhibits a mostly complementary profile of advantages-disadvantages. In fact, it decreases the sampling efficiency of the modelling methods (S2), by leading to a potentially very large number of models to be subsequently discarded on the mere basis of their violations of MS-derived restraints (F2a), and by providing no *ab initio* limitations to the available conformational/configurational space of the system. Furthermore, it may still require programming activity if available restraint analysis tools (F2a) are lacking or inefficient in the case of the implemented information. However, this approach warrants the maximum freedom to the user in the choice of the sampling program; this may result very useful in those cases where the peculiar features of a specific program are strongly required to model the investigated system. In addition, a compared analysis of both structural features and scoring function values between models accepted and rejected on the basis of MS-based data may allow the identification of potential issues in the selected models and the corresponding data sets (steps F2c-X).

2.2 Integration of MS-based data into modelling procedures

Although an ever-increasing number of MS-based strategies has been developed, they provide essentially two information classes for model building: i) surface accessible residues, from chemical/isotopic labelling or limited proteolysis experiments (Renzone et al., 2007a); ii) couples of residues whose relative distances span a prefixed range, from crosslinking experiments (Sinz, 2006; Renzone et al., 2007a). Details on the nature of the combined biochemical and MS approaches used to generate these data and the experimental procedures adopted in these cases is provided in the exhaustive reviews reported above.

2.2.1 Surface-related information (selective proteolysis and chemical labelling)

Although many structural generation approaches include surface-dependant terms, usually they are not exposed to the user; thus, direct implementation of accessibility information is always indirect and ranges from very difficult to impossible. In some docking programs, surface residue patches can be excluded from the exploration, thus restricting the region of space to be sampled (Section 3.2). This information is generally exploited through programs that build and evaluate different kinds of molecular surfaces, applied during the model validation stages. In this view, the main available programs and their usage will be described in the section dedicated to model validation (Section 3.3.2).
In the case of modelling procedures based on sequence alignment with templates of known 3D structure, surface-dependent data can be employed both to validate alignments before modelling (early steps in S1 stage), and to filter the structures resulting from the different steps of a traditional model building procedure (stages F1 or F2a, and FF).

2.2.2 Crosslinks

Cross-linking information often directly contribute to the model building procedure (under the form of distance restraints or direct linker addition to the simulated systems) (stage S1 in Fig.1), in addition to their model validation/interpretation role (stages F1, F2a, FF).

Whenever information from crosslinking experiments is integrated within the modelling procedure, the most common approach recurring in literature is its translation into distance constraints (i.e. "hard", fixed distances) or restraints (i.e. variable within an interval and/or around a fixed distance with a given tolerance) involving atoms, in a full-atomistic representation, or higher-order units, such as residues, secondary structure (SS) elements, or domains, in coarse-grained models. A less common approach consists in the explicit inclusion of the crosslinker atoms in the simulation.

2.2.2.1 Distance restraints

Distance restraints (DRs) are usually implemented by adding a penalty term to the scoring function used to generate, classify or select the models, whenever the distance between specified atom pairs exceeds a threshold value. In this way, associated experimental information can be introduced rather easily and with moderate computational overheads in all the molecular modelling and simulation approaches based on scoring functions. However, since crosslinking agents are molecules endowed with well-defined and specific conformational and interaction properties, both internal and with crosslinked molecules, accurate theoretical and experimental estimates of distance ranges associated with the corresponding cross-link agents only qualitatively correspond to experimentally-detected distances between pairs of cross-linked residues (Green et al., 2001; Leitner et al., 2010). Steric bumps, specific favourable or unfavourable electrostatic interactions, presence of functional groups capable of promoting/hampering the crosslinking reaction and changes in crosslinker conformational population under the effects of macromolecule are all possible causes for observed discrepancies.

2.2.2.2 Explicit linkers

Explicit inclusion of crosslinkers in the systems, although potentially allowing to overcome the limits of DRs, presently suffers from several drawbacks that limit its usage to either final selection/validation stages, or to cases where a limited number of totally independent and simultaneously holding crosslinks are observed. In fact, when many crosslinks are detected in a system by MS analysis, they very often correspond to mixtures of different patterns, because crosslinks can interfere each other either by direct steric hindrance, or by competition for one of the macromolecule reacting groups, or by inducing deformation in the linked system, thus preventing further reactions. However, the added information from explicit crosslinkers may: i) allow disambiguation between alternative predicted binding modes, ii) provide more realistic and strict estimates of the linker length to be used in further stages of DR-based calculations, iii) help modelling convergence, iv) substantially contribute to model validation.

An attempt to reproduce by an implicit approach at least the geometrical constraints associated with a physical linker has been performed by developing algorithms to identify minimum-length paths on protein surfaces (Potluri et al., 2004). This approach provides upper/lower bounds to possible crosslinking distances on static structures but it only worked on static structures as a post-modelling validation tool, and no further applications have been reported so far.

3. Available computational approaches in MS3D

MS-based data can be used to obtain structural information on different classes of problems:

a. single conformational states (e.g. the overall fold);

b. conformational changes upon mutations/environmental modifications;
c. macromolecular aggregation (multimerization);
d. binding of small ligands to macromolecules.

Sampling efficiency and physical soundness of the scoring functions used during sampling (stages S1/S2 of Fig. 1) and to select computed structures (stages F1/F2b and FF) generally represent the main current limitations of 3D structure prediction and simulation methods. In this view, introduction of experimental data represents a powerful approach to reduce the geometrical space to be explored during sampling, and also an independent criterion to evaluate the quality of selected models.

From a computational point of view, structural problems a)-d) translate into system-dependent proper combinations of:

A. fold identification and characterization;
B. docking;
C. structural refinement and characterization of dynamic properties and of changes under the effects of local or environmental perturbations.

Since the optimal combination of methods for a given problem depends upon a large number of system- and data-dependent parameters, and the number of programs developed for biomolecular simulations is huge, an exhaustive description and compared analysis of methods for biomolecular structure generation/refinement is practically impossible. However, we will try to offer a general overview of the main approaches to generate, refine and select 3D structures in MS3D applications, with a special attention to possible ways of introducing MS-based data and exploiting their full information content.

3.1 Fold identification and characterization

The last CASP (Critical Assessment of techniques for protein Structure Prediction) experiment call (CASP9, 2010) classified modelling methods in two main categories: "Template Based Modelling" (TBM) and "Template Free Modelling" (TFM), depending if meaningful homology can be identified or not before modelling between the target sequence and those of proteins/domains whose 3D structures are known (templates).

TFM represents the most challenging task because it requires the exploration of the widest conformational space and heavily relies on scoring methods inspired by those principles of physics governing protein folding (*de novo* or *ab initio* methods), eventually integrated by statistical predictions, such as probabilities of interresidue contacts, surface accessibility of single residues or local patches and SS occurrence. When number and quality of these information increase, together with the extent of target sequence for which they are available, "folding recognition" and "threading" techniques can be used, including a broad range of methods at the interface between TFM and TBM. In these approaches, several partial 3D structure "seeds" are generated by statistical prediction or distant homology relationships, and their relative arrangements are subsequently optimized by strategies deriving from *de novo* methods.

The most typical TBM approach, "comparative" or "homology" modelling (HM), uses experimentally elucidated structures of related protein family members as "templates" to model the structure of the protein under investigation (the "target"). Target sequence can either be fully covered by one or more templates, exhibiting good homology over most of the target sequence, or can require a "patchwork" of different templates, each best covering a different region of the target.

A further group of approaches, presently under active development and already exhibiting good performances in CASP and other benchmark and testing experiments, is formed by the "integrative" or "hybrid" methods. They combine information from a varied set of computational and experimental sources, often acting as/based on "metaservers", i.e. servers that submit a prediction request to several other servers, then averaging their results to provide a consensus that in many cases is more reliable than the single predictions from which it originated. Some metaservers use the consensus as input to their own prediction algorithms to further elaborate the models.

In order to provide some guidelines for structural prediction/refinement tasks in the presence of MS-based data, a general procedure will be outlined for protein fold/structure modelling. The starting step in protein modelling is usually represented by a search for already structurally-characterized similar sequences. Sensitive methods for sequence homology detection and alignment have been developed, based on iterative profile searches, e.g. PSI-Blast (Altschul et al., 1997), Hidden Markov Models, e.g. SAM (K. Karplus et al. 1998), HMMER (Eddy, 1998), or profile-profile alignment such as FFAS03 (Jaroszewski et al., 2005), profile.scan (Marti-Renom et al., 2004), and HHsearch (Soding, 2005).

When homology with known templates is over 40%, HM programs can be used rather confidently. In this case, especially when alignments to be used in modelling have already been obtained, local programs represent a more viable alternative to web-based methods than in TFM processes. If analysis is limited to most popular programs and web services capable of implementing user MS-based restraints (strategy S1 in Fig. 1), the number of possible candidates considerably decreases. Among web servers, on the basis of identified homologies with templates, Robetta is automatically capable of switching from *ab initio* to comparative modelling, while I-TASSER requires user-provided alignment or templates to activate comparative modelling mode. A very powerful, versatile and popular HM program, available both as a standalone application, and as a web service, and embedded in many modelling servers, is MODELLER (http://www.salilab.org/modeller/). It include routines for template search, sequence and structural alignments, determination of homology-derived restraints, model building, loop modelling, model refinement and validation. MS-based distance restraints can be added to those produced from target-template alignments, as well as to other restraints enforcing secondary structures, symmetry or part of the structure that must not be allowed to change upon modelling. However, some scripting ability is required to fully exploit MODELLER versatility.

The overall accuracy of HM models calculated from alignments with sequence identities of 40% or higher is almost always good (typical root mean square deviations (RMSDs) from corresponding experimental structures less than 2Å). The frequency of models deviating by more than 2Å RMSD from experimental structures rapidly increases when target-template sequence identity falls significantly below 30–40%, the so-called "twilight zone" of HM (Blake & Cohen, 2001; Melo & Sali, 2007). In such cases, the quality of resulting modelled structures significantly increases by combining additional information, both of statistical origin, such as SS prediction profiles, and from sparse experimental data (low resolution NMR or chemical crosslinking, limited proteolysis, chemical/isotopical labelling coupled with MS).

If the search does not produce templates with sufficient homology and/or covering of the target sequence, TFM or mixed TFM/TBM methods must be used. Many programs based on *ab initio*, fold recognition and threading methods are presently offered as web services; this is because very often they use a metaserver approach for some steps, need extensive

searches in large databases, require huge computational resources, or to better protect underlying programs and algorithms, currently under very active development. Although this may offer some advantages, especially to users less-experienced in biocomputing or endowed with limited computing facilities, it may also imply strong limitations in the full exploitation of the features implemented in the different methods, with particularly serious implications in MS3D. Only few servers either include a NMR structure determination module (not always suitable for MS-based data), or explicitly allow the optional usage of user-provided distance restraints in the main input form. Fortunately, two of the most used and versatile servers, Robetta (http://robetta.bakerlab.org/) and I-TASSER (http://zhanglab.ccmb.med.umich.edu/I-TASSER/), good performers at the last CASP rounds (http://predictioncenter.org/), allow input of distance restraints in the modelling procedure, via a NMR-dedicated service for Robetta (Rosetta-NMR, suitable for working with sparse restraint) (Bowers et al., 2000), or directly in the main prediction submission page (I-TASSER). Other servers can still allow the implementation of MS-based information in the model generation step if they can save intermediate results, such as sequence alignments, SS or fold predictions. These latter, after addition of MS-based restraints, can be then included into suitable modelling programs, to be run either locally or on web servers.
A successful examples of modelling with MS-based information in a low-homology case is Gadd45β. A model was built, despite the low sequence identity (<20%) with template identified by fold recognition programs, through the introduction of additional SS restraints, which were based on SS profiles and experimental data from limited proteolysis and alkylation reactions combined with MS analysis (Papa et al., 2007). Model robustness was confirmed by comparison with the homolog Gadd45γ structure solved later (Schrag JD et al., 2008), where the only divergence in SS profiles was the occurrence of two short 3_{10} helices (three residues each long) and an additional two-residues β-strand in predicted loop regions (Fig. 2). Furthermore, this latter β-strand is so distorted that only a few SS assignment programs could identify it, and the corresponding sequence in Gadd45β, predicted unstructured and outside the template alignment, was not modelled at all.

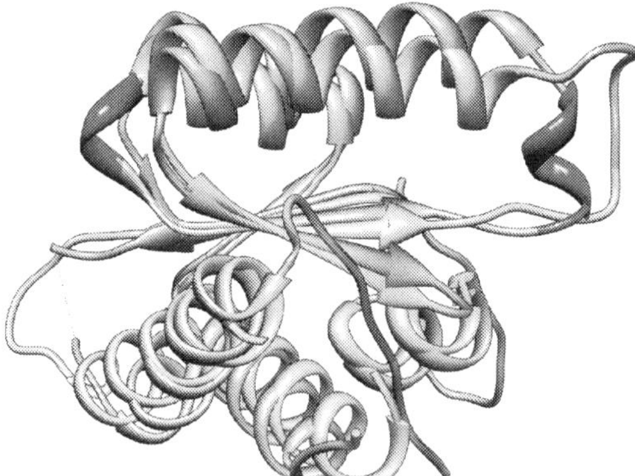

Fig. 2. Comparison between the MS3D model of Gadd45β (light green) and the crystallographic structure of its homolog Gadd45γ (light blue). Sequences with different SS profiles are painted green in Gadd45β and magenta in Gadd45γ.

3.2 Docking

Usually, methods for protein docking involve a six-dimensional search of the rotational and translational space of one protein with respect to the other where the molecules are treated as rigid or semirigid-bodies. However, during protein-protein association, the interface residues of both molecules may undergo conformational changes that sometimes involve not only side-chains, but also large backbone rearrangements. To manage at least in part these conformational changes, protein docking protocols have introduced some degree of protein flexibility by either use of "soft" scoring functions allowing some steric clash, or explicit inclusion of domain movement/side chain flexibility. Biological information from experimental data on regions or residues involved in complexation can guide the search of complex configurations or filter out wrong solutions. Among the programs most frequently used for protein-protein docking, recently reviewed by Moreira and colleagues (Moreira et al., 2010), some of them can manage biological information and will be discussed in this context.

In the Attract program (http://www.t38.physik.tu-muenchen.de/08475.htm), proteins are represented with a reduced model (up to 3 pseudoatoms per amino acid) to allow the systematic docking minimization of many thousand starting structures. During the docking, both partner proteins are treated as rigid-body and the protocol is based on energy minimization in translational and rotational degrees of freedom of one protein with respect to the other. Flexibility of critical surface side-chains as well as large loop movements are introduced in the calculation by using a multiple conformational copy approach (Bastard et al., 2006). Experimental data can be taken into account at various stages of the docking procedure.

The 3D-Dock algorithm (http://www.sbg.bio.ic.ac.uk/docking/) performs a global scan of translational and rotational space of the two interacting proteins, with a scoring function based on shape complementarity and electrostatic interaction. The protein is described at atomic level, while the side-chain conformations are modelled by multiple copy representation using a rotamer library. Biological information can be used as distance restraints to filter final complexes.

HADDOCK (http://www.nmr.chem.uu.nl/haddock/) makes use of biochemical or biophysical interaction data, introduced as ambiguous intermolecular distance restraints between all residues potentially involved in the interaction. Docking protocol consists of four steps: 1) topology and structure generation; 2) randomization of orientations and rigid body energy minimization; 3) semi-flexible simulated annealing (SA) in torsion angle space; 4) flexible refinement in Cartesian space with explicit solvent (water or DMSO). The final structures are clustered using interface backbone RMSD and scored by their average interaction energy and buried interface area. Recently, also explicit inclusion of water molecules at the interface was incorporated in the protocol.

Molfit (http://www.weizmann.ac.il/Chemical_Services/molfit/) represents each molecule involved in docking process by a 3-dimensional grid of complex numbers and estimates the extent of geometric and chemical surface complementarity by correlating the grids using Fast Fourier Transforms (FFT). During the search, contacts involving specified surface regions of either one or both molecules are up- or down-weighted, depending on available structural and biochemical data or sequence analysis (Ben-Zeev et al., 2003). The solutions are sorted by their complementarity scores and the top ranking solutions are further refined by small rigid body rotations around the starting position.

PatchDock (http://bioinfo3d.cs.tau.ac.il/PatchDock/) is based on shape complementarity. First, the surfaces of interacting molecules are divided according to the shape in concave, convex and flat patches; then, complementarity among patches are identified by shape-matching techniques. The algorithm is a rigid body docking, but some flexibility is indirectly considered by allowing some steric clashes. The resulting complexes are ranked on the basis of the shape complementarity score. PatchDock allows integration of external information by a list of binding site residues, thus restricting the matching stage to their corresponding patches.

RosettaDock (http://rosettadock.graylab.jhu.edu/) try to mimics the two stages of a docking process, recognition and binding, as hypothesized in Camacho & Vajda, 2001. Recognition is simulated by a low resolution phase in which a coarse-grained representation of proteins, with side chains replaced by single pseudoatoms, undergoes a rigid body Monte Carlo (MC) search on translations and rotations. Binding is emulated by a high-resolution refinement phase where explicit sidechains are added by using a backbone-dependent rotamer packing algorithm. The sampling problem is handled by supercomputing clusters to ensure a very large number of decoys that are discriminated by scoring functions at the end of both stages of docking. The docking search problem can be simplified when biological information is available on the binding region of one or both interacting proteins. The reduction of conformational space to be sampled could be pursued by: i) opportunely pre-orienting the partner proteins, or ii) reducing docking sampling to the high-affinity domain, in the case of multidomain proteins, or iii) using loose distance constraints.

ZDOCK (http://zdock.bu.edu/) is a rigid body docking program based on FFT algorithm and an energy function that combines shape complementarity, electrostatics and desolvation terms. RDOCK (http://zdock.bu.edu/) is a refinement program to minimize and rerank the solutions found by ZDOCK. The complexes are minimized by CHARMm (Brooks et al., 1983) to remove clashes and improve energies, then electrostatic and desolvation terms are recalculated in a more accurate fashion with respect to ZDOCK. Biological information can be used either to avoid undesirable contacts between certain residues during ZDOCK calculations or to filter solutions after RDOCK.

As in protein folding, also for docking the use of MS-based information allowed the modelling of several complexes even in the lack of suitable templates with high homology. The fold of prohibitin proteins PHB1 and PHB2 was predicted (Back et al., 2002) by SS and fold recognition algorithms, while crosslinking allowed to model the relative spatial arrangement of the two proteins in their 1:1 complex. Another example of combined use of SS information, chemical crosslinking, limited proteolysis and MS analysis results with a low sequence identity (~ 20%) template is the modelling of porcine aminoacylase 1 dimer; in this case, standard modelling procedures based on automatic alignment had failed to produce a dimeric model consistent with experimental data (D'Ambrosio et al., 2003).

In the case of protein-small ligand docking, the conformational space to be explored is reduced by the small size of the ligand, whose full flexibility can usually be allowed, and by the limited fraction of protein surface to be sampled, corresponding to the binding site, often already known. Among the programs for ligand-flexible docking that allow protein side-chains flexibility, Autodock is one of most popular (http://autodock.scripps.edu/). AutoDock combines a grid-based method with a Lamarckian Genetic Algorithm to allow a rapid evaluation of the binding energy. A simulated annealing method and a traditional genetic algorithm are also available in Autodock4.

In general, MS-based data can be used to limit the protein region to be sampled (Kessl et al., 2009) or can be explicitly considered in the docking procedure, as in the case of the mapping of Sso7d ATPase site (Renzone et al., 2007b). In this case, three independent approaches for molecular docking/MD studies were followed, considering both FSBA-derivatives and the ATP-Sso7d non-covalent complex: i) unrestrained MD, starting from a full-extended, external conformation for Y7-FSBA and K39-FSBA residue sidechains, and from several random orientations for ATP, with an initial distance of 20 Å from Sso7d surface, in regions not involved in protein binding; ii) restrained MD, by gradually imposing distance restraints corresponding to a H-bond between adenine NH_2 group and each accessible (i.e., within a distance lower or equal to the maximum length of the corresponding FSBA-derivative) donor sidechain; iii) rigid ligand docking, by calculating 2000 ZDOCK models of the non-covalent complex of Sso7d with an adenosine molecule. The rigid ligand docking reproduced only in part features from other approaches, as rigid docking correctly predicted the anchoring point for adenosine ring, but failed to achieve a correct position for the ribose moiety, due to the required concerted rearrangement of two Sso7d loops involved in the binding. This latter feature represents one of the main advantages of modelling strategies involving MD (in particular, in cartesian coordinates) because MD-based simulation techniques are the best or the only approaches that reproduce medium-to-large scale concerted rearrangements of non-contiguous regions.

3.3 Model simulation, refinement and validation

Refinement (R stage in Fig.1) and validation of final models (FF stage) represent very important steps, especially in cases of low homologies with known templates and when fine details of the models are used to predict or explain functional properties of the investigated system. In addition, very often the modelled structures are aimed at understanding the structural effects of point mutations or other local sequence alterations (sequence deletions/insertions, addition or deletion of disulphide bridges, formation of covalent constructs between two molecules and post-translational modifications), or of changes in environmental parameters (temperature, pressure, salt concentration and pH). In these cases, techniques are required to simulate the static or dynamic behaviour of the investigated system in its perturbed and unperturbed states.

3.3.1 Computational techniques and programs for model simulation and refinement

Model refinement, when not implemented in the modelling procedure, can be performed by energy minimization (EM) or, better, by different molecular simulation methods, mostly based on variants of molecular dynamics (MD) or Monte Carlo (MC) techniques. They are also commonly used to characterize dynamic properties and structural changes upon local or environmental perturbations.

Structures deriving from folding or docking procedures need, in general, at least a structural regularization by EM before final validation steps, to avoid meaningless results from many methods. Scoring functions of the latter evaluate the probity of parameters, such as dihedral angle distributions, presence and distribution of steric bumps, voids in the molecular core, specific nonbonded interactions (H-bonds, hydrophobic clusters). Representing a mandatory step in most MC/MD protocols, EM programs are included in all the molecular simulation packages, and they share with MC/MD most input files and part of the setup parameters. Thus, unless they are be explicitly discussed, all system- and restraint-related features or issues illustrated for simulation methods also implicitly held for EM.

As we are mostly interested in techniques implementing experimentally-derived constraints or restraints, some of the most popular methods for constraints-based modelling will be briefly described. These methods have been developed and optimized mainly to identify and refine 3D structures consistent with spatial constraints from diffraction and resonance experiments (de Bakker et al., 2006). They have also been extensively applied to both TBM (Fiser & Sali, 2003) and free modelling prediction and simulation (Bradley et al., 2005; Schueler-Furman et al., 2005), and are often used to refine/validate models produced in TFM and TBM approaches described in sections 3.1 and 3.2. There are two main categories of constraint-based modelling algorithms: i) distance geometry embedding, which uses a metric matrix of distances from atomic coordinates to their collective centroid, to project distance space to 3D space (Havel et al. 1983; Aszodi et al. 1995, 1997); ii) minimization, which incorporates distance constraints in variable energy optimization procedures, such as molecular dynamics (MD) and Monte Carlo (MC). For both MD and MC, it is possible to work both in full cartesian coordinates, or in the restricted torsion angle (TA) space, with covalent structure parameters kept fixed at their reference values, thus originating the Torsional Angle MD (TAMD) and Torsional Angle MC (TAMC) approaches. They are currently implemented in several modelling and refinement packages, developed for structural refinement of X-ray or NMR structures (Rice & Brünger, 1994; Stein et al. 1997; Güntert et al., 1997), folding prediction (Gray et al., 2003), or more general packages (Mathiowetz et al., 1994; Vaidehi et al., 1997). Standard MC/MD methods are only useful for structural refinement, local exploration and to characterize limited global rearrangements. However, they are also widely used as sampling techniques in folding/docking approaches, although in those cases enhanced sampling extensions of both methods are employed. Simulated annealing (SA) (Kirkpatrick et al., 1983) and replica exchange (RE) approaches (Nymeyer et al., 2004) are the most common examples of these MC/MD enhancements, both potentially overcoming the large energy barriers required for sampling the wide conformational and configurational spaces to be explored in folding and docking applications, respectively.

A non-exhaustive list of the most diffused simulation packages including a more-than-basic treatment of distance-related restraints and also exhibiting good versatility (i.e. implementation of different algorithms, approaches, force fields and solvent representations), may include at least: AMBER (http://ambermd.org/), CHARMM (http://www.charmm.org/), DESMOND (http://deshawresearch.com/resources.html), GROMACS (http://www.gromacs.org/) and TINKER (http://dasher.wustl.edu/tinker). CYANA (http:www.cyana.org) and XPLOR/CNS (http://cns-online.org/v1.3/), although originally more specialized for structural determination and refinement from NMR and NMR/X-ray data, respectively, have been recently included in several TFM and TBM protocols, thanks to their efficient implementations of TAMD and distance or torsional angle restraints. The choice of a simulation program should ideally keep into account several criteria, ranging from computational efficiency, to support of sampling or refinement algorithms, to integration with other tools for TFM or TBM applications.

The main problems associated with simulation methods having relevant potential implications on MS3D are: i) insufficient sampling; ii) inaccuracy in the potential energy functionals driving the simulations; iii) influence of the approach used to implement experimentally-derived information on final structure sets.

Sampling problem can be approached both by increasing the sampling efficiency with MC/MD variations like SA and RE, and by decreasing the size of the space to be explored.

This latter result can be reached by reducing the overall number of degrees of freedom to be explicitly sampled and/or by reducing the number of possible values per variable to a small, finite number (discretization, like in grid-based methods), and/or by restraining acceptable variable ranges. Reduction of the total number of degrees of freedom can be accomplished by switching to coarse-grained representations of the system, where a number of explicit atoms, ranging from connected triples, to amino acid sidechains, to whole residues, up to full protein subdomains, are replaced by a single particle. This method is frequently used in initial stages of *ab initio* folding modelling, or in the simulation of very large systems, such as giant structural proteins of huge protein aggregates.

Another possible way to reduce the number of degrees of freedom is the aforementioned TA approach, requiring for a N atom system only N/3 torsional angles compared with 3N coordinates in atomic cartesian space (Schwieters & Clore, 2001). Moreover, as the high frequency motions of bending and stretching are removed, TAMD can use longer time steps in the numerical integration of equations of motion than that required for a classical molecular dynamics in cartesian space. Its main limitation may derive from neglecting covalent geometry variations (in particular, bending centred on protein Cα atoms) that are known to be associated with conformational variations (Berkholz et al., 2009), for instance from α-helix to β-strand, and that can be important in concerted transitions or in large structures with extensive and oriented SS regions. Discretization is mostly employed in the initial screening of computationally intensive problems, such as *ab initio* modelling. Restraining variable value ranges in MS3D is usually associated with either predictive methods (SS, H-bond pattern, residue exposure), or to homology analysis, or to experimentally-derived information. Origin, nature and form of these restraints have already been discussed in previous sections, while some more detail on the implementation of distance-related information into simulation programs will be given at the end of this section.

While the implementation of restraints can be very variable in methods where the scoring function does not intend to mimic or replicate a physical interaction between involved entities, in methods based on physically-sounding molecular potential functions (forcefields) have DRs implemented by a more limited number of approaches. At its simplest, a DR will be represented as a harmonic restraint, for which only the target distance and the force constant need to be specified in input. This functional form is present in practically all most common programs, but either requires a precise knowledge of the target distance, or it will result in a very loose restraint if the force constant is lowered too much to account for low-precision target values, the usual case in MS-based data. In a more complex and useful form, implemented with slight variations in several programs (AMBER, CHARMM, GROMACS, XPLOR/CNS, DESMOND, TINKER), the restraint is a well with a square bottom with parabolic sides out to a defined distance, and then linear beyond that on both (AMBER) or just the upper limit side (CHARMM, GROMACS, XPLOR/CNS, DESMOND). In some programs (CHARMM, AMBER, XPLOR/CNS), it is possible to select an alternative behaviour when a distance restraint gets very large (Nilges et al,1988b) by "flattening out" the potential, thus leading to no force for large violations; this allows for errors in constraint lists, but might tend to ignore constraints that should be included to pull a bad initial structure towards a more correct one.

Other forms for less-common applications can also be available in the programs or be implemented by an user. However, the most interesting additional features of versatile DR implementations are the different averages that can be used to describe DRs: i) complex restraints can involve atom groups rather than single atoms at either or both restraint sides; ii) time-averaged DRs, where target values are satisfied on average within a given time lapse rather than instantaneously; iii) ambiguous DRs, averaged on different distance pairs. The latter two cases are very useful when the overall DRs are not fully consistent each other, because they are observed in the presence of conformational equilibria and, as such, they are associated with different microstates of the system. In addition, complex and versatile protocols can be simply developed in those programs where different parameters can be smoothly varied during the simulation (AMBER).

3.3.2 Programs for model validation

A validation of the final models, very often included in part in the available automated modelling protocols, represents a mandatory step, especially for more complex (low-homology, few experimental data) modelling tasks. A huge number of protein and nucleic acid structural analysis and validation tools exists, based on many different criteria, and subjected to continuous development and testing; thus, even a CASP section is dedicated to structural assessment tools (http://www.predictioncenter.org/), and the "Bioinformatics Links Directory" site alone currently reports 76 results matching "3-D structural analysis" (Brazas et al., 2010). Being outside the scope of the present report, information on 3D structural validation tools can be searched on specialized sites such as http://bioinformatics.ca/links_directory/ . However, similarly to what stated on prediction metaservers, a general principle for validation is to possibly use several tools, based on different criteria, looking for emergent properties and consensus among the results.
Specific parameters associated with MS-based data can be usually analysed with available tools. Distance restraints and their violations can be analysed both on single structures and on ensembles (sets of possible solutions of prediction methods, frames from molecular dynamics trajectories) with several graphic or textual programs, the most specialized obviously being those tools developed for the analysis of NMR-derived structures.
Surface information can be analysed by programs like:
DSSP (http://swift.cmbi.ru.nl/gv/dssp/),
NACCESS (http://www.bioinf.manchester.ac.uk/naccess/),
GETAREA (http://curie.utmb.edu/getarea.html/),
ASA-VIEW (http://gibk26.bse.kyutech.ac.jp/jouhou/shandar/netasa/asaview/)
that calculate different kinds of molecular surfaces, such as van der Waals, accessible, or solvent excluded surfaces for overall systems and contact surfaces for complexes are used. However, differently from distance restraints, available programs usually work on a single input structure at a time, thus making structure filtering and analysis on the large ensembles of models potentially produced by conformational prediction, molecular simulation or docking calculations, a painful or impossible task. In these cases, scripts or programs to automate the surface calculations and to average or filter the results must be developed.

4. Modelling with sparse experimental restraints

In the previous section many of the computational methods that can concur to produce structural models in MS3D applications have been outlined, together with different ways to integrate MS-based experimental information into them. Here we will refocus on the overall computational approach in MS3D, to illustrate some of its peculiar features and issues, its present potentialities and the variety of possible combinations of data and protocols that can be devised to optimally handle different types of structural problems. Depending on nature and quantity of available experimental information and on previous knowledge of the investigated system, different combinations of the methods mentioned in previous sections can be optimally employed. We will start illustrating examples of methods for *de novo* protein folding, a frontier application of modelling with sparse restraints, because it is based on minimal additional information on the system under investigation.

The MONSSTER program (Skolnick et al. 1997) only makes use of SS profiles and a limited number of long-distance restraints. By employing system discretization and coarse-graining to reduce required sampling, a protein is represented by a lattice-based Cα-backbone trace with single interaction center rotamers for the side-chains. By using N/7 (N is the protein length) long-range restraints, this method is able to produce folds of moderate resolution, falling in the range from 4 to 5.5 Å of RMSD for Cα-traces with respect to the native conformation for all α and α/β proteins, whereas β-proteins require, for the same resolution, N/4 restraints. A more recent method for *de novo* protein modelling (Latek et al., 2007) adopts restrained folding simulations supported by SS predictions, reinforced with sparse experimental data. Authors focused on NMR chemical-shift-based restraints, but also sparse restraints from different sources can be employed. A significant improvement of model quality was already obtained by using a number of DRs equal to N/12.

As already stated by Latek and colleagues, the introduction of DRs in protein folding protocol represents a critical step that in principle could negatively affect the sampling of conformational space. In fact, restraint application at too early stages of calculations can trap the protein into local minima, where restraints are satisfied, but the native conformation is not reached. In addition to the number, even the specific distribution of long-range restraints along the sequence can affect the sampling efficiency. To test the influence of data sets in folding problem, we applied a well-tested protocol of SA, developed for AMBER program and mainly oriented to NMR structure determination, to the folding simulation of bovine pancreatic trypsin inhibitor (BPTI), by using different sets of ten long-distance restraints, randomly selected from available NMR data (Berndt et al., 1992), with optional inclusion of a SS profile. Fig. 3 shows representative structures for each restraint set.

The four native BPTI disulphide bridges were taken into account by additional distance and angle restraints. BPTI represents a typical benchmark for this kind of studies, due to its peculiar topology (an α/β fold with long connection loops, stabilized by disulphide bonds) still associated with a limited size (58 residues), and to the availability of both X-ray and NMR accurate structures. SA cycles of 50 structures each were obtained and compared for four combinations of three sets (S1-3) of ten long distance restraints, totally non-redundant among different sets and SS profiles: a) S1+SS profile; b) S1 alone; c) S2+SS profile; d) S3+SS profile. S1 set performed definitely better than the other two, its best model exhibiting a RMSD value of 2.4 Å on protein backbone of residues 3-53 from the representative NMR structure.

This set was also able to provide a reasonable low-resolution fold even in the absence of SS restraints (b). S3 resulted in almost correctly folded models, but with significantly worse RMSD values than S1 (c). In S3 pseudomirror images (d) of the BPTI fold occurred several times and only one model out of 50 was correctly folded (not shown).

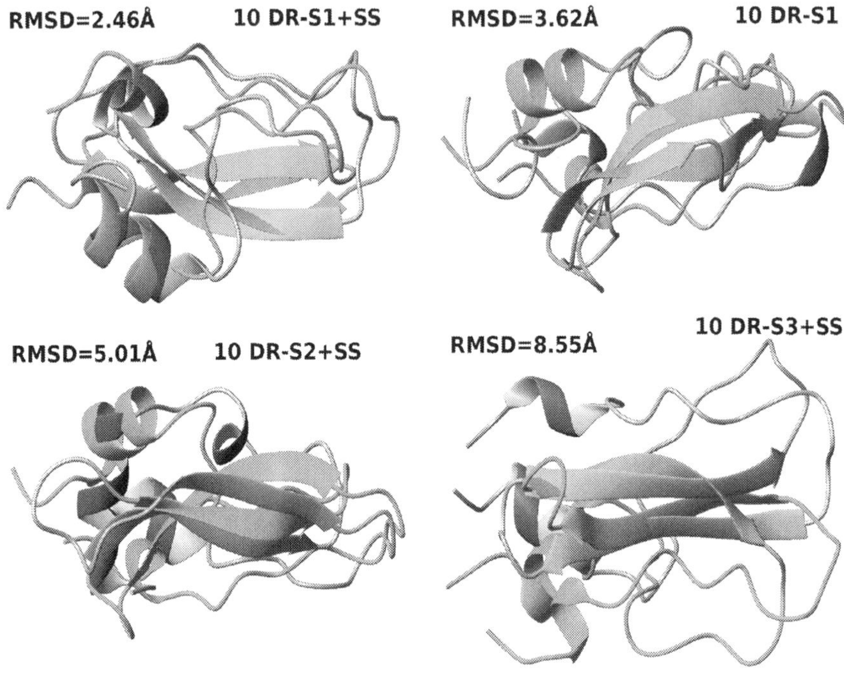

Fig. 3. *Ab initio* modelling from sparse restraints of BPTI. Representative models from different restraint sets (S1, S2, S3), with optional SS dihedral restraints are shown in ribbon representation, coloured in red/yellow (helices), cyan (strands) and grey (loops). Models are best-fitted on Cα atoms of residues 3-53 to the representative conformation of the NMR high-resolution structure (PDB code: 1pit) (green), except for the S3+SS set, where superposition with β-sheet only is shown, to better illustrate pseudo-mirroring effect, although RMSD values are calculated on the same 3-53 residue range as other models.

These results suggest a strong dependency of results upon both the exact nature of experimental data used in structure determination, and the protocol followed for model building. Thus, the number of restraints estimated in the aforementioned studies as necessary/sufficient for a reliable structural prediction should be prudently interpreted for practical purposes. If a proper protocol is adopted, increasing quantity, quality and distribution homogeneity of data should decrease this dependency, but the problem still remains severe when using very sparse restraints, such as those associated with many MS3D applications. A careful validation of the models and, possibly, execution of more modelling cycles with variations in different protocol parameters, can help to identify and solve these kinds of problems.
However, in spite of these potential issues, *ab initio* MS3D can provide a precious insight into systems that are impossible to study with other structural methods. In addition to increases in the number of experimental data, also homology-based information and other statistically-derived constraints can substantially increase the reliability of MS3D predictions. Thus, suitable combinations of experimental data, predictive methods and computational approaches have allowed the modelling of many different proteins and protein complexes spanning a wide range of sizes and complexity. The illustrative examples shown in Table 1 represents just a sample of systems affordable with current computational MS3D techniques and a guideline to select possible approaches for different problem classes. Heterogeneity of reported systems, data and methods, while suggesting the enormous potentialities of MS3D approaches, practically prevents any really meaningful critical comparison among methods, whose description in applicative papers is often incomplete. A standardization of MS3D computational methods is still far from being achieved, since it requires considerable computational effort to tackle with the considerable number of strategies and parameters that should be tested in a truly exhaustive analysis. Furthermore, the extreme sensitivity of modelling with sparse data to constraint distribution, as seen in the example shown in Fig. 3, either introduces some degree of arbitrariness in comparative analyses, or make them even more computationally-intensive, by requiring the use of more subsets for each system setup to be sampled.
Advancements in MS3D experimental approaches continuously change the scenarios for computational procedures, by substantially increasing the number of data, as well as the types of crosslinking or labelling agents and proteolytic enzymes. The large number of crosslinks obtained for apolipoproteins (Silva et al., 2005; Tubb et al., 2008) or CopA copper ATPase (Lübben et al., 2009) represent good examples of these trends (Table 1).

5. Conclusion

As already stated in the preceding section, the compared analysis of computational approaches involved in MS3D is still considerably limited, because of the complexity both of the systems to be investigated, and of the methods themselves, especially when they are used in combination with restraints as sparse as those usually available in MS3D studies. The continuous development in all involved experimental and computational techniques considerably accelerates the obsolescence of the results provided by any accurate methodological analysis, thus representing a further disincentive to these usually very time consuming studies. In this view, rather than strict prescriptions, detailed recipes or sharp critical compared analysis of available approaches, this study was meant to provide an

System	Exp. data[a]	Use of exp. data[b]	Available info[c]	Additional data[d]	Modelling techniques[e]/ Refinement[e]	Software	Reference
Annexin A2/p11 complex	**3 XL, CL**	**XL:PSF, CL:PSF**	**XR**: ANXA2 (part); $(p11)_2$-(ANXA2 N-terminus)$_2$ complex	**SSp** for missing residues	multistep **PPD / NA**	Rosetta	Shultz et al., 2007
Apolipoprotein (apo) A-I	**17 XL**	**XL:IIS**	**XR**: 4 templates	Helix amphipatic profiles	**HM**(regions); **XL / EM** with **DRs**	MOE-AMBER	Silva et al., 2005
Apolipoprotein (apo) A-IV	**21 XL**	**XL:IIS**	**XR**: res. 1-241: 1 template	**SSp:** residues 242-378	**HM** of 1-241; merged with **SSp** by **XL / EM** with **DRs**	MOE, I-TASSER	Tubb et al., 2008
Calmodulin (CaM)-Munc13 peptides complexes	**XL** (6 x Munc13; 3x ubMunc13-2); **PL**	**XL,PL: IIS,PSF**	**XR**: different CaM-peptides complexes	**SSp** for Munc13 peptides	**FP** of Munc13 peptides; multistep **PPD** with **IIS** & **PSF/NA**	Bhageerath program, Patch-Dock, Rosetta Dock	Dimova et al., 2009
CopA copper ATPase	**18 XL**	**XL:IIS, PSF**	**XR**: different templates for the diverse domains	TM-helix predictions	**HM**; Multistep **PPD** with **PSF** and **IIS / EM**	SwissModel, 3Dgarden, Haddock, Xplor-NIH	Lübben et al., 2009
Ffh-FtsY complex	**9 XL**	**XL:IIS**	**XR**: isolated proteins	NA	Multistep **PPD** with **IIS / SA-MD, EM**	CHARMM, Dock, Multidock	Chu et al., 2004
G-actin–cofilin complex	**2 XL**	**XL:PSF**	**XR**: isolated proteins, templates for 5 N-ter residues of cofilin	NA	**HM** of missing regions; **PPD** with **PSF** / NA	InsightII, Autodock 3.0	Grintsevich et al., 2008
Latexin–carboxypeptidase A complex	**3 XL**	**XL:PSF**	**XR**: isolated proteins	NA	**PPD** with **PSF** / NA	In house rigid PD with hydrophobic energy score	Mouradov et al., 2006
MutL - MutH complex	**4 XL**	**XL:PSF**	**XR**: isolated proteins		**PPD** with **PSF** / NA	BIGGER	Giron-Monzon et al., 2004

Urease complex	**2 XL**	**XL:PSF**	**XR**: template	NA	**HM** with **PSF** / NA	MODELLER	Carlsohn et al., 2004
GTPases Rab4 and Rab5	**LP**	**LP:PSF**	**XR**: templates	NA	**HM** with **PSF / EM**	QUANTA/ CHARMm, DISCOVER/ AMBER	Nikolova et al., 1998
ERK2 - PTP-SL complex	**XL**	**XL:IIS**	**XR**: templates	NA	**HM, aiM, PPD** with **IIS / SA-MD, EM**	InsightII, HMMSTR/ Rosetta, RosettaDock, CHARMm	Balasu et al., 2009
Aminoacylase 1	**1 XL, CL, LP**	**XL,CL: PSF**	**XR**: templates	**SSp**	**HM** with **PSF / EM**	MODELLER, AMBER	D'Ambrosio et al., 2003
Sso7d-melittin / ATP complex	**LP, PL**	**LP:PSF, PL: IIS**	**XR**: isolated proteins	NA	**PPD, PLD** with **IIS** & **PSF / MD, EM**	ZDOCK, AMBER	Renzone et al., 2007b
Gadd45β-MKK7 complex	**LP, AA**	**LP,AA: IIS**	**XR**: isolated proteins	**SSp**	**HM** with **IIS, PPD** with **PSF/MD,EM**	ZDOCK, AMBER	Papa et al., 2007
Calmodulin-Melittin complex	**LP,CL**	**LP,CL: PSF**	**XR**: CaM, CaM-peptides complexes	**NA**	**MD/EM**	AMBER	Scaloni et al., 1998

Table 1. Some examples of MS3D studies from literature. The following abbreviations have been used (in **bold**, non standard abbreviations): [a]**XL**: crosslinking, **CL**: chemical labelling, **PL**: photoaffinity labeling, **LP**: limited proteolysis; **AA**: alkylation analyses, [b]see [a], **PSF**: post-sampling filtering with experimental data, **IIS**: esperimental data integrated in sampling; [c]**XR**: x-ray crystallography; [d]**SSp**: secondary structure prediction, [e]see [a,b], **HM**: Homology modeling, **aiM**: *ab-initio* modeling, **FP**: fold prediction, **MD**: molecular dynamics, **SA**: Simulate Annealing, **EM**: energy minimization, **PPD**: protein-protein docking, **PLD**: protein-small ligand docking; **DR**: distance restraints, TM: trans-membrane; NA: not available.

overall and as wide as possible picture of the state-of-art approaches in MS3D computational techniques and their potential application fields. However, in spite of these limitations, some general conclusions can still be drawn.

For predictive methods that stay behind the most ambitious MS3D applications (*ab initio* folding, folding prediction, threading), at least when used in the absence of experimental data, metaservers exhibit on average best performances than the single employed servers, as also shown by the results of the last CASP rounds on automatic servers (http://predictioncenter.org/). This suggests two distinct considerations: 1) the accuracy of sampling and scoring exhibited by each single method, as well the rationale behind them, are still so limited to prevent reliable predictions on best performing methods in any given case; 2) nevertheless, most methods tend to locate correct solutions, or, in general, groups of solutions including the correct one or a close analogue. Therefore, a consensus among the predictions from different servers generally improves the final solutions, by smoothing down both extreme results and random fluctuations associated with each single approach. Well consolidated metaservers, such as Robetta or I-TASSER, can be regarded as reasonable starting guesses for general folding problems, also considering that they both include distance-related restraints in their available options. However, special classes of systems (e.g. transmembrane proteins or several enzyme families) can instead benefit from employing specifically-devised approaches.

In comparing server-based applications to standalone programs (often available in alternative for a given approach), potential users should also consider that the former require less computational skill and resources, but are intrinsically less flexible than the latter, and that legal and secrecy issues may arise, because several servers consider submitted prediction requests and the corresponding results as public data, as usually clearly stated in submission pages. In addition to possible information "leakage" in projects, the public status of the models would prevent their use in patents.

When considering more specifically MS3D procedures, it has been shown that even a small number of MS-based restraints can significantly help in restricting the overall space to be explored and in identifying the correct fold/complexation mode, especially if they are introduced in early modelling stages of a computational procedure optimized to deal with both the investigated system and the available data. Thus, experimental restraints can allow the use of a single model generation procedure, rather than a multiple/metaserver approach, at least in non-critical cases. In fact, they should filter out all wrong solutions deriving from the biases of the modelling method, leaving only those close to the "real" one, if it is included in the sampled set. In particular, since the lowest energy structure should ideally also be associated with a minimum violation of experimentally-derived restraints, the coincidence of minimum energy structures with least violated restraints should be suggestive of correct modelling convergence and evaluation of experimental data. However, particular care must be adopted not only in the choice of the overall computational procedure, but especially of the protocol used to introduce experimental information, because a too abrupt build up of the restraints can easily bring to local minima far from the correct solution. Comparison of proper scoring functions other than energy between experimentally-restrained and unrestrained solutions may provide significant help in identifying potential issues in data or protocols. Estimates of the sensitivity of solutions to changes in protocols may also enforce the reliability of best converged cases. In particular, when other restraints are also present, the relative strength and/or introduction order of the different sets could play an important role in the final result; thus, their weight should be carefully evaluated by performing more modelling runs with different setups.

When evaluating the overall modelling procedures, their corresponding caveats and performance issues, the importance of many details in setup and validation of MS3D computational procedures fully emerges, thus suggesting that they still requires a considerable human skill, although many full automated programs and servers allow in principle the use of MS3D protocols even to inexperienced users. This is also demonstrated for the pure *ab initio* modelling stage by the still superior performances obtained by human-guided predictions in CASP rounds, when compared to fully automated servers.

Future improvements in MS3D are expected as a natural consequence of continuous development in biochemical/MS techniques for experimental data, and in hardware/software for molecular simulations and predictive methods. However, some specific, less expensive and, possibly, quicker evolution in MS3D could be propelled by targeted development of computational approaches more directly related to the real nature of the experimental information on which MS3D is based, notably algorithms implementing surface-dependent contributions and more faithful representations of crosslinkers than straight distance restraints.

6. References

Aebersold, R. & Mann, M. (2003). Mass spectrometry-based proteomics. *Nature*, Vol.422, pp. 198-207.

Altschul, S.F.; Madden, T.L.; Schaffer, A.A.; Zhang, J.; Zhang, Z.; Miller, W. & Lipman, D.J. (1997). Gapped BLAST and PSI-BLAST: A new generation of protein database search programs. *Nucleic Acids Research*, Vol.25, pp.3389-3402.

Aszodi, A.; Gradwell, M.J. & Taylor, W.R.(1995). Global fold determination from a small number of distance restraints. *Journal of Molecular Biology* Vol.251, pp.308-326.

Aszodi, A.; Munro, R.E. & Taylor, W.R.(1997). Protein modeling by multiple sequence threading and distance geometry. *Proteins*, Vol. 29, pp.38-42.

Back, J.W.; de Jong, L.; Muijsers, A.O. & de Koster, C.G. (2003). Chemical crosslinking and mass spectrometry for protein structural modeling. *Journal of Molecular Biology*, Vol.331,pp.303-313.

Back, J.W.; Sanz, M.A.; De Jong, L.; De Koning, L.J.; Nijtmans, L.G.; De Koster, C.G.; Grivell, L.A.; Van Der Spek, H. & Muijsers, A.O.(2002). A structure for the yeast prohibitin complex: Structure prediction and evidence from chemical crosslinking and mass spectrometry. *Protein Science*, Vol. 11, pp.2471-2478.

Balasu, M.C.; Spiridon, L.N.; Miron, S. ; Craescu, C.T.; Scheidig, A.J., Petrescu, A.J. & Szedlacsek, S.E. (2009). Interface Analysis of the Complex between ERK2 and PTP-SL. *Plos one*, Vol. 4, pp. e5432.

Bastard, K.; Prévost, C. & Zacharias, M. (2006). Accounting for loop flexibility during protein-protein docking. *Proteins*, Vol.62, pp. 956-969.

Ben-Zeev, E. & Eisenstein, M. (2003). Weighted geometric docking: incorporating external information in the rotation-translation scan. *Proteins*, Vol.52, pp. 24-27.

Berndt, K.D.; Güntert, P.; Orbons, L.P. & Wüthrich, K. (1992). Determination of a high-quality nuclear magnetic resonance solution structure of the bovine pancreatic trypsin inhibitor and comparison with three crystal structures. *Journal of Molecular Biology*, Vol.227, pp.757-775.

Blake, J.D. & Cohen, F.E. (2001). Pairwise sequence alignment below the twilight zone. *Journal of Molecular Biology*, Vol. 307, pp. 721-735.

Bowers, P.M.; Strauss, C.E.M. & Baker, D. (2000). De novo protein structure determination using sparse NMR data. *Journal of Biomolecular NMR*, Vol.18, pp.311-318.

Brazas, M.D.; Yamada, J.T. & Ouellette, B.F.F. (2010). Providing web servers and training in Bioinformatics: 2010 update on the Bioinformatics Links Directory . *Nucleic Acids Research*, Vol. 38, pp.W3-W6.

Brooks, B.R.; Bruccoleri, R.E.; Olafson, B.D.; States, D.J.; Swaminathan, S. & Karplus M.(2003). CHARMM: A Program for Macromolecular Energy, Minimization, and Dynamics Calculations. *Journal of Computational Chemistry*, Vol.4, pp.187-217.

Camacho, C. J. & Vajda, S. (2001). Protein docking along smooth association pathways. *PNAS USA*, Vol.98, pp.10636-10641.

Carlsohn, E.; Ångström, J. ; Emmett, M.R.; Marshall, A.G. & Nilsson, C.L. (2004). Chemical cross-linking of the urease complex from Helicobacter pylori and analysis by Fourier transform ion cyclotron resonance mass spectrometry and molecular modeling *International Journal of Mass Spectrometry*, Vol.234, pp. 137-144.

Chu, F.; Shan, S.; Moustakas, D.T.; Alber, F.; Egea, P.F.; Stroud, R.M.; Walter, P. & Burlingame A.L. (2004). Unraveling the interface of signal recognition particle and its receptor by using chemical cross-linking and tandem mass spectrometry. *PNAS*, Vol.101, pp. 16454-16459.

D'Ambrosio, C.; Talamo, F.; Vitale, R.M.; Amodeo, P.; Tell, G.; Ferrara, L. & Scaloni, A. (2003). Probing the Dimeric Structure of Porcine Aminoacylase 1 by Mass Spectrometric and Modeling Procedures. *Biochemistry*, Vol. 42, pp. 4430-4443.

de Bakker, P.I.; Furnham, N.; Blundell, T.L. & DePristo, M.A. (2006). Conformer generation under restraints. *Current Opinion in Structural Biology*, Vol. 16, pp.160-165.

Dimova, K; Kalkhof, S.; Pottratz, I.; Ihling, C.; Rodriguez-Castaneda, F.; Liepold, T.; Griesinger, C.; Brose, N.; Sinz, A. & Jahn, O. (2009). Structural Insights into the Calmodulin-Munc13 Interaction Obtained by Cross-Linking and Mass Spectrometry. *Biochemistry*, Vol.48, pp. 5908-5921.

Eddy, S.R. (1998). Profile hidden Markov models. *Bioinformatics*, Vol.14, pp.755-763.

Fabris, D. & Yu, E.T. (2010). The collaboratory for MS3D:a new cyberinfrastructure for the structural elucidation of biological macromolecules and their assemblies using mass spectrometry-based approaches. *Journal Proteome Research*, Vol.7, pp. 4848-4857.

Fiser, A. & Sali, A. (2003). Modeller: generation and refinement of homology base protein structure models. *Methods in Enzymology*, Vol. 374, pp.461-491.

Förster, F.; Webb, B.; Krukenberg, K.A.; Tsuruta, H.; Agard, D.A. & Sali A.(2008). Integration of Small-Angle X-Ray Scattering Data into Structural Modeling of Proteins and Their Assemblies. *Journal of Molecular Biology*, Vol.382, pp.1089-1106.

Friedhoff, P. (2005). Mapping protein-protein interactions by bioinformatics and crosslinking.. *Analitycal & Bioanalitycal Chemistry*, Vol.381,pp.78-80.

Giron-Monzon, L.; Manelyte, L.; Ahrends, R.; Kirsch, D.; Spengler, B. & Friedhoff, P. (2004). Mapping Protein-Protein Interactions between MutL and MutH by Cross-linking. *The Journal of Biochemical Chemistry*, Vol.279, pp. 49338-49345.

Gray, J.J.; Moughon, S.; Wang, C.; Schueler-Furman, O.; Kuhlman, B.; Rohl, C.A. & Baker, D. (2003). Protein-protein docking with simultaneous optimization of rigid-body displacement and side-chain conformations. *Journal of Molecular Biology*, Vol.331, pp.281-299.

Green, N.S.; Reisler, E. & Houk, K.N. (2001). Quantitative evaluation of the lengths of homobifunctional protein cross-linking reagents used as molecular rulers. *Protein Science*, Vol.10, pp.1293-1304.

Grintsevich, E.E.; Benchaar, S.A.; Warshaviak, D.; Boontheung, P.; Halgand, F.; Whitelegge, J.P.; Faull, K.F.; Ogorzalek Loo, R.R; Sept, D.; Loo, J.A. & Reisler, E. (2008). Mapping the Cofilin Binding Site on Yeast G-Actin by Chemical Cross-Linking. *Journal of Molecular Biology*, Vol.377, pp. 395-409.

Güntert, P.; Mumenthaler, C. & Wüthrich, K. (1997). Torsion angle dynamics for NMR structure calculation with the new program Dyana. *Journal of Molecular Biology*, Vol. 273, pp. 283-298.

Havel, T.F.; Kuntz, I.D. & Crippen, G.M.(1983). The combinatorial distance geometry method for the calculation of molecular conformation. I. A new approach to an old problem. *Journal of Theoretical Biology*, Vol. 310, pp.638-642.

Jaroszewski, L.; Rychlewski, L.; Li, Z.; Li, W. & Godzik, A. (2005). FFAS03: a server for profile- profile sequence alignments. *Nucleic Acids Research*, Vol.33, pp.W284-288.

Karplus, K.; Barrett, C. & Hughey R. (1998). Hidden Markov models for detecting remote protein homologies. *Bioinformatics*, Vol.14, pp.846-856.

Kessl, J.J.; Eidahl, J.O.; Shkriabai, N.; Zhao, Z.; McKee, C.J.; Hess, S.; Burke, T.R. Jr & Kvaratskhelia, M. (2009). An allosteric mechanism for inhibiting HIV-1 integrase with a small molecule.*Molecular Pharmacology*, Vol. 76, pp.824-832.

Kirkpatrick, S.; Gelatt, C.D. Jr. & Vecchi, M.P. (1983). Optimization by Simulated Annealing. *Science*, Vol 220,pp. 671-680.

Latek, D.; Ekonomiuk, D. & Kolinski , A.(2007). Protein structure prediction: combining de novo modeling with sparse experimental data. *Journal of Computational Chemistry*, Vol. 28, pp.1668-1676.

Leitner, A.; Walzthoeni, T.; Kahraman, A.; Herzog, F.; Rinner, O.; Beck, M. & Aebersolda, R. (2010). Probing Native Protein Structures by Chemical Cross-linking, Mass Spectrometry, and Bioinformatics. *Molecular & Cellular Proteomics*, Vol.24, pp. 1634-1649.

Lin, M.; Lu, H.M.; Rong Chen,R. & Liang, J.(2008). Generating properly weighted ensemble of conformations of proteins from sparse or indirect distance constraints. *The Journal of Chemical Physics*, Vol.129, pp.094101-094114.

Marti-Renom, M.A.; Madhusudhan, M.S. & Sali, A. (2004). Alignment of protein sequences by their profiles. *Protein Science*, Vol.13, pp.1071-1087.

Lübben, M.; Portmann, R.; Kock, G.; Stoll, R.; Young, M.M. & Solioz, M. (2009). Structural model of the CopA copper ATPase of Enterococcus hirae based on chemical cross-linking. *Biometals*, Vol.22, pp. 363-375.

Mathiowetz, A.M.; Jain, A.; Karasawa, N. & Goddard, W.A. III. (1994). Protein simulation using techniques suitable for very large systems: The cell multipole method for nonbond interactions and the Newton-Euler inverse mass operator method for internal coordinate dynamics. *Proteins*, Vol. 20, pp. 227-247.

Melo, F. & Sali, A. (2007). Fold assessment for comparative protein structure modeling. *Protein Science*, Vol. 16, pp. 2412-2426.

Millevoi, S.; Thion, L.; Joseph, G.; Vossen, C.; Ghisolfi-Nieto, L. & Erard, M. (2001). Atypical binding of the neuronal POU protein N-Oct3 to noncanonical DNA targets. Implications for heterodimerization with HNF-3b. *European Journal Biochemistry*, Vol.268, pp. 781-791.

Moreira, I.S.; Fernandes, P.A. & Ramos, M.J. (2010). Protein-protein docking dealing with the unknown. *Journal of Computational Chemistry*, Vol. 31, pp.317-342.

Mouradov, D.; Craven, A.; Forwood, J.K.; Flanagan, J.U.; García-Castellanos, R.; Gomis-Rüth, F.X.; Hume, D.A.; Martin, J.L.; Kobe, B. & Huber, T. (2006). Modelling the structure of latexin-carboxypeptidase. A complex based on chemical cross-linking and molecular docking. *Protein Engineering, Design & Selection*, Vol.19, pp. 9-16.

Nikolova, L.; Soman, K. ; Nichols, J.C.; Daniel, D.S., Dickey, B.F. & Hoffenberg, S. (1998). Conformationally variable Rab protein surface regions mapped by limited proteolysis and homology modelling. *Biochemical Journal*, Vol.336, pp. 461-469.

Nilges, M.; Clore, G.M. & Gronenborn, A.M.(1988a). Determination of three dimensional structures of proteins from interproton distance data by hybrid distance geometry-dynamical simulated annealing calculations. *FEBS Letters*, Vol.229, pp.317-324.

Nilges, M.; Gronenborn, A.M.; Brünger, A.T. & Clore, G.M. (1988b). Determination of three- dimensional structures of proteins by simulated annealing with interproton distance restraints: application to crambin, potato carboxypeptidase inhibitor and barley serine proteinase inhibitor 2. *Protein Engineering*, Vol.2, pp.27-38.

Nymeyer, H.; Gnanakaran, S. and García, A.E. (2004). Atomic simulations of protein folding using the replica exchange algorithm. *Methods in Enzymology*, Vol.383, pp.111-149.

Papa, S.; Monti, S.M.; Vitale, R.M.; Bubici, C.; Jayawardena, S.; Alvarez, K.; De Smaele, E.; Dathan, N.; Pedone, C.; Ruvo M. & Franzoso, G. (2007). Insights into the structural basis of the GADD45beta-mediated inactivation of the JNK kinase, MKK7/JNKK2.. *Journal of Biological Chemistry*, Vol. 282, pp. 19029-19041.

Potluri, S.; Khan, A.A.; Kuzminykh, A.; Bujnicki, J.M., Friedman, A.M. & Bailey-Kellogg, C. (2004). Geometric Analysis of Cross-Linkability for Protein Fold Discrimination. *Pacific Symposium on Biocomputing*, Vol.9, pp.447-458.

Renzone, G.; Salzano, A.M.; Arena, S.; D'Ambrosio, C. & Scaloni, A.(2007a). Mass Spectrometry-Based Approaches for Structural Studies on Protein Complexes at Low-Resolution. *Current Proteomics*, Vol. 4, pp. 1-16.

Renzone, G.; Vitale, R.M.; Scaloni, A.; Rossi, M., Amodeo, P. & Guagliardi A. (2007b). Structural Characterization of the Functional Regions in the Archaeal Protein Sso7d. *Proteins: Structure, Function, and Bioinformatics*, Vol. 67, pp. 189-197.

Rice, L.M & Brünger, A.T. (1994). Torsion angle dynamics: Reduced variable conformational sampling enhances crystallographic structure refinement. *Proteins*, Vol. 19, pp. 277-290.

Russell, R.B.; Alber, F.; Aloy, P.; Davis, F.P.; Korkin, D.;Pichaud, M; Topf, M. & Sali, A. (2004). A structural perspective on protein-protein interactions. *Current Opinion in Structural Biology*, Vol.14, pp. 313-324.

Scaloni, A; Miraglia, N.; Orrù, S.; Amodeo, P.; Motta, A.; Marino, G. & Pucci, P.(1998). Topology of the calmodulin-melittin complex. *Journal of Molecular Biology*, Vol. 277, pp.945-958.

Schrag, J.D.; Jiralerspong, S.; Banville, M; Jaramillo, M.L. & O'Connor-McCourt, M.D. (2007). The crystal structure and dimerization interface of GADD45gamma. *PNAS*, Vol. 105, pp. 6566-6571.

Schueler-Furman, O.; Wang, C.; Bradley, P.; Misura, K. & Baker, D. (2005). Progress in modeling of protein structures and interactions. *Science*, Vol. 310, pp.638-642.

Schulz,D.M.; Kalkhof, S.; Schmidt, A.; Ihling, C.; Stingl, C.; Mechtler, K.; Zschörnig, O & Sinz, A. (2007). Annexin A2/P11 interaction: New insights into annexin A2 tetramer structure by chemical crosslinking, high-resolution mass spectrometry, and computational modeling. *Proteins: Structure Function & Bioinformatics*, Vol.69, pp. 254-269.

Schwieters, C.D. & Clore, G.M. (2001). Internal Coordinates for Molecular Dynamics and Minimization in Structure Determination and Refinement. *Journal of Magnetic Resonance*, Vol. 152, pp.288-302.

Silva, R.A.G.D.; Hilliard, G.M.; Fang, J.; Macha, S. & Davidson, W.S. (2005). A Three-Dimensional Molecular Model of Lipid-Free Apolipoprotein A-I Determined by Cross-Linking/Mass Spectrometry and Sequence Threading. *Biochemistry*, Vol.44, pp. 2759-2769.

Singh, P.; Panchaud, A. & Goodlett, D.R. (2010) Chemical Cross-Linking and Mass Spectrometry As a Low-Resolution Protein Structure Determination Technique. *Analytical Chemistry*, Vol. 82, pp. 2636-2642

Sinz, A. (2006). Chemical cross-linking and mass spectrometry to map three dimensional protein structures and protein-protein interactions. *Mass Spectrometry Reviews*, Vol.25, pp. 663-682.

Skolnick,J.; Kolinski, A. & Ortiz, A.R.(1997). MONSSTER: A Method for Folding Globular Proteins with a Small Number of Distance Restraints. *Journal of Molecular Biology*, Vol. 265 pp.217-241.

Soding, J. (2005). Protein homology detection by HMM-HMM comparison. *Bioinformatics*, Vol.21, pp.951-960.

Stein, E.G.; Rice, L.M & Brünger, A.T. (1997). Torsion-angle molecular dynamics as a new efficient tool for NMR structure calculation. *Journal of Magnetic Resonance*, Vol. 124, pp. 154-164.

Tubb, M.R.; Silva, R.A.G.D.; Fang, J.; Tso, P. & Davidson, W.S. (2008). A Three-dimensional Homology Model of Lipid-free Apolipoprotein A-IV Using Cross-linking and Mass Spectrometry. *The Journal of Biochemical Chemistry*, Vol.283, pp. 17314--17323.

Vaidehi, N., Jain, A. & Goddard, W.A. III (1996). Constant temperature constrained molecular dynamics: The Newton–Euler inverse mass operator method. *Journal of Physical Chemistry*, Vol. 100, pp.10 508–10517.

Van Dijk, A.D.J.; Boelens, R. & Bonvin, A.M.J.J. (2005). Data-driven docking for the study of biomolecular complexes. *FEBS Journal*, Vol.272, pp.293–312.

Young, M.M.; Tang, N.; Hempel, J.C.; Oshiro, C.M.; Taylor, E.W.; Kuntz, I.D.; Gibson, B.W. & Dollinger G. (2000). High throughput protein fold identification by using experimental constraints derived from intramolecular cross-links and mass spectrometry. *PNAS*, Vol.97, pp. 5802-2806.

Zheng, X.; Wintrode, P.L. & Chance M.R. (2007). Complementary Structural Mass Spectrometry Techniques Reveal Local Dynamics in Functionally Important Regions of a Metastable Serpin. *Structure*, Vol.16, pp. 38-51.

8

Synthetic Biology & Bioinformatics Prospects in the Cancer Arena

Lígia R. Rodrigues and Leon D. Kluskens
IBB - Institute for Biotechnology and Bioengineering, Centre of Biological Engineering, University of Minho, Campus de Gualtar, Braga Portugal

1. Introduction

Cancer is the second leading cause of mortality worldwide, with an expected 1.5-3.0 million new cases and 0.5-2.0 million deaths in 2011 for the US and Europe, respectively (Jemal et al., 2011). Hence, this is an enormously important health risk, and progress leading to enhanced survival is a global priority. Strategies that have been pursued over the years include the search for new biomarkers, drugs or treatments (Rodrigues et al., 2007). Synthetic biology together with bioinformatics represents a powerful tool towards the discovery of novel biomarkers and the design of new biosensors.

Traditionally, the majority of new drugs has been generated from compounds derived from natural products (Neumann & Neumann-Staubitz, 2010). However, advances in genome sequencing together with possible manipulation of biosynthetic pathways, constitute important resources for screening and designing new drugs (Carothers et al. 2009). Furthermore, the development of rational approaches through the use of bioinformatics for data integration will enable the understanding of mechanisms underlying the anti-cancer effect of such drugs (Leonard et al., 2008; Rocha et al., 2010).

Besides in biomarker development and the production of novel drugs, synthetic biology can also play a crucial role in the level of specific drug targeting. Cells can be engineered to recognize specific targets or conditions in our bodies that are not naturally recognized by the immune system (Forbes, 2010).

Synthetic biology is the use of engineering principles to create, in a rational and systematic way, functional systems based on the molecular machines and regulatory circuits of living organisms or to re-design and fabricate existing biological systems (Benner & Sismour, 2005). The focus is often on ways of taking parts of natural biological systems, characterizing and simplifying them, and using them as a component of a highly unnatural, engineered, biological system (Endy, 2005). Virtually, through synthetic biology, solutions for the unmet needs of humankind can be achieved, namely in the field of drug discovery. Indeed, synthetic biology tools enable the elucidation of disease mechanisms, identification of potential targets, discovery of new chemotherapeutics or design of novel drugs, as well as the design of biological elements that recognize and target cancer cells. Furthermore, through synthetic biology it is possible to develop economically attractive microbial production processes for complex natural products.

Bioinformatics is used in drug target identification and validation, and in the development of biomarkers and tools to maximize the therapeutic benefit of drugs. Now that data on cellular signalling pathways are available, integrated computational and experimental projects are being developed, with the goal of enabling *in silico* pharmacology by linking the genome, transcriptome and proteome to cellular pathophysiology. Furthermore, sophisticated computational tools are being developed that enable the modelling and design of new biological systems. A key component of any synthetic biology effort is the use of quantitative models (Arkin, 2001). These models and their corresponding simulations enable optimization of a system design, as well as guiding their subsequent analysis. Dynamic models of gene regulatory and reaction networks are essential for the characterization of artificial and synthetic systems (Rocha et al., 2008). Several software tools and standards have been developed in order to facilitate model exchange and reuse (Rocha et al., 2010).

In this chapter, synthetic biology approaches for cancer diagnosis and drug development will be reviewed. Specifically, examples on the design of RNA-based biosensors, bacteria and virus as anti-cancer agents, and engineered microbial cell factories for the production of drugs, will be presented.

2. Synthetic biology: tools to design, build and optimize biological processes

Synthetic biology uses biological insights combined with engineering principles to design and build new biological functions and complex artificial systems that do not occur in Nature (Andrianantoandro et al., 2006). The building blocks used in synthetic biology are the components of molecular biology processes: promoter sequences, operator sequences, ribosome binding sites (RBS), termination sites, reporter proteins, and transcription factors. Examples of such building blocks are given in Table 1.

Great developments of DNA synthesis technologies have opened new perspectives for the design of very large and complex circuits (Purnick & Weiss, 2009), making it now affordable to synthesize a given gene instead of cloning it. It is possible to synthesize *de novo* a small virus (Mueller et al., 2009), to replace the genome of one bacterium by another (Lartigue et al., 2007) and to make large chunks of DNA coding for elaborate genetic circuits. Software tools to simulate large networks and the entire panel of omics technologies to analyze the engineered microorganism are available (for details see section 3). Finally, the repositories of biological parts (e.g. Registry of Standard Biological Parts (http://partsregistry.org/)) will increase in complexity, number and reliability of circuits available for different species.

Currently, the design and synthesis of biological systems are not decoupled. For example, the construction of metabolic pathways or any circuit from genetic parts first requires a collection of well characterized parts, which do not yet fully exist. Nevertheless, this limitation is being addressed through the development and compilation of standard biological parts (Kelly et al., 2009). When designing individual biological parts, the base-by-base content of that part (promoter, RBS, protein coding region, terminator, among others) is explicitly dictated (McArthur IV & Fong, 2010). Rules and guidelines for designing genetic parts at this level are being established (Canton et al., 2008). Particularly, an important issue when designing protein-coding parts is codon optimization, encoding the same amino acid sequence with an alternative, preferred nucleotide sequence. Although a particular sequence, when expressed, may be theoretically functional, its expression may be far from optimal or even completely suppressed due to codon usage bias in the heterologous host.

Genetic part	Examples	Rationale
Transcriptional control		
Constitutive promoters	*lacIq, SV40, T7, sp6*	"Always on" transcription
Regulatory regions	*tetO, lacO, ara, gal4, rhl* box	Repressor and activator sites
Inducible promoters	*ara, ethanol, lac, gal, rhl, lux, fdhH, sal, glnK, cyc1*	Control of the promoter by induction or by cell state
Cell fate regulators	GATA factors	Control cell differentiation
Transcriptional control		
RNA interference (RNAi)	Logic functions, RNAi repressor	Genetic switch, logic evaluation and gene silencing
Riboregulators	Ligand-controlled ribozymes	Switches for detection and actuation
Ribosome binding site	Kozak consensus sequence mutants	Control the level of translation
Post-transcriptional control		
Phosphorylation cascades	Yeast phosphorylation pathway	Modulate genetic circuit behavior
Protein receptor design	TNT, ACT and EST receptors	Control detection thresholds and combinatorial protein function
Protein degradation	Ssra tags, peptides rich in Pro, Glu, Ser and Thr	Protein degradation at varying rates
Localization signals	Nuclear localization, nuclear export and mitochondrial localization signals	Import or export from nucleus and mitochondria
Others		
Reporter genes	GFP, YFP, CFP, LacZ	Detection of expression
Antibiotic resistance	ampicilin, chloramphenicol	Selection of cells

Table 1. Genetic elements used as components of synthetic regulatory networks (adapted from McArthur IV & Fong, 2010 and Purnick & Weiss, 2009). Legend: CFP, cyan fluorescent protein; GFP, green fluorescent protein; YFP, yellow fluorescent protein.

Codon optimization of coding sequences can be achieved using freely available algorithms such as Gene Designer (see section 3). Besides codon optimization, compliance with standard assembly requirements and part-specific objectives including activity or specificity modifications should be considered. For example, the BioBrick methodology requires that parts exclude four standard restriction enzyme sites, which are reserved for use in assembly (Shetty et al., 2008). Extensive collections of parts can be generated by using a naturally occurring part as a template and rationally modifying it to create a library of that particular genetic part. Significant progress in this area has been recently demonstrated for promoters and RBS (Ellis et al., 2009; Salis et al., 2009). Ellis and co-workers (2009) constructed two promoter libraries that can be used to tune network behavior a *priori* by fitting mathematical promoter models with measured parameters. By using this model-guided design approach the authors were able to limit the variability of the system and increase predictability.

However, it is well-known that noisy or leaky promoters can complicate the system design. In these cases a finer control over expression can be established by weakening the binding strength of the downstream gene (Ham et al., 2006), or by using two promoter inputs to drive transcription of an output via a modular AND gate (Anderson et al., 2006). Additionally, modular and scalable RNA-based devices (aptamers, ribozymes, and transmitter sequences) can be engineered to regulate gene transcription or translation (Win & Smolke, 2007).

Design at the pathway level is not only concerned with including the necessary parts, but also with controlling the expressed functionality of those parts. Parts-based synthetic metabolic pathways will require tunable control, just as their natural counterparts which often employ feedback and feed-forward motifs to achieve complex regulation (Purnick & Weiss, 2009). Using a synthetic biology approach, the design of DNA sequences encoding metabolic pathways (e.g. operons) should be relatively straightforward. Synthetic scaffolds and well-characterized families of regulatory parts have emerged as powerful tools for engineering metabolism by providing rational methodologies for coordinating control of multigene expression, as well as decoupling pathway design from construction (Ellis et al., 2009). Pathway design should not overlook the fact that exogenous pathways interact with native cellular components and have their own specific energy requirements. Therefore, modifying endogenous gene expression may be necessary in addition to balancing cofactor fluxes and installing membrane transporters (Park et al., 2008).

After designing parts, circuits or pathways, the genomic constructs ought to be manufactured through DNA synthesis. Nucleotide's sequence information can be outsourced to synthesis companies (e.g. DNA2.0, GENEART or Genscript, among others). The convenience of this approach over traditional cloning allows for the systematic generation of genetic part variants such as promoter libraries. Also, it provides a way to eliminate restriction sites or undesirable RNA secondary structures, and to perform codon optimization. The ability to make large changes to DNA molecules has resulted in standardized methods for assembling basic genetic parts into larger composite devices, which facilitate part-sharing and faster system-level construction, as demonstrated by the BioBrick methodology (Shetty et al., 2008) and the Gateway cloning system (Hartley, 2003). Other approaches based on type II restriction enzymes, such as Golden Gate Shuffling, provide ways to assemble many more components together in one step (Engler et al., 2009). A similar one-step assembly approach, circular polymerase extension cloning (CPEC), avoids the need for restriction-ligation, or single-stranded homologous recombination altogether (Quan & Tian, 2009). Not only is this useful for cloning single genes, but also for assembling parts into larger sequences encoding entire metabolic pathways and for generating combinatorial part libraries. On a chromosomal level, disruption of genes in *Escherichia coli* and other microorganisms has become much faster with the development of RecBCD and lambda RED-assisted recombination systems (Datsenko & Wanner, 2000), allowing the insertion, deletion or modification by simply using linear gene fragments. Additionally, multiplex automated genome engineering (MAGE) has been introduced as another scalable, combinatorial method for producing large-scale genomic diversity (Wang et al., 2009). This approach makes chromosomal modification easier by simultaneously mutating target sites across the chromosome. Plasmid-based expression and chromosomal integration are the two common vehicles for implementing synthetic metabolic pathways. Recently, the chemically inducible chromosomal evolution (CIChE) was proposed as a long-term expression alternative method (Tyo et al., 2009). This new method avoids complications associated with plasmid replication and segregation, and can be used to integrate multiple copies of genes into the genome. All these techniques will provide technical platforms for the rapid synthesis of parts and subsequent pathways.

The majority of the synthetic biology advances has been achieved purely *in vitro* (Isalan et al., 2008), or in microorganisms involving the design of small gene circuits without a direct practical application, although scientifically very exciting. These studies have offered fundamental insight into biological processes, like the role and sources of biological noise; the existence of biological modules with defined properties; the dynamics of oscillatory behavior; gene transcription and translation; or cell communication (Alon, 2003; Kobayashi et al., 2004). An interesting example of a larger system that has been redesigned is the refactoring of the T7 bacteriophage (Chan et al., 2005). Another successful example has been the production of terpenoid compounds in *E. coli* (Martin et al., 2003) and *Saccharomyces cerevisiae* (Ro et al., 2006) that can be used for the synthesis of artemisinin. Bacteria and fungi have long been used in numerous industrial microbiology applications, synthesizing important metabolites in large amounts. The production of amino acids, citric acid and enzymes are examples of other products of interest, overproduced by microorganisms. Genetic engineering of strains can contribute to the improvement of these production levels. Altogether, the ability to engineer biological systems will enable vast progress in existing applications and the development of several new possibilities. Furthermore, novel applications can be developed by coupling gene regulatory networks with biosensor modules and biological response systems. An extensive RNA-based framework has been developed for engineering ligand-controlled gene regulatory systems, called ribozyme switches. These switches exhibit tunable regulation, design modularity, and target specificity and could be used, for example, to regulate cell growth (Win & Smolke, 2007). Engineering interactions between programmed bacteria and mammalian cells will lead to exciting medical applications (Anderson et al., 2006). Synthetic biology will change the paradigm of the traditional approaches used to treat diseases by developing "smart" therapies where the therapeutic agent can perform computation and logic operations and make complex decisions (Andrianantoandro et al., 2006). There are also promising applications in the field of living vectors for gene therapy and chemical factories (Forbes, 2010; Leonard et al., 2008).

3. Bioinformatics: a rational path towards biological behavior predictability

In order to evolve as an engineering discipline, synthetic biology cannot rely on endless trial and error methods driven by verbal description of biomolecular interaction networks. Genome projects identify the components of gene networks in biological organisms, gene after gene, and DNA microarray experiments discover the network connections (Arkin, 2001). However, these data cannot adequately explain biomolecular phenomena or enable rational engineering of dynamic gene expression regulation. The challenge is then to reduce the amount and complexity of biological data into concise theoretical formulations with predictive ability, ultimately associating synthetic DNA sequences to dynamic phenotypes.

3.1 Models for synthetic biology

The engineering process usually involves multiple cycles of design, optimization and revision. This is particularly evident in the process of constructing gene circuits (Marguet et

al., 2007). Due to the large number of participating species and the complexity of their interactions, it becomes difficult to intuitively predict a design behavior. Therefore, only detailed modeling can allow the investigation of dynamic gene expression in a way fit for analysis and design (Di Ventura et al., 2006). Modeling a cellular process can highlight which experiments are likely to be the most informative in testing model hypothesis, and for example allow testing for the effect of drugs (Di Bernardo et al., 2005) or mutant phenotypes (Segre et al., 2002) on cellular processes, thus paving the way for individualized medicine.
Data are the precursor to any model, and the need to organize as much experimental data as possible in a systematic manner has led to several excellent databases as summarized in Table 2. The term "model" can be used for verbal or graphical descriptions of a mechanism underlying a cellular process, or refer to a set of equations expressing in a formal and exact manner the relationships among variables that characterize the state of a biological system (Di Ventura et al., 2006). The importance of mathematical modeling has been extensively demonstrated in systems biology (You, 2004), although its utility in synthetic biology seems even more dominant (Kaznessis, 2009).

Name	Website
BIND (Biomolecular Interaction Network Database)	http://www.bind.ca/
Brenda (a comprehensive enzyme information system)	http://www.brenda.uni-koeln.de/
CSNDB (Cell Signaling Networks Database)	http://geo.nihs.go.jp/csndb/
DIP (Database of Interacting Proteins)	http://dip.doe-mbi.ucla.edu/
EcoCyc/Metacyc/BioCyc (Encyclopedia of E. coli genes and metabolism)	http://ecocyc.org/
EMP (Enzymes and Metabolic Pathways Database)	http://www.empproject.com/
GeneNet (information on gene networks)	http://wwwmgs.bionet.nsc.ru/mgs/systems/genenet/
Kegg (Kyoto Encyclopedia of Genes and Genomes)	http://www.genome.ad.jp/kegg/kegg.html
SPAD (Signaling Pathway Database)	http://www.grt.kyushu-u.ac.jp/eny-doc/
RegulonDB (*E. coli* K12 transcriptional network)	http://regulondb.ccg.unam.mx/
ExPASy-beta (Bioinformatics Resource Portal)	http://beta.expasy.org/

Table 2. Databases of molecular properties, interactions and pathways (adapted from Arkin, 2001).

Model-driven rational engineering of synthetic gene networks is possible at the level of topologies or at the level of molecular components. In the first one, it is considered that molecules control the concentration of other molecules, e.g. DNA-binding proteins regulate the expression of specific genes by either activation or repression. By combining simple regulatory interactions, such as negative and positive feedback and feed-forward loops, one may create more complex networks that precisely control the production of protein molecules (e.g. bistable switches, oscillators, and filters). Experimentally, these networks can be created using existing libraries of regulatory proteins and their corresponding operator sites. Examples of these models are the oscillator described by Gardner *et al* (2000) and repressilator by Elowitz and Leibler (2000). In the second level, the kinetics and strengths of molecular interactions within the system are described. By altering the characteristics of the components, such as DNA-binding proteins and their corresponding DNA sites, one can modify the system dynamics without modifying the network topology. Experimentally, the DNA sequences that yield the desired characteristics of each component can be engineered to achieve the desired protein-protein, protein-RNA, or protein-DNA binding constants and enzymatic activities. For example, Alon and co-workers (2003) showed how simple mutations on the DNA sequence of the lactose operon can result in widely different phenotypic behavior.

Various mathematical formulations can be used to model gene circuits. At the population level, gene circuits can be modeled using ordinary differential equations (ODEs). In an ODE formulation, the dynamics of the interactions within the circuit are deterministic. That is, the ODE formulation ignores the randomness intrinsic to cellular processes, and is convenient for circuit designs that are thought to be less affected by noise or when the impact of noise is irrelevant (Marguet et al., 2007). An ODE model facilitates further sophisticated analyses, such as sensitivity analysis and bifurcation analysis. Such analyses are useful to determine how quantitative or qualitative circuit behavior will be impacted by changes in circuit parameters. For instance, in designing a bistable toggle switch, bifurcation analysis was used to explore how qualitative features of the circuit may depend on reaction parameters (Gardner et al., 2000). Results of the analysis were used to guide the choice of genetic components (genes, promoters and RBS) and growth conditions to favor a successful implementation of designed circuit function. However, in a single cell, the gene circuit's dynamics often involve small numbers of interacting molecules that will result in highly noisy dynamics even for expression of a single gene. For many gene circuits, the impact of such cellular noise may be critical and needs to be considered (Di Ventura et al., 2006). This can be done using stochastic models (Arkin, 2001). Different rounds of simulation using a stochastic model will lead to different results each time, which presumably reflect aspects of noisy dynamics inside a cell. For synthetic biology applications, the key of such analysis is not necessarily to accurately predict the exact noise level at each time point. This is not possible even for the simplest circuits due to the "extrinsic" noise component for each circuit (Elowitz et al., 2002). Rather, it is a way to determine to what extent the designed function can be maintained and, given a certain level of uncertainty or randomness, to what extent additional layers of control can minimize or exploit such variations. Independently of the model that is used, these can be evolved *in silico* to optimize designs towards a given function. As an example, genetic algorithms were used by Francois and Hakim (2004) to design gene regulatory networks exhibiting oscillations.

In most attempts to engineer gene circuits, mathematical models are often purposefully simplified to accommodate available computational power and to capture the qualitative behavior of the underlying systems. Simplification is beneficial partially due to the limited quantitative characterization of circuit elements, and partially because simpler models may better reveal key design constraints. The limitation, however, is that a simplified model may fail to capture richer dynamics intrinsic to a circuit. Synthetic models combine features of mathematical models and model organisms. In the engineering of genetic networks, synthetic biologists start from mathematical models, which are used as the blueprints to engineer a model out of biological components that has the same materiality as model organism but is much less complex. The specific characteristics of synthetic models allow one to use them as tools in distinguishing between different mathematical models and evaluating results gained in performing experiments with model organisms (Loettgers, 2007).

3.2 Computational tools for synthetic biology

Computational tools are essential for synthetic biology to support the design procedure at different levels. Due to the lack of quantitative characterizations of biological parts, most design procedures are iterative requiring experimental validation to enable subsequent refinements (Canton et al., 2008). Furthermore, stochastic noise, uncertainty about the cellular environment of an engineered system, and little insulation of components complicate the design process and require corresponding models and analysis methods (Di Ventura et al., 2006). Many computational standards and tools developed in the field of systems biology (Wierling et al., 2007) are applicable for synthetic biology as well.

As previously discussed, synthetic gene circuits can be constructed from a handful of basic parts that can be described independently and assembled into interoperating modules of different complexity. For this purpose, standardization and modularity of parts at different levels is required (Canton et al., 2008). The Registry of Standard Biological Parts constitutes a reference point for current research in synthetic biology and it provides relevant information on several DNA-based synthetic or natural building blocks. Most computational tools that specifically support the design of artificial gene circuits use information from the abovementioned registry. Moreover, many of these tools share standardized formats for the input/output files. The System Biology Markup Language (SBML) (http://sbml.org) defines a widely accepted, XML-based format for the exchange of mathematical models in biology. It provides a concise representation of the chemical reactions embraced by a biological system. These can be translated into systems of ODEs or into reaction systems amenable to stochastic simulations (Alon, 2003). Despite its large applicability to simulations, SBML currently lacks modularity, which is not well aligned with parts registries in synthetic biology. Alternatively, synthetic gene systems can be described according to CellML language which is more modular (Cooling et al., 2008).

One important feature to enable the assembly of standard biological parts into gene circuits is that they share common inputs and outputs. Endy (2005) proposed RNA polymerases and ribosomes as the molecules that physically exchange information between parts. Their fluxes, measured in PoPS (Polymerase Per Second) and in RiPS (Ribosomes Per Second) represent biological currents (Canton et al., 2008). This picture, however, does not seem sufficient to describe all information exchanges even in simple engineered gene circuits, since other signal carriers like transcription factors and environmental "messages" should be explicitly introduced and not indirectly estimated by means of PoPS and RiPS (Marchisio & Stelling, 2008). Based on the assumption that parts share common input/output signals, several computational tools have been proposed for gene circuit design, as presented in Table 3. Comparing these circuit design tools it is obvious that we are still far from an ideal solution. The software tools differ in many aspects such as scope of parts and circuit descriptions, the mode of user interaction, and the integration with databases or other tools.

Circuit design and simulation	
Biojade	http://web.mit.edu/jagoler/www/biojade/
Tinkercell	http://www.tinkercell.com/Home
Asmparts	http://soft.synth-bio.org/asmparts.html
ProMoT	http://www.mpimagdeburg.mpg.de/projects/promot
GenoCAD	http://www.genocad.org/genocad/
GEC	http://research.microsoft.com/gec
TABASCO	http://openwetware.org/wiki/TABASCO
Circuit optimization	
Genetdes	http://soft.synth-bio.org/genetdes.html
RoVerGeNe	http://iasi.bu.edu/~batt/rovergene/rovergene.htm
DNA and RNA design	
Gene Designer	https://www.dna20.com/index.php?pageID=220
GeneDesign	http://www.genedesign.org/
UNAFold	http://www.bioinfo.rpi.edu/applications/hybrid/download.php
Vienna RNA package	http://www.tbi.univie.ac.at/~ivo/RNA/
Zinc Finger Tools	http://www.scripps.edu/mb/barbas/zfdesign/zfdesignhome.php
Protein Design	
Rosetta	http://www.rosettacommons.org/main.html
RAPTOR	http://www.bioinformaticssolutions.com/products/raptor/index.php
PFP	http://dragon.bio.purdue.edu/pfp/
Autodock 4.2	http://autodock.scripps.edu/
HEX 5.1	http://webloria.loria.fr/~ritchied/hex/
Integrated workflows	
SynBioSS	http://synbioss.sourceforge.net/
Clotho	http://biocad-server.eecs.berkeley.edu/wiki/index.php/Tools
Biskit	http://biskit.sf.net/

Table 3. Computational design tools for synthetic biology (adapted from Marchisio & Stelling, 2009; Matsuoka et al., 2009; and Prunick & Weiss, 2009)

Biojade was one of the first tools being reported for circuit design (Goler, 2004). It provides connections to both parts databases and simulation environments, but it considers only one kind of signal carrier (RNA polymerases). It can invoke the simulator TABASCO (Kosuri et al., 2007), thus enabling genome scale simulations at single base-pair resolution. CellDesigner (Funahashi et al., 2003) has similar capabilities for graphical circuit composition. However, parts modularity and consequently circuit representation do not appear detailed enough. Another tool for which parts communicate only by means of PoPS, but not restricted to a single mathematical framework, is the Tinkercell. On the contrary, in Asmparts (Rodrigo et al., 2007a) the circuit design is less straightforward and intuitive because the tool lacks a Graphic User Interface. Nevertheless, each part exists as an independent SBML module and the model kinetics for transcription and translation permit to limit the number of parameters necessary for a qualitative system description. Marchisio and Stelling (2008) developed a new framework for the design of synthetic circuits where each part is modeled independently following the ODE formalism. This results in a set of composable parts that communicate by fluxes of signal carriers, whose overall amount is constantly updated inside their corresponding pools. The model also considers transcription factors, chemicals and small RNAs as signal carriers. Pools are placed among parts and devices: they store free signal carriers and distribute them to the whole circuit. Hence, polymerases and ribosomes have a finite amount; this permits to estimate circuit scalability with respect to the number of parts. Mass action kinetics is fully employed and no approximations are required to depict the interactions of signal carriers with DNA and mRNA. The authors implemented the corresponding models into ProMoT (Process Modeling Tool), software for the object-oriented and modular composition of models for dynamic processes (Mirschel et al., 2009). GenoCAD (Czar et al., 2009) and GEC (Pedersen & Phillips, 2009) introduce the notions of a grammar and of a programming language for genetic circuit design, respectively. These tools use a set of rules to check the correct composition of standard parts. Relying on libraries of standard parts that are not necessarily taken from the Registry of Standard Biological Parts, these programs can translate a circuit design into a complete DNA sequence. The two tools differ in capabilities and possible connectivity to other tools.

The ultimate goal of designing a genetic circuit is that it works, i.e. that it performs a given function. For that purpose, optimization cycles to establish an appropriate structure and a good set of kinetic parameters values are required. These optimization problems are extremely complex since they involve the selection of adequate parts and appropriate continuous parameter values. Stochastic optimization methods (e.g. evolutionary algorithms) attempt to find good solutions by biased random search. They have the potential for finding globally optimal solutions, but optimization is computationally expensive. On the other hand, deterministic methods (e.g. gradient descent) are local search methods, with less computational cost, but at the expense of missing good solutions.

The optimization problem can be tackled by tools such as Genetdes (Rodrigo et al., 2007b) and OptCircuit (Dasika & Maranas, 2008). They rely on different parts characterizations and optimization algorithms. Genetdes uses a stochastic method termed "Simulated Annealing" (Kirkpatrick et al., 1983), which produces a single solution starting from a random circuit configuration. As a drawback, the algorithm is more likely to get stuck in a local minimum than an evolutionary algorithm. OptCircuit, on the contrary, treats the circuit design problem with a deterministic method (Bansal et al., 2003), implementing a procedure towards a "local' optimal solution. Each of these optimization algorithms requires a very simplified model for gene dynamics where, for instance, transcription and translation are treated as a single step process. Moreover, the current methods can cope only with rather small circuits. Another tool that has been described by Batt and co-workers (2007), RoVerGeNe, addresses the problem of parameter estimation more specifically. This tool permits to tune the performance and to estimate the robustness of a synthetic network with a known behavior and for which the topology does not require further improvement.

Detailed design of synthetic parts that reproduce the estimated circuit kinetics and dynamics is a complex task. It requires computational tools in order to achieve error free solutions in a reasonable amount of time. Other than the placement/removal of restriction sites and the insertion/deletion of longer motifs, mutations of single nucleotides may be necessary to tune part characteristics (e.g. promoter strength and affinity toward regulatory factors). Gene Designer (Villalobos et al., 2006) is a complete tool for building artificial DNA segments and codon usage optimization. GeneDesign (Richardson et al., 2006) is another tool to design long synthetic DNA sequences. Many other tools are available for specific analysis of the DNA and RNA circuit components. The package UNAFold (Markham & Zuker, 2008) predicts the secondary structure of nucleic acid sequences to simulate their hybridizations and to estimate their melting temperature according to physical considerations. A more accurate analysis of the secondary structure of ribonucleic acids can be performed through the Vienna RNA package (Hofacker, 2003). Binding sites along a DNA chain can be located using Zinc Finger Tools (Mandell & Barbas, 2006). These tools allows one to search DNA sequences for target sites of particular zinc finger proteins (Kaiser, 2005), whose structure and composition can also be arranged. Thus, gene control by a class of proteins with either regulation or nuclease activity can be improved. Furthermore, tools that enable promoter predictions and primers design are available, such as BDGP and Primer3. Another relevant task in synthetic biology is the design and engineering of new proteins. Many tools have been proposed for structure prediction, homology modeling, function prediction, docking simulations and DNA-protein interactions evaluation. Examples include the Rosetta package (Simons et al., 1999); RAPTOR (Xu et al., 2003); PFP (Hawkins et al., 2006); Autodock 4.2 (Morris et al., 2009) and Hex 5.1 (Ritchie, 2008).

Further advance in computational synthetic biology will result from tools that combine and integrate most of the tasks discussed, starting with the choice and assembly of biological parts to the compilation and modification of the corresponding DNA sequences. Examples of such tools comprise SynBioSS (Hill et al., 2008); Clotho and Biskit (Grunberg et al., 2007). Critical elements are still lacking, such as tools for automatic information integration (literature and databases), and tools that re-use standardized model entities for optimal circuit design. Overall, providing an extended and integrated information technology infrastructure will be crucial for the development of the synthetic biology field.

4. A roadmap from design to production of new drugs

Biological systems are dynamic, that is they mutate, evolve and are subject to noise. Currently, the full knowledge on how these systems work is still limited. As previously discussed, synthetic biology approaches involve breaking down organisms into a hierarchy of composable parts, which is useful for conceptualization purposes. Reprogramming a cell involves the creation of synthetic biological components by adding, removing, or changing genes and proteins. Nevertheless, it is important to notice that assembly of parts largely depends on the cellular context (the so-called chassis), thus restraining the abstraction of biological components into devices and modules, and their use in design and engineering of new organisms or functions.

One level of abstraction from the DNA synthesis and manipulation is parts production, which optimization can be accomplished through either rational design or directed evolution. Applying rational design to parts alteration or creation is advantageous, in that it cannot only generate products with a defined function, but it can also produce biological insights into how the designed function comes about. However, it requires prior structural knowledge of the part, which is frequently unavailable. Directed evolution is an alternative method that can effectively address this limitation. Many synthetic biology applications will require parts for genetic circuits, cell–cell communication systems, and non-natural metabolic pathways that cannot be found in Nature, simply because Nature is not in need of them (Dougherty & Arnold, 2009). In essence, directed evolution begins with the generation of a library containing many different DNA molecules, often by error-prone DNA replication, DNA shuffling or combinatorial synthesis (Crameri et al., 1998). The library is next subjected to high-throughput screening or selection methods that maintain a link between genotype and phenotype in order to enrich the molecules that produce the desired function. Directed evolution can also be applied at other levels of biological hierarchy, for example to evolve entire gene circuits (Yokobayashi et al., 2002). Rational design and directed evolution should not be viewed as opposing methods, but as alternate ways to produce and optimize parts, each with their own unique strengths and weaknesses. Directed evolution can complement this technique, by using mutagenesis and subsequent screening for improved synthetic properties (Brustad & Arnold, 2010). In addition, methods have been developed to incorporate unnatural amino acids in peptides and proteins (Voloshchuk & Montclare, 2009). This will expand the toolbox of protein parts, and add beneficial effects, such as increased *in vivo* stability, when incorporated in proteinaceous therapeutics. Also, the development of *de novo* enzymes has seen a significant increase lately. The principle of computational design uses the design of a model, capable of stabilizing the transition state of a reaction. From there on, individual amino acids are positioned around it to create a catalytic site that stabilizes the transition state. The mRNA display technique resembles phage display and is a technique for the *in vitro* selection and evolution of proteins. Translated proteins are associated with their mRNA via a puromycin linkage. Selection occurs by binding to an immobilized substrate, after which a reverse transcriptase step will reveal the cDNA and thus the nucleotide sequence (Golynskiy & Seelig, 2010). If the selection step includes measurement of product formation from the substrate, novel peptides with catalytic properties can be selected.

For the design, engineering, integration and testing of new synthetic gene networks, tools and methods derived from experimental molecular biology must be used (for details see section 2). Nevertheless, progress on these tools and methods is still not enough to guarantee the complete success of the experiment. As a result, design of synthetic biological systems has become an iterative process of modeling, construction, and experimental testing that continues until a system achieves the desired behavior (Purnick & Weiss, 2009). The process begins with the abstract design of devices, modules, or organisms, and is often guided by mathematical models (Koide et al., 2009). Afterwards, the newly constructed systems are tested experimentally. However, such initial attempts rarely yield fully functional implementations due to incomplete biological information. Rational redesign based on mathematical models improves system behavior in such situations (Koide et al., 2009; Prather & Martin, 2008). Directed evolution is a complimentary approach, which can yield novel and unexpected beneficial changes to the system (Yokobayashi et al., 2002). These retooled systems are once again tested experimentally and the process is repeated as needed. Many synthetic biological systems have been engineered successfully in this fashion because the methodology is highly tolerant to uncertainty (Matsuoka et al., 2009). Figure 1 illustrates the above mentioned iterative approach used in synthetic biology.

Since its inception, metabolic engineering aims to optimize cellular metabolism for a particular industrial process application through the use of directed genetic modifications (Tyo et al., 2007). Metabolic engineering is often seen as a cyclic process (Nielsen, 2001), where the cell factory is analyzed and an appropriate target is identified. This target is then experimentally implemented and the resulting strain is characterized experimentally and, if necessary, further analyses are conducted to identify novel targets. The application of synthetic biology to metabolic engineering can potentially create a paradigm shift. Rather than starting with the full complement of components in a wild-type organism and piecewise modifying and streamlining its function, metabolic engineering can be attempted from a bottom-up, parts-based approach to design by carefully and rationally specifying the inclusion of each necessary component (McArthur IV & Fong, 2010). The importance of rationally designing improved or new microbial cell factories for the production of drugs has grown substantially since there is an increasing need for new or existing drugs at prices that can be affordable for low-income countries. Large-scale re-engineering of a biological circuit will require systems-level optimization that will come from a deep understanding of operational relationships among all the constituent parts of a cell. The integrated framework necessary for conducting such complex bioengineering requires the convergence of systems and synthetic biology (Koide et al., 2009). In recent years, with advances in systems biology (Kitano, 2002), there has been an increasing trend toward using mathematical and computational tools for the *in silico* design of enhanced microbial strains (Rocha et al., 2010).

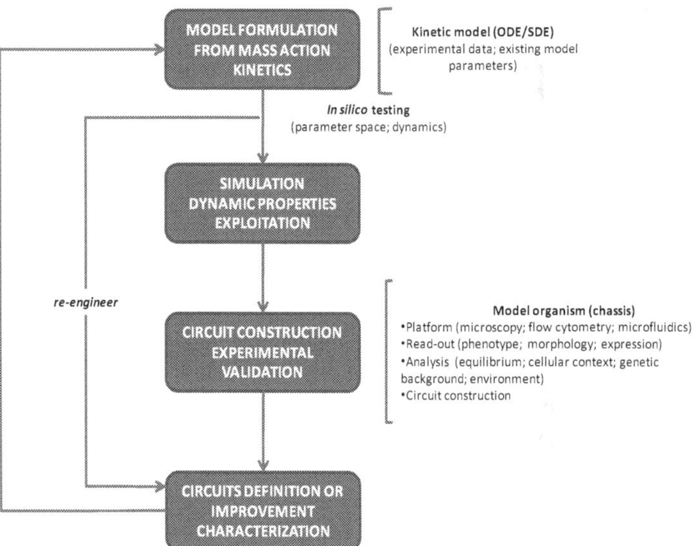

Fig. 1. The iterative synthetic biology approach to design a given biological circuit/system.

Current models in both synthetic and systems biology emphasize the relationship between environmental influences and the responses of biological networks. Nevertheless, these models operate at different scales, and to understand the new paradigm of rational systems re-engineering, synthetic and systems biology fields must join forces (Koide et al., 2009). Synthetic biology and bottom-up systems biology methods extract discrete, accurate, quantitative, kinetic and mechanistic details of regulatory sub-circuits. The models generated from these approaches provide an explicit mathematical foundation that can ultimately be used in systems redesign and re-engineering. However, these approaches are confounded by high dimensionality, non-linearity and poor prior knowledge of key dynamic parameters (Fisher & Henzinger, 2007) when scaled to large systems.

Consequently, modular sub-network characterization is performed assuming that the network is isolated from the rest of the host system. The top-down systems biology approach is based on data from high-throughput experiments that list the complete set of components within a system in a qualitative or semi-quantitative manner. Models of overall systems are similarly qualitative, tending toward algorithmic descriptions of component interactions. Such models are amenable to the experimental data used to develop them, but usually sacrifice the finer kinetic and mechanistic details of the molecular components involved (Price & Shmulevich, 2007). Bridging systems and synthetic biology approaches is being actively discussed and several solutions have been suggested (Koide et al., 2009).

A typical synthetic biology project is the design and engineering of a new biosynthetic pathway in a model organism (chassis). Generally, *E. coli* is the preferred chassis since it is well-studied, easy to manipulate, and its reproduction in biological cultures is very handy. Initially, relevant databases like Kegg and BioCyc (Table 2) can be consulted for identifying all the possible metabolic routes that allow the production of a given drug from metabolites that exist in native *E. coli*. Then, for each reaction, the species that are known to possess the corresponding enzymes/genes must be identified. This step is relevant, since most often the same enzyme exhibits different kinetic behavior among different species. Information on sequences and kinetic parameters can be extracted from the above-mentioned sources, relevant literature and also from Brenda and Expasy databases. Afterwards, the information collected is used to build a family of dynamic models (Rocha et al., 2008, 2010) that enable the simulation of possible combinations regarding pathway configuration and the origin of the enzymes (associated with varying kinetic parameters). Optflux tool can be used for simulations and metabolic engineering purposes (http://www.optflux.org/). Using the same input (a fixed amount of precursor) it is possible to select the configuration that allows obtaining higher drug yields. Furthermore, through the use of genome-scale stoichiometric models coupled with the dynamic model it is possible to understand the likely limitations regarding the availability of the possible precursors. In fact, if the precursor for the given drug biosynthesis is a metabolic intermediate, possible limitations in its availability need to be addressed in order to devise strategies to cope with it. Based on this information, the next step involves the construction of the enzymatic reactions that will lead to the production of the drug from a metabolic precursor in *E. coli*. The required enzymes are then synthesized based on the gene sequences previously selected from the databases. The cloning strategy may include using a single plasmid with two different promoters; using two different plasmids, with different copy numbers and/or origins of replication; or ultimately integrating it into the genome, in order to allow fine tuning of the expression of the various enzymes necessary. Finally, a set of experiments using the engineered bacterium needs to be performed to evaluate its functionality, side-product formation and/or accumulation, production of intermediate metabolites and final product (desired drug). In the case of the previously mentioned artemisinin production, DNA microarray analysis and targeted metabolic profiling were used to optimize the synthetic pathway, reducing the accumulation of toxic intermediates (Kizer et al., 2008). These types of methodologies enable the validation of the drug production model and the design of strategies to further improve its production.

5. Novel strategies for cancer diagnosis and drug development

Cancer is a main issue for the modern society and according to the World Health Organization it is within the top 10 of leading causes of death in middle- and high-income countries. Several possibilities to further improve existing therapies and diagnostics, or to develop novel alternatives that still have not been foreseen, can be drawn using synthetic biology approaches. Promising future applications include the development of RNA-based biosensors to produce a desired response *in vivo* or to be integrated in a cancer diagnosis device; the design and engineering of bacteria that can be programmed to target a tumor and release a therapeutic agent *in situ*; the use of virus as a tool for recognizing tumors or for gene therapy; and the large scale production of complex chemotherapeutic agents, among others.

5.1 RNA-based biosensors

Synthetic biology seeks for new biological devices and systems that regulate gene expression and metabolite pathways. Many components of a living cell possess the ability to carry genetic information, such as DNA, RNA, proteins, among others. RNA has a critical role in several functions (genetic translation, protein synthesis, signal recognition of particles) due to its functional versatility from genetic blueprint (e.g. mRNA, RNA virus genomes. Its catalytic function as enzyme (e.g. ribozymes, rRNA) and regulator of gene expression (e.g. miRNA, siRNA) makes it stand out among other biopolymers with a more specialized scope (e.g. DNA, proteins) (Dawid et al., 2009). Therefore, non-coding RNA molecules enable the formation of complex structures that can interact with DNA, other RNA molecules, proteins and other small molecules (Isaacs et al., 2006).

Natural biological systems contain transcription factors and regulators, as well as several RNA-based mechanisms for regulating gene expression (Saito & Inoue, 2009). A number of studies have been conducted on the use of RNA components in the construction of synthetic biologic devices (Topp & Gallivan, 2007; Win & Smolke, 2007). The interaction of RNA with proteins, metabolites and other nucleic acids is affected by the relationship between sequence, structure and function. This is what makes the RNA molecule so attractive and malleable to engineering complex and programmable functions.

5.1.1 Riboswitches

One of the most promising elements are the riboswitches, genetic control elements that allow small molecules to regulate gene expression. They are structured elements typically found in the 5′-untranslated regions of mRNA that recognize small molecules and respond by altering their three-dimensional structure. This, in turn, affects transcription elongation, translation initiation, or other steps of the process that lead to protein production (Beisel & Smolke, 2009; Winkler & Breaker, 2005). Biological cells can modulate gene expression in response to physical and chemical variations in the environment allowing them to control their metabolism and preventing the waste of energy expenditure or inappropriate physiological responses (Garst & Batey, 2009). There are currently at least twenty classes of riboswitches that recognize a wide range of ligands, including purine nucleobases (purine riboswitch), amino acids (lysine riboswitch), vitamin cofactors (cobalamin riboswitch), amino sugars, metal ions (mgtA riboswitch) and second messenger molecules (cyclic di-GMP riboswitch) (Beisel & Smolke, 2009). Riboswitches are typically composed of two distinct domains: a metabolite receptor known as the aptamer domain, and an expression platform whose secondary structure signals the regulatory response. Embedded within the aptamer domain is the switching sequence, a sequence shared between the aptamer domain and the expression platform (Garst & Batey, 2009). The aptamer domain is part of the RNA and forms precise three-dimensional structures. It is considered a structured nucleotide pocket belonging to the riboswitch, in the 5′-UTR, which when bound regulates downstream gene expression (Isaacs et al., 2006). Aptamers specifically recognize their corresponding target molecule, the ligand, within the complex group of other metabolites, with the appropriate affinity, such as dyes, biomarkers, proteins, peptides, aromatic small molecules, antibiotics and other biomolecules. Both the nucleotide sequence and the secondary structure of each aptamer remain highly conserved (Winkler & Breaker, 2005). Therefore, aptamer domains are the operators of the riboswitches.

A strategy for finding new aptamer sequences is the use of SELEX (Systemic Evolution of Ligands by Exponential enrichment method). SELEX is a combinatorial chemistry technique for producing oligonucleotides of either single-stranded DNA or RNA that specifically bind to one or more target ligands (Stoltenburg et al., 2007). The process begins with the synthesis of a very large oligonucleotide library consisting of randomly generated sequences of fixed length flanked by constant 5' and 3' ends that serve as primers. The sequences in the library are exposed to the target ligand and those that do not bind the target are removed, usually by affinity chromatography. The bound sequences are eluted and amplified by PCR to prepare for subsequent rounds of selection in which the stringency of the elution conditions is increased to identify the tightest-binding sequences (Stoltenburg et al., 2007). SELEX has been used to evolve aptamers of extremely high binding affinity to a variety of target ligands. Clinical uses of the technique are suggested by aptamers that bind tumor markers (Ferreira et al., 2006). The aptamer sequence must then be placed near to the RBS of the reporter gene, and inserted into *E. coli* (chassis), using a DNA carrier (i.e. plasmid), in order to exert its regulatory function.

Synthetic riboswitches represent a powerful tool for the design of biological sensors that can, for example, detect cancer cells, or the microenvironment of a tumor, and in the presence of a given molecule perform a desired function, like the expression *in situ* of a therapeutic agent. Several cancer biomarkers have been identified in the last decade; therefore there are many opportunities of taking these compounds as templates to design adequate riboswitches for their recognition. Alternatively, the engineering goal might be the detection of some of these biomarkers in biological samples using biosensors with aptamers as the biological recognition element, hence making it a less invasive approach. The development of aptamer-based electrochemical biosensors has made the detection of small and macromolecular analytes easier, faster, and more suited for early detection of protein biomarkers (Hianik & Wang, 2009). Multi-sensor arrays that provide global information on complex samples (e.g. biological samples) have deserved much interest recently. Coupling an aptamer to these devices will increase its specificity and selectivity towards the selected target(s). The selected target may be any serum biomarker that when detected in high amounts in biological samples can be suggestive of tumor activity.

5.2 Bacteria as anti-cancer agents

Bacteria possess unique features that make them powerful candidates for treating cancer in ways that are unattainable by conventional methods. The moderate success of conventional methods, such as chemotherapy and radiation, is related to its toxicity to normal tissue and inability to destroy all cancer cells. Many bacteria have been reported to specifically target tumors, actively penetrate tissue, be easily detected and/or induce a controlled cytotoxicity. The possibility of engineering interactions between programmed bacteria and mammalian cells opens unforeseen progresses in the medical field. Emerging applications include the design of bacteria to produce therapeutic agents (*in vitro* or *in vivo*) and the use of live bacteria as targeted delivery systems (Forbes, 2010; Pawelek et al., 2003). An impressive example of these applications is described by Anderson and co-workers (2006). The authors have successfully engineered *E. coli* harboring designed plasmids to invade cancer cells in an environmentally controlled way, namely in a density-dependent manner under anaerobic growth conditions and arabinose induction. Plasmids were built containing the *inv* gene from *Yersinia pseudotuberculosis* under control of the Lux promoter, the hypoxia-responsive fdhF promoter, and the arabinose-inducible araBAD promoter. This is significant because the tumor environment is often hypoxic and allows for high bacterial cell densities due to depressed immune function in the tumor. Therefore, this work demonstrated, as a "proof of concept", that one can potentially use engineered bacteria to target diseased cells without significantly impacting healthy cells.

Ideally, an engineered bacterium for cancer therapy would specifically target tumors enabling the use of more toxic molecules without systemic effects; be self-propelled enabling its penetration into tumor regions that are inaccessible to passive therapies; be responsive to external signals enabling the precise control of location and timing of cytotoxicity; be able to sense the local environment allowing the development of responsive therapies that can make decisions about where and when drugs are administered; and be externally detectable, thus providing information about the state of the tumor, the success of localization and the efficacy of treatment (Forbes, 2010). Indeed some of these features naturally exist in some bacteria, e.g. many genera of bacteria have been shown to preferentially accumulate in tumors, including *Salmonella, Escherichia, Clostridium* and *Bifidobacterium*. Moreover, bacteria have motility (flagella) that enable tissue penetration and chemotactic receptors that direct chemotaxis towards molecular signals in the tumor microenvironment. Selective cytotoxicity can be engineered by transfection with genes for therapeutic molecules, including toxins, cytokines, tumor antigens and apoptosis-inducing factors. External control can be achieved using gene promoter strategies that respond to small molecules, heat or radiation. Bacteria can be detected using light, magnetic resonance imaging and positron emission tomography. At last, genetic manipulation of bacteria is easy, thus enabling the development of treatment strategies, such as expression of anti-tumor proteins and including vectors to infect cancer cells (Pawelek et al., 2003). To date, many different bacterial strategies have been implemented in animal models (e.g. *Salmonella* has been tested for breast, colon, hepatocellular, melanoma, neuroblastoma, pancreatic and prostate cancer), and also some human trials (e.g. *C. butyricum* M-55 has been tested for squamous cell carcinoma, metastic, malignant neuroma and melanoma) have been carried out (Forbes, 2010).

Ultrasound is one of the techniques often used to treat solid tumors (e.g. breast cancer); however, this technique is not always successful, as sometimes it just heats the tumor without destroying it. Therefore, we are currently engineering the heat shock response machinery from *E. coli* to trigger the release of a therapeutic agent *in situ* concurrent with ultrasound treatment. For that purpose, several modeling and engineering steps are being implemented. The strategy being pursued is particularly useful for drugs that require *in situ* synthesis because of a poor bioavailability, thereby avoiding repetitive oral doses to achieve sufficient concentration inside the cells. The use of live bacteria for therapeutic purposes naturally poses some issues (Pawelek et al., 2003), but currently the goal is to achieve the proof-of-concept that an engineered system will enable the production of a cancer-fighting drug triggered by a temperature increase.

5.3 Alternative nanosized drug carriers

The design of novel tumor targeted multifunctional particles is another extremely interesting and innovative approach that makes use of the synthetic biology principles. The modest success of the traditional strategies for cancer treatment has driven research towards the development of new approaches underpinned by mechanistic understanding of cancer progression and targeted delivery of rational combination therapy.

5.3.1 Viral drug delivery systems

The use of viruses, in the form of vaccines, has been common practice ever since its first use to combat smallpox. Recently, genetic engineering has enlarged the applications of viruses, since it allows the removal of pathogen genes encoding virulence factors that are present in the virus coat. As a result, it can elicit immunity without causing serious health effects in humans. In the light of gene therapy, the use of virus-based entities hold a promising future, since by nature, they are being delivered to human target cells, and can be easily manipulated genetically. As such, they may be applied to target and lyse specific cancer cells, delivering therapeutics *in situ*. Bacteriophages are viruses that specifically and only infect bacteria. They have gained more attention the last decades, mainly in phage display technology. In anti-cancer therapy, this technique has contributed enormously to the identification of new tumor-targeting molecules (Brown, 2010). *In vivo* phage display technology identified a peptide exhibiting high affinity to hepatocellular carcinoma cells (Du et al., 2010). In a different approach, a phage display-selected ligand targeting breast cancer cells was incorporated in liposomes containing siRNA. The delivered liposomes were shown to significantly downregulate the *PRDM14* gene in the MCF7 target cells (Bedi et al., 2011). In addition, the option to directly use bacteriophages as drug-delivery platforms has been explored. A recent study described the use of genetically modified phages able to target tumor cell receptors via specific antibodies resulting in endocytosis, intracellular degradation, and drug release (Bar et al., 2008). Using phage display a variety of cancer cell binding and internalizing ligands have been selected (Gao et al., 2003). Bacteriophages can also be applied to establish an immune response. Eriksson and co-workers (2009) showed that a tumor-specific M13 bacteriophage induced regression of melanoma target cells, involving tumor-associated macrophages and being Toll-like receptor-dependent. Finally, marker molecules or drugs can be chemically conjugated onto the phage surface, making it a versatile imaging or therapy vehicle that may reduce costs and improve life quality (Steinmetz, 2010). An M13 phage containing cancer cell-targeting motifs on the surface was chemically modified to conjugate with fluorescent molecules, resulting in both binding and imaging of human KB cancer cells (Li et al., 2010). Besides being genetically part of the virus, anti-tumor compounds can also be covalently linked to it. We are currently, by phage display, selecting phages that adhere and penetrate tumor cells. Following this selection, we will chemically conjugate anti-cancer compounds (e.g. doxorubicin) to bacteriophages, equipped with the cancer cell-recognizing peptides on the phage surface. We anticipate that such a multifunctional nanoparticle, targeted to the tumor using a tumor "homing" peptide, will enable a significant improvement over existing anti-cancer approaches.

5.4 Microbial cell factories for the production of drugs

On a different perspective, as exemplified in section 4, synthetic biology approaches can be used for the large scale production of compounds with pharmaceutical applications. One of the easily employable approaches to develop synthetic pathways is to combine genes from different organisms, and design a new set of metabolic pathways to produce various natural and unnatural products. The host organism provides precursors from its own metabolism, which are subsequently converted to the desired product through the expression of the heterologous genes (see section 4). Existing examples of synthetic metabolic networks make use of transcriptional and translational control elements to regulate the expression of enzymes that synthesize and breakdown metabolites. In these systems, metabolite concentration acts as an input for other control elements (Andrianantoandro et al., 2006). An entire metabolic pathway from *S. cerevisiae*, the mevalonate isoprenoid pathway for synthesizing isopentyl pyrophosphate, was successfully transplanted into *E. coli*. In combination with an inserted synthetic amorpha-4, 11-diene synthase, this pathway produced large amounts of a precursor to the anti-malarial drug artemisinin. This new producing strain is very useful since a significant decrease in the drug production time and costs could be achieved (Martin et al., 2003). In addition to engineering pathways that produce synthetic metabolites, artificial circuits can be engineered using metabolic pathways connected to regulatory proteins and transcriptional control elements (Andrianantoandro et al., 2006). One study describes such a circuit based on controlling gene expression through acetate metabolism for cell–cell communication (Bulter et al., 2004). Metabolic networks may embody more complex motifs, such as an oscillatory network. A recently constructed metabolic network used glycolytic flux to generate oscillations through the signaling metabolite acetyl phosphate (Fung et al., 2005). The system integrates transcriptional regulation with metabolism to produce oscillations that are not correlated with the cell division cycle. The general concerns of constructing transcriptional and protein interaction-based modules, such as kinetic matching and optimization of reactions for a new environment, apply for metabolic networks as well. In addition, the appropriate metabolic precursors must be present. For this purpose, it may be necessary to include other enzymes or metabolic pathways that synthesize precursors for the metabolite required in a synthetic network (Leonard et al., 2008; McArthur IV & Fong, 2010).

Many polyketides and nonribosomal peptides are being used as antibiotic, anti-tumor and immunosuppressant drugs (Neumann & Neumann-Staubitz, 2010). In order to produce them in heterologous hosts, assembly of all the necessary genes that make up the synthetic pathways is essential. The metabolic systems for the synthesis of polyketides are composed of multiple modules, in which an individual module consists of either a polyketide synthase or a nonribosomal peptide synthetase. Each module has a specific set of catalytic domains, which ultimately determine the structure of the metabolic product and thus its function. Recently, Bumpus et al. (2009) presented a proteomic strategy to identify new gene clusters for the production of polyketides and nonribosomal peptides, and their biosynthetic pathways, by adapting mass-spectrometry-based proteomics. This approach allowed identification of genes that are used in the production of the target product in a species, for which a complete genome sequence is not available. Such newly identified pathways can then be copied into a new host strain that is more suitable for producing polyketides and nonribosomal peptides at an industrial scale. This exemplifies that the sources of new pathways are not limited to species with fully sequenced genomes.

The use of synthetic biology approaches in the field of metabolic engineering opens enormous possibilities, especially toward the production of new drugs for cancer treatment. Our goal is to design and model a new biosynthetic pathway for the production of natural drugs in *E. coli*. Key to this is the specification of gene sequences encoding enzymes that catalyze each reaction in the pathway, and whose DNA sequences can be incorporated into devices that lead to functional expression of the molecules of interest (Prather & Martin, 2008). Partial pathways can be recruited from independent sources and co-localized in a single host (Kobayashi et al., 2004). Alternatively, pathways can be constructed for the production of new, non-natural products by engineering existing routes (Martin et al., 2003).

6. Conclusion

Despite all the scientific advances that humankind has seen over the last centuries, there are still no clear and defined solutions to diagnose and treat cancer. In this sense, the search for innovative and efficient solutions continues to drive research and investment in this field. Synthetic biology uses engineering principles to create, in a rational and systematic way, functional systems based on the molecular machines and regulatory circuits of living organisms, or to re-design and fabricate existing biological systems. Bioinformatics and newly developed computational tools play a key role in the improvement of such systems. Elucidation of disease mechanisms, identification of potential targets and biomarkers, design of biological elements for recognition and targeting of cancer cells, discovery of new chemotherapeutics or design of novel drugs and catalysts, are some of the promises of synthetic biology. Recent achievements are thrilling and promising; yet some of such innovative solutions are still far from a real application due to technical challenges and ethical issues. Nevertheless, many scientific efforts are being conducted to overcome these limitations, and undoubtedly it is expected that synthetic biology together with sophisticated computational tools, will pave the way to revolutionize the cancer field.

7. References

Alon, U. (2003). Biological networks: the tinkerer as an engineer. *Science*, Vol.301, No.5641, (September 2003), pp. 1866-1867, ISSN 0036-8075

Anderson, J.C.; Clarke, E.J.; Arkin, A.P. & Voigt, C.A. (2006). Environmentally controlled invasion of cancer cells by engineered bacteria. *Journal of Molecular Biology*, Vol.355, No.4, (January 2006), pp. 619–627, ISSN 00222836

Andrianantoandro, E.; Basu, S.; Karig, D.K. & Weiss, R. (2006). Synthetic biology: new engineering rules for an emerging discipline. *Molecular Systems Biology* Vol.2, No. 2006.0028, (May 2006), pp. 1-14, ISSN 1744-4292

Arkin, A.P. (2001). Synthetic cell biology. *Current Opinion in Biotechnology* Vol.12, No.6, (December 2001), pp. 638-644, ISSN 0958-1669

Bansal, V.; Sakizlis, V.; Ross, R.; Perkins, J.D. & Pistikopoulos, E.N. (2003). New algorithms for mixed-integer dynamic optimization. *Computers and Chemical Engineering* Vol.27, No.5, (May 2003), pp. 647-668, ISSN 0098-1354

Bar, H.; Yacoby, I. & Benhar, I. (2008). Killing cancer cells by targeted drug-carrying phage nanomedicines. *BMC Biotechnology* Vol.8, No.37, (April 2008), pp. 1-14, ISSN 1472-6750

Batt, G., Yordanov, B., Weiss, R. & Belta, C. (2007). Robustness analysis and tuning of synthetic gene networks. *Bioinformatics* Vol.23, No.18, (July 2007), pp. 2415-2422, ISSN 1367-4803

Bedi, D.; Musacchio, T.; Fagbohun, O.A.; Gillespie, J.W.; Deinnocentes, P.; Bird, R.C.; Bookbinder, L.; Torchilin, V.P. & Petrenko, V.A. (2011). Delivery of siRNA into breast cancer cells via phage fusion protein-targeted liposomes. *Nanomedicine: Nanotechnology, Biology, and Medicine*, doi:10.1016/j.nano.2010.10.004, ISSN 1549-9634

Beisel, C.L. & Smolke, C.D. (2009). Design principles for riboswitch function. *PLoS Computational Biology* Vol.5, No.4, (April 2009), e1000363, pp. 1-14, ISSN 1553-734X

Benner, S.A. & Sismour, A.M. (2005). Synthetic biology. *Nature Reviews Genetics* Vol.6, No.7, (July 2005), pp. 533–543, ISSN 1471-0056

Brown, K.C. (2010). Peptidic tumor targeting agents: the road from phage display peptide selections to clinical applications. *Current Pharmaceutical Design* Vol.16, No.9, (March 2010), pp. 1040-1054, ISSN 1381-6128

Brustad, E.M. & Arnold, F.H. (2010). Optimizing non-natural protein function with directed evolution. *Current Opinion in Chemical Biology* Vol.15, No.2, (April 2010), pp. 1-10, ISSN 1367-5931

Bulter, T.; Lee, S.G.; Wong, W.W.; Fung, E.; Connor, M.R. & Liao, J.C. (2004). Design of artificial cell–cell communication using gene and metabolic networks. *PNAS* Vol.101, No.8 (February 2004), pp. 2299–2304, ISSN 0027-8424

Bumpus, S.; Evans, B.; Thomas, P.; Ntai, I. & Kelleher, N. (2009). A proteomics approach to discovering natural products and their biosynthetic pathways. *Nature Biotechnology* Vol.27, No.10, (September 2009), pp. 951-956, ISSN 1087-0156

Canton, B.; Labno, A. & Endy, D. (2008). Refinement and standardization of synthetic biological parts and devices. *Nature Biotechnology* Vol.26, No. 7, (July 2008), pp. 787-793, ISSN 1087-0156

Carothers, J.M.; Goler, J.A. & Keasling, J.D. (2009). Chemical synthesis using synthetic biology. *Current Opinion in Biotechnology* Vol.20, No.4, (August 2009), pp. 498-503, ISSN 0958-1669

Chan, L.Y.; Kosuri, S. & Endy, D. (2005). Refactoring bacteriophage T7. *Molecular Systems Biology* Vol.1, No. 2005.0018, (September 2005), pp. 1-10, ISSN 1744-4292

Cooling, M.T.; Hunter, P. & Crampin, E.J. (2008). Modelling biological modularity with CellML. *IET Systems Biology* Vol.2, No.2, (March 2008), pp. 73-79, ISSN 1751-8849

Crameri, A.; Raillard, S.A.; Bermudez, E. & Stemmer, W.P. (1998). DNA shuffling of a family of genes from diverse species accelerates directed evolution. *Nature* Vol.391, No.6664, (January 1998), pp. 288-291, ISSN 0028-0836

Czar, M.J.; Cai, Y. & Peccoud, J. (2009). Writing DNA with GenoCAD. *Nucleic Acids Research* Vol.37, No.2, (May 2009), pp. W40-W47, ISSN 0305-1048

Dasika, M.S. & Maranas, C.D. (2008). OptCircuit: an optimization based method for computational design of genetic circuits. *BMC Systems Biology* Vol.2, No.24, pp. 1-19, ISSN 1752-0509

Datsenko, K.A. & Wanner, B.L. (2000). One-step inactivation of chromosomal genes in Escherichia coli K-12 using PCR products. *PNAS* Vol.97, No. 12, (June 2000), pp. 6640-6645, ISSN 0027-8424

Dawid, A.; Cayrol, B. & Isambert, H. (2009). RNA synthetic biology inspired from bacteria: construction of transcription attenuators under antisense regulation. *Physical Biology* Vol.6, No. 025007, (July 2009), pp. 1-10, ISSN 1478-3975

Di Bernardo, D.; Thompson, M.J.; Gardner, T.S.; Chobot, S.E.; Eastwood, E.L.; Wojtovich, A.P.; Elliott, S.E.; Schaus, S.E. & Collins, J.J. (2005). Chemogenomic profiling on a genome-wide scale using reverse-engineered gene networks. *Nature Biotechnology* Vol.23, No.3, (March 2005), pp. 377-383, ISSN 1087-0156

Di Ventura, B.; Lemerle, C.; Michalodimitrakis, K. & Serrano, L. (2006). From *in vivo to in silico* biology and back. *Nature* Vol.443, No.7111, (October 2006), pp. 527-533, ISSN 0028-0836

Dougherty, M.J. & Arnold, F.H. (2009). Directed evolution: new parts and optimized function. *Current Opinion in Biotechnology* Vol.20, No.4, (August 2009), pp. 486-491, ISSN 0958-1669

Du B, Han H, Wang Z, Kuang L, Wang L, Yu L, Wu M, Zhou Z, Qian M. (2010). Targeted drug delivery to hepatocarcinoma in vivo by phage-displayed specific binding peptide. *Molecular Cancer Research* Vol.8, No.2, (February 2010), pp.135-144, ISSN 1541-7786

Ellis, T.; Wang, X. & Collins, J.J. (2009). Diversity-based, model guided construction of synthetic gene networks with predicted functions. *Nature Biotechnology* Vol.27, No.5, (May 2009), pp. 465- 471, ISSN 1087-0156

Elowitz, M.B. & Leibler, S. (2000). A synthetic oscillatory network of transcriptional regulators. *Nature* Vol.403, No.6767, (January 2000), pp. 335-338, ISSN 0028-0836

Elowitz, M.B.; Levine, A.J.; Siggia, E.D. & Swain, P.S. (2002). Stochastic gene expression in a single cell. *Science* Vol.297, No.5584, (August 2002), pp. 1183–1186, ISSN 0036-8075

Endy, D. (2005). Foundations for engineering biology. *Nature* Vol.438, No. 7067, (November 2005), pp. 449-453, ISSN 0028-0836

Engler, C.; Gruetzner, R.; Kandzia, R. & Marillonnet, S. (2009). Golden gate shuffling: a one pot DNA shuffling method based on type ils restriction enzymes. *PLoS ONE* Vol.4, No.5, (May 2009), e5553, pp. 1-9, ISSN 1932-6203

Eriksson, F.; Tsagozis, P.; Lundberg, K.; Parsa, R.; Mangsbo, S.M.; Persson, M.A.; Harris, R.A. & Pisa, P.J. (2009). Tumor-specific bacteriophages induce tumor destruction through activation of tumor-associated macrophages. *The Journal of Immunology* Vol.182, No.5, (March 2009), pp. 3105-3111, ISSN 0022-1767

Ferreira, C.S.; Matthews, C.S. & Missailidis, S. (2006). DNA aptamers that bind to MUC1 tumour marker: design and characterization of MUC1-binding single-stranded DNA aptamers. *Tumour Biology* Vol.27, No.6, (October 2006), pp. 289-301, ISSN 1010-4283

Fisher, J. & Henzinger, T.A. (2007). Executable cell biology. *Nature Biotechnology* Vol.25, No.11, (November 2007), pp. 1239–1249, ISSN 1087-0156

Forbes, N.S. (2010). Engineering the perfect (bacterial) cancer therapy. *Nature Reviews Cancer* Vol.10, No.11, (October 2010), pp. 785-794, ISSN 1474-175X

Francois, P. & Hakim, V. (2004). Design of genetic networks with specified functions by evolution in silico. *PNAS* Vol.101, No.2, (January 2004), pp. 580–585, ISSN 0027-8424

Funahashi, A.; Morohashi, M. & Kitano, H. (2003). CellDesigner: a process diagram editor for gene-regulatory and biochemical networks. *BIOSILICO* Vol.1, No.5, (November 2003),pp. 159–162, ISSN 1478-5382

Fung, E.; Wong, W.W.; Suen, J.K.; Bulter, T.; Lee, S.G. & Liao, J.C. (2005). A synthetic gene metabolic oscillator. *Nature* Vol.435, No.7038, (May 2005), pp. 118–122, ISSN 0028-0836

Gao, C.; Mao, S.; Ronca, F.; Zhuang, S.; Quaranta, V.; Wirsching, P. & Janda, K.D. (2003). De novo identification of tumor-specific internalizing human antibody-receptor pairs by phage-display methods. *Journal of Immunological Methods.* Vol.274, No.1-2, (March 2003), pp. 185-197, ISSN 0022-1759

Gardner, T.S.; Cantor, C.R. & Collins, J.J. (2000). Construction of a genetic toggle switch in Escherichia coli. *Nature* Vol.403, No.6767, (January 2000), pp. 339-342, ISSN 0028-0836

Garst, A.D. & Batey, R.T. (2009). A switch in time: detailing the life of a riboswitch. *Biochimica Biophysica Acta* Vol.1789, No.9-10, (September-October 2009), pp. 584-591, ISSN 0006-3002

Goler, J.A. (2004). A design and simulation tool for synthetic biological systems. Cambridge, MA: MIT

Golynskiy, M.V. & Seelig, B. (2010). De novo enzymes: from computational design to mRNA display. *Trends in Biotechnology* Vol.28, No.7, (July 2010), pp. 340-345, ISSN 0167-7799

Grunberg, R.; Nilges, M. & Leckner, J. (2007). Biskit —a software platform for structural bioinformatics. *Bioinformatics* Vol.23, No.6, (March 2007), pp. 769-770, ISSN 1367-4803

Ham, T.S.; Lee, S.K.; Keasling, J.D. & Arkin, A.P. (2006). A tightly regulated inducible expression system utilizing the fim inversion recombination switch. *Biotechnology and Bioengineering* Vol.94, No.1, (May 2006), pp. 1–4, ISSN 0006-3592

Hartley, J.L. (2003). Use of the gateway system for protein expression in multiple hosts. *Current Protocols in Protein Science* Chap5: Unit 5.17

Hawkins, T.; Luban, S. & Kihara, D. (2006). Enhanced automated function prediction using distantly related sequences and contextual association by PFP. *Protein Science* Vol.15, No.6, (June 2006), pp. 1550-1556, ISSN 1469-896X

Hianik, T. & Wang, J. (2009). Electrochemical aptasensors - recent achievements and perspectives. *Electroanalysis* Vol.21, No.11, (June 2009), pp. 1223-1235, ISSN 1521-4109

Hill, A.D.; Tomshine, J.R.; Weeding, E.M.; Sotiropoulos, V. & Kaznessis, Y.N. (2008). SynBioSS: the synthetic biology modeling suite. *Bioinformatics* Vol.24, No.21, (November 2008),pp. 2551-2553, ISSN 1367-4803

Hofacker, I.L. (2003). Vienna RNA secondary structure server. *Nucleic Acids Research* Vol.31, No.13, (July 2003), pp. 3429-3431, ISSN 0305-1048

Isaacs, F.J.; Dwyer, D.J. & Collins, J.J. (2006). RNA synthetic biology. *Nature Biotechnology* Vol.24, No.5, (May 2006), pp. 545-554, ISSN 1087-0156

Isalan, M.; Lemerle, C.; Michalodimitrakis, K.; Horn, C.; Beltrao, P.; Raineri, E.; Garriga-Canut, M.; Serrano, L. (2008). Evolvability and hierarchy in rewired bacterial gene networks. *Nature* Vol.452, No.7189, (April 2008), pp. 840–845, ISSN 0028-0836

Jemal, A.; Bray, F.; Center, M.M.; Ferlay, J.; Ward, E. & Forman, D. (2011). Global cancer statistics. *CA Cancer Journal for Clinicians* Vol.61, No.2, (March-April 2011),pp. 69-90, ISSN 0007-9235

Kaiser, J. (2005). Gene therapy. Putting the fingers on gene repair. *Science* Vol.310, No.5756, (December 2005), pp.1894-1896, ISSN 0036-8075

Kaznessis, Y.N. (2009). Computational methods in synthetic biology. *Biotechnology Journal* Vol.4, No.10, (October 2009), pp.1392-1405, ISSN 1860-7314

Kelly, J.; Rubin, A.J.; Davis, J. II; Ajo-Franklin, C.M.; Cumbers, J.; Czar, M.J.; de Mora, K.; Glieberman, A.I.; Monie, D.D. & Endy, D. (2009). Measuring the activity of BioBrick promoters using an in vivo reference standard. *Journal of Biological Engineering* Vol.3, No.4, (March 2009), pp. 1-13, ISSN 1754-1611

Kirkpatrick, S.; Gelatt, C.D. Jr. & Vecchi, M.P. (1983). Optimization by Simulated Annealing. *Science* Vol.220, No.4598, (May 1983), pp. 671-680, ISSN 0036-8075

Kitano, H. (2002). Systems biology: a brief overview. *Science* Vol.295, No.5560, (March 2002), pp. 1662-1664, ISSN 0036-8075

Kizer, L.; Pitera, D.J.; Pfleger, B.F. & Keasling, J.D. (2008). Application of functional genomics to pathway optimization for increased isoprenoid production. *Applied and Environmental Microbiology* Vol.74, No.10, (May 2008), pp. 3229–3241, ISSN 0099-2240

Kobayashi, H.; Kaern, M.; Araki, M.; Chung, K.; Gardner, T.S.; Cantor, C.R. & Collins, J.J. (2004). Programmable cells: interfacing natural and engineered gene networks. *PNAS* Vol.101, No.22, (June 2004), pp. 8414–8419, ISSN 0027-8424

Koide, T.; Pang, W.L. & Baliga, N.S. (2009). The role of predictive modeling in rationally reengineering biological systems. *Nature Reviews Microbiology* Vol.7, No.4, (April 2009), pp. 297-305, ISSN 1740-1526

Kosuri, S.; Kelly, J.R. & Endy, D. (2007). TABASCO: a single molecule, base pair resolved gene expression simulator. BMC *Bioinformatics* Vol.8, No.480, (December 2007), pp. 1-15, ISSN 1471-2105

Lartigue, C.; Glass, J.I.; Alperovich, N.; Pieper, R.; Parmar, P.P.; Hutchison III, C.A.; Smith, H.O. & Venter, J.C. (2007). Genome transplantation in bacteria: changing one species to another. *Science* Vol.317, No.5838, (August 2007), pp. 632-638, ISSN 0036-8075

Leonard, E.; Nielsen, D.; Solomon, K. & Prather, K.J. (2008). Engineering microbes with synthetic biology frameworks. *Trends in Biotechnology* Vol.26, No.12, (December 2008), pp. 674-681, ISSN 0167-7799

Li K, Chen Y, Li S, Nguyen HG, Niu Z, You S, Mello CM, Lu X, Wang Q. (2010). Chemical modification of M13 bacteriophage and its application in cancer cell imaging. *Bioconjugate Chemistry* Vol.21, No.7, (December 2010), pp. 1369-1377, ISSN 1043-1802

Loettgers, A. (2007). Model organisms and mathematical and synthetic models to explore regulation mechanisms. *Biological Theory Vol.*2, No.2, (December 2007), pp. 134-142, ISSN 1555-5542

Mandell, J.G. & Barbas, C.F. III (2006). Zinc Finger Tools: custom DNA binding domains for transcription factors and nucleases. *Nucleic Acids Research Vol.*34, (July 2006), pp. W516-523, ISSN 0305-1048

Marchisio, M.A. & Stelling, J. (2008). Computational design of synthetic gene circuits with composable parts. *Bioinformatics Vol.*24, No.17, (September 2008), pp.1903–1910, ISSN 1367-4803

Marchisio, M.A. & Stelling, J. (2009). Computational design tools for synthetic biology. *Current Opinion in Biotechnology Vol.*20, No.4, (August 2009), pp. 479-485, ISSN 0958-1669

Marguet, P.; Balagadde, F.; Tan, C. & You, L. (2007). Biology by design: reduction and synthesis of cellular components and behavior. *Journal of the Royal Society Interface* Vol.4, No.15, (August 2007), pp. 607-623, ISSN 1742-5689

Markham, N.R. & Zuker, M. (2008). UNAFold: software for nucleic acid folding and hybridization. *Methods Molecular Biology Vol.*453, No.I, pp. 3-31, ISSN 1064-3745

Martin, V.J.; Pitera, D.J.; Withers, S.T.; Newman, J.D. & Keasling, J.D. (2003). Engineering a mevalonate pathway in Escherichia coli for production of terpenoids. *Nature Biotechnology Vol.*21, No.7, (July 2003), pp.796–802, ISSN 1087-0156

Matsuoka, Y.; Ghosh, S. & Kitano, H. (2009). Consistent design schematics for biological systems: standardization of representation in biological engineering. *Journal of the Royal Society Interface Vol.*6, No.4, (August 2009), pp. S393-S404, ISSN 1742-5689

McArthur IV, G.H. & Fong, S.S. (2010). Toward engineering synthetic microbial metabolism. *Journal of Biomedicine and Biotechnology* doi:10.1155/2010/459760, ISSN 1110-7243

Mirschel, S.; Steinmetz, K.; Rempel, M.; Ginkel, M. & Gilles, E.D. (2009). PROMOT: modular modeling for systems biology. *Bioinformatics Vol.*25, No.5, (March 2009), pp. 687-689, ISSN 1367-4803

Morris, G.M.; Huey, R.; Lindstrom, W.; Sanner, M.F.; Belew, R.K.; Goodsell, D.S. & Olson, A.J. (2009). AutoDock4 and AutoDockTools4: automated docking with selective receptor flexibility. *Journal of Computational Chemistry Vol.*30, No.16, (December 2009), pp. 2785-2791, ISSN 0192-8651

Mueller, S.; Coleman, J.R. & Wimmer, E. (2009). Putting synthesis into biology: a viral view of genetic engineering through de novo gene and genome synthesis. *Chemistry & Biology Vol.*16, No.3, (March 2009), pp. 337-347, ISSN 1074-5521

Neumann, H. & Neumann-Staubitz, P. (2010). Synthetic biology approaches in drug discovery and pharmaceutical biotechnology. *Applied Microbiology and Biotechnology* Vol.87, No.1, (June 2010), pp. 75-86, ISSN 0175-7598

Nielsen, J. (2001). Metabolic engineering. *Applied Microbiology and Biotechnology Vol.*55, No.3, (April 2001), pp. 263-283, ISSN 0175-7598

Park, J.H.; Lee, S.Y.; Kim, T.Y. & Kim, H.U. (2008). Application of systems biology for bioprocess development. *Trends in Biotechnology Vol.*26, No.8, (August 2008), pp. 404–412, ISSN 0167-7799

Pawelek, J.; Low, K. & Bermudes, D. (2003). Bacteria as tumour-targeting vectors. *Lancet Oncology Vol.*4, No.9, (September 2003), pp. 548–556, ISSN 1470-2045

Pedersen, M. & Phillips, A. (2009). Towards programming languages for genetic engineering of living cells. *Journal of the Royal Society Interface Vol.*6, No.4, (August 2009), pp. S437-S450, ISSN 1742-5689

Prather, K. & Martin, C.H. (2008). De novo biosynthetic pathways: rational design of microbial chemical factories. *Current Opinion in Biotechnology Vol.*19, No.5, (October 2008), pp. 468–474, ISSN 0958-1669

Price, N.D. & Shmulevich, I. (2007). Biochemical and statistical network models for systems biology. *Current Opinion in Biotechnology Vol.*18, No.4, (August 2007), pp. 365–370, ISSN 0958-1669

Purnick, P.E. & Weiss, R. (2009). The second wave of synthetic biology: from modules to systems. *Nature Reviews Molecular Cell Biology Vol.*10, No.6, (June 2009), pp. 410-422, ISSN 1471-0072

Quan, J. & Tian, J. (2009). Circular polymerase extension cloning of complex 1 gene libraries and pathways. *PLoS ONE* Vol.4, No.7, (July 2009), e6441, pp.1-6, ISSN 1932-6203

Richardson, S.M.; Wheelan, S.J.; Yarrington, R.M. & Boeke, J.D. (2006). GeneDesign: rapid, automated design of multikilobase synthetic genes. *Genome Research* Vol.16, No.4, (April 2006), pp. 550-556, ISSN 1088-9051

Ritchie, D.W. (2008). Recent progress and future directions in protein-protein docking.*Current Protein & Peptide Science* Vol.9, No.1, (February 2008), pp. 1-15, ISSN 1389-2037

Ro, D.K.; Paradise, E.M.; Ouellet, M.; Fisher, K.J.; Newman, K.L.; Ndungu, J.M.; Ho, K.A.; Eachus, R.A.; Ham, T.S.; Kirby, J.; Chang, M.C.; Withers, S.T.; Shiba, Y.; Sarpong, R. & Keasling, J.D. (2006). Production of the antimalarial drug precursor artemisinic acid in engineered yeast. *Nature* Vol.440, No.7086, (April 2006), pp. 940–943, ISSN 0028-0836

Rocha, I.; Forster, J. & Nielsen, J. (2008). Design and application of genome-scale reconstructed metabolic models. *Methods in Molecular Biology* Vol.416, No.IIIB, pp. 409-431, ISSN 1064-3745

Rocha, I.; Maia, P.; Evangelista, P.; Vilaça, P.; Soares, S.; Pinto, J.P.; Nielsen, J.; Patil, K.R.; Ferreira, E.C. & Rocha, M. (2010). OptFlux: an open-source software platform for in silico metabolic engineering. *BMC Systems Biology* Vol.4, No. 45, pp. 1-12, ISSN 1752-0509

Rodrigo, G.; Carrera, J. & Jaramillo, A. (2007a). Asmparts: assembly of biological model parts. *Systems and Synthetic Biology* Vol.1, No.4, (December 2007), pp. 167-170, ISSN 1872-5325

Rodrigo, G.; Carrera, J. & Jaramillo, A. (2007b). Genetdes: automatic design of transcriptional networks. *Bioinformatics* Vol.23, No.14, (July 2007), pp.1857-1858, ISSN 1367-4803

Rodrigues, L.R.; Teixeira, J.A.; Schmitt, F.; Paulsson, M. & Lindmark Måsson, H. (2007). The role of osteopontin in tumour progression and metastasis in breast cancer. *Cancer Epidemology Biomarkers & Prevention* Vol.16, No.6, (June 2007), pp. 1087–1097, ISSN 1055-9965

Saito, H. & Inoue, T. (2009). Synthetic biology with RNA motifs. *The International Journal of Biochemistry & Cell Biology* Vol.41, No.2, (February 2009), pp. 398-404, ISSN 1357-2725

Salis, H.M., Mirsky, E.A. & Voigt, C.A. (2009). Automated design of synthetic ribosome binding sites to control protein expression. *Nature Biotechnology* Vol.27, No.10, (October 2009), pp. 946–950, ISSN 1087-0156

Segre, D.; Vitkup, D. & Church, G.M. (2002). Analysis of optimality in natural and perturbed metabolic networks. *PNAS* Vol.99, No.23, (November 2001), pp. 15112-15117, ISSN 0027-8424

Shetty, R.P.; Endy, D. & Knight, T.F. Jr. (2008). Engineering BioBrick vectors from BioBrick parts. *Journal of Biological Engineering* Vol.2, No.1, (April 2008), pp.5-17 ISSN 1754-1611

Simons, K.T.; Bonneau, R.; Ruczinski, I. & Baker, D. (1999). Ab initio protein structure prediction of CASP III targets using ROSETTA. *Proteins* Vol.3, pp. 171-176, ISSN 0887-3585

Steinmetz, N.F. (2010).Viral nanoparticles as platforms for next-generation therapeutics and imaging devices. *Nanomedicine: Nanotechnology, Biology, and Medicine* Vol.6, No.5, (October 2010), pp. 634-641, ISSN 1549-9634

Stoltenburg, R.; Reinemann, C. & Strehlitz, B. (2007). SELEX—A (r)evolutionary method to generate high-affinity nucleic acid ligands. *Biomolecular Engineering* Vol.24, No.4, (October 2007), pp. 381–403, ISSN 1389-0344

Topp, S. & Gallivan, J.P. (2007). Guiding bacteria with small molecules and RNA. *Journal of the American Chemical Society* Vol.129, No.21, (May 2007), pp. 6807-6811, ISSN 0002-7863

Tyo, K.E.; Alper, H.S. & Stephanopoulos, G. (2007). Expanding the metabolic engineering toolbox: more options to engineer cells. *Trends in Biotechnology* Vol.25, No.3, (March 2007), pp. 132-137, ISSN 0167-7799

Tyo, K.E.J.; Ajikumar, P.K. & Stephanopoulos, G. (2009). Stabilized gene duplication enables long-term selection-free heterologous pathway expression. *Nature Biotechnology* Vol.27, No.8, (August 2009), pp. 760–765, ISSN 1087-0156

Villalobos, A.; Ness, J.E.; Gustafsson, C.; Minshull, J. & Govindarajan, S. (2006). Gene Designer: a synthetic biology tool for constructing artificial DNA segments. *BMC Bioinformatics* Vol.7, No.285, (June 2006), pp.1-8, ISSN 1471-2105

Voloshchuk, N. & Montclare, J.K. (2010). Incorporation of unnatural amino acids for synthetic biology. *Molecular Biosystems* Vol.6, No.1, (January 2010), pp. 65-80, ISSN 1742-2051

Wang, H.H.; Isaacs, F.J.; Carr, P.A.; Sun, Z.Z.; Xu, G.; Forest, C.R. & Church, G.M. (2009). Programming cells by multiplex genome engineering and accelerated evolution. *Nature* Vol.460, No.7257, (August 2009), pp. 894–898, ISSN 0028-0836

Wierling, C.; Herwig, R. & Lehrach, H. (2007). Resources, standards and tools for systems biology. *Briefings in Functional Genomics and Proteomics* Vol.6, No.3, (September 2007), pp. 240-251, ISSN 1473-9550

Win, M.N. & Smolke, C.D. (2007). From the cover: a modular and extensible RNA-based gene-regulatory platform for engineering cellular function. *PNAS* Vol.104, No.36, (September 2007), pp. 14283–14288, ISSN 0027-8424

Winkler, W.C. & Breaker, R.R. (2005). Regulation of bacterial gene expression by riboswitches. *Annual Review of Microbiology* Vol.59, pp. 487-517, ISSN 0066-4227

Xu, J.; Li, M.; Kim, D. & Xu, Y. (2003). RAPTOR: optimal protein threading by linear programming. *Journal of Bioinformatics and Computational Biology* Vol.1, No.1, (April 2003), pp. 95-117, ISSN 0219-7200

Yokobayashi, Y.; Weiss, R. & Arnold, F.H. (2002). Directed evolution of a genetic circuit. *PNAS* Vol.99, No.26, (September 2002), pp. 16587–16591, ISSN 0027-8424

You, L. (2004). Toward computational systems biology. *Cell Biochemistry and Biophysics* Vol.40, No.2, pp. 167–184, ISSN 1085-9195

9

An Overview of Hardware-Based Acceleration of Biological Sequence Alignment

Laiq Hasan and Zaid Al-Ars
TU Delft
The Netherlands

1. Introduction

Efficient biological sequence (proteins or DNA) alignment is an important and challenging task in bioinformatics. It is similar to string matching in the context of biological data and is used to infer the evolutionary relationship between a set of protein or DNA sequences. An accurate alignment can provide valuable information for experimentation on the newly found sequences. It is indispensable in basic research as well as in practical applications such as pharmaceutical development, drug discovery, disease prevention and criminal forensics. Many algorithms and methods, such as, dot plot (Gibbs & McIntyre, 1970), *Needleman-Wunsch (N-W)* (Needleman & Wunsch, 1970), *Smith-Waterman (S-W)* (Smith & Waterman, 1981), FASTA (Pearson & Lipman, 1985), BLAST (Altschul et al., 1990), HMMER (Eddy, 1998) and ClustalW (Thompson et al., 1994) have been proposed to perform and accelerate sequence alignment activities. An overview of these methods is given in (Hasan et al., 2007). Out of these, S-W algorithm is an optimal sequence alignment method, but its computational cost makes it inappropriate for practical purposes. To develop efficient and optimal sequence alignment solutions, the S-W algorithm has recently been implemented on emerging accelerator platforms such as *Field Programmable Gate Arrays (FPGAs)*, *Cell Broadband Engine (Cell/B.E.)* and *Graphics Processing Units (GPUs)* (Buyukkur & Najjar, 2008; Hasan et al., 2010; Liu et al., 2009; 2010; Lu et al., 2008). This chapter aims at providing a broad overview of sequence alignment in general with particular emphasis on the classification and discussion of available methods and their comparison. Further, it reviews in detail the acceleration approaches based on implementations on different platforms and provides a comparison considering different parameters. This chapter is organized as follows:
The remainder of this section gives a classification, discussion and comparison of the available methods and their hardware acceleration. Section 2 introduces the S-W algorithm which is the focus of discussion in the succeeding sections. Section 3 reviews CPU-based acceleration. Section 4 provides a review of FPGA-based acceleration. Section 5 overviews GPU-based acceleration. Section 6 presents a comparison of accelerations on different platforms, whereas Section 7 concludes the chapter.

1.1 Classification

Sequence alignment aims at identifying regions of similarity between two DNA or protein sequences (the query sequence and the subject or database sequence). Traditionally, the methods of pairwise sequence alignment are classified as either global or local, where pairwise means considering only two sequences at a time. Global methods attempt to match as many characters as possible, from end to end, whereas local methods aim at identifying short stretches of similarity between two sequences. However, in some cases, it might also be needed to investigate the similarities between a group of sequences, hence multiple sequence alignment methods are introduced. Multiple sequence alignment is an extension of pairwise alignment to incorporate more than two sequences at a time. Such methods try to align all of the sequences in a given query set simultaneously. Figure 1 gives a classification of various available sequence alignment methods.

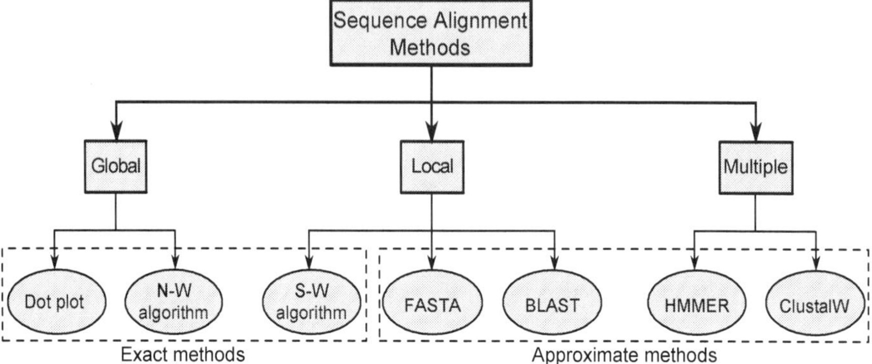

Fig. 1. Various methods for sequence alignment

These methods are categorized into three types, i.e. global, local and multiple, as shown in the figure. Further, the figure also identifies the exact methods and approximate methods. The methods shown in Figure 1 are discussed briefly in the following subsection.

1.2 Discussion of available methods

Following is a brief description of the available methods for sequence alignment.

Global methods

Global methods aim at matching as many characters as possible, from end to end between two sequences i.e. the *query sequence (Q)* and the *database sequence (D)*. Methods carrying out global alignment include dot plot and N-W algorithm. Both are categorized as exact methods. The difference is that dot plot is based on a basic search method, whereas N-W on *dynamic programming (DP)* (Giegerich, 2000).

Local methods

In contrast to global methods, local methods attempt to identify short stretches of similarity between two sequences i.e. Q and D. These include exact method like S-W and heuristics based approximate methods like FASTA and BLAST.

Multiple alignment methods

It might be of interest in some cases to consider the similarities between a group of sequences. Multiple sequence alignment methods like HMMER and ClustalW are introduced to handle such cases.

1.3 Comparison

The alignment methods can be compared on the basis of their temporal and spatial complexities and parameters like alignment type and search procedure. A summary of the comparison is shown in Table 1. It is interesting to note that all the global and local sequence alignment methods essentially have the same computational complexity of $O(L_Q L_D)$, where L_Q and L_D are the lengths of the query and database sequences, respectively. Yet despite this, each of the algorithms has very different running times, with BLAST being the fastest and dynamic programming algorithms being the slowest. In case of multiple sequence alignment methods, ClustalW has the worst time complexity of $O(L_Q^2 L_D^2)$, whereas HMMER has a time complexity of $O(L_Q L_D^2)$. The space complexities of all the alignment methods are also essentially identical, around $O(L_Q L_D)$ space, except BLAST, the space complexity of which is $O(20^w + L_Q L_D)$. In the exact methods, dot plot uses a basic search method, whereas N-W and S-W use DP. On the other hand, all the approximate methods are heuristic based. It is also worthy to note that FASTA and BLAST have to make sacrifices on sensitivity to be able to achieve higher speeds. Thus, a trade off exists between speed and sensitivity and we must come to a compromise to be able to efficiently align sequences in a biologically relevant manner in a reasonable amount of time.

Method	Type	Accuracy	Search	Time complexity	Space complexity
Dot plot	Global	Exact	Basic	$O(L_Q L_D)$	$O(L_Q L_D)$
N-W	Global	Exact	DP	$O(L_Q L_D)$	$O(L_Q L_D)$
S-W	Local	Exact	DP	$O(L_Q L_D)$	$O(L_Q L_D)$
FASTA	Local	Approximate	Heuristic	$O(L_Q L_D)$	$O(L_Q L_D)$
BLAST	Local	Approximate	Heuristic	$O(L_Q L_D)$	$O(20^w + L_Q L_D)$
HMMER	Multiple	Approximate	Heuristic	$O(L_Q L_D^2)$	$O(L_Q L_D)$
ClustalW	Multiple	Approximate	Heuristic	$O(L_Q^2 L_D^2)$	$O(L_Q L_D)$

Table 1. Comparison of various sequence alignment methods

1.4 Hardware platforms

Work has been done on accelerating sequence alignment methods, by implementing them on various available hardware platforms. Following is a brief discussion about such platforms.

CPUs

CPUs are well known, flexible and scalable architectures. By exploiting the *Streaming SIMD Extension (SSE)* instruction set on modern CPUs, the running time of the analyses is decreased significantly, thereby making analyses of data intensive problems like sequence alignment feasible. Also emerging CPU technologies like multi-core combines two or more independent processors into a single package. The *Single Instruction Multiple Data-stream (SIMD)* paradigm is heavily utilized in this class of processors, making it appropriate for data parallel applications like sequence alignment. SIMD describes CPUs with multiple processing elements that perform the same operation on multiple data simultaneously. Thus, such machines exploit data level parallelism. The SSE instruction set extension in modern CPUs contains 70 new SIMD instructions. This extension greatly increases the performance when exactly the same operations are to be performed on multiple data objects, making sequence alignment a typical application.

FPGAs

FPGAs are reconfigurable data processing devices on which an algorithm is directly mapped to basic processing logic elements. To take advantage of using an FPGA, one has to implement massively parallel algorithms on this reconfigurable device. They are thus well suited for certain classes of bioinformatics applications, such as sequence alignment. Methods like the ones based on systolic arrays are used to accelerate such applications.

GPUs

Initially stimulated by the need for real time graphics in video gaming, GPUs have evolved into powerful and flexible vector processors, ideal for accelerating a variety of data parallel applications. GPUs have in the last couple of years developed themselves from a fixed function graphics processing unit into a flexible platform that can be used for *high performance computing (HPC)*. Applications like bioinformatics sequence alignment can run very efficiently on these architectures.

2. Smith-Waterman algorithm

In 1981, Smith and Waterman described a method, commonly known as the *Smith-Waterman (S-W)* algorithm (Smith & Waterman, 1981), for finding common regions of local similarity. S-W method has been used as the basis for many subsequent algorithms, and is often quoted as a benchmark when comparing different alignment techniques. When obtaining the local S-W alignment, a matrix H is constructed using the following equation.

$$H_{i,j} = \max \begin{cases} 0 \\ H_{i-1,j-1} + S_{i,j} \\ H_{i-1,j} - d \\ H_{i,j-1} - d \end{cases} \quad (1)$$

Where $S_{i,j}$ is the similarity score and d is the penalty for a mismatch. The algorithm can be implemented using the following pseudo code.

Initialization:

```
H(0,j) = 0
H(i,0) = 0
```

Matrix Fill:

```
for each i,j = 1 to M,N
    {
        H(i,j) = max(0,
                     H(i-1,j-1) + S(i,j),
                     H(i-1,j) - d,
                     H(i,j-1) - d)
    }
```

Traceback:

```
H(opt) = max(H(i,j))
traceback(H(opt))
```

The H matrix is constructed with one sequence lined up against the rows of a matrix, and another against the columns, with the first row and column initialized with a predefined value (usually zero) i.e. if the sequences are of length M and N respectively, then the matrix for the alignment algorithm will have $(M+1) \times (N+1)$ dimensions. The matrix fill stage scores each cell in the matrix. This score is based on whether the two intersecting elements of each sequence are a match, and also on the score of the cell's neighbors to the left, above, and diagonally upper left. Three separate scores are calculated based on all three neighbors, and the maximum of these three scores (or a zero if a negative value would result) is assigned to the cell. This is done for each cell in the matrix resulting in $O(MN)$ complexity for the matrix fill stage. Even though the computation for each cell usually only consists of additions, subtractions, and comparisons of integers, the algorithm would nevertheless perform very poorly if the lengths of the query sequences become large. The traceback step starts at the cell with the highest score in the matrix and ends at a cell when the similarity score drops below a certain predefined threshold. For doing this, the algorithm requires to find the maximum cell which is done by traversing the entire matrix, making the time complexity for the traceback $O(MN)$. It is also possible to keep track of the cell with the maximum score, during the matrix filling segment of the algorithm, although this will not change the overall complexity. Thus, the total time complexity of the S-W algorithm is $O(MN)$. The space complexity is also $O(MN)$.

In order to reduce the $O(MN)$ complexity of the matrix fill stage, multiple entries of the H matrix can be calculated in parallel. This is however complicated by data dependencies, whereby each $H_{i,j}$ entry depends on the values of three neighboring entries $H_{i,j-1}$, $H_{i-1,j}$ and $H_{i-1,j-1}$, with each of those entries in turn depending on the values of three neighboring entries, which effectively means that this dependency extends to every other entry in the region $H_{x,y} : x \leq i,\ y \leq j$. This implies that it is possible to simultaneously compute all the elements in each anti-diagonal, since they fall outside each other's data dependency regions. Figure 2 shows a sample H matrix for two sequences, with the bounding boxes indicating the elements that can be computed in parallel. The bottom-right cell is highlighted to show that its data dependency region is the entire remaining matrix. The dark diagonal arrow indicates the direction in which the computation progresses. At least 9 cycles are required for this computation, as there are 9 bounding boxes representing 9 anti-diagonals and a maximum of 5 cells may be computed in parallel.

The degree of parallelism is constrained to the number of elements in the anti-diagonal and the maximum number of elements that can be computed in parallel are equal to the number of elements in the longest anti-diagonal (l_d), where,

$$l_d = \min(M, N) \tag{2}$$

Theoretically, the lower bound to the number of steps required to calculate the entries of the H matrix in a parallel implementation of the S-W algorithm is equal to the number of anti-diagonals required to reach the bottom-right element, i.e. $M + N - 1$ (Liao et al., 2004).

Figure 3 shows the logic circuit to compute an element of the H matrix. The logic contains three adders, a sequence comparator circuit (*SeqCmp*) and three max operators (*MAX*). The sequence comparator compares the corresponding characters of two input sequences and outputs a match/mismatch score, depending on whether the two characters are equal or not. Each max operator finds the maximum of its two inputs. The time to compute an element is 4 cycles, assuming that the time for each cycle is equal to the latency of one add or compare operation.

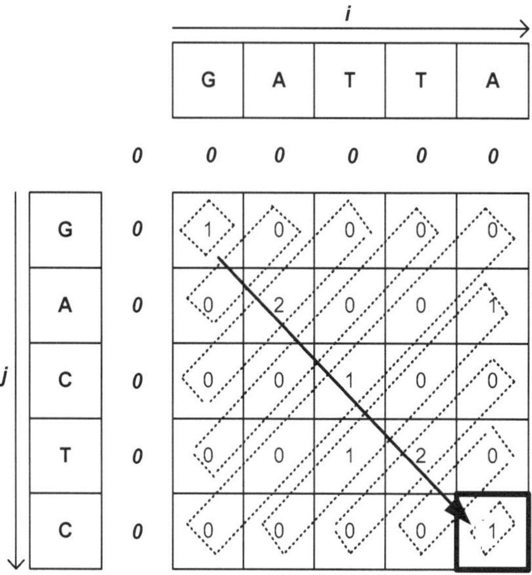

Fig. 2. Sample H matrix, where the dotted rectangles show the elements that can be computed in parallel

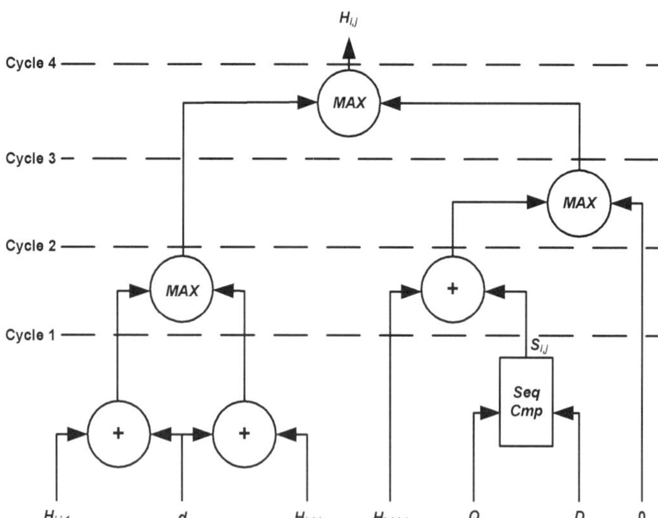

Fig. 3. Logic circuit to compute cells in the H matrix, where + is an adder, *MAX* is a max operator and *SeqCmp* is the sequence comparator that generates match/mismatch scores

3. CPU-based acceleration

In this section CPU-based acceleration of the S-W algorithm is reviewed. Furthermore, an estimation of the performance for top-end and future systems is made.

3.1 Recent implementations

The first CPU implementations used a sequential way of calculating all the matrix values. These implementations were slow and therefore hardly used. In 2007, Farrar introduced a SSE implementation for S-W (Farrar, 2007). His work used SSE2 instructions for an Intel processor and was up to six times faster than existing S-W implementations. Two years later, a *Smith-Waterman implementation on Playstation 3 (SWPS3)* was introduced (Szalkowski et al., 2009), which was based on a minor adjustment to Farrar's implementation. SWPS3 is a vectorized implementation of the Smith-Waterman local alignment algorithm optimized for both the IBM Cell/B.E. and Intel x86 architectures. A SWPS3 version optimized for multi threading has been released recently (Aldinucci et al., 2010). The SSE implementations can be viewed as being semi parallel, as they constantly calculate sixteen, eight or less values at the same time, while discarding startup and finish time. Table 2 presents the performance achieved by these implementations on various CPU platforms.

Implementation	Peak performance	Benchmark hardware	Peak performance (per thread)
(Farrar, 2007)	2.9 GCUPS	2.0 GHz, Xeon Core 2 Duo single thread	3.75 GCUPS
(Szalkowski et al., 2009)	15.7 GCUPS	2.4 GHz Core 2 Quad Q6600, 4 threads	4.08 GCUPS
(Aldinucci et al., 2010)	35 GCUPS	2.5 GHz, 2x Xeon Core Quad E5420, 8 threads	4.38 GCUPS

Table 2. Performance achieved by various S-W CPU implementations (Vermij, 2011)

3.2 Performance estimations for top-end and future CPUs

With the data from Table 2, we make an estimate of the performance on the current top-end CPUs and take a look into the future. Table 3 gives the estimated peak performances based on the SIMD register width, the number of cores, clock speed and the known speed per core. We assumed linear scaling in the number of cores as suggested in Table 2, and the given performances may therefore not be reliable. Non-ideal inter-core communication, memory bandwidth limitations and shared caches could lead to a lower peak performance. Furthermore, no distinction in performance is made between Intel and AMD processors. Hence, Table 3 must be used as an indication to where the S-W performance could go on in current and future CPUs (Vermij, 2011).

4. FPGA-based acceleration

FPGAs are programmable logic devices. To map an application on flexible FPGA platforms, a program is written in a hardware description language like VHDL. The flexibility, difficulty

System	Released	SIMD register width	Cores (threads)	Clock speed	Peak performance (estimated)
Xeon Beckton	2010	128	8 (16)	2.26 GHz	32 GCUPS
Opteron Magny-Cours	2010	128	12 (12)	2.3 GHz	48 GCUPS
Opteron Interlagos	2011	128	16 (16)	2.3 GHz	64 GCUPS

Table 3. Estimated peak performance for current top-end and future CPUs (Vermij, 2011)

of design as well as the performance of FPGA implementations fall typically somewhere between pure software running on a CPU and an *Application Specific Integrated Circuit (ASIC)*. FPGAs are widely used to accelerate applications like S-W based sequence alignment. Implementations rely on the ability to create building blocks called *processing elements (PEs)* that can update one matrix cell every clock cycle. Furthermore, multiple PEs can be linked together in a two dimensional or linear systolic arrays to process huge data in parallel. This section provides a brief description of traditional systolic arrays followed by a discussion of existing and future FPGA-based S-W implementations.

4.1 Systolic arrays

Systolic array is an arrangement of processors in an array, where data flows synchronously across the array between neighbors, usually with data flowing in a specific direction (Kung & Leiserson, 1979), (Quinton & Robert, 1991). Each processor at each step takes in data from one or more neighbors (e.g. North and West), processes it and, in the next step, outputs results to the opposite neighbors (South and East). Systolic arrays can be implemented in rectangular or *2-dimensional (2D)* and linear or *1-dimensional (1D)* fashion. Figure 4 gives a pictorial view of both implementation types.

They best suit compute-intensive applications like biological sequence alignment. The disadvantage is that being highly specialized processors type, they are difficult to implement and build.

In (Pfeiffer et al., 2005), a concept to accelerate S-W algorithm on the basis of linear systolic array is demonstrated. The reason for choosing this architecture is outlined by demonstrating the efficiency and simplicity in combination with the algorithm. Nevertheless, there are two key methodologies to speedup this massively parallel system. By turning the processing from bit-parallel to bit-serial, the actual improvement is enabled. This change is performance neutral, but in combination with the early maximum detection, a considerable speedup is possible. Another effect of this improvement is a data dependant execution time of the processing elements. Here, the second acceleration prevents idle times to exploit the hardware and speeds up the computation. This can be accomplished by a globally asynchronous timing representing a self-timed linear systolic array. The authors have provided no performance estimation due to the initial stage of their work, that is why it cannot be compared with other related work.

In (Vermij, 2011), the working of a *linear systolic array (LSA)* is explained. Such an array works like the SSE unit in a modern CPU. But instead of having a fixed length of lets say 16, the FPGA based array can have any length.

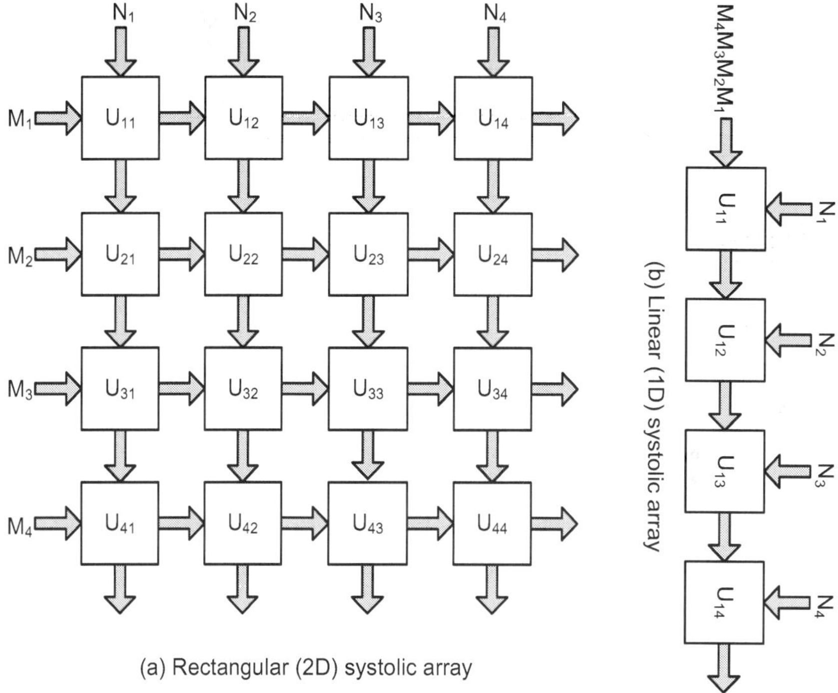

Fig. 4. Pictorial view of systolic array architectures

4.2 Existing FPGA implementations

In Section 3, we discussed some existing S-W implementations running on a CPU. A comparable analysis for FPGAs is rather hard. There are very few real, complete implementations that give usable results. Most research implementations only discuss synthetic tests, giving very optimistic numbers for implementations that are hardly used in practice. Furthermore, there is a great variety in the types of FPGAs used. Since every FPGA series has a different way of implementing circuitry, it is hard to make a fair comparison. In addition, the performance of the implementations relies heavily on the data widths used. Smaller data widths lead to smaller PEs, which lead to faster implementations. These numbers are not usually published. The first, third and fourth implementations shown in Table 4 make this clear, where the performance is given in terms of *Giga Cell Updates Per Second (GCUPS)*. Using the same FPGA device, these three implementations differ significantly in performance. The most reliable numbers are from Convey and SciEngines, as shown in the last two entries of Table 4. These implementations work the same in practice for real cases and are build for maximal performance (Vermij, 2011).

Reference	FPGA	Frequency	PEs	Performance (per FPGA)	Performance (per system)
(Puttegowda et al., 2003)	Virtex2 XC2V6000	180 MHz	7000	1260 GCUPS	—
(Yu et al., 2003)	Virtex2 XCV1000-6	—	4032	742 GCUPS	—
(Oliver et al., 2005)	Virtex2 XC2V6000	55 MHz	252	13.9 GCUPS	—
(Gok & Yilmaz, 2006)	Virtex2 XC2V6000	112 MHz	482	54 GCUPS	—
(Altera, 2007)	Stratix2 EP2S180	66.7 MHz	384	25.6 GCUPS	—
(Cray, 2010)	Virtex4	200 MHz	120	24.1 GCUPS	—
(Convey, 2010)	Virtex5 LX330	150 MHz	1152	172.8 GCUPS	691.2 GCUPS
(SciEngines, 2010)	Spartan6 LX150	—	—	47 GCUPS	6046 GCUPS

Table 4. Performance of various FPGA implementations (Vermij, 2011)

4.3 Future FPGA implementations

The performance of S-W implementations on FPGA can foremost be increased by using larger and faster FPGAs. Larger FPGAs can contain more PEs and therefore deliver higher performance in terms of GCUPS. The largest Xilinx Virtex 6 FPGA device has roughly 2.5 times more area than the largest Virtex 5 FPGA, so the peak performance of the former can be estimated at $2.5 \times 172.8 = 432$ GCUPS (using the numbers from the Convey implementation) (Vermij, 2011).

5. GPU-based acceleration

The parallelization capabilities of GPUs can be best exploited for accelerating biological sequence alignment applications. This section provides some brief background information about GPUs. Furthermore, it presents the current GPU implementations for S-W based sequence alignment.

5.1 GPU background

Compute Unified Device Architecture (CUDA) is the hardware and software architecture that enables NVIDIA GPUs (Fermi™, 2009) to execute programs written in C, C++, Fortran, OpenCL, DirectCompute and other languages. A CUDA program calls kernels that run on the GPU. A kernel executes in parallel across a set of threads, where a thread is the basic unit in the programming model that executes an instance of the kernel, and has access to registers and per thread local memory. The programmer organizes these threads in grids of thread blocks, where a thread block is a set of concurrently executing threads and has a shared memory for communication between the threads. A grid is an array of thread blocks that execute the same kernel, read inputs from and write outputs to global memory, and synchronize between interdependent kernel calls. Figure 5 gives a block diagram description of the GPU architecture.

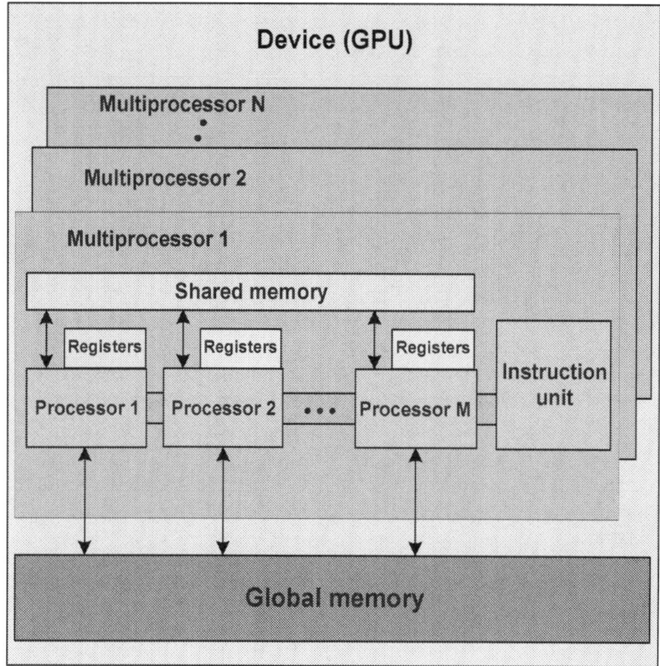

Fig. 5. Block diagram description of a GPU architecture

5.2 Current implementations

The first known implementations of S-W based sequence alignment on a GPU are presented in (Liu, Schmidt, Voss, Schroder & Muller-Wittig, 2006) and (Liu, Huang, Johnson & Vaidya, 2006). These approaches are similar and use the OpenGL graphics API to search protein databases. First the database and query sequences are copied to GPU texture memory. The score matrix is then processed in a systolic array fashion, where the data flows in anti-diagonals. The results of each anti-diagonal are again stored in texture memory, which are then used as inputs for the next pass. The implementation in (Liu, Schmidt, Voss, Schroder & Muller-Wittig, 2006) searched 99.8% of Swiss-Prot (almost 180,000 sequences) and managed to obtain a maximum speed of 650 MCUPS compared to around 75 for the compared CPU version. The implementation discussed in (Liu, Huang, Johnson & Vaidya, 2006) offers the ability to run in two modes, i.e. one with and one without traceback. The version with no traceback managed to perform at 241 MCUPS, compared to 178 with traceback and 120 for the compared CPU implementation. Both implementations were benchmarked using a Geforce GTX 7800 graphics card.

The first known CUDA implementation, 'SW-CUDA', is discussed in (Manavski & Valle, 2008). In this approach, each of the GPU's processors performs a complete alignment instead of them being used to stream through a single alignment. The advantage of this is that no communication between processing elements is required, thereby reducing memory reads and writes. This implementation managed to perform at 1.9 GCUPS on a single Geforce GTX 8800 graphics card when searching Swiss-Prot, compared to around 0.12 GCUPS for the compared

CPU implementation. Furthermore, it is shown to scale almost linearly with the amount of GPUs used by simply splitting up the database.
Various improvements have been suggested to the approach presented in (Manavski & Valle, 2008), as shown in (Akoglu & Striemer, 2009; Liu et al., 2009). In (Liu et al., 2009), for sequences of more than 3,072 amino acids an 'inter-task parallelization' method similar to the systolic array and OpenGL approaches is used as this, while slower, requires less memory. The 'CUDASW++' solution presented in (Liu et al., 2009) manages a maximum speed of about 9.5 GCUPS searching Swiss-Prot on a Geforce GTX 280 graphics card. An improved version, 'CUDASW++ 2.0' has been published recently (Liu et al., 2010). Being the fastest Smith-Waterman GPU implementation to date, 'CUDASW++ 2.0' managed 17 GCUPS on a single GTX 280 GPU, outperforming CPU-based BLAST in its benchmarks.
In (Kentie, 2010), an enhanced GPU implementation for protein sequence alignment using database and memory access optimizations is presented. Each processing element in this implementation is used to independently generate a complete alignment between a query sequence and a database sequence. This eliminates the need for inter-processor communication and results in efficient resource utilization. The GPU used for implementation (i.e. NVIDIA GTX 275) contains 240 processors, while the latest release of Swiss-Prot contains more than 500,000 protein sequences. Hence, it is possible to keep all processors well occupied while aligning query sequences with the sequences in the Swiss-Prot database. The results demonstrate that the implementation presented in (Kentie, 2010) achieves a performance of 21.4 GCUPS on an NVIDIA GTX 275 graphics card. Table 5 summarizes these GPU implementations. Besides NVIDIA, ATI/AMD (AMD, 2011) also produces graphics cards but to our knowledge no S-W implementations on such cards are available.

Implementation	Device	Database searched	Performance
(Liu, Schmidt, Voss, Schroder & Muller-Wittig, 2006)	GTX 7800	Swiss-Prot	650 MCUPS
(Liu, Huang, Johnson & Vaidya, 2006)	GTX 7800	983 protein sequences	241 MCUPS
(Manavski & Valle, 2008)	GTX 8800	Swiss-Prot	1.9 GCUPS
(Liu et al., 2009)	GTX 280	Swiss-Prot	9.5 GCUPS
(Liu et al., 2010)	GTX 280	Swiss-Prot	17 GCUPS
(Kentie, 2010)	GTX 275	Swiss-Prot	21.4 GCUPS

Table 5. Summary of the existing GPU implementations

6. Comparison of acceleration on different platforms

This section compares the performance of S-W implementations on various platforms like CPUs, FPGAs and GPUs. The comparison is based on parameters like cost, energy consumption, flexibility, scalability and future prospects. For the CPU, we consider a four way, 48 core Opteron machine. For GPUs, a fast PC with 4 high end graphics cards, and for FPGAs the fastest S-W system known, the one from SciEngines (SciEngines, 2010). The results are shown in Figure 6. Following is a discussion per metric (Vermij, 2011).

Performance/Euro

FPGAs can deliver the best amount of GCUPS per Euro, followed closely by GPUs. The gap between GPUs and CPUs can be explained by the extra money needed for a 4 way CPU system, while plugging 4 GPUs on a commodity motherboard is free. This result explains

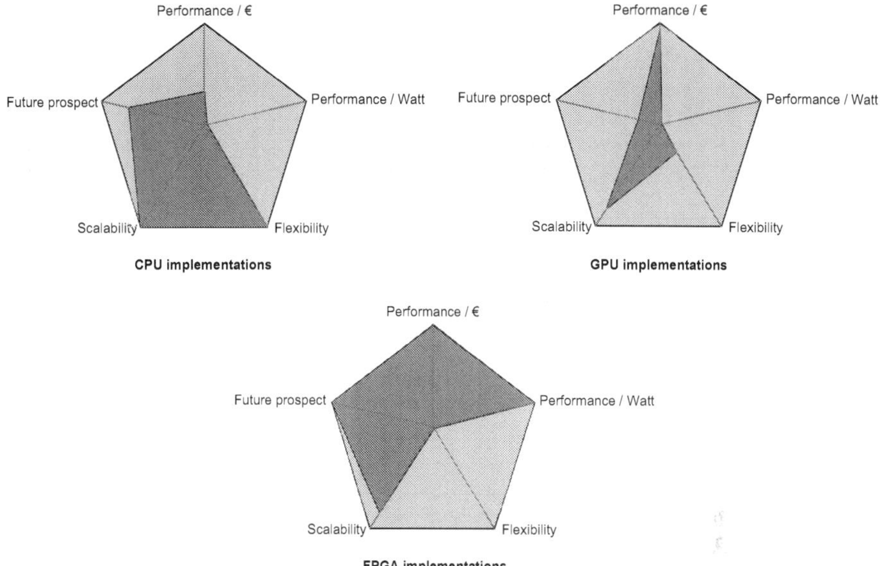

Fig. 6. Analysis of various S-W metrics for implementations on different platforms (Vermij, 2011)

why FPGAs are used for high performance computing. S-W might not be the algorithm of choice to show a major performance per Euro gain from using FPGAs. Nevertheless, it shows the trend.

Performance/Watt

It is clear that here, in contrast to the previous metric, FPGAs are the absolute winner. Full systems can deliver thousands of GCUPS for around 1000 Watts. This is another important reason for using FPGAs for sequence alignment. Note that, while not visible in the graphs, CPUs score around twice as good as GPUs.

Flexibility

This metric represents the effort needed to change a fast linear gap implementation to use affine gaps. A skilled engineer would manage to do this in a day for a CPU implementation, in a couple of days for GPU implementations, and many weeks for their FPGA counterpart.

Scalability

Here we took CPUs as baseline. Given a suitable problem, CPUs are very scalable as they can be connected together using standard networking solutions. By the same token, GPUs are also scalable, as they can take advantage of the scalability of CPUs, but will introduce some extra latency. Therefore they score a bit lower than CPUs. Depending on the used platform, FPGAs can also be made scalable.

Future prospect

In the past few years, there is a trend for CPUs to replace GPUs in high speed systems. This trend is expected to continue, thereby reducing the market share of GPUs in favour of CPUs. CPUs therefore score highly on this metric while GPUs score rather low. In the very specific HPC areas, however, where memory bandwidth requirements are low and the problem is very composable, FPGAs will likely continue to be the best choice. S-W partially lies in this category.

7. Conclusions

This chapter provided a classification, discussion and comparison of the available sequence alignment methods and their acceleration on various available hardware platforms. A detailed introduction about the S-W algorithm, its pseudo code, data dependencies in the H matrix and logic circuit to compute values of the cells in the H matrix are provided. A review of CPU-based acceleration of S-W algorithm was presented. Recent CPU implementations were discussed and compared. Further, performance estimations for top end current and future CPUs were provided. FPGA-based acceleration of S-W algorithm and a discussion about systolic arrays was given. Existing FPGA implementations were discussed and compared. Further, an insight into the future FPGA implementations was touched upon. GPU-based acceleration of S-W algorithm and GPU background were presented. Current GPU implementations were discussed and compared. Furthermore, this chapter presented a comparison of S-W accelerations on different hardware platforms. The comparison was based on the following parameters.

- Performance per euro
- Performance per unit watt
- Flexibility
- Scalability
- Future prospects

8. References

Akoglu, A. & Striemer, G. M. (2009). Scalable and highly parallel implementation of Smith-Waterman on graphics processing unit using CUDA, *Cluster Computing* Vol. 12(No. 3): 341–352.

Aldinucci, M., Meneghin, M. & Torquati, M. (2010). Efficient Smith-Waterman on multi-core with fastflow, *Proceedings of the 2010 IEEE International Symposium on Parallel and Distributed Processing*, IEEE, Pisa, Italy, pp. 195–199.

Altera (2007). Implementation of the Smith-Waterman algorithm on a reconfigurable supercomputing platform, *Altera White Paper*, Altera, pp. 1–18.

Altschul, S. F., Gish, W., Miller, W., Myers, E. W. & Lipman, D. J. (1990). A basic local alignment search tool, *Journal of Molecular Biology* Vol. 215: 403–410.

AMD (2011). ATI/AMD.
URL: *http://www.amd.com/us/products/Pages/graphics.aspx*

Buyukkur, A. B. & Najjar, W. (2008). Compiler generated systolic arrays for wavefront algorithm acceleration on FPGAs, *Proceedings of International Conference on Field Programmable Logic and Applications (FPL08)*, Heidelberg, Germany, pp. 1–4.

Convey (2010). Convey HC1.
URL: *http://www.convey.com*

Cray (2010). Cray XD1.
URL: *http://www.cray.com*

Eddy, S. R. (1998). Profile hidden morkov models, *Bioinformatics Review* Vol. 14: 755-763.

Farrar, M. (2007). Striped Smith-Waterman speeds database searches six times over other SIMD implementations, *Bioinformatics* Vol. 23(2): 156–161.

Fermi™(2009). Nvidia's next generation cuda™ compute architecture, *White paper NVIDIA corporation* .

Gibbs, A. J. & McIntyre, G. A. (1970). The diagram, a method for comparing sequences, its use with amino acid and nucleotide sequences, *European Journal of Biochemistry* Vol. 16(No. 22): 1–11.

Giegerich, R. (2000). A systematic approach to dynamic programming in bioinformatics, *Bioinformatics* Vol. 16: 665–677.

Gok, M. & Yilmaz, C. (2006). Efficient cell designs for systolic Smith-Waterman implementation, *Proceedings of International Conference on Field Programmable Logic and Applications (FPL06)*, Madrid, Spain, pp. 1–4.

Hasan, L., Al-Ars, Z. & Taouil, M. (2010). High performance and resource efficient biological sequence alignment, *Proceedings of* 32nd *Annual International Conference of the IEEE EMBS*, Buenos Aires, Argentina, pp. 1767–1770.

Hasan, L., Al-Ars, Z. & Vassiliadis, S. (2007). Hardware acceleration of sequence alignment algorithms - an overview, *Proceedings of International Conference on Design & Technology of Integrated Systems in Nanoscale Era (DTIS'07)*, Rabat, Morocco, pp. 96–101.

Kentie, M. (2010). Biological sequence alignment using graphics processing units, *M.Sc. Thesis CE-MS-2010-35*, Computer Engineering Laboratory, TU Delft, The Netherlands, 2010.

Kung, H. T. & Leiserson, C. E. (1979). Algorithms for VLSI processor arrays, *in: C. Mead, L. Conway (eds.):* Introduction to VLSI Systems; Addison-Wesley.

Liao, H. Y., Yin, M. L. & Cheng, Y. (2004). A parallel implementation of the Smith-Waterman algorithm for massive sequences searching, *Proceedings of* 26th *Annual International Conference of the IEEE EMBS*, San Francisco, CA, USA, pp. 2817–2820.

Liu, W., Schmidt, B., Voss, G., Schroder, A. & Muller-Wittig, W. (2006). Bio-sequence database scanning on a GPU, *Parallel and Distributed Processing Symposium*, IEEE, Rhodes Island, pp. 1–8.

Liu, Y., Huang, W., Johnson, J. & Vaidya, S. (2006). GPU accelerated Smith-Waterman, *Proceedings of International Conference on Computational Science, ICCS 2006*, Springer, Reading, UK, pp. 1–8.

Liu, Y., Maskell, D. & Schmidt, B. (2009). CUDASW++: Optimizing Smith-Waterman sequence database searches for CUDA-enabled graphics processing units, *BMC Research Notes* Vol. 2(No. 1:73).

Liu, Y., Schmidt, B. & Maskell, D. (2010). CUDASW++2.0: Enhanced Smith-Waterman protein database search on CUDA-enabled GPUs based on SIMT and virtualized SIMD abstractions, *BMC Research Notes* Vol. 3(No. 1:93).

Lu, J., Perrone, M., Albayraktaroglu, K. & Franklin, M. (2008). HMMER-cell: High performance protein profile searching on the Cell/B.E. processor, *Proceedings of IEEE International Symposium on Performance Analysis of Systems and Software (ISPASS-2008)*, Austin, Texas, USA, pp. 223–232.

Manavski, S. A. & Valle, G. (2008). CUDA compatible GPU cards as efficient hardware accelerators for Smith-Waterman sequence alignment, *BMC Bioinformatics* Vol. 9(No. 2:S10).

Needleman, S. & Wunsch, C. (1970). A general method applicable to the search for similarities in the amino acid sequence of two proteins, *Journal of Molecular Biology* Vol. 48(No. 3): 443–453.

Oliver, T., Schmidt, B. & Maskell, D. (2005). Hyper customized processors for bio-sequence database scanning on FPGAs, *Proceedings of FPGA'05*, ACM, Monterey, California, USA, pp. 229–237.

Pearson, W. R. & Lipman, D. J. (1985). Rapid and sensitive protein similarity searches, *Science* Vol. 227: 1435–1441.

Pfeiffer, G., Kreft, H. & Schimmler, M. (2005). Hardware enhanced biosequence alignment, *Proceedings of International Conference on METMBS*, pp. 1–7.

Puttegowda, K., Worek, W., Pappas, N., Dandapani, A. & Athanas, P. (2003). A run-time reconfigurable system for gene-sequence searching, *Proceedings of 16th International Conference on VLSI Design*, IEEE, USA, pp. 561–566.

Quinton, P. & Robert, Y. (1991). *Systolic Algorithms and Architectures*, Prentice Hall Int.

SciEngines (2010). Sciengines rivyera.
URL: *http://www.sciengines.com*

Smith, T. F. & Waterman, M. S. (1981). Identification of common molecular subsequences, *Journal of Molecular Biology* Vol. 147: 195–197.

Szalkowski, A., Ledergerber, C., Krhenbhl1, P. & Dessimoz, C. (2009). SWPS3 - A fast multi-threaded vectorized Smith-Waterman for IBM Cell/B.E. and x86/SSE2, *BMC Research Notes* Vol. 1(No. 1:107).

Thompson, J. D., Higgins, D. G. & Gibson, T. J. (1994). ClustalW: Improving the sensitivity of progressive multiple sequence alignment through sequence weighting, position-specific gap penalties and weight matrix choice, *Nucleic Acids Research* Vol. 22(No. 22): 4673–4680.

Vermij, E. (2011). Genetic sequence alignment on a supercomputing platform, *M.Sc. Thesis*, Computer Engineering Laboratory, TU Delft, The Netherlands, 2011.

Yu, C. W., Kwong, K. H., Lee, K. H. & Leong, P. H. W. (2003). A Smith-Waterman systolic cell, *International Workshop on Field Programmable Logic and Applications (FPL03)*, Springer, pp. 375–384.

Part 2

Case Studies

10

Retrieving and Categorizing Bioinformatics Publications through a MultiAgent System

Andrea Addis[1], Giuliano Armano[1], Eloisa Vargiu[1] and Andrea Manconi[2]
[1]*University of Cagliari - Dept. of Electrical and Electronic Engineering*
[2]*Institute for Biomedical Technologies, National Research Council*
Italy

1. Introduction

The huge and steady increase of available digital documents, together with the corresponding volume of daily updated contents, makes the problem of retrieving and categorizing documents and data a challenging task. To this end, automated content-based document management systems have gained a main role in the field of intelligent information access (Armano et al., 2010).
Web retrieval is highly popular and presents a technical challenge due to the heterogeneity and size of the Web, which is continuously growing (see (Huang, 2000), for a survey). In particular, it becomes more and more difficult for Web users to select contents that meet their interests, especially if contents are frequently updated (e.g., news aggregators, newspapers, scientific digital archives, RSS feeds, and blogs). Supporting users in handling the huge and widespread amount of Web information is becoming a primary issue.
Among other kinds of information, let us concentrate on publications and scientific literature, largely available on the Web for any topic. As for bioinformatics, it can be observed that the steady work of researchers, in conjunction with the advances in technology (e.g., high-throughput technologies), has arisen in a growing amount of known sequences. The information related with these sequences is daily stored in the form of scientific articles. Digital archives like BMC Bioinformatics[1], PubMed Central[2] and other online journals and resources are more and more searched for by bioinformaticians and biologists, with the goal of downloading articles relevant to their scientific interests. However, for researchers, it is still very hard to find out which publications are in fact of interest without an explicit classification of the relevant topics they describe.
Traditional filtering techniques based on keyword search are often inadequate to express what the user is really searching for. This principle is valid also in the field of scientific publications retrieval, where researchers could obtain a great benefit from the adoption of automated tools able to search for publications related with their interests.
To be effective in the task of selecting and suggesting to a user only relevant publications, an automated system should at least be able (i) to extract the required information and (ii) to encode and process it according to a given set of categories. Personalization could also be provided according to user needs and preferences.

---

[1] http://www.biomedcentral.com/bmcbioinformatics/
[2] http://www.pubmedcentral.gov/

In this chapter, we present PUB.MAS, a multiagent system able to retrieve and categorize bioinformatics publications from selected Web sources. The chapter extends and revises our previous work (Armano et al., 2007). The main extensions consist of a more detailed presentation of the information extraction task, a deep explanation of the adopted hierarchical text categorization technique, and the description of the prototype that has been implemented. Built upon X.MAS (Addis et al., 2008), a generic multiagent architecture aimed at retrieving, filtering and reorganizing information according to user interests, PUB.MAS is able to: (i) extract information from online digital archives; (ii) categorize publications according to a given taxonomy; and (iii) process user's feedback. As for information extraction, PUB.MAS provides specific wrappers able to extract publications from RSS-based Web pages and from Web Services. As for categorization, PUB.MAS performs Progressive Filtering (PF), the effective hierarchical text categorization technique described in (Addis et al., 2010). In its simplest setting, PF decomposes a given rooted taxonomy into pipelines, one for each existing path between the root and each node of the taxonomy, so that each pipeline can be tuned in isolation. To this end, a threshold selection algorithm has been devised, aimed at finding a sub-optimal combination of thresholds for each pipeline. PUB.MAS provides also suitable strategies to allow users to express what they are really interested in and to personalize search results accordingly. Moreover, PUB.MAS provides a straightforward approach to user feedback with the goal of improving the performance of the system depending on user needs and preferences.

The prototype allows users to set the sources from which publications will be extracted and the topics s/he is interested in. As for the digital archives, the user can choose between BMC Bioinformatics and PubMed Central. As for the topics of interest, the user can select one or more categories from the adopted taxonomy, which is taken from the TAMBIS ontology (Baker et al., 1999).

The overall task begins with agents able to handle the selected digital archives, which extract the candidate publications. Then, all agents that embody a classifier trained on the selected topics are involved to perform text categorization. Finally, the system supplies the user with the selected publications through suitable interface agents.

The chapter is organized as follows. First, we give a brief survey of relevant related work on: (i) scientific publication retrieval; (ii) hierarchical text categorization; and (iii) multiagent systems in information retrieval. Subsequently, we concentrate on the task of retrieving and categorizing bioinformatics publications. Then, PUB.MAS is illustrated together with its performances and the implemented prototype. Conclusions end the chapter.

2. Background

Information Retrieval (IR) is the task of representing, storing, organizing, and accessing information items. IR has changed considerably in recent years with the expansion of the World Wide Web and the advent of modern and inexpensive graphical user interfaces and mass storage (Baeza-Yates & Ribeiro-Neto, 1999). The most relevant IR issues that help to clarify the contextual setting of this chapter are: (i) the work done on scientific publication retrieval, (ii) the work done on Hierarchical Text Categorization (HTC), and (iii) the work done on multiagent systems (MAS) for information retrieval.

2.1 Scientific publication retrieval

In the academic area, online search engines are used to find out scientific resources, as journals and conference proceedings. However, finding and selecting appropriate information on the

Web is still difficult. To simplify this process, several frameworks and systems have been developed to retrieve scientific publications from the Web.
Bollacker et al. (2000) developed CiteSeer[3], the well-known automatic generator of digital libraries of scientific literature. Being aimed at eliminating most of the manual effort of finding useful publications on the Web, CiteSeer uses sophisticated acquisition, parsing, and presentation methods. In particular, CiteSeer uses a three-stage process: database creation and feature extraction; personalized filtering of new publications; and personalized adaptation and discovery of interesting research and trends. These functions are interdependent: information filtering affects what is discovered, whereas useful discoveries tune the information filtering. In (McNee et al., 2002), the authors study how to recommend research papers using the citation between papers to create the user-item matrix. In particular, they test the ability of collaborative filtering to recommend citations that could be additional references for a target research paper. Janssen & Popat (2003) developed UpLib, a personal digital library system that consists of a full-text indexed repository accessed through an active agent via a Web interface. UpLib is mainly concerned with the task of collecting personal collections comprising tens of thousands of documents. In (Mahdavi et al., 2009), the authors start from the assumption that trend detection in scientific publication retrieval systems helps scholars to find relevant, new and popular special areas by visualizing the trend of the input topic. To this end, they developed a semi-automatic system based on a semantic approach.
As for the specific task of retrieving information in the field of bioinformatics, a lot of work has been done –some of it being recalled hereafter. Tanabe et al. (1999) developed MedMiner, an Internet-based hypertext program able to filter and organize large amounts of textual and structured information returned from public search engines –like GeneCards and PubMed. Craven & Kumlien (1999) applied machine learning techniques to automatically map information from text sources into structured representations, such as knowledge bases. Friedman et al. (2001) propose GENIES, a system devised to extract information about cellular pathways from the biological literature in accordance with a given domain knowledge. Ramu (2003) developed a Web Server for SIR (Ramu, 2001), a simple indexing and retrieval system that combines sequence motif search with keyword search. The Web Server, called SIRW, is a generic tool used by the bioinformatics community for searching and analyzing biological sequences of interest. Rocco & Critchlow (2003) propose a system aimed at finding classes of bioinformatics data sources and integrating them behind a unified interface. The main goal of the system is to eliminate the human effort required to maintain a repository of information sources. Kiritchenko et al. (2004) propose a system aimed at retrieving Medline articles that mention genes. After being retrieved, articles are categorized according to the Gene Ontology (GO) codes. Delfs et al. (2004) developed and implemented GoPubMed, a system that allows to submit keywords to PubMed, extracts GO terms from the retrieved abstracts, and supplies the user with the relevant ontology for browsing. Corney et al. (2004) propose BioRAT (Biological Research Assistant for Text mining), a new information extraction tool specifically tailored for biomedical tasks. Able to access and analyze both abstracts and full-length papers, it incorporates a domain specific document search ability.

2.2 Hierarchical text categorization

In recent years several researchers have investigated the use of hierarchies for text categorization.
Until the mid-1990s researchers mostly ignored the hierarchical structure of categories that occur in several domains. In 1997, Koller & Sahami (1997) carry out the first proper study

[3] http://citeseer.ist.psu.edu/

on HTC on the Reuters-22173 collection. Documents are classified according to the given hierarchy by filtering them through the single best-matching first-level class and then sending them to the appropriate second level. This approach shows that hierarchical models perform well when a small number of features per class is used, as no advantages were found using the hierarchical model for large numbers of features. McCallum et al. (1998) propose a method based on naïve Bayes. The authors compare two techniques: (i) exploring all possible paths in the given hierarchy and (ii) greedily selecting at most two branches according to their probability, as done in (Koller & Sahami, 1997). Results show that the latter is more error prone while computationally more efficient. Mladenić & Grobelnik (1998) use the hierarchical structure to decompose a problem into a set of subproblems, corresponding to categories (i.e., the nodes of the hierarchy). For each subproblem, a naïve Bayes classifier is generated, considering examples belonging to the given category, including all examples classified in its subtrees. The classification applies to all nodes in parallel; a document is passed down to a category only if the posterior probability for that category is higher than a user-defined threshold. D'Alessio et al. (2000) propose a system in which, for a given category, the classification is based on a weighted sum of feature occurrences that should be greater than the category threshold. Both single and multiple classifications are possible for each document to be tested. The classification of a document proceeds top-down possibly through multiple paths. An innovative contribution of this work is the possibility of restructuring a given hierarchy or building a new one from scratch. Dumais & Chen (2000) use the hierarchical structure for two purposes: (i) training several SVMs, one for each intermediate node and (ii) classifying documents by combining scores from SVMs at different levels. The sets of positive and negative examples are built considering documents that belong to categories at the same level, and different feature sets are built, one for each category. Several combination rules have also been assessed. In the work of Ruiz & Srinivasan (2002), a variant of the Hierarchical Mixture of Experts model is used. A hierarchical classifier combining several neural networks is also proposed in (Weigend et al., 1999). Gaussier et al. (2002) propose a hierarchical generative model for textual data, i.e., a model for hierarchical clustering and categorization of co-occurrence data, focused on documents organization. In (Rousu et al., 2005), a kernel-based approach for hierarchical text classification in a multi-label context is presented. The work demonstrates that the use of the dependency structure of microlabels (i.e., unions of partial paths in the tree) in a Markovian Network framework leads to improved prediction accuracy on deep hierarchies. Optimization is made feasible by utilizing decomposition of the original problem and making incremental conditional gradient search in the subproblems. Ceci & Malerba (2007) present a comprehensive study on hierarchical classification of Web documents. They extend a previous work (Ceci & Malerba, 2003) considering hierarchical feature selection mechanisms, a naïve Bayes algorithm aimed at avoiding problems related to different document lengths, the validation of their framework for a probabilistic SVM-based classifier, and (iv) an automated threshold selection algorithm. More recently, in (Esuli et al., 2008), the authors propose a multi-label hierarchical text categorization algorithm consisting of a hierarchical variant of ADABOOST.MH, a well-known member of the family of "boosting" learning algorithms. Bennett & Nguyen (2009) study the problem of the error propagation under the assumption that the higher the node in the hierarchy is the worse is the mistake, as well as the problem of dealing with increasingly complex decision surfaces. Brank et al. (2010) deal with the problem of classifying textual documents into a topical hierarchy of categories. They construct a coding matrix gradually, one column at a time, each new column being defined in a way that the corresponding binary classifier attempts to correct the most common mistakes of the current ensemble of binary classifiers. The goal is to achieve good performance while keeping reasonably low the number of binary classifiers.

2.3 MultiAgent Systems in information retrieval

Autonomous agents and MAS have been successfully applied to a number of problems and have been largely used in different application domains (Wooldridge & Jennings, 1995).
As for MAS in IR, in the literature, several centralized agent-based architectures aimed at performing IR tasks have been proposed. Among others, let us recall NewT (Sheth & Maes, 1993), Letizia (Lieberman, 1995), WebWatcher (Armstrong et al., 1995), and SoftBots (Etzioni & Weld, 1995). NewT is composed by a society of information-filtering interface agents, which learn user preferences and act on her/his behalf. These information agents use a keyword-based filtering algorithm, whereas adaptive techniques are relevance feedback and genetic algorithms. Letizia is an intelligent user-interface agent able to assist a user while browsing the Web. The search for information is performed through a cooperative venture between the user and the software agent: both browse the same search space of linked Web documents, looking for interesting ones. WebWatcher is an information search agent that follows Web hyperlinks according to user interests, returning a list of links deemed interesting. In contrast to systems for assisted browsing or IR, SoftBots accept high-level user goals and dynamically synthesize the appropriate sequence of Internet commands according to a suitable ad-hoc language.
Despite the fact that a centralized approach could have some advantages, in IR tasks it may encompass several problems, in particular how to scale up the architectures to large numbers of users, how to provide high availability in case of constant demand of the involved services, and how to provide high trustability in case of sensitive information, such as personal data. To overcome the above drawbacks, suitable MAS devoted to perform IR tasks have been proposed. In particular, Sycara et al. (2001) propose Retsina, a MAS infrastructure applied in many domains. Retsina is an open MAS infrastructure that supports communities of heterogeneous agents. Three types of agents have been defined: (i) *interface agents*, able to display the information to the users; (ii) *task agents*, able to assist the user in the process of handling her/his information; and (iii) *information agents*, able to gather relevant information from selected sources.
Among other MAS, let us recall IR-agents (Jirapanthong & Sunetnanta, 2000), CEMAS (Bleyer, 1998) and the cooperative multiagent system for Web IR proposed in (Shaban et al., 2004). IR-agents implement an XML-based multiagent model for IR. The corresponding framework is composed of three kinds of agents: (i) *managing agents*, aimed at extracting the semantics of information and at performing the actual tasks imposed by coordinator agents, (ii) *interface agents*, devised to interact with the users, and (iii) *search agents*, aimed at discovering relevant information on the Web. IR-agents do not take into account personalization, while providing information in a structured form without the adoption of specific classification mechanisms. In CEMAS, Concept Exchanging MultiAgent System, the basic idea is to provide specialized agents for exchanging concepts and links, representing the user, searching for new relevant documents matching existing concepts, and supporting agent coordination. Although CEMAS provides personalization and classification mechanisms based on a semantic approach, and it is mainly aimed at supporting scientists while looking for comprehensive information about their research interests. Finally, in (Shaban et al., 2004) the underlying idea is to adopt intelligent agents that mimic everyday-life activities of information seekers. To this end, agents are also able to profile the user in order to anticipate and achieve her/his preferred goals. Although interesting, the approach is mainly focused on cooperation among agents rather than on IR issues.

3. The proposed approach

A system for information retrieval must take into account several issues, the most relevant being:

1. how to deal with different information sources and to integrate new information sources without re-writing significant parts of it;
2. how to suitably encode data in order to put into evidence the informative content useful to discriminate among categories;
3. how to control the imbalance between relevant and irrelevant articles (the latter being usually much more numerous than the former);
4. how to allow the user to specify her/his preferences;
5. how to exploit the user's feedback to improve the overall performance of the system.

The above issues are typically strongly interdependent in state-of-the-art systems. To better concentrate on these aspects separately, we adopted a layered multiagent architecture, able to promote the decoupling among all aspects deemed relevant.
To perform the task of retrieving scientific publications, the actual system –sketched in Figure 1– involves three main activities: extracting the required information from selected online sources, categorizing it according to a given taxonomy while taking into account also users preferences, and providing suitable feedback mechanisms.

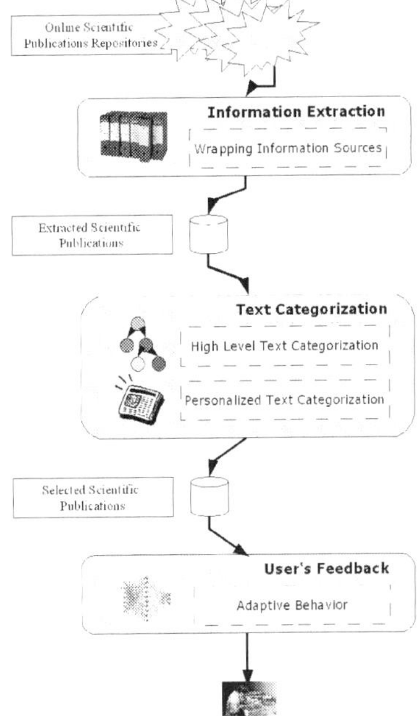

Fig. 1. PUB.MAS: the multiagent system devised for classifying bioinformatics publications.

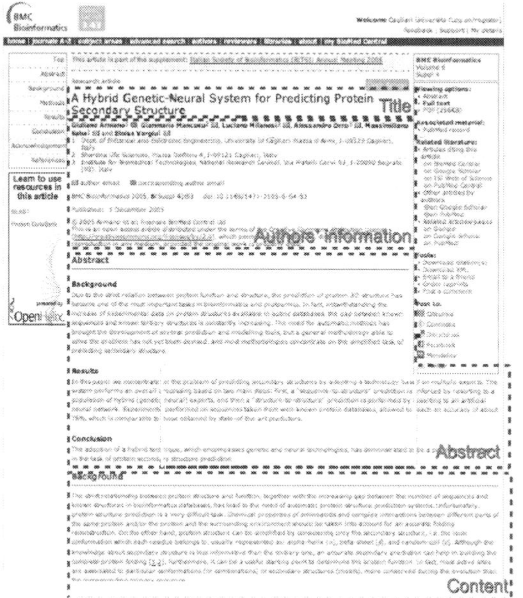

Fig. 2. The structure of a BMC Bioinformatics page.

3.1 Information extraction

This phase is devoted to deal with the huge amount of information provided by information sources. To this end suitable wrappers have been implemented, able to handle the structure of a document by saving the information about the corresponding metadata. In general, given a Web source, a specific wrapper must be implemented, able to map each Web page, designed according to the constraints imposed by the Web source, to a suitable description, which contains relevant data in a structured form –such as title, text content, and references. To make this point clearer, let us consider the structure of the BMC Bioinformatics page of the paper "A Hybrid Genetic-Neural System for Predicting Protein Secondary Structure" (Armano et al., 2005) reported in Figure 2. In this case, it is quite easy to implement the mapping function, since, for each description field, a corresponding tag exists, making it very simple to process the pages.

A suitable encoding of the text content has also been enforced during this phase: all non-informative words such as prepositions, conjunctions, pronouns and very common verbs are deleted using a stop-word list; after that, a standard stemming algorithm (Porter, 1980) removes the most common morphological and inflexional suffixes. The subsequent step requires the adoption of suitable domain knowledge. For each category of the underlying taxonomy, feature selection (based on the information-gain heuristics) has been adopted to reduce the dimensionality of the feature space.

3.2 Text categorization

Scientific publications are classified according to a high-level taxonomy, which is independent from the specific user. To this end, classifiers are combined according to the links that hold within the taxonomy, giving rise to "vertical" and "horizontal" combinations of classifiers.

3.2.1 Vertical combination

The Approach

Vertical combination is currently performed by resorting to Progressively Filtering (PF), a simple categorization technique framed within the local classifier per node approach, which admits only binary decisions. In PF, each classifier is entrusted with deciding whether the input in hand can be forwarded or not to its children. The first proposals in which sequential boolean decisions are applied in combination with local classifiers per node can be found in (D'Alessio et al., 2000), (Dumais & Chen, 2000), and (Sun & Lim, 2001). In Wu et al. (2005), the idea of mirroring the taxonomy structure through binary classifiers is clearly highlighted; the authors call this technique "binarized structured label learning".

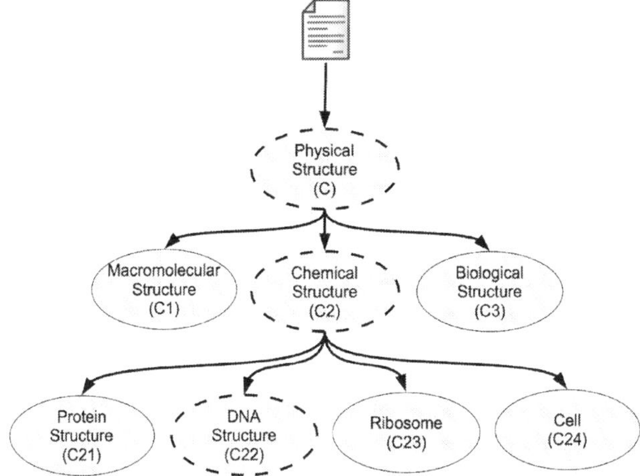

Fig. 3. An example of PF (highlighted with bold-dashed lines).

In PF, given a taxonomy, where each node represents a classifier entrusted with recognizing all corresponding positive inputs (i.e., interesting documents), each input traverses the taxonomy as a "token", starting from the root. If the current classifier recognizes the token as positive, it is passed on to all its children (if any), and so on. A typical result consists of activating one or more branches within the taxonomy, in which the corresponding classifiers have accepted the token. Figure 3 gives an example of how PF works. A theoretical study of the approach is beyond the scope of this chapter, the interested reader could refer to (Armano, 2009) for further details.

A simple way to implement PF consists of unfolding the given taxonomy into pipelines of classifiers, as depicted in Figure 4. Each node of the pipeline is a binary classifier able to recognize whether or not an input belongs to the corresponding class (i.e., to the corresponding node of the taxonomy). Partitioning the taxonomy in pipelines gives rise to a set of new classifiers, each represented by a pipeline.

Finally, let us note that the implementation of PF described in this chapter performs a sort of "flattening" though *preserving* the information about the hierarchical relationships embedded in a pipeline (Addis et al., 2010). For instance, the pipeline $\langle C, C2, C21 \rangle$ actually represents the classifier C21, although the information about the existing subsumption relationships (i.e., $C21 \leq C2 \leq C$) is preserved.

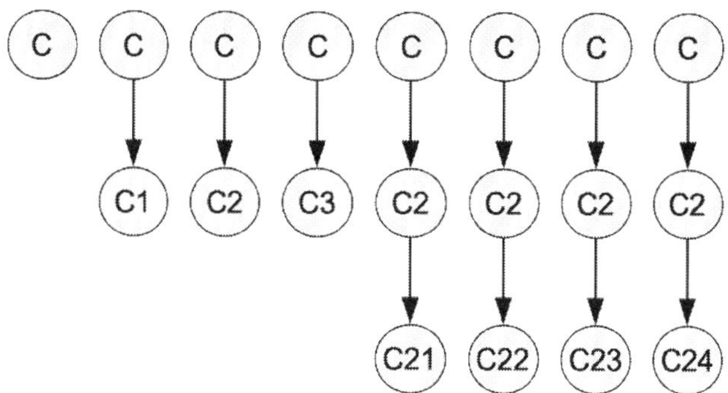

Fig. 4. The pipelines corresponding to the taxonomy in Figure 3.

The Threshold Selection Algorithm

As we know from classical text categorization, given a set of documents D and a set of labels C, a function $CSV_i : D \rightarrow [0,1]$ exists for each $c_i \in C$. We assume that the behavior of c_i is controlled by a threshold θ_i, responsible for relaxing or restricting the acceptance rate of the corresponding classifier. Given $d \in D$, $CSV_i(d) \geq \theta_i$ permits to categorize d under c_i, whereas $CSV_i(d) < \theta_i$ is interpreted as a decision not to categorize d under c_i.

In PF, let us still assume that CSV_i exists for each $c_i \in C$, with the same semantics adopted in the classical case. Considering a pipeline π, composed of n classifiers, the acceptance policy strictly depends on the vector $\theta_\pi = \langle \theta_1, \theta_2, \cdots, \theta_n \rangle$ that embodies the thresholds of all classifiers in π. In order to categorize d under π, the following constraint must be satisfied: $\forall k = 1 \ldots n$, $CSV_i(d) \geq \theta_k$; otherwise, d is not categorized under c_i.

A further simplification of the problem consists of allowing a classifier to have different behaviors, depending on which pipeline it is embedded in. Each pipeline can be considered in isolation from the others. For instance, given $\pi_1 = \langle C, C2, C21 \rangle$ and $\pi_2 = \langle C, C2, C22 \rangle$, the classifier C is not compelled to have the same threshold in π_1 and in π_2 (the same holds for C2).

Given a utility function[4], we are interested in finding an effective and computationally "light" way to reach a sub-optimum in the task of determining the best vector of thresholds. Unfortunately, finding the best acceptance thresholds is a difficult task. Exhaustively trying each possible combination of thresholds (brute-force approach) is unfeasible, the number of thresholds being virtually infinite. However, the brute-force approach can be approximated by defining a granularity step that requires to check only a finite number of points in the range $[0,1]$, in which the thresholds are permitted to vary with step δ. Although potentially useful, this "relaxed" brute force algorithm for calibrating thresholds (RBF for short) is still too heavy from a computational point of view. On the contrary, the threshold selection algorithm described in this chapter is characterized by low time complexity while maintaining the capability of finding near-optimum solutions.

Bearing in mind that the lower the threshold the less restrictive is the classifier, we adopt the greedy bottom-up algorithm for selecting decision threshold that relies on two functions described in (Addis et al., 2011):

[4] Different utility functions (e.g., precision, recall, F_β, user-defined) can be adopted, depending on the constraints imposed by the underlying scenario.

- *Repair* ($\mathcal{R}$), which operates on a classifier C by increasing or decreasing its threshold –i.e., $\mathcal{R}(up, C)$ and $\mathcal{R}(down, C)$, respectively– until the selected utility function reaches and maintains a local maximum.
- *Calibrate* ($\mathcal{C}$), which operates going downwards from the given classifier to its offspring. It is intrinsically recursive and, at each step, calls $\mathcal{R}$ to calibrate the current classifier.

Given a pipeline $\pi = \langle C_1, C_2, \ldots, C_L \rangle$, *TSA* is defined as follows (all thresholds are initially set to 0):

$$TSA(\pi) := for\ \ k = L\ \ downto\ \ 1\ \ do\ \ \mathcal{C}(up, C_k) \tag{1}$$

which asserts that $\mathcal{C}$ is applied to each node of the pipeline, starting from the leaf ($k = L$).
The *Calibrate* function is defined as follows:

$$\begin{aligned} \mathcal{C}(up, C_k) &:= \mathcal{R}(up, C_k), \quad k = L \\ \mathcal{C}(up, C_k) &:= \mathcal{R}(up, C_k); \mathcal{C}(down, C_{k+1}), \quad k < L \end{aligned} \tag{2}$$

and

$$\begin{aligned} \mathcal{C}(down, C_k) &:= \mathcal{R}(down, C_k), \quad k = L \\ \mathcal{C}(down, C_k) &:= \mathcal{R}(down, C_k); \mathcal{C}(up, C_{k+1}), k < L \end{aligned} \tag{3}$$

where the ";" denotes a sequence operator, meaning that in "*a;b*" action *a* is performed *before* action *b*. The reason why the direction of threshold optimization changes at each call of Calibrate (and hence of Repair) lies in the fact that increasing the threshold θ_{k-1} is expected to forward less false positives to C_k, which allows to decrease θ_k. Conversely, decreasing the threshold θ_{k-1} is expected to forward more false positives to C_k, which must react by increasing θ_k.
It is worth pointing out that, as also noted in (Lewis, 1995), the sub-optimal combination of thresholds depends on the adopted dataset, hence it needs to be recalculated for each dataset.

3.2.2 Horizontal combination

To express what the user is really interested in, we implemented suitable horizontal composition strategies by using extended boolean models (Lee, 1994). In fact, a user is typically not directly concerned with topics that coincide with classes of the given taxonomy. Rather, a set of arguments of interest can be obtained by composing such generic topics with suitable logical operators (i.e., *and*, *or*, and *not*). For instance, a user might be interested in being kept informed about all articles that involve both "cell" *and* "nucleus". This compound topic can be dealt with by composing the *cell* and the *nucleus* classifiers. To address this issue, we adopted a soft boolean perspective, in which the combination is evaluated using *P*-norms (Golub & Loan, 1996).

3.3 Users' feedback

So far, a simple solution based on the *k*-NN technology has been implemented and experimented to deal with the problem of supporting the user's feedback. When a irrelevant article is evidenced by the user, it is immediately embedded in the training set of the *k*-NN classifier that implements the feedback. A check performed on this training set after inserting the negative example allows to trigger a procedure entrusted with keeping the number of negative and positive examples balanced. In particular, when the ratio between negative and positive examples exceeds a given threshold (by default set to 1.1), some examples are randomly extracted from the set of "true" positive examples and embedded in the above training set.

4. PUB.MAS

To retrieve and categorize scientific publications, we customized X.MAS (Addis et al., 2008), a generic multiagent architecture built upon JADE (Bellifemine et al., 2007) devised to facilitate the implementation of information retrieval and information filtering applications. The motivation for adopting a MAS lies in the fact that a centralized classification system might be quickly overwhelmed by a large and dynamic document stream, such as daily-updated online publications. Furthermore, the Web is intrinsically a pervasive system and offers the opportunity to take advantage of distributed computing paradigms and spread knowledge resources.

4.1 The system

PUB.MAS, is organized in the three "virtual" layers depicted in Figure 1, by customizing X.MAS as follows:

- *Information Level.* Agents at this level are devoted to deal with the selected information sources. A wrapper able to deal with the RSS (Really Simple Syndication) format has been implemented, aimed at extracting publications from BMC Bioinformatics. It is worth pointing out that the RSS format allows to easily process any given page, since a corresponding RSS tag exists for each relevant item. Furthermore, the growing amount of Web Services providing scientific publications requires the implementation of wrappers explicitly devoted to extract information from them. In order to invoke Web Services from our multiagent system, required to access the PubMed Central Web Service, we implemented an ad-hoc wrapper by adopting WSIG (Greenwood & Calisti, 2004).
- *Filter Level.* Filter agents are devoted to select information deemed relevant to the users, and to cooperate to prevent information from being overloaded and redundant. A suitable encoding of the text content has been enforced at this level to facilitate the work of agents belonging to the task level. As already pointed out, all non-informative words such as prepositions, conjunctions, pronouns and very common verbs are removed using a stop-word list. After that, a standard stemming algorithm removes the most common morphological and inflexional suffixes. Then, for each category, feature selection, based on the information-gain heuristics, has been adopted to reduce the dimensionality of the feature space.
- *Task Level.* Task agents are devoted to identify relevant scientific publications, depending on user interests. Agents belonging to this architectural level are devoted to perform two kinds of actions: to classify any given input in accordance with the selected set of categories, and to decide whether it may be of interest to the user or not. Each task agent has been trained by resorting to a state-of-the-art technique, i.e. k-NN, in its *"weighted"* variant (Cost & Salzberg, 1993). The choice of adopting weighted k-NN stems from the fact that it does not require specific training and is very robust with respect to the impact of noisy data. Furthermore, the adoption of weighted k-NN is related with the choice of P-norms for implementing the "and" operation, as P-norms combination rules require values in [0,1].
- *Interface Level.* Interface agents are devoted to perform the feedback that originates from the users –which can be exploited to improve the overall ability of discriminating relevant from irrelevant inputs. To this end, PUB.MAS uses the k-NN solution previously described.

4.2 The prototype

Since our primary interest consists of classifying scientific articles for bioinformaticians or biologists, a high-level *is-a* taxonomy has been extracted from the TAMBIS ontology (Baker et al., 1999). A fragment of the taxonomy is depicted in Figure 5.

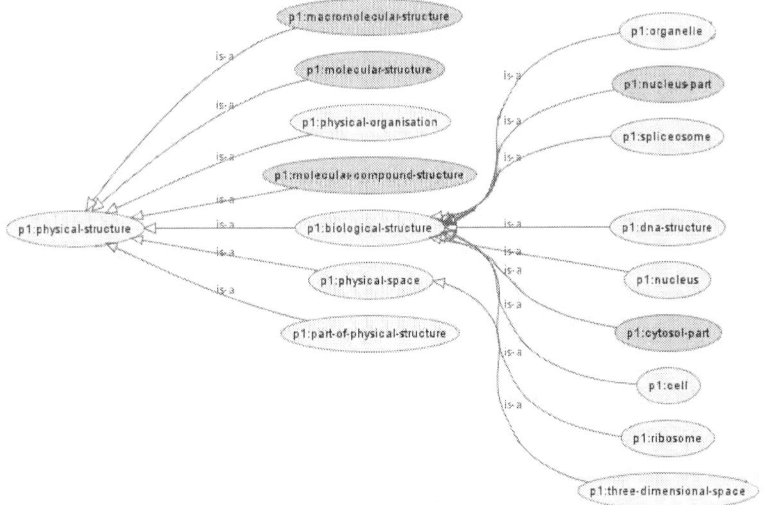

Fig. 5. A fragment of the adopted taxonomy

Through a suitable user interface (see Figure 6), the user can set the sources from which publications will be extracted and the topics s/he is interested in. As for the digital archives, the user can choose between BMC Bioinformatics and PubMed Central. As for the topics of interest, the user can select one or more categories in accordance with the adopted taxonomy and compose them in order to build her/his personal document collection. For instance, in the example reported in Figure 6, the user queries the system on *(cell AND nucleus) OR (organelle)*. The search for relevant documents is activated by clicking on the *Start Search* button. First, information agents devoted to handle Web sources extract the documents. Then, all agents that embody a classifier trained on the selected topics are involved to perform text categorization. Finally, the system supplies the user with the selected articles through suitable interface agents (see Figure 7).

4.3 Experimental results

To assess the system, different kinds of tests have been performed, each aimed at highlighting (and getting information about) a specific issue. First, we estimated the *normalized* confusion matrix for each classifier belonging to the highest level of the taxonomy. Afterwards, we tested the importance of defining user's interests by resorting to a relaxation of the logical operators. Finally, we assessed the solution devised for implementing user's feedback, based on the *k*-NN technique.

Tests have been performed using selected publications extracted from the BMC Bioinformatics site and from the PubMed Central digital archive. Publications have been classified by an expert of the domain according to the proposed taxonomy. For each item of the taxonomy, a

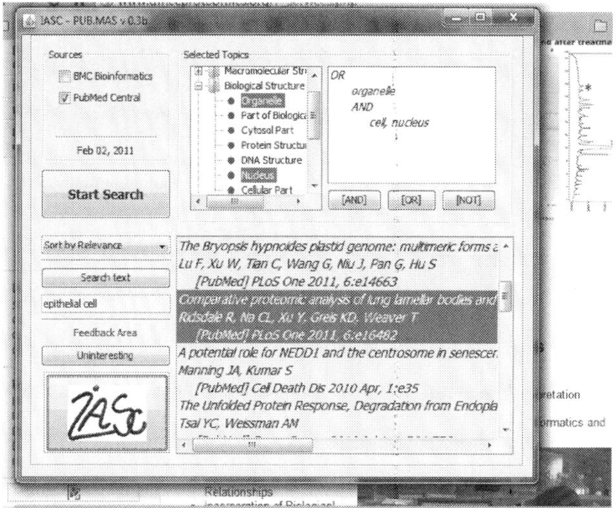

Fig. 6. The user interface

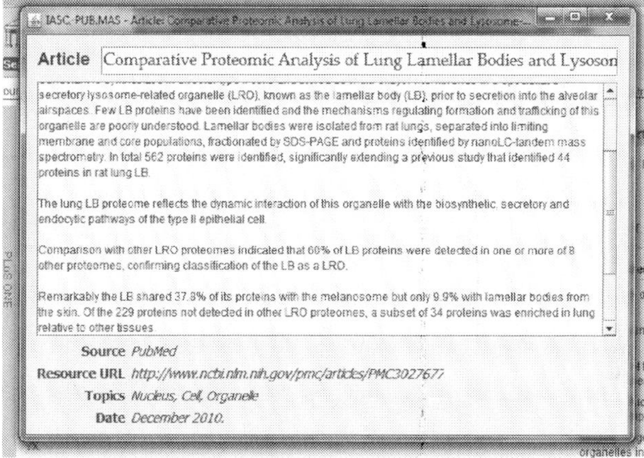

Fig. 7. A publication retrieved by PUB.MAS

set of about 100-150 articles has been selected to train the corresponding k-NN classifier, and 300-400 articles have been used to test it.

As for the estimation of the normalized confusion matrices (one for each classifier), we fed classifiers with balanced sets of positive and negative examples. Given a classifier, we performed several runs to obtain an averaged confusion matrix. Normalization has been imposed row by row on the averaged confusion matrix. In particular, true negatives and false positives are divided by the number of negative examples; conversely, the number of false negatives and true positives are divided by the number of positive examples. In so doing,

we obtain an estimation of the conditional probability $P(\hat{c}(x)|c(x))$, where x is the input to be classified, $\hat{c}(x)$ is the output of the classifier, and $c(x)$ is the category of x
To assess the impact of exploiting a taxonomy over precision and recall, we selected some relevant samples of three classifiers in pipeline. They have been tested by imposing randomly-selected relevant and irrelevant inputs, their ratio being set to $1/100$, to better approximate the expected behavior of the pipelines in real-world conditions. Averaging the results obtained in all experiments in which a pipeline of three classifiers was involved, PF allowed to reach an accuracy of 95%, a precision of 80%, and a recall of 44%.
Figure 8 and 9 report experimental results focused on average precision and recall, respectively. Experimental results are compared with those derived theoretically. Let us note that results show that the filtering effect of a pipeline is not negligible. In particular, in presence of imbalanced inputs, a pipeline of three classifiers is able to counteract a lack of equilibrium of about 10 irrelevant articles vs. one relevant article. Since, at least in principle, the filtering activity goes with the power of the number of classifiers involved in the pipeline, it is easy to verify that PF could also counteract a ratio between irrelevant and relevant articles with an order of magnitude of hundreds or thousands, provided that the number of levels of the underlying taxonomy is deep enough (at least 3-4).

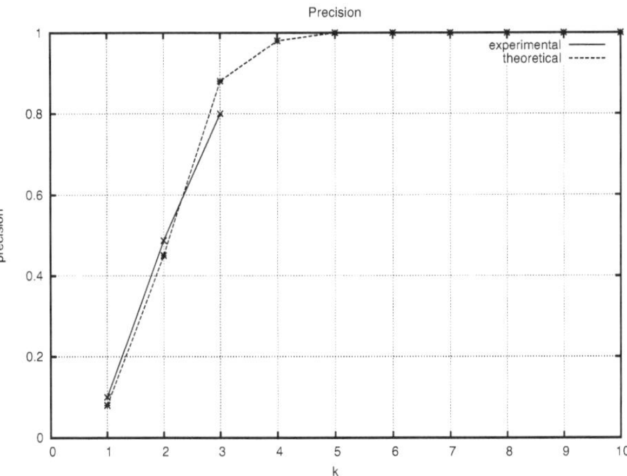

Fig. 8. Average precision using three classifiers in pipeline

To test the validity of the horizontal composition mechanisms, the system has been tested on 20 selected users. The behavior of the system has been monitored over a two-week period by conducting regular interviews with each user to estimate her/his satisfaction and the correctness of the process. All users stated their satisfaction with the system after just one or two days.
As for the user's feedback, we obtained an improvement of about 0.3% on the precision of the system by populating a k-NN classifier with examples selected as relevant by the system, taking care of balancing true positives with false positives.

5. Conclusions

It becomes more and more difficult for Web users to search for, find, and select contents according to their preferences. The same happens when researchers surf the Web searching

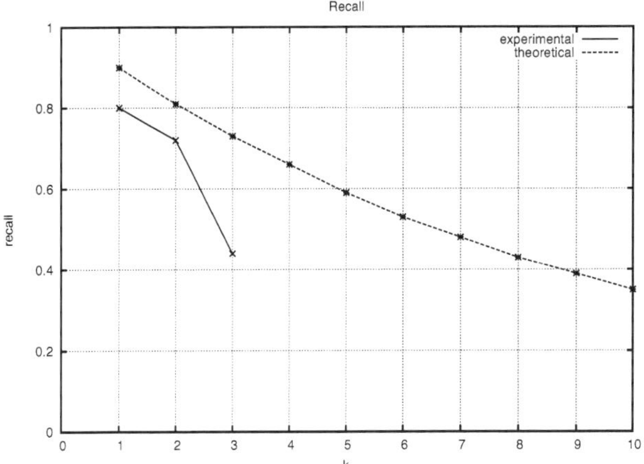

Fig. 9. Average recall using three classifiers in pipeline

for scientific publication of interest. Hence, supporting users in the task of dealing with the information provided by the Web is a primary issue. In this chapter, we focused on automatically retrieving and categorizing scientific publications and presented PUB.MAS, a system devoted to provide personalized search results in terms of bioinformatics publications. The system encompasses three main tasks: extracting scientific publications from online repositories, classifying them using hierarchical text categorization, and providing suitable feedback mechanisms. To validate the approach, we performed several experiments. Results show that the approach is effective and can be adopted in practice.

6. Acknowledgements

This work has been supported by the Italian Ministry of Education under the projects: FIRB ITALBIONET (RBPR05ZK2Z), Bioinformatics analysis applied to Populations Genetics (RBIN064YAT 003) and SHIWA.

7. References

Addis, A., Armano, G. & Vargiu, E. (2008). From a generic multiagent architecture to multiagent information retrieval systems, *AT2AI-6, Sixth International Workshop, From Agent Theory to Agent Implementation*, pp. 3–9.

Addis, A., Armano, G. & Vargiu, E. (2010). Assessing progressive filtering to perform hierarchical text categorization in presence of input imbalance, *Proceedings of International Conference on Knowledge Discovery and Information Retrieval (KDIR 2010)*, pp. 14–23.

Addis, A., Armano, G. & Vargiu, E. (2011). A comparative experimental assessment of a threshold selection algorithm in hierarchical text categorization, *Advances in Information Retrieval. The 33rd European Conference on Information Retrieval (ECIR 2011)*, pp. 32–42.

Armano, G. (2009). On the progressive filtering approach to hierarchical text categorization, *Technical report*, DIEE - University of Cagliari.

Armano, G., de Gemmis, M., Semeraro, G. & Vargiu, E. (2010). *Intelligent Information Access*, Springer-Verlag, Studies in Computational Intelligence series.

Armano, G., Manconi, A. & Vargiu, E. (2007). A multiagent system for retrieving bioinformatics publications from web sources, *IEEE TRANSACTIONS ON NANOBIOSCIENCE* 6(2): 104–109. Special Session on GRID, Web Services, Software Agents and Ontology Applications for Life Science.

Armano, G., Mancosu, G., Milanesi, L., Orro, A., Saba, M. & Vargiu, E. (2005). A hybrid genetic-neural system for predicting protein secondary structure, *BMC BIOINFORMATICS* 6 (suppl. 4):s3.

Armstrong, R., Freitag, D., Joachims, T. & Mitchell, T. (1995). Webwatcher: A learning apprentice for the world wide web, *AAAI Spring Symposium on Information Gathering*, pp. 6–12.

Baeza-Yates, R. A. & Ribeiro-Neto, B. (1999). *Modern Information Retrieval*, Addison-Wesley Longman Publishing Co., Inc., Boston, MA, USA.

Baker, P., Goble, C., Bechhofer, S., Paton, N., Stevens, R. & Brass, A. (1999). An ontology for bioinformatics applications, *Bioinformatics* 15(6): 510–520.

Bellifemine, F., Caire, G. & Greenwood, D. (2007). *Developing Multi-Agent Systems with JADE (Wiley Series in Agent Technology)*, John Wiley and Sons.

Bennett, P. N. & Nguyen, N. (2009). Refined experts: improving classification in large taxonomies, *SIGIR '09: Proceedings of the 32nd international ACM SIGIR conference on Research and development in information retrieval*, ACM, New York, NY, USA, pp. 11–18.

Bleyer, M. (1998). *Multi-Agent Systems for Information Retrieval on the World Wide Web*, PhD thesis, University of Ulm, Germany.

Bollacker, K. D., Lawrence, S. & Giles, C. L. (2000). Discovering relevant scientific literature on the web, *IEEE Intelligent Systems* 15(2): 42–47.

Brank, J., Mladenić, D. & Grobelnik, M. (2010). Large-scale hierarchical text classification using svm and coding matrices, *Large-Scale Hierarchical Classification Workshop*.

Ceci, M. & Malerba, D. (2003). Hierarchical classification of HTML documents with WebClassII, *in* F. Sebastiani (ed.), *Proceedings of ECIR-03, 25th European Conference on Information Retrieval*, Berlin Heidelberg NewYork: Springer, pp. 57–72.

Ceci, M. & Malerba, D. (2007). Classifying web documents in a hierarchy of categories: a comprehensive study, *Journal of Intelligent Information Systems* 28(1): 37–78.

Corney, D. P. A., Buxton, B. F., Langdon, W. B. & Jones, D. T. (2004). Biorat: Extracting biological information from full-length papers, *Bioinformatics* 20(17): 3206–3213.

Cost, W. & Salzberg, S. (1993). A weighted nearest neighbor algorithm for learning with symbolic features, *Machine Learning* 10: 57–78.

Craven, M. & Kumlien, J. (1999). Constructing biological knowledge-bases by extracting information from text sources, *Proceedings of the Seventh International Conference on Intelligent Systems for Molecular Biology*, Germany, pp. 77–86.

D'Alessio, S., Murray, K. & Schiaffino, R. (2000). The effect of using hierarchical classifiers in text categorization, *Proceedings of of the 6th International Conference on Recherche dŠInformation Assistée par Ordinateur (RIAO)*, pp. 302–313.

Delfs, R., Doms, A., Kozlenkov, A. & Schroeder, M. (2004). Gopubmed: ontology-based literature search applied to gene ontology and pubmed, *Proc. of German Bioinformatics Conference*, pp. 169–178.

Dumais, S. T. & Chen, H. (2000). Hierarchical classification of Web content, *in* N. J. Belkin, P. Ingwersen & M.-K. Leong (eds), *Proceedings of SIGIR-00, 23rd ACM International Conference on Research and Development in Information Retrieval*, ACM Press, New York, US, Athens, GR, pp. 256–263.

Esuli, A., Fagni, T. & Sebastiani, F. (2008). Boosting multi-label hierarchical text categorization, *Inf. Retr.* 11(4): 287–313.

Etzioni, O. & Weld, D. (1995). Intelligent agents on the internet: fact, fiction and forecast, *IEEE Expert* 10(4): 44–49.

Friedman, C., Kra, P., Yu, H., Krauthammer, M. & Rzhetsky, A. (2001). Genies: a natural-language processing system for the extraction of molecular pathways from journal articles, *Bioinformatics* 17: S74–S82.

Gaussier, É., Goutte, C., Popat, K. & Chen, F. (2002). A hierarchical model for clustering and categorising documents, *in* F. Crestani, M. Girolami & C. J. V. Rijsbergen (eds), *Proceedings of ECIR-02, 24th European Colloquium on Information Retrieval Research*, Springer Verlag, Heidelberg, DE, Glasgow, UK, pp. 229–247.

Golub, G. & Loan, C. V. (1996). *Matrix Computations*, Baltimore: The Johns Hopkins University Press.

Greenwood, D. & Calisti, M. (2004). An automatic, bi-directional service integration gateway, *IEEE Systems, Cybernetics and Man Conference*.

Huang, L. (2000). A survey on web information retrieval technologies, *Technical report, ECSL*.

Janssen, W. C. & Popat, K. (2003). Uplib: a universal personal digital library system, *DocEng '03: Proceedings of the 2003 ACM symposium on Document engineering*, ACM Press, New York, NY, USA, pp. 234–242.

Jirapanthong, W. & Sunetnanta, T. (2000). An xml-based multi-agents model for information retrieval on www, *Proceedings of the 4th National Computer Science and Engineering Conference (NCSEC2000)*.

Kiritchenko, S., Matwin, S. & Famili, A. F. (2004). Hierarchical text categorization as a tool of associating genes with gene ontology codes, *2nd European Workshop on Data Mining and Text Mining for Bioinformatics*, pp. 26–30.

Koller, D. & Sahami, M. (1997). Hierarchically classifying documents using very few words, *in* D. H. Fisher (ed.), *Proceedings of ICML-97, 14th International Conference on Machine Learning*, Morgan Kaufmann Publishers, San Francisco, US, Nashville, US, pp. 170–178.

Lee, J. (1994). Properties of extended boolean models in information retrieval, *Proceedings of the 17th Annual international ACM SIGIR Conference on Research and Development in information Retrieval*, pp. 182–190.

Lewis, D. D. (1995). Evaluating and optimizing autonomous text classification systems, *SIGIR '95: Proceedings of the 18th annual international ACM SIGIR conference on Research and development in information retrieval*, ACM, New York, NY, USA, pp. 246–254.

Lieberman, H. (1995). Letizia: An agent that assists web browsing, *in* C. S. Mellish (ed.), *Proceedings of the Fourteenth International Joint Conference on Artificial Intelligence (IJCAI-95)*, Morgan Kaufmann publishers Inc.: San Mateo, CA, USA, Montreal, Quebec, Canada, pp. 924–929.

Mahdavi, F., Ismail, M. A. & Abdullah, N. (2009). Semi-automatic trend detection in scholarly repository using semantic approach, *Proceedings of World Academy of Science, Engineering and Technology*, pp. 224–226.

McCallum, A. K., Rosenfeld, R., Mitchell, T. M. & Ng, A. Y. (1998). Improving text classification by shrinkage in a hierarchy of classes, *in* J. W. Shavlik (ed.), *Proceedings of ICML-98, 15th International Conference on Machine Learning*, Morgan Kaufmann Publishers, San Francisco, US, Madison, US, pp. 359–367.

McNee, S. M., Albert, I., Cosley, D., Gopalkrishnan, P., Lam, S. K., Rashid, A. M., Konstan, J. A. & Riedl, J. (2002). On the recommending of citations for research papers, *Proceedings of the 2002 ACM conference on Computer supported cooperative work*, CSCW '02, ACM, New York, NY, USA, pp. 116–125.

Mladenić, D. & Grobelnik, M. (1998). Feature selection for classification based on text hierarchy, *Text and the Web, Conference on Automated Learning and Discovery CONALD-98*.

Porter, M. (1980). An algorithm for suffix stripping, *Program* 14(3): 130–137.

Ramu, C. (2001). SIR: a simple indexing and retrieval system for biological flat file databases, *Bioinformatics* 17(8): 756–758.

Ramu, C. (2003). SIRW: A web server for the simple indexing and retrieval system that combines sequence motif searches with keyword searches., *Nucleic Acids Res* 31(13): 3771–3774.

Rocco, D. & Critchlow, T. (2003). Automatic discovery and classification of bioinformatics web sources, *Bioinformatics* 19(15): 1927–1933.

Rousu, J., Saunders, C., Szedmak, S. & Shawe-Taylor, J. (2005). Learning hierarchical multi-category text classification models, *ICML '05: Proceedings of the 22nd international conference on Machine learning*, ACM, New York, NY, USA, pp. 744–751.

Ruiz, M. E. & Srinivasan, P. (2002). Hierarchical text categorization using neural networks, *Information Retrieval* 5(1): 87–118.

Shaban, K., Basir, O. & Kamel, M. (2004). Team consensus in web multi-agents information retrieval system, *Team Consensus in Web Multi-agents Information Retrieval System*, pp. 68–73.

Sheth, B. & Maes, P. (1993). Evolving agents for personalized information filtering, *Proceedings of the 9th Conference on Artificial Intelligence for Applications (CAIA-93)*, pp. 345–352.

Sun, A. & Lim, E. (2001). Hierarchical text classification and evaluation, *ICDM '01: Proceedings of the 2001 IEEE International Conference on Data Mining*, IEEE Computer Society, Washington, DC, USA, pp. 521–528.

Sycara, K., Paolucci, M., van Velsen, M. & Giampapa, J. (2001). The RETSINA MAS infrastructure, *Technical Report CMU-RI-TR-01-05*, Robotics Institute Technical Report, Carnegie Mellon.

Tanabe, L., Scherf, U., Smith, L. H., Lee, J. K., Hunter, L. & Weinstein, J. N. (1999). Medminer: an internet text-mining tool for biomedical information, *BioTechniques* 27: 1210–1217.

Weigend, A. S., Wiener, E. D. & Pedersen, J. O. (1999). Exploiting hierarchy in text categorization, *Information Retrieval* 1(3): 193–216.

Wooldridge, M. J. & Jennings, N. R. (1995). Agent Theories, Architectures, and Languages: A Survey, *in* M. J. Wooldridge & N. R. Jennings (eds), *Workshop on Agent Theories, Architectures & Languages (ECAI'94)*, Vol. 890 of *Lecture Notes in Artificial Intelligence*, Springer-Verlag, Amsterdam, The Netherlands, pp. 1–22.

Wu, F., Zhang, J. & Honavar, V. (2005). Learning classifiers using hierarchically structured class taxonomies, *Proc. of the Symp. on Abstraction, Reformulation, and Approximation*, Vol. 3607, Springer Verlag, pp. 313–320.

11

GRID Computing and Computational Immunology

Ferdinando Chiacchio and Francesco Pappalardo
University of Catania
Italy

1. Introduction

Biological function emerges from the interaction of processes acting across a range of spatio-temporal scales. Therefore understanding disease and developing potential therapeutic strategies requires studies that bridge across multiple levels. This requires a systems biology approach and the tools used must be based on effective mathematical algorithms and built by combining experimental and theoretical approaches, addressing concrete problems and clearly defined questions.
Technological revolutions in both biotechnology and information technology have produced enormous amounts of data and are accelerating the extension of our knowledge of biological systems. These advances are changing the way biomedical research, development and applications are done. Mathematical and computational models are increasingly used to help interpret data produced by high-throughput genomics and proteomics projects, and through advances in instrumentation. Advanced applications of computer models that enable the simulation of biological processes are used to generate hypotheses and plan experiments.
Computational modeling of immune processes has emerged as a major support area for immunology and vaccinology research. Computational models have been developed for the simulation of immune processes at the molecular, cellular, and system levels. Computer models are used to complement or replace actual testing or experimentation. They are commonly used in situations where experimentation is expensive, dangerous, or impossible to perform. Models of the immune system fall into two categories:

1. molecular interactions (such as peptide binding to receptors), and
2. system-level interactions (such as models of immune response, or cellular models of immune system).

Until recently, models of molecular level immune processes have been successful in supporting immunology research, such as antigen processing and presentation. Computational models of the complete immune system have mainly been developed in the domain of theoretical immunology and were used to offer possible explanations of the overall function of the immune system, but were usually not applied in practice.
Newer computational tools that focus on immune interactions include numerous methods used for the mapping of T-cell epitopes (targets of immune responses). Computational methods for the simulation of antigen processing include prediction of proteasomal cleavage and peptide binding to transporters associated with antigen processing (TAP). The basic methods for the prediction of antigen processing and presentation have been extended to more sophisticated computational models for identification of promiscuous

Human Leukocyte Antigens (HLA)-restricted T-cell epitopes, identification of Major Histocompatibility Complex (MHC) supermotifs and T-cell epitope hot-spots.
Computational models of cellular or higher level processes or interactions have a longer history than those focusing on molecular processes, but are also more complex. These include models of T- cell responses to viruses, analysis of MHC diversity under host-pathogen co-evolution, B-cell maturation, or even the dynamic model of the immune system that can simulate both cellular and humoral immune responses.
The main problems that prevented the use of these models in practical applications, such as design of vaccines and optimization of immunization regimens are: *a)* large combinatorial complexity of the human immune system that could not be supported by existing computational infrastructures, *b)* lack of understanding of specific molecular interactions that resulted in an idealization of representation of molecular interactions as binary strings, and *c)* lack of experimental model data and correlation of model parameters to real-life measurements. Recent developments provide remedies to these problems and we are in the position to address each of these issues.
Grid computing brought powerful computational infrastructure and the capacity that can match the complexity of the real human immune system. Models of molecular interactions have reached high accuracy and we are routinely using prediction methods of antigen processing and presentation to identify the best targets for vaccine constructs. Finally, experimental models of immune responses to tumors and infectious diseases have been successfully modeled computationally.
In this work, we present two different experiences in which we show successful stories in using computational immunology approaches that have been implemented using GRID infrastructure:

- modeling atherosclerosis, a disease affecting arterial blood vessels, that is one of most common disease of the developed countries;
- a biological optimization problem on the Grid, i.e. an optimal protocol search algorithm based on Simulated Annealing (SA) capable to suggest optimal Triplex vaccine dosage used against mammary carcinoma induced lung metastasis.

2. Modeling atherogenesis using GRID computing

Atherosclerosis is, in large part, due to the deposition of low density lipoproteins (LDLs), i.e., plasma proteins carrying cholesterol and triglycerides, that determine the formation of multiple plaques within the arteries (Hanson, 2002; Ross, 1999). The origin of atherosclerosis is still not fully understood. However there are risk factors which increase the probability of developing atherosclerosis in humans. Some of these risk factors are beyond a person's control (smoking, obesity), others seem to have genetic origin (familial hypercholesterolemia, diabetes, hypertension) (Romero-Corral et al., 2006). Common denominator in all the form of atherosclerosis is the elevated level of LDL, which is subject to oxidation becoming oxidized low density lipoproteins (ox-LDL), that promotes an inflammatory response and immune activation in the artery walls (Berliner et al., 1996). The formation of atherosclerotic plaques in the artery reduces both the internal diameter of vessels and the blood flux leading to a number of serious pathologies (Vinereanu, 2006). Early studies demonstrated that ox-LDL can induce activation of monocytes/macrophages, endothelial cells and T cells. Ox-LDLs engulfed by macrophages form the so called foam cells (Steinberg, 1997). These cells represent the nucleus

of the plaques formation. Ox-LDL promotes also immune activation of B cells inducing the production of specific anti ox-LDL antibody (OLAB).
Atherosclerosis and their anatomical consequences cause severe problems. Stenosis (narrowing) and aneurysm of the artery are chronic, slowly progressing and cumulative effects indicating the progression of atherosclerotic disease. In both case the result is an insufficient blood supply to the organ fed by the artery. Most commonly, soft plaque suddenly ruptures, causes the formation of a thrombus that will rapidly slow or stop blood flow, leading to death of the tissues fed by the artery. This catastrophic event is called infarction and is not predictable. The most common event is thrombosis of the coronary artery causing infarction (a heart attack): However, since atherosclerosis is a body wide process, similar events also occur in the arteries of the brain (stroke attack), intestines, kidneys, etc. Those atherosclerosis associated events often cause of dead or serious invalidating diseases and require preventive treatments. Vaccine research for atherosclerosis is a hot pharmaceutical topic.
Recently we proposed a model based on the Agent Based Model (ABM) paradigm (Pappalardo et al., 2008) which reproduces clinical and laboratory parameters associated to atherogenesis. The model and its computer implementation (SimAthero simulator) considers all the relevant variables that play an important role in atherogenesis and its induced immune response, i.e., LDL, ox-LDL, OLAB, chitotriosidase and the foam cells generated in the artery wall.
We present three different situations over a time scale of two years. The standard normal patients where no foam cells are formed; patients having high level of LDL but who delay to apply appropriate treatments and finally patients who may have many events of high level of LDL but takes immediately appropriate treatments.

2.1 Description of the model

2.1.1 The biological scenario

Exogenous and endogenous factors induce in humans a very small, first oxidative process of blood circulating native LDLs (minimally modified LDLs or mm-LDLs). In endothelium mm-LDLs are extensively oxidized from intracellular oxidative products and then recognized by the macrophage scavenger receptor. High level and persistent in time LDLs lead to macrophages engulfment and their transformation in foam cells. Contrary, low level of LDLs and their oxidized fraction, lead to the internalization of the oxidized low density lipoproteins and subsequent presentation by major histocompatibility complex class II at the macrophages surface. Recognition of ox-LDL by macrophages and naive B cells, leads, by T helper lymphocytes cooperation, to the activation of humoral response and production of OLAB. When the OLAB/ox-LDL immune complexes are generated in the vascular wall, the macrophages catch them by the Fc receptor or via phagocytosis and destroy ox-LDL in the lysosome system. During this process, the activated macrophage releases chitotriosidase enzyme, that is then used as a marker of macrophage activation.

2.1.2 The model

To describe the above scenario one needs to include all the crucial entities (cells, molecules, adjuvants, cytokines, interactions) that biologists and medical doctors recognize as relevant in the game. The model described in (Pappalardo et al., 2008) contains entities and interactions which both biologist and MD considered relevant to describe the process.
Atherosclerosis is a very complex phenomenon which involves many components some of them not fully understood. In the present version of the simulator we considered only in

the immune system processes that control the atherogenesis. These processes may occur in immune system organs like lymph nodes or locally in the artery endothelium. To describe the Immune processes we considered both cellular and molecular entities.

Cellular entities can take up a state from a certain set of suitable states and their dynamics is realized by means of state-changes. A state change takes place when a cell interacts with another cell or with a molecule or both of them. We considered the relevant lymphocytes that play a role in the atherogenesis-immune system response, B lymphocytes and helper T lymphocytes. Monocytes are represented as well and we take care of macrophages. Specific entities involved in atherogenesis are present in the model: low density lipoproteins, oxidized low density lipoproteins, foam cells, auto antibodies anti oxidized low density lipoproteins and chitotriosidase enzyme. Cytotoxic T lymphocytes are not taken into consideration because they are not involved in the immune response (only humoral response is present during atherogenesis).

Molecular entities The model distinguishes between simple small molecules like interleukins or signaling molecules in general and more complex molecules like immunoglobulins and antigens, for which we need to represent the specificity. We only represent interleukin 2 that is necessary for the development of T cell immunologic memory, one of the unique characteristics of the immune system, which depends upon the expansion of the number and function of antigen-selected T cell clones. For what is related to the immunoglobulins, we represent only type immunoglobulins of class G (IgG). This just because at the actual state we don't need to represent other classes of Ig and because IgG is the most versatile immunoglobulin since it is capable of carrying out all of the functions of immunoglobulins molecules. Moreover IgG is the major immunoglobulin in serum (75% of serum Ig is IgG) and IgG is the major Ig in extra vascular spaces.

The actual model does not consider multi-compartments processes and mimics all processes in a virtual region in which all interactions take place. Our physical space is therefore represented by a 2D domain bounded by two opposite rigid walls and left and right periodic boundaries. This biological knowledge is represented using an ABM technique. This allows to describe, in a defined space, the immune system entities with their different biological states and the interactions between different entities. The system evolution in space and in time is generated from the interactions and diffusion of the different entities. Compared to the complexity of the real biological system our model is still very naive and it can be extended in many aspects. However, the model is sufficiently complete to describe the major aspects of the atherogenesis-immune system response phenomenon.

The computer implementation of the model (SimAthero hereafter) has two main classes of parameters: the first one refers to values known from standard immunology literature (Abbas et al., 2007; Celada et al., 1996; Goldspy et al., 2000; Klimov et al., 1999); the second one collects all the parameters with unknown values which we arbitrarily set to plausible values after performing a series of tests (*tuning phase*).

The simulator takes care of the main interactions that happens during an immune response against atherogenesis.

Physical proximity is modeled through the concept of lattice-site. All interactions among cells and molecules take place within a lattice-site in a single time step, so that there is no correlation between entities residing on different sites at a fixed time. The simulation space is represented as a $L \times L$ hexagonal (or triangular) lattice (six neighbors), with periodic boundary conditions to the left and right side, while the top and bottom are represented by rigid walls. All entities are allowed to move with uniform probability between neighboring lattices in the

grid with equal diffusion coefficient. In the present release of the simulator chemotaxis is not implemented.
LDLs values can be fixed in order to simulate different patients both in normolipidic condition and in hypercholesterolemic condition. The same applies to ox-LDLs. However human habits change with time and personal life style. A normolipidic patient can change its attitude becoming an hypercholesterolemic one and vice versa. For this reason we allow the simulator to accept varying life style conditions and preventive actions to decrease risk factors.

2.2 Results

The model described include the possibility of mimicking biological diversity between patients. The general behavior of a class of virtual patients arise from the results of a suitable set of patients, i.e., the mean values of many runs of the simulator of different patients under the same conditions. The class of virtual patients described by the model were tuned against *human* data data collected by (Brizzi et al., 2003; 2004) where different conditions, normal and hypercholesterolemic diabetic patients were analyzed.
In this section we analyze the behavior of the same patients in three broad class of clinical conditions to show how SimAthero could be used in order to analyze and predict the effects of various LDL levels in blood. The normal patient simulation is used as control experiment for the other simulations. The differences among these four clinical conditions depend on the LDL level and the time interval which occurs between the time in which concentration of LDL rise above normal level and the time in which the patient take appropriate treatments (lifestyle o drug) to reduce it to normal level.
Jobs were launched using the SimAthero simulator on the COMETA Grid. The submission process was done through the web interface of the ImmunoGrid project (http://www.immunogrid.eu).
A patient with a LDL level of roughly 950-970 ng/μl of blood is considered normal in clinical practice and he has with very low risk of atheroslerotique plaque. The results of SimAthero for a virtual normal patient (Figure 1) show that he will not support the formation of foam cells and, as a consequence, the beginning of atherogenesis process is absent.
We then simulated a scenario in which a patient, due to several reasons (diet, life style, oxidative agents and so on, so forth) leads its LDL level at 1300 ng/μl, taking it up to 1700 ng/mul. Looking at figure 2 one can observe about 12 foam cells per μl at the end of in silico follow up. This leads to a small atherogenesis process due to the high level of LDL.
Lastly (figure 3), we analyzed a virtual patient that initially takes its LDL level to small peaks, causing no damage. After that, he takes its LDL level to a hypercholesterolemic behavior, generating a small damage, as shown. This shows that small LDL alteration are completely taken under control by the normal behavior of the organism, but high LDL peaks lead to foam cells formation and then to the beginning of the atherogenesis process.

2.3 Remarks on atherosclerosis modeling using GRID computing

Atherosclerosis is a pathology where the immune control plays a relevant role. We presented studies on the increased atherosclerosis risk using an ABM model of atherogenesis and its induced immune system response in humans. Very few mathematical models (Cobbold et al., 2002; Ibragimov et al., 2005) and (to our best knowledge) no computational models of atherogenesis have been developed to date.
It is well known that the major risk in atherosclerosis is persistent high level of LDL concentration. However it is not known if short period of high LDL concentration can cause

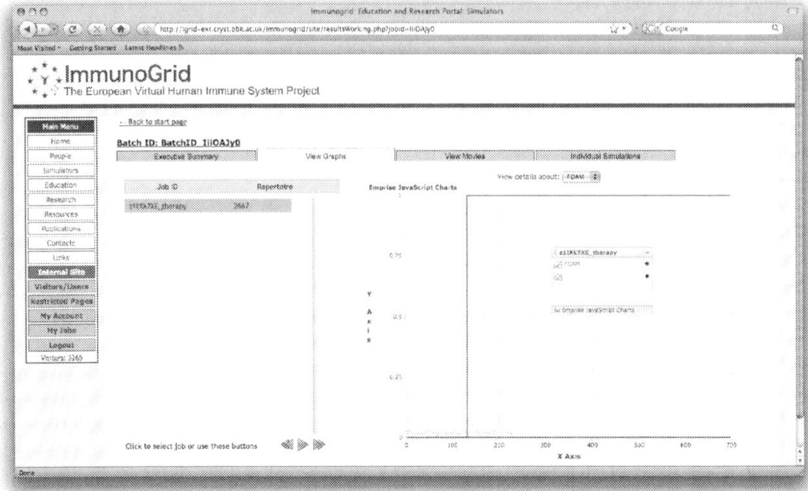

Fig. 1. Simulation results of a virtual patient with level of LDL considered normal. The follow-up period is two years. The figure shows that foam cells formation is absent in this patient.

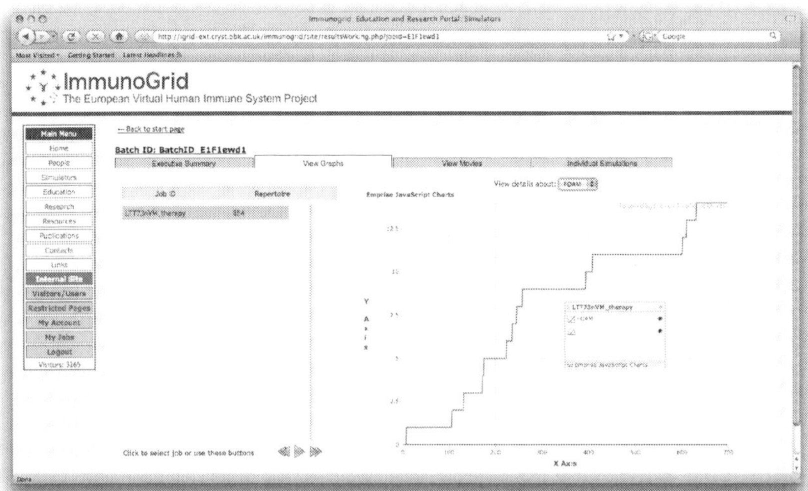

Fig. 2. Simulation results of a virtual patient with level of LDL considered at high risk. The follow-up period is two years. The figure shows that foam cells formation is present, leading to an atherogenesis process.

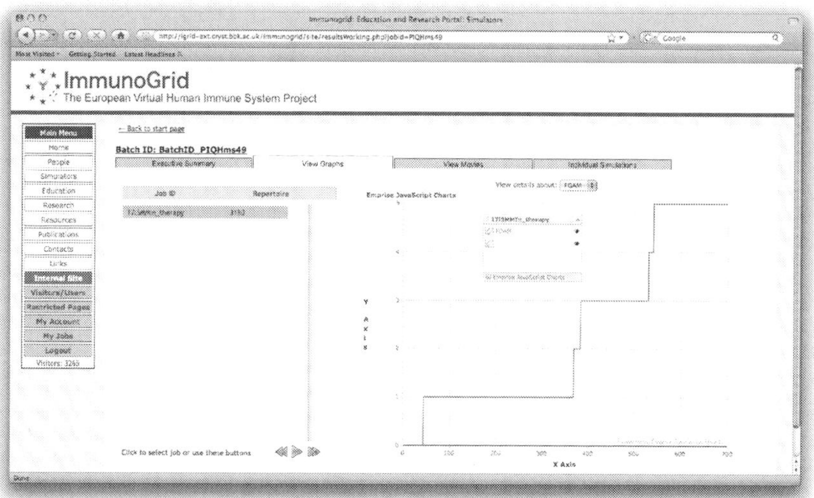

Fig. 3. Simulation results of a virtual patient with level of LDL considered quasi-normal at the beginning and then at high risk. The follow-up period is two years. The figure shows that foam cells formation is negligible in the first time, but becomes important soon after.

irreversible damage and if reduction of the LDL concentration (either by lfe style or drug) can drastically or partially reduce the already acquired risk.
Using an ABM cellular model describing the initial phase of plaque formation (atherogenesis) we are able to simulate the effect of life style which increases the risk of atherosclerosis.

3. Vaccine dosage optimization using GRID

As a second application, we show an example of how a *physiological* model in conjunction with some optimization techniques can be used to speed-up the research of an opitmal vaccination schedule for an immunopreventive vaccine, using the powerful of the GRID computing paradigm.
A vaccination schedule is usually designed empirically using a combination of immunological knowledge, vaccinological experience from previous endeavors, and practical constraints. In subsequent trials, the schedule of vaccinations is then renewed on the basis of the protection elicited in the first batch of subjects and their immunological responses e.g. kinetics of antibody titers, cell mediated response, etc. The problem of defining optimal schedules is particularly important in cancer immunopreventive approaches, which requires a sequence of vaccine administrations to keep a high level of protective immunity against the constant generation of cancer cells over very long periods, ideally for the entire lifetime of the host.
The Triplex vaccine represents a clear example of such immunopreventive approaches. It has been designed to raise the immune response against the breast cancer for the prevention of the mammary carcinoma formation in HER-2/Neu mouse models using a *Chronic schedule* in a follow up time between 52 and 57 weeks.
However it is not known if the *Chronic schedule* schedule is minimal, i.e. if it can guarantee survival for the mice population avoiding unnecessary/redundant vaccine administrations.

Shorter heuristic protocols failed, in *in vivo* experiments, in fulfilling this requirement, but between the Chronic and the shorter schedules there is still a huge number of possibilities which remain yet unexplored.
The SimTriplex () is a *physiological* computational model developed with the aim to answer to this question. It demonstrated able to reproduce *in silico* the *in vivo* Immune System (IS) - breast cancer competition elicted by the Triplex vaccine.
Optimal search strategy was biologically guided. Considering that Chronic proved to be effective for tumor control, the optimal search tried to find a protocol with minimum number of vaccine administrations able to reproduce, in silico, the time evolution of the chronic schedule. This strategy was used in the GA optimal search. It is well known that GA are slowly converging algorithms; the GA optimal search required several days using 32 nodes of an High Performance Computing infrastructure.
To this end we decided to investigate the applicability of **Simulated Annealing** (SA), a global optimization algorithm, widely tested and known for its computational speed and ability to achieve optimal solutions. (Interested readers can found an extended description of SA algorithm in (Van Laarhoven at al., 1987)). The combination of the Simulated Annealing algorithm with biologically driven heuristic strategies, leads to a much faster algorithm and better results for the optimal vaccination schedule problem for Triplex vaccine. In this context we remark how the COMETA grid infrastructure demostrated an excellent framework for protocol search and validation. We first executed the SA algorithm on a subset of the virtual mice population using MPI jobs. We therefore checked the protocol quality calculating (with simple jobs) the survivals of the entire population.

3.1 The algorithm

The work done by Kirkpatrick (Kirkpatrick et al., 1983) opened the path to a deep analogy between Statistical Mechanics (the behavior of systems with many degrees of freedom in thermal equilibrium at a finite temperature) and Combinatorial Optimization (the method of finding the minimum, if any, of a given function with respect to many parameters). There is a close similarity, indeed, between the procedure of *annealing* in solids and the framework required for optimization of complex systems.

3.1.1 The optimal vaccination schedule search problem

The SimTriplex model (Pappalardo et al., 2005) has been created with the aim to mimic the behavior of the immune system stimulated by the Triplex vaccine. It simulates all the major interactions of cells and molecules of the immune system in vaccinated as well as naive HER-2/neu mice. *In silico* experiments showed an excellent agreement with the *in vivo* ones.
As previously said, a protocol is said to be optimal if it can maintain efficacy with a minimum number of vaccine administrations. As in standard drug administration, the vaccination protocol must assure survival for a high percentages of patients. Schedule design is usually achieved using *medical consensus*, i.e. a public statement on a particular aspect of medical knowledge available at the time it was written, and that is generally agreed upon as the evidence-based, state-of-the-art (or state-of-science) knowledge by a representative group of experts in that area. Our goal is to improve medical consensus, helping biologists in design vaccine protocols with simulators and optimization techniques.
Let us consider a time interval $[0, T]$, in which we study the action of the vaccine on a set of virtual mice S. We then discretize the given time interval in $N-1$ equally spaced subintervals of width Δt (=8 hours), i.e. $\{t_1 = 0,\ t_2,\ \ldots, t_i,\ \ldots,\ t_N = T\}$.

Let $\mathbf{x} = \{x_1, x_2, \ldots, x_i, \ldots x_N\}$ be a binary vector representing a vaccine schedule, where $x_i = 0/1$ means respectively administration/no administration of the vaccine at time t_i. The number of vaccine administrations is given by $n = \sum_{i=1}^{N} x_i$. With $T = 400$ days, and $\Delta t = 8$ hours the search space D has cardinality 2^{400}, excluding any exhaustive search.
One of the wet biologists requirements imposes no more than two administrations a week (monday and thursday) because this is already considered a very intensive schedule from an immunological point of view. This reduces the cardinality of the search space D, from 2^{400} ($\sim 10^{120}$) to 2^{114} ($\sim 10^{34}$). We still have no chance for an exhaustive search.
The time of the carcinoma in situ (CIS) formation is computed through SimTriplex simulator. It is defined by $\tau(\mathbf{x}, \lambda_j)$, which is a function of the vaccination schedule $\mathbf{x}$ administered to the mouse $j \in S$ and a parameter λ_j which represents the biological diversity. The vaccine will be obviously effective if $\tau \geq T$.
As pointed out in § 1, any optimal protocol should try to reproduce , *in silico*, the chronic time evolution of cancer cells. This leads us to the need to use two thresholds on the allowed maximum number of cancer cells.
We also note here that, due to biological variability, a schedule found for a single mouse is not immunologically effective as it usually does not reveal able to protect high percentages of the treated patients. Having this in mind, we formulate our optimization problem as follows.
Let $\{j_1, j_2, \ldots, j_m\} \subset S$, with $m = 8$, a random chosen subset of *in silico* mice, the problem is defined as:

$$\begin{cases} \tau(\bar{\mathbf{x}}, \lambda_{j_1}) = \max(\tau(\mathbf{x}, \lambda_{j_1})) \\ \tau(\bar{\mathbf{x}}, \lambda_{j_2}) = \max(\tau(\mathbf{x}, \lambda_{j_2})) \\ \quad \vdots \qquad \vdots \\ \tau(\bar{\mathbf{x}}, \lambda_{j_m}) = \max(\tau(\mathbf{x}, \lambda_{j_m})) \\ n(\bar{\mathbf{x}}) = \min(n(\mathbf{x})) \\ \text{subject to:} \\ M_1(\bar{\mathbf{x}}) \leq \gamma_1, \; t \in [0, T_{in}] \\ M_2(\bar{\mathbf{x}}) \leq \gamma_2, \; t \in [T_{in}, T] \end{cases} \tag{1}$$

where $M_1(\mathbf{x})$ and $M_2(\mathbf{x})$ are the maximum number of cancer cells in $[0, T_{in}]$ (cellular-mediated controlled phase) and in $[T_{in}, T]$ (humoral-mediated controlled phase) respectively, and $T_{in} \sim T/3$, while γ_1 and γ_2 represent cancer cells threshold in $[0, T_{in}]$ and in $[T_{in}, T]$, respectively.
We deal with a multi-objective discrete and constrained optimization problem.
We modified this last formulation of the problem grouping all the $\tau(\mathbf{x}, \lambda_{j_h})$ $(h = 1, \ldots, m)$ by a proper statistical indicator. We chose the harmonic mean H of the survivals:

$$H(\mathbf{x}, \lambda_{j_1}, \ldots, \lambda_{j_m}) = \frac{m}{\sum_{i=1}^{m} \frac{1}{\tau(\mathbf{x}, \lambda_{j_i})}} \tag{2}$$

since it is very frequently used when statistic measurements of time are involved.
Therefore, the system (1) translates as:

$$\begin{cases} H(\bar{\mathbf{x}}, \vec{\lambda}) = \max(H(\mathbf{x}, \vec{\lambda})) \\ n(\bar{\mathbf{x}}) = \min(n(\mathbf{x})) \\ \text{subject to:} \\ M_1(\bar{\mathbf{x}}) \leq \gamma_1, \; t \in [0, T_{in}] \\ M_2(\bar{\mathbf{x}}) \leq \gamma_2, \; t \in [T_{in}, T] \end{cases} \tag{3}$$

with $\vec{\lambda} = (\lambda_{j_1}, \ldots, \lambda_{j_m})$.

3.1.2 Simulated annealing

An acclaimed Monte Carlo method, commonly referred as the *Metropolis criterion*, has been designed by Metropolis (Metropolis et al., 1953) with the aim to mimic the evolution of the complex systems towards equilibrium at a fixed temperature. This method randomly perturbates the position of the particles of a solid modifying its *configuration*. If the energy difference, ΔE, between the unperturbed and perturbed configurations is negative, the new configuration has lower energy and it's considered as the new one. Otherwise the probability of acceptance of the new configuration is given by the *Boltzmann factor* (which expresses the "probability" of a state with energy E relative to the probability of a state of zero energy).
After a large number of perturbations the probability distribution of the states should approach the Boltzmann distribution.
The Metropolis algorithm can also be used in combinatorial optimization problems to generate sequences of configurations of a system using a *cost function* C and control parameter c respectively as the energy and temperature in the physical annealing.
The SA algorithm can be represented as a sequence of Metropolis algorithms evaluated at a sequence of decreasing values of the control parameter c. A generalization of the method is given as follows: a generation mechanism is defined so that, given a configuration i, another configuration j can be obtained by choosing at random an element in the neighborhood of i. If $\Delta C_{ij} = C(j) - C(i) \leq 0$, then the probability that the next configuration is j is given by 1; if $\Delta C_{ij} > 0$ the probability is given by $e^{-\Delta C_{ij}/c}$ (Metropolis criterion).
This process is continued until the probability distribution, P, of the configurations approaches the Boltzmann distribution, which translates as:

$$P\{configuration = i\} = \frac{1}{Q(c)} \cdot e^{-C(i)/c} \tag{4}$$

where $Q(c)$ is a normalization constant depending on the control parameter c.
The control parameter is then lowered in steps, with the system being allowed to reach equilibrium by generating a chain of configurations at each step. The algorithm stops for some fixed value of the control parameter c where virtually no deteriorations can be accepted anymore.
At the end the final *frozen* configuration is assumed as a solution of the problem.

3.1.3 Implementation

In our *in silico* experiment, we select a population of 200 virtual mice and a simple random sample of $k = 8$ mice. To use the SA algorithm for the optimal vaccine schedule search problem we tried to define the SA relevant concepts (the solid configuration, the temperature, the energy and the semi-equilibrium condition) in terms of vaccine protocol and to describe the main protocol elements (the number of injections, the mean survival age, the time distribution of injections) in terms of a cooling process.
As previously said, we describe any candidate protocol as a binary vector $\mathbf{x}$ of cardinality $V = 114$. The total number of vaccine administrations n and the total number of possible schedules with n vaccine administrations $\mathcal{M}$ are given by:

$$n = \sum_{j=1}^{V} x_n^l(j)$$

$$\mathcal{M} = V!/[n!\ (V-n)!]$$

The configuration distribution is defined by $\mathbf{x}_{in}$, n_{in}, and the initial energy E_{in} as defined later on.

The *temperature* is slowly but constantly lowered to reach a state with minimum energy. We coupled this entity with n, the *number of vaccine administrations* of a semi-equilibrium configuration.

At a given temperature, a semi-equilibrium configuration is reached when its *Energy* is minimal. Since we want to maximize survival times of a mice sample set, the concept of Energy can be easily associated with the *harmonic mean* H of the survival times τ_i, $H(\bar{\mathbf{x}}, \vec{\lambda})$ (i.e. $E \propto H$). As a matter of fact H decreases when the cumulative survival time of the sample increases, in perfect accord with the energy definition.

The perturbation of a protocol has been initially implemented as random *1* bits reallocation (Pennisi et al., 2008b). This perturbation has been heuristically improved using *biological* knowledge. As we want to optimize mice survival, scheduling many vaccine administrations after the death of a mouse makes nonsense. So, in moving from $\mathbf{x}_i^j$ to $\mathbf{x}_i^{j+1}$, we improve random bits reallocation also moving some "1" at a suitable time $t < min\{\tau_i\}$, $i = 1, k$.

The SA algorithm for protocol optimization. *i*) starts from a randomly chosen initial vaccine distribution and finds the initial semi-equilibrium configuration n_{in}, $\mathbf{x}_{n_{in}}^{l_{in}}$, $E_{n_{in}}^{l_{in}}$

ii) Decrease the number of injections of 1 unit; *iii*) find a semi-equilibrium configuration $\mathbf{x}_i$, E_i according to Metropolis algorithm; *iv*) cycle on (*ii*). The algorithm stops when, once the algorithm control parameter, i.e. the number of vaccine administrations, is decreased from n to $n-1$, the Metropolis algorithm is not able to find a semi-equilibrium configuration, i.e. an acceptable value of survivals, in λ iterations. The accepted protocol is the last found at temperature n.

3.2 Computational results and conclusions

In silico optimal protocol search is a two-step process: *search* and *validation*. During the search step, the optimization tecnique tries to find an optimal protocol. As pointed out in section 3.1.1, optimal search stragies have to be executed on a representative subset of the population in order to guarantee significative survival percentages for the population.

The search technique therefore needs to test simultaneously every candidate protocol on the mice subset to compute its fitness function. This process usually requires a relatively small number of nodes with high communication throughputs, representing a typical massive parrallel application. In our case it has been implemented using the MPI (Message passing Interface) libraries.

Validation represents a completely different process. Here the protocol found by the search technique is tested over the entire population to compute mean survival rates, requiring an high number of CPUs with almost no need of communication. The *"search and validation"* process is represented in figure 4.

In this context the Cometa grid revealed itself as an excellent tool for our needs. It demonstrated to be highly flexible, giving us the opportunity to execute these so-different processes on the same infrastructure. We only needed to define the first process as an *MPI*

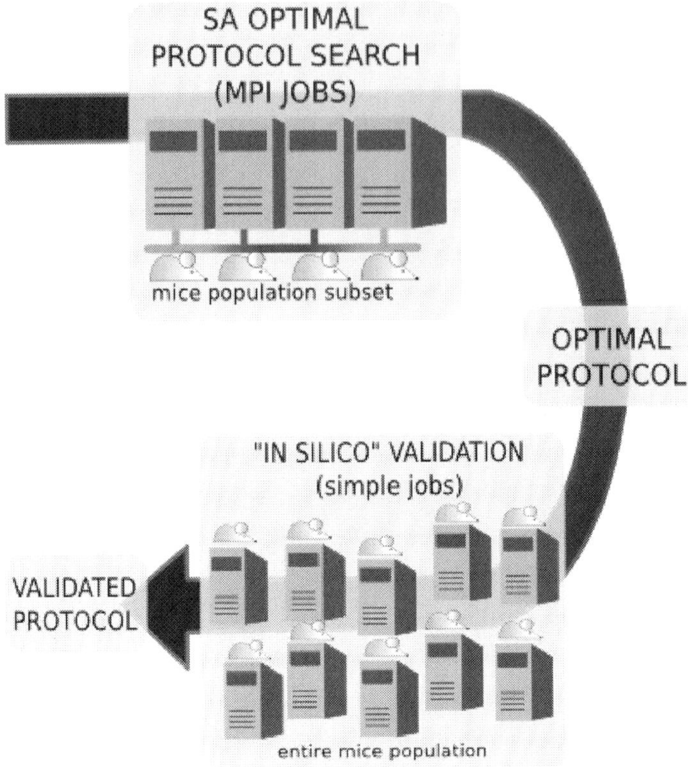

Fig. 4. The "in silico" *search* and *validation* optimal protocol representation.

job and the second as a sequence of *simple jobs* in a trasparent way, without worrying of the hidden the architectures behind the job submission interface.
To compare the results with those obtained using GA optimization technique (Lollini et al., 2006), we executed the SA algorithm on the same 8 random selected virtual mice sample used by GA. The protocol was then validated on the same population set (200 virtual mice).
The SA *in silico* tumor free percentages of the mice population show no substantial difference with GA results (87% for GA vs 86,5% for SA). Figure 5 shows the mean number of cancer cells, computed on the 200-mice set, for the GA-protocol (up lhs) and the SA-protocol (down lhs). Only the SA-protocol is able to totally fulfill the safety threshold conditions (shown in red).
Moreover the SA algorithm required a computational effort of about 2 hrs on a 8 processor unit to find a protocol with 37 vaccine administrations, showing speed-up factor of $\sim 1.4 \cdot 10^2$ in respect to previous GA experiments (Pennisi et al., 2008a).

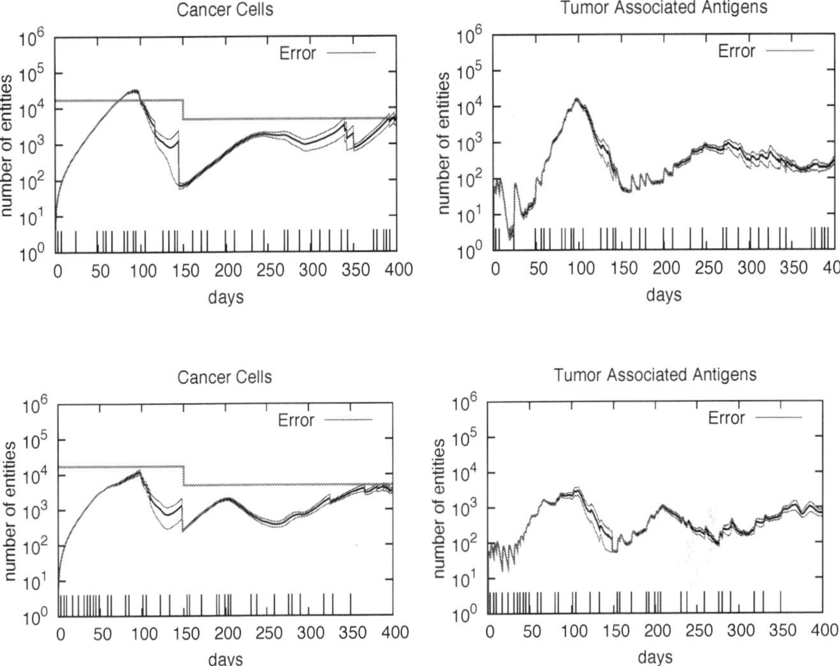

Fig. 5. Cancer cells behaviors and thresholds in GA (top) and SA (bottom). Small vertical bars on the x-axis represent vaccine administration times. Broken-line graphs on the lhs represent safety thresholds.

4. Conclusions

Mathematical models and Cellular Automata are mostly used for cellular level simulations, while a range of statistical modeling applications are suitable for the analysis of sequences at molecular level of the immune system.
Grid computing technology brings the possibility of simulating the immune system at the natural scale. In our opinion, a Grid solution is only as good as the interface provided to the users. We presented two successful stories in which we have shown successful stories in using computational immunology approaches that have been implemented using GRID infrastructure.
We like to conclude by stressing the interdisciplinary nature of the experiences described above and by noting that the contribution of life scientists needs to go beyond the only data supply, as it is extremely important in defining the biological scenario and ultimately construct a robust and validated mathematical or computational model. Only through a common effort of life and computer scientists it is possible to turn software into a valuable tools in life sciences.

5. Acknowledgments

This work makes use of results produced by the PI2S2 Project managed by the Consorzio COMETA, a project co-funded by the Italian Ministry of University and Research (MIUR) within the Piano Operativo Nazionale "Ricerca Scientifica, Sviluppo Tecnologico, Alta Formazione" (PON 2000-2006). More information is available at: http://www.pi2s2.it and http://www.consorzio-cometa.it.

6. References

Abbas, A.K.; Lichtman, A.H.; Pillai, S. (2007) Cellular and molecular immunology, *Saunders*, 6th edition.

Artieda, M.; Cenarro, A.; Gañàn, A.; Jericó, I.; Gonzalvo, C.; Casado, J.M.; Vitoria, I.; Puzo, J.; Pocoví,M.; Civeira,F. (2003) Serum chitotriosidase activity is increased in subjects with atherosclerosis disease, *Arterioscler. Thromb. Vasc. Biol.*, 23, 1645-1652.

Artieda, M.; Cenarro, A.; Gañàn, A.; Lukic, A.; Moreno, E.; Puzo, J.; Pocoví, M.; Civeira, F. (2007) Serum chitotriosidase activity, a marker of activated macrophages, predicts new cardiovascular events independently of C-Reactive Protein, *Cardiology*, 108, 297-306.

Berliner, J.A.; Heinecke, J.W. (1996) The role of oxidized lipoproteins in atherogenesis, *Free Radic. Biol. Med.*, 20(5), 707-727.

Brizzi, P.; Isaja, T.; D'Agata, A.; Malaguarnera, A.; Malaguarnera, M.; Musumeci, S. (2002) Oxidized LDL antibodies (OLAB) in patients with beta-thalassemia major, *J. Atheroscler. Thromb.*, 9(3), 139-144.

Brizzi, P.; Tonolo, G.; Carusillo, F.; Malaguarnera, M.; Maioli, M.; Musumeci, S. (2003) Plasma Lipid Composition and LDL Oxidation, *Clin. Chem. Lab. Med.*, 41(1), 56-60.

Brizzi, P.; Tonolo, G.; Bertrand, G.; Carusillo, F.; Severino, C.; Maioli, M.; Malaguarnera, L.; Musumeci, S. (2004) Autoantibodies against oxidized low-density lipoprotein (ox-LDL) and LDL oxidation status, *Clin. Chem. Lab. Med.*, 42(2), 164-170.

Celada, F.; Seiden, P.E. (1996) Affinity maturation and hypermutation in a simulation of the humoral immune response, *Eur. J. Immunol.*, 26, 1350.

Cobbold, C.A.; Sherratt, J.A.; Maxwell, S.R.J. (2002) Lipoprotein Oxidation and its Significance for Atherosclerosis: a Mathematical Approach, *Bulletin of Mathematical Biology*, 64, 65-95.

Goldsby, R.A. et al. (2000) In Austen,K.F., Frank,M.M., Atkinson,J.P. and Cantor,H. (eds.) Kuby Immunology. W.H. Freeman and Company, New York.

Hanson, G.K. (2002) Inflammation, atherosclerosis, and coronary artery disease, *N. Engl. J. Med.*, 352(16), 1685-1695.

Ibragimov, A.I.; McNeal, C.J.; Ritter, L.R.; Walton,J.R. (2005) A mathematical model of atherogenesis as an inflammatory response, *Math Med Biol.*, 22(4), 305-333.

Klimov, A.N.; Nikul'cheva, N.G. (1999) Lipid and Lipoprotein Metabolism and Its Disturbances, St. Petersburg: Piter Kom.

Lollini,P.-L. (2008) Private communication.

Orem, C.; Orem, A.; Uydu, H.A.; Celik, S.; Erdol, C.; Kural,B.V. (2002) The effects of lipid-lowering therapy on low-density lipoprotein auto-antibodies: relationship with low-density lipoprotein oxidation and plasma total antioxidant status, *Coron. Artery Dis.*, 13(1), 56-71.

Pappalardo, F.; Musumeci, S.; Motta, S. (2008) Modeling immune system control of atherogenesis, *Bioinformatics*, 24:15, 1715-1721.

Romero-Corral, A.; Somers, V.K.; Korinek, J.; Sierra-Johnson, J.; Thomas, R.J.; Allison, T.G.; Lopez-Jimenez,F. (2006) Update in prevention of atherosclerotic heart disease: management of major cardiovascular risk factors, *Rev. Invest. Clin.*, 58(3), 237-244.

Ross,R. (1999) Atherosclerosis–an inflammatory disease, *N. Engl. J. Med.*, 340(2), 115-126.

Tonolo,G. (2008) Private communication.

Shaw, P.X.; Hörkkö, S.; Tsimikas, S.; Chang, M.K.; Palinski, W.; Silverman, G.J.; Chen, P.P.; Witztum, J.L. (2001) Human-derived anti-oxidized LDL autoantibody blocks uptake of oxidized LDL by macrophages and localizes to atherosclerotic lesions in vivo, *Arterioscler. Thromb. Vasc. Biol.*, 21(8), 1333-1339.

Shoji, T.; Nishizawa, Y.; Fukumoto, M.; Shimamura, K.; Kimura, J; Kanda, H.; Emoto, M.; Kawagishi, T.; Morii,H. (2000) Inverse relationship between circulating oxidized low density lipoprotein (oxLDL) and anti-oxLDL antibody levels in healthy subjects, *Atherosclerosis*, 148(1), 171-177.

Steinberg,D. (1997) Low density lipoprotein oxidation and its pathobiological significance, *J. Biol. Chem.*, 272(34), 20963-20966.

Tinahones, F.J.; Gomez-Zumaquero, J.M.; Rojo-Martinez, G.; Cardona, F.; Esteva de Antonio, I.E.; Ruiz de Adana, M.S.; Soriguer, F.K. (2002) Increased levels of anti-oxidized low-density lipoprotein antibodies are associated with reduced levels of cholesterol in the general population, *Metabolism.*, 51(4), 429-431.

Tinahones, F.J.; Gomez-Zumaquero, J.M.; Garrido-Sanchez, L.; Garcia-Fuentes, E.; Rojo-Martinez, G.; Esteva, I.; Ruiz de Adana, M.S.; Cardona, F.; Soriguer,F. (2005) Influence of age and sex on levels of anti-oxidized LDL antibodies and anti-LDL immune complexes in the general population, *J. Lipid Res.*, 46(3), 452-457.

Vinereanu,D. (2006) Risk factors for atherosclerotic disease: present and future, *Herz*, Suppl. 3, 5-24.

Metropolis, N.; Rosenbluth, A.W.; Rosenbluth, M.N.; Teller, A.H.; Teller, E. (1953) "Equation of State Calculations by Fast Computing Machines", *Journal of Chemical Physics*, 21, 1087-1092.

Kirkpatrick, S.; Gelatt, C. D.; Vecchi, Jr., M. P. Optimization by simulated Annealing Science, Vol. 220, No. 4598, 671-680, 1983

Van Laarhoven, P.J.M.; Aarts, E.H.L. (1987) Simulated Annealing: Theory and Applications, *Springer Edition.*

Kirschner, D.; Panetta, J.C. (1998) Modeling immunotherapy of the tumor-immune interaction. *J. Math. Biol.*, (37), 235-252.

Nani,F.,Freedman,S. (2000) A mathematical model of cancer treatment by immunotherapy, *Math. Biosci.*, 163, 159-199.

Pappalardo, F.; Lollini, P.L.; Castiglione, F; Motta, S. Modeling and simulation of cancer immunoprevention vaccine. Bioinformatics. 21, 2891-2897

Agur, Z.; Hassin, R.; Levy, S. Optimizing chemotherapy scheduling using local search heuristics, Operations Research, 54:5, 829-846.

Kumar, N.; Hendriks, B.S.; Janes, K.A.; De Graaf, D.; Lauffenburger D.A. Applying computational modeling to drug discovery and development Drug Discovery Today Volume 11, Issues 17-18, September 2006, Pages 806-811

Lollini, P.L.; Motta, S.; Pappalardo, F. Discovery of cancer vaccination protocols with a genetic algorithm driving an agent based simulator. BMC. Bioinformatics. 7, 352.

Davies, M.N.; Flower, D.R. (2007) Harnessing bioinformatics to discover new vaccines, *Drug Discovery Today*, 12(9-10), 389-395.

Castiglione, F.; Piccoli, B. (2007) Cancer immunotherapy, mathematical modeling and optimal control, *J. Theo. Biol.*, 247, 723-732.

Pennisi, M.; Catanuto, R.; Pappalardo, F.; Motta, S. (2008) Optimal vaccination schedules using Simulated Annealing, *Bioinformatics*, 24:15, 1740-1742.

Pennisi, M.; Catanuto, R.; Mastriani, E.; Cincotti, A.; Pappalardo,F.; Motta,S. (2008) Simulated Annealing And Optimal Protocols, Journal of Circuits Systems, and Computers, 18:8, 1565-1579.

A Comparative Study of Machine Learning and Evolutionary Computation Approaches for Protein Secondary Structure Classification

César Manuel Vargas Benítez, Chidambaram Chidambaram, Fernanda Hembecker and Heitor Silvério Lopes
Bioinformatics Laboratory, Federal University of Technology Paraná (UTFPR)
Curitiba - PR - Brazil

1. Introduction

Proteins are essential to life and they have countless biological functions. Proteins are synthesized in the ribosome of cells following a template given by the messenger RNA (mRNA). During the synthesis, the protein folds into a unique three-dimensional structure, known as native conformation. This process is called protein folding. The biological function of a protein depends on its three-dimensional conformation, which in turn, is a function of its primary and secondary structures.

It is known that ill-formed proteins can be completely inactive or even harmful to the organism. Several diseases are believed to result from the accumulation of ill-formed proteins, such as Alzheimer's disease, cystic fibrosis, Huntington's disease and some types of cancer. Therefore, acquiring knowledge about the secondary structure of proteins is an important issue, since such knowledge can lead to important medical and biochemical advancements and even to the development of new drugs with specific functionality.

A possible way to infer the full structure of an unknown protein is to identify potential secondary structures in it. However, the pattern formation rules of secondary structure of proteins are still not known precisely.

This paper aims at applying Machine Learning and Evolutionary Computation methods to define suitable classifiers for predicting the secondary structure of proteins, starting from their primary structure (that is, their linear sequence of amino acids).

The organization of this paper is as follows: in Section 2 we introduce some basic concepts and some important aspects of molecular biology, computational methods for classification tasks and the protein classification problem. Next, in Sections 3 and 4, we present, respectively, a review of the machine learning and evolutionary computation methods used in this work. In Section 6, we describe the methodology applied to develop the comparison of different classification algorithms. Next, Section 7, the computational experiments and results are detailed. Finally, in the last Section 8, discussion about results, conclusions and future directions are pointed out.

2. Background

2.1 Molecular biology

Proteins are considered the primary components of living beings and they have countless biological functions. Finding the proteins that make up an organism and understanding their function is the foundation of molecular Biology (Hunter, 1993).
All proteins are composed by a chain of amino acids (also called residues) that are linked together by means of peptide bonds. Each amino acid is characterized by a central carbon atom (also known as $C\alpha$) to which are attached (as shown in Figure 1) a hydrogen atom, an amino group (NH_2), a carboxyl group (COOH) and a side-chain that gives each amino acid a distinctive function (also known as radical R). Two amino acids form a peptide bond when the carboxyl group of one molecule reacts with the amino group of the other. This process of amino acids aggregation is known as dehydration by releasing a water molecule (Griffiths et al., 2000). All amino acids have the same backbone and they differ from each other by the side-chain, which can range from just a hydrogen atom (in glycine) to a complex heterocyclic group (in tryptophan). The side-chain defines the physical and chemical properties of the amino acids of a protein (Cooper, 2000; Nelson & Cox, 2008).

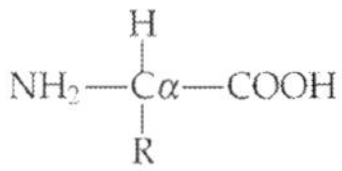

$$
\begin{array}{c}
H \\
| \\
NH_2 - C\alpha - COOH \\
| \\
R
\end{array}
$$

Fig. 1. General structure of an α-amino acid. The side-chain (R element) attached to the $C\alpha$ defines the function of the amino acid

There are numerous amino acids in the nature, but only 20 are proteinogenic. They are shown in Table 1. The first to be discovered was asparagine, in 1806. The last, threonine, was identified in 1938 (Nelson & Cox, 2008).
To understand the structures and functions of proteins, it is of fundamental importance to have knowledge about the properties of the amino acids, defined by their side-chain. Thus, the amino acids can be grouped into four categories: hydrophobic (also called non-polar), hydrophilic (also called polar), neutral, basic and acid (Cooper, 2000). Kyte & Doolittle (1982) proposed an hydrophobicity scale for all 20 amino acids. See the Table 1 for detailed information about the proteinogenic amino acids.
Polar amino acids can form hydrogen bonds with water and tend to be positioned preferably outwards of the protein, i.e., they are capable to interact with the aqueous medium (which is polar). On the other hand, hydrophobic amino acids tend to group themselves in the inner part of the protein, in such a way to get protected from the aqueous medium by the polar amino acids.
According to this behavior in aqueous solution, one can conclude that the polarity of the side chain directs the process of protein structures formation (Lodish et al., 2000).
From the chemical point of view, proteins are structurally complex and functionally sophisticated molecules. The structural organization of proteins is commonly described into four levels of complexity (Cooper, 2000; Griffiths et al., 2000; Lodish et al., 2000; Nelson & Cox, 2008), in which the upper cover the properties of lower: primary, secondary, tertiary and quaternary structures.
The primary structure is the linear sequence of amino acids. This is the simplest level of organization, it represents only the peptide bonds between amino acids.

Amino acid name	symbol	K&D	Normalised Value	Type
Isoleucine	ILE	+4.5	10.00	Hydrophobic
Valine	VAL	+4.2	9.68	Hydrophobic
Leucine	LEU	+3.8	9.26	Hydrophobic
Phenylalanine	PHE	+2.8	8.20	Hydrophobic
Cysteine	CYS	+2.5	7.88	Hydrophobic
Methionine	MET	+1.9	7.25	Hydrophobic
Alanine	ALA	+1.8	7.14	Hydrophobic
Glycine	GLY	-0.4	4.81	Neutral
Threonine	THR	-0.7	4.49	Neutral
Serine	SER	-0.8	4.39	Neutral
Tryptophan	TRP	-0.9	4.28	Neutral
Tyrosine	TYR	-1.3	3.86	Neutral
Proline	PRO	-1.6	3.54	Neutral
Histidine	HIS	-3.2	1.85	Hydrophilic
Glutamine	GLN	-3.5	1.53	Hydrophilic
Asparagine	ASN	-3.5	1.53	Hydrophilic
Glutamic acid	GLU	-3.5	1.53	Hydrophilic
Aspartic acid	ASP	-3.5	1.53	Hydrophilic
Lysine	LYS	-3.9	1.10	Hydrophilic
Arginine	ARG	-4.0	1.00	Hydrophilic

Table 1. Kyte and Doolittle (K & D) hydrophobicity scale and the normalized scale used in the computational experiments

The secondary structure of a protein refers to the local conformation of some part of a three-dimensional structure. There are, basically, three main secondary structures: α-helices (Pauling et al., 1951a), β-sheets (Pauling et al., 1951b) and turns (Lewis et al., 1973). In the structure of an α-helix, the backbone is tightly turned around an imaginary helix (spiral) and the side-chains of the amino acids protrude outwards the backbone (Figure 2(a)). The β-sheet is formed by two or more polypeptide segments of the same molecule, or different molecules, arranged laterally and stabilized by hydrogen bonds between the NH and CO groups (Figure 2(b)). Adjacent polypeptides in a β-sheet can have same direction (parallel β-sheet) or opposite directions (antiparallel β-sheet). Functionally, the antiparallel β-sheets are present in various types of proteins, for example, enzymes, transport proteins, antibodies and cell-surface proteins (Branden & Tooze, 1999). Turns are composed by a small number of amino acids and they are usually located in the surface of proteins forming folds that redirect the polypeptide chain into the protein. They allow large proteins to fold in highly compact structures.

Secondary structures can be associated through side-chain interactions to motifs (Branden & Tooze, 1999; Griffiths et al., 2000; Nölting, 2006). Motifs are patterns often found in three-dimensional structures that perform specific functions. For instance, the helix-turn-helix motif is important in DNA-protein interactions.

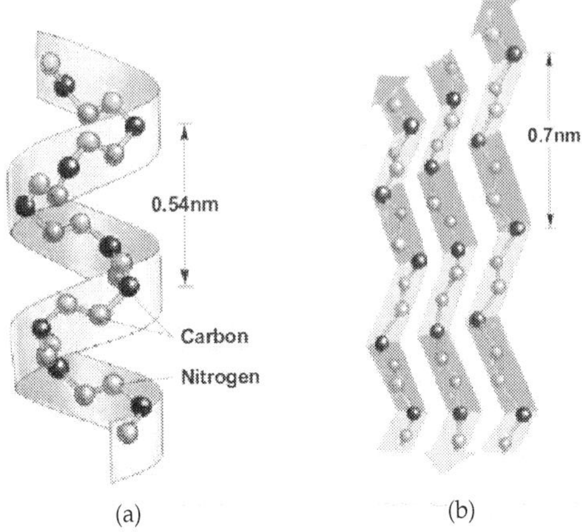

Fig. 2. α-helix (a) and β-sheet (b) structures. Adapted from (Alberts et al., 2002)

The tertiary structure represents the conformation of a polypeptide chain, i.e. the three-dimensional arrangement of the amino acids. The tertiary structure is the folding of a polypeptide as a result of interactions between the side chains of amino acids that are in different regions of the primary structure. Figure 3(a) shows an example of tertiary structure, where one can observe the presence of two secondary structures: α-helix and β-sheet.
Finally, the arrangement of three-dimensional structures constitutes quaternary structure (as shown in Figure 3(b)). Figures 3(a) and 3(b) were drawn using RasMol [1] from PDB files (see Section 2.2).
The Proteins can be classified into two major groups, considering higher levels of structure (Nelson & Cox, 2008): fibrous and globular proteins. Both groups are structurally different: fibrous proteins consist of a single type of secondary structure; globular proteins have a nonrepetitive sequence and often contain several types of secondary structure. Helices are the most abundant form of secondary structure in globular proteins, followed by sheets, and in the third place, turns (Nölting, 2006).

2.2 Protein databases and classification

Finding protein functions has been, since long ago, an important topic in the Bioinformatics community. As mentioned in Section 2, the function of a protein is directly related to its structure. Due to its great importance for Medicine and Biochemistry, many research has been done about proteins (including the many genome sequencing projects) and, consequently, many information is available. There are many resources related to protein structure and function. Table 2 lists some protein databases. Basically, the protein databases can be classified into two classes: sequence and structure databases.

[1] RasMol is a molecular visualization software. Available at http://www.rasmol.org

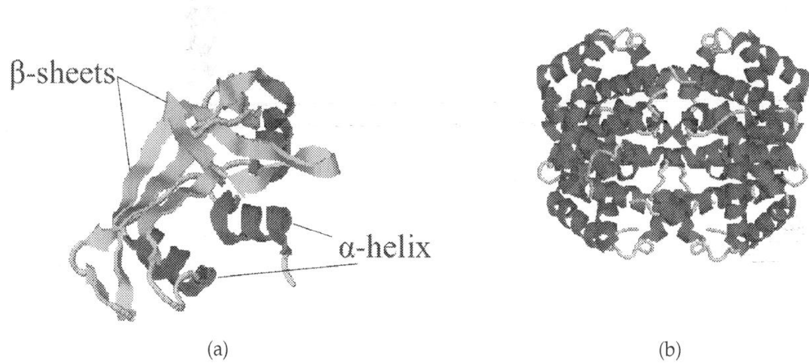

Fig. 3. Tertiary structure of Ribonuclease-A (a) and quaternary structure of Hemoglobin (b).

Database	Description	Web Address
PDB	repository of protein structures	http://www.pdb.org
UniProtKB/TrEMBL	repository of protein amino acid sequences, name/description, taxonomic data and citation information	http://www.uniprot.org/
PIR	protein sequence databases	http://pir.georgetown.edu/
PROSITE	documentation entries describing sequence motif definitions, protein domains, families and functional patterns	http://www.expasy.org/prosite/
PRINTS	Fingerprints information on protein sequences	http://www.bioinf.man.ac.uk/dbbrowser/
BLOCKS	Multiple-alignment blocks	http://blocks.fhcrc.org/
eMOTIF	protein motif database, derived from PRINT and BLOCKS	http://motif.stanford.edu/emotif/
PRODOM	protein domain databases	http://protein.toulouse.inra.fr/prodom.html
InterPro	protein families and domains	http://www.ebi.ac.uk/interpro/

Table 2. Some important protein databases

In this work we used the Protein Data Bank (PDB) (Berman et al., 2000; Bernstein et al., 1977) that is an international repository of three-dimensional structure of biological macromolecules. The data is, typically, obtained by X-ray crystallography (Drenth, 1999; Sunde & Blake, 1997) or NMR spectroscopy (Nuclear Magnetic Resonance) (Jaroniec et al., 2004; Wüthrich, 1986). Almost all protein structures known today are stored in the PDB (Bernstein et al., 1977).

Despite the growing number of protein sequences already discovered, only a few portion of them have their three-dimensional and secondary structures unveiled. For instance, the UniProtKB/TrEMBL repository (The UniProt Consortium, 2010) of protein sequences has currently around 13,89 million records (as in March/2011), and the PDB registers the structure of only 71,794 proteins. This fact is due to the cost and difficulty in unveiling the structure of proteins, from the biochemical and biological point of view. Therefore, computer science has an important role here, proposing models and methods for studying the Protein Structure Prediction problem (PSP).

There are two basic approaches that are used to the prediction of protein functions: prediction of the protein structure and then prediction of function from the structure, or else, classifying proteins and supposing that similar sequences will have similar functions (Tsunoda et al., 2011). Several approaches to solve the PSP exist, each addressing the problem by using a computational method to obtain optimal or quasi-optimal solutions, such as Molecular Dynamics with *ab initio* model (Hardin et al., 2002; Lee et al., 2001), neural nets (Yanikoglu & Erman, 2002) and evolutionary computation methods with lattice (Benítez & Lopes, 2010; Lopes, 2008; Scapin & Lopes, 2007; Shmygelska & Hoos, 2005) and off-lattice (Kalegari & Lopes, 2010) models.

However, there is no consensus about protein classification which is done using different properties of proteins through several approaches. Many computational techniques have been used to classify proteins into families, such as structural transformations (Ohkawa et al., 1996), data compression (Chiba et al., 2001), genetic programming (Koza, 1996), Markov chains (Durbin et al., 1998) and neural networks (Wang & Ma, 2000; Weinert & Lopes, 2004).

Other papers focus on motifs discovery, as a starting step for protein classification. For instance, (Tsunoda & Lopes, 2006) present a system based on a genetic algorithm that was conceived to discover motifs that occur very often in proteins of a given family but rarely occur in proteins of other families. Also, Tsunoda et al. (2011) present an evolutionary approach for motif discovery and transmembrane protein classification, named GAMBIT. Techniques of clustering for sequences analysis were presented by (Manning et al., 1997). Chang et al. (2004) proposed a Particle Swarm Optimization (PSO) approach to motif discovery using two protein families from the PROSITE.

Wolstencroft et al. (2006) describes the addition of an ontology that captures human understanding of recognizing members of protein phosphatases family by domain architecture as an ontology. G.Mirceva & Davcev (2009) present an approach based on Hidden Markov Models (HMMs) and the Viterbi algorithm to domain classification of proteins. Davies et al. (2007) present an hierarchical classification of G protein-coupled receptors (GPCRs) The GPCRs are a common target for therapeutic drugs (Klabunde & Hessler, 2002).

3. Machine learning methods

In this section, we focus on the Machine Learning (ML) algorithms applied to the classification of a secondary protein data set. Most of the methods discussed in this section can be considered as important supervised ML techniques. It is frequently cited in the literature that the efficiency of ML algorithms can be quite different from data set to data set. Therefore, it is usual to test the data set with many different ML algorithms. For this purpose,

Weka[2] (Witten et al., 2011) is an appropriate and a flexible tool which contains a collection of state-of-the-art ML algorithms and data preprocessing tools. It allows users to test the algorithms with new data sets quickly.
It is not an easy task to select which algorithm is the most suitable for a specific domain or problem. Also, as there are many algorithms that use numerical data, an expert is needed to tune the data. In this context, the comparison of many different algorithms is not an easy task. Such a comparison can be performed by statistical analysis of the accuracy rate obtained from trained classifiers on some specific data set (Kotsiantis, 2007). To support this comparison and to evaluate the quality of the methods, three main measures can be considered (Mitchell, 1997):

- The classification rate of the training data
- The correct classification rate of some test data
- The average performance using cross validation

The most well-known classification algorithms are grouped in the Weka workbench, including Bayesian classifiers, Neural networks, Meta-learners, Decision trees, Lazy classifiers and rule-based classifiers (Witten et al., 2011). Bayesian methods include Naïve Bayes, complement Naïve Bayes (CNB), multinomial Naïve Bayes, Bayesian networks and AODE. Both decision trees and rule-based classifiers belong to the category of logic-based supervised classification algorithms (Kotsiantis, 2007). Decision trees algorithms include several variants, such as NBTree, ID3, J48 (also known as C4.5) and alternative decision trees (ADTree). PART, decision tables, Rider, JRip, and NNge are included in the group of rule learners. Lazy Bayesian rules (LBR), Instance-Based learning schemes (IB1 and IBk), Kstar, and Locally-Weighted Learning (LWL) are part of the lazy learning algorithms.
Besides these basic classification learning algorithms, there are some meta-learning schemes such as LogitBoost, MultiBoostAB, and ADABoost. These boosting algorithms enable users to combine instances of one or more of the above-mentioned algorithms. In addition to these all algorithms, Weka workbench includes neural networks methods such as Multiplayer Perceptron (MLP), Sequential Minimal Optimization algorithm (SMO), Radial Basis Function (RBF) network, and logistic and voted perceptron (Witten et al., 2011). In the following subsections, we discuss specific aspects and present a brief comparison of some classification methods mentioned previously and used in the protein classification task of this work.

3.1 Bayesian methods

Bayesian methods are the most practical approaches amongst many learning algorithms and those that provide learning based on statistical approaches. These methods are characterized by induction of probabilistic networks from the training data (Kotsiantis, 2007). In the Weka workbench there are several methods, such as: CNB, NaïveBayes, NaïveBayesSimple and AODE. Bayesian methods emerged as alternative approaches for decision trees and neural networks and, at the same time, being competitive with them for real-world applications (John & Langley, 1995). However, it is known that, to apply such methods, prior knowledge of many probabilities is required (Mitchell, 1997). Naïve Bayes (NB) is one of the learning algorithms widely applied to classification tasks. The NB classifier is the most effective algorithm of the Bayesian group of algorithms, and it can easily handle unknown or missing values (Mitchell, 1997). The main advantage of the NB is that it does not require a long

[2] Available at: *http://www.cs.waikato.ac.nz/ml/weka/*

time to train the classifier with the data set. This classifier is based on the assumption of attribute independence. Some Bayesian variants have tried to alleviate this assumption, such as the NBTree (Zheng & Webb, 2000) algorithm, which has a high computational overhead. To improve the prediction accuracy without increasing the computational cost, Webb et al. (2005) developed a variant algorithm of NB called AODE (Aggregating One-Dependence Estimators) that uses a weaker attribute independence assumption than NB. NB classifier traditionally relies on the assumption that the numerical attributes are generated by a single Gaussian distribution. However, this is not always the best approximation for the density estimation on continuous attributes. Following this direction, John & Langley (1995) proposed a new approach described as flexible Bayes, in which a variety of nonparametric density estimation methods were used instead of Gaussian distribution. This approach is implemented through NaïveBayesUpdateable class in the Weka tool. Bayesian networks (BN) methods use several search algorithms that works under conditions of uncertainty. All BN learners are considered as slow and not suitable for large data sets (Cheng et al., 2002). Unlike decision trees and neural networks, the main aspect of this method is that it obtains prior information of the data set from the structural relationships among the features (Kotsiantis, 2007).

3.2 Decision trees

Basically, the methods of this category classify data by building decision trees, in which each node represents a feature in an instance and each branch represents the value of a node. They are heuristic, non-incremental, and do not use any world knowledge. In this category, there are many learning methods, amongst of which, ID3, J48 and C4.5 are most well-known (Kotsiantis, 2007). Not only these algorithms are based on decision trees. There are also some hybrid versions like NBTree, LMTree, RandomForest and ADTree. In general, these versions have competitive performance with the traditional C4.5 algorithm. The earliest version of ID3 was developed by Quinlan (1993) and its improved version is C4.5 (Martin, 1995). ID3 allows to work with errors in the training data, as well as missing attribute values. This method uses a hill-climbing strategy to search in the hypothesis space, so as to find out a decision tree that correctly classifies the training data. This learning method can handle noisy training data and is less sensitive to errors of individual training examples (Mitchell, 1997). From the decision trees, during the learning process, a set of rules in the disjunctive form are obtained by traversing the different possible paths in the entire tree. A limitation of the ID3 algorithm is that it cannot guarantee the optimal solution. Another improved version of the algorithm, J48, implements Quinlan's C4.5 algorithm by generating a pruned or an unpruned decision tree (Witten et al., 2011). The NBTree algorithm was proposed by Kohavi (Kohavi, 1996) to overcome the accuracy problem encountered with both Naïve-Bayes and decision trees in small data sets. NBTree is a hybrid algorithm which induces a mix of decision-tree and Naïve-Bayes classifiers. Eventually, this algorithm outperforms both C4.5 and Naïve-Bayes. To predict numeric quantities, Landwehr et al. (2005) introduced a new learning method combining tree induction methods and logistic regression models into decision trees. This method is denominated logistic model trees (LMTree). LMTree produces a single tree which is not easily interpretable, but better than multiple trees. This algorithm achieved performance higher than decision trees with C4.5 and logistic regression.

RandomForest is a classifier consisting of a collection of tree-structured classifiers in which each tree depends on the values of a random vector, sampled independently and with the same distribution for all trees in the forest (Breiman, 2001).

Freund & Schapire (1999) presented a new type of classification algorithm, combining decision trees and boosting, called alternative decision trees (ADTree). An important feature of the ADTree algorithm is the measure of confidence or classification margin. This learning algorithm was compared with other improved version of C4.5 with boosting, denominated as C5.0. According to the experiments, it can be realized that the ADTree is competitive with C5.0 and generate easily interpretable small rules.

3.3 Neural Networks

Neural Network (NN) learning methods are robust to errors in the training data. They provide a good approximation to real-valued, discrete-valued, and vector-valued target functions. High accuracy and high speed rate of classification are some relevant aspects of NN classifiers (Kotsiantis, 2007). Algorithms like Multiplayer Perceptron (MLP), SMO, RBFNetwork and Logistic are part of the Weka workbench. A RBF neural networks is a particular NN constructed from spatially localized kernel functions, and uses a different error estimation and gradient descent function called the radial basis function (RBF). This method uses a cross-validation technique to stop the training which is not present in other NN algorithms (Mitchell, 1997). Platt (1998) proposed an improved algorithm for training support vector machines (SVMs) called Sequential Minimal Optimization SMO. The results obtained for real-world test sets showed that the SMO is 1200 times faster than linear SVMs and 15 times faster than non-linear SVMs.

3.4 Meta-learning methods

Most of the methods of this category such as boosting, bagging and wagging are common committee learning approaches that reduce the classification error from learned classifiers. Boosting is the one of the most important recent advancements in classification algorithms, since it can improve dramatically their performance (Friedman et al., 2000). It is also known as machine learning meta-algorithm or discrete AdaBoost. AdaBoost, as a boosting algorithm, can efficiently convert a weak learning algorithm into a short learning algorithm. Furthermore, it is an adaptive behavior with error rates of the individual weak hypotheses (Freund & Schapire, 1999). AdaBoost calls a given weak learning algorithm repeatedly in a series of rounds and trains the classifiers by over-weighting the training samples that were misclassified in the next iteration. A complete algorithm description can be found in Friedman et al. (2000). One of the important properties of the ADABoost algorithm is the identification of outliers which are either mislabeled in the training data or inherently ambiguous and hard to categorize (Freund & Schapire, 1999). Many other developments on AdaBoost resulted in variants such as Discrete AdaBoost, Real AdaBoost, LPBoost and LogitBoost. Both ADABoost and bagging generic techniques can be employed together with any baseline classification technique. Wagging is variant of bagging, that requires a base learning algorithm capable of using training cases with differing weights (Webb, 2000). MultiBoosting is an extension technique of AdaBoost with wagging. It offers a potential computational advantage over AdaBoost (Webb, 2000).

3.5 Rule-based methods

Rules can be extracted from the data set using many different machine learning algorithms. The IF-THEN rules is a convenient way to represent the underlying knowledge present in the data set, which can be easily understood by domain experts. Among a variety of rule-based methods that were investigated, decision trees and separate-and-conquer strategy are considered the most important (Frank & Witten, 1998). Based on these approaches, emerged two dominant rule-based learning methods: C4.5 (Quinlan, 1993) and RIPPER (Cohen, 1995). In fact, RIPPER is an optimized algorithm which is very efficient in large samples with noisy data, and produce error rates lower than or equivalent to C4.5 (Cohen, 1995). In the Weka workbench, rule-based classifiers have many learning algorithms, including PART, DecisionTable, Ridor, JRip and NNge. Decision tables, which are simple space hypotheses, are represented by DTM (Decision Table Majority). Kohavi (1995) evaluated the power of decision tables through Inducer DTM (IDTM). IDTM, on some data sets, obtained comparable performance accuracy as C4.5. JRip implements a propositional rule learner called Repeated Incremental Pruning to Produce Error Reduction (RIPPER), proposed by Cohen (1995). Frank & Witten (1998) proposed a new approach, by combining the paradigms of decision trees and separate-and-conquer, called PART. PART is based on the repeated generation of partial decision trees in a separate-and-conquer manner. Not only accuracy of a learning algorithm, but also the size of a rule set resulted from the learning process is important. The size of a rule strongly influences on the degree of comprehensibility. Rule sets produced by PART are generally smaller as C4.5 and more accurate than RIPPER (Frank & Witten, 1998). Another algorithm that is part of Weka workbench is the Non-Nested Generalized Exemplars (NNge), proposed by Martin (1995). NNge generalizes exemplars without nesting (exemplars contained within one another) or overlaping. By generalization, examples in the data set that belongs to the same class are grouped together. When tested against domains containing both large and small disjuncts, NNge performs better than C4.5. NNge performs well on data sets that combine small and large disjuncts, but it performs poorly in domains with a high degree of noise (Martin, 1995).

3.6 Lazy methods

Instance-based methods are considered as lazy learning methods because the classification or induction process is done only after receiving a new instance or a training example. Learning process will be started on the stored examples only after when a new query instance is encountered (Mitchell, 1997). Nearest Neighbor algorithm is the one of the most basic instance-based methods. The instance space is defined in terms of Euclidean distance. However, since Euclidean distance is inadequate for many domains, several improvements were proposed to the instance-based nearest neighbor algorithm which are known as IB1 to IB5 (Martin, 1995). Zheng & Webb (2000) proposed a novel algorithm called Lazy Bayesian Rule learning algorithm (LBR) in which lazy learning techniques are applied to Bayesian tree induction. Error minimization was maintained as an important criteria in this algorithm. Experiments done with different domains showed that, on average, LBR overcomes mostly all other algorithms including Naïve Bayes classifier and C4.5. The Locally Weighted Learning algorithm (LWL) is similar to other lazy learning methods, however it behaves differently when classifying a new instance. LWL algorithm constructs a new Naïve Bayes model using a weighted set of training instances (Frank et al., 2003). It empirically outperforms both standard Naïve Bayes as well as nearest-neighbor methods on most data sets tested by those authors.

4. Gene expression programming and GEPCLASS

Gene expression programming (GEP) is proposed by Ferreira (2001), and it has features of both genetic algorithms (GAs) and genetic programming (GP). The basic difference between the three algorithms is the way the individuals are defined in each algorithm. In the traditional GAs individuals are called chromosomes and are represented as linear binary strings of fixed length. On the other side, in GP, individuals or trees are non-linear entities of different sizes and shapes. On the other hand, in GEP, the individuals are encoded as linear strings of fixed length (chromosomes) which are afterwards expressed as non-linear entities of different sizes and shapes (simple diagram representations or expression trees (ETs)) (Ferreira, 2001). In fact, in GEP, chromosomes are simple, compact, linear and relatively small entities, that are manipulated by means of special genetic operators (replication, mutation, recombination and transposition). ETs, in turn, are the phenotypical representation of the chromosome. Unlike genetic algorithms, selection operates over ETs (over the phenotype, not the genotype). During the reproduction phase, only chromosomes are generated, modified and transmitted to the next generations. The interplay of chromosomes and ETs allows to translate the language of chromosomes into the language of ETs. The use of varied set of genetic operators introduce genetic diversity in GEP populations always producing valid ETs. In the same way as other evolutionary algorithms, in GEP, the initial population must be defined either randomly or using some previous knowledge collected from the problem. Next, chromosomes are expressed into ETs which will be then evaluated according to the definition of the problem resulting in a fitness measure. During the iteration process, the best individual(s) is(are) kept and the rest are submitted to a fitness-based selection procedure. Selected individuals naturally go through modifications by means of genetic operators leading to a new generation of individuals. The whole process is repeated until a stopping criterion is met (Weinert & Lopes, 2006). The structural organization of GEP genes are based on Open Reading Frames (ORF), a biological terminology. Although the length of genes is constant, the length of ORFs are not. GEP genes are composed of a head that contain symbols that represent both function and terminals, and a tail that contains only terminals. GEP chromosomes are basically formed by more than one gene of different lengths. Unlike GP, in which an individual of the population is modified by only one operator at a time, in GEP, an individual may be changed by one or several genetic operators. Besides the genetic operations like replication, mutation and recombination, GEP includes operations based on transpositions and insertion of elements.

In this work, we used the GEPCLASS system[3] that was specially developed for data classification with some modifications regarding the original GEP algorithm (Weinert & Lopes, 2006). Prior implementing data classification using evolutionary algorithms, it is necessary to definet whether an individual represents a single rule (Michigan approach) or a complete solution composed by a set of rules (Pittsburg approach) (Freitas, 1998). This system can implement both approaches, either by an explicit decision of a user, or allowing the algorithm decide by itself which one is the most suitable for a given classification task, during the evolutionary process. GEPCLASS can manage both continuous and categorical (nominal) attributes in the data set. If a given attribute is nominal, GEPCLASS uses only $=$ or $\neq$ as relational operators. Otherwise, if the attribute is continuous or ordered categorical, all relational operators can be used.

Figure 4 shows a 2-genes chromosome with different lengths for heads and tails. In the chromosome part, upward arrows show the points delimiting the coding sequence of each gene. This chromosome transformed into an ET and, later, to a candidate rule. GEPCLASS uses variable-length chromosomes that can have one or more genes. Genes within a given chromosome are of the same size. Using chromosomes with different lengths in the population can introduce healthy genetic diversity during the search.

[3] Freely available at: *http://bioinfo.cpgei.ct.utfpr.edu.br/en/softwares.htm*

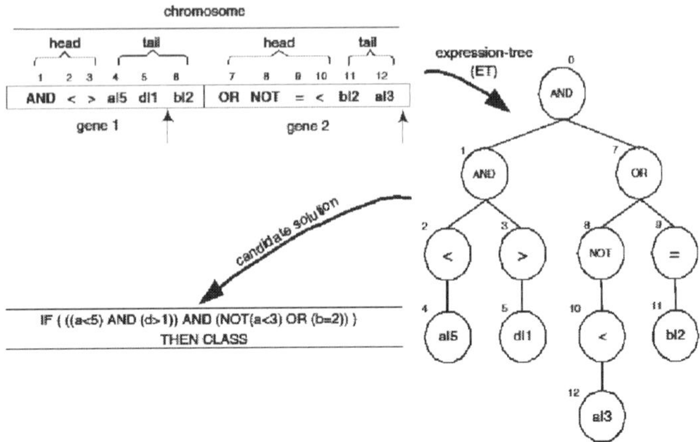

Fig. 4. Example of chromosome structure, expression tree (ET) and rule in GEPCLASS. Adapted from (Weinert & Lopes, 2006).

5. Hidden Markov models

A hidden Markov model (HMM) has an underlying stochastic process that is not observable (it is hidden), but can only be observed through another set of stochastic processes that produce the sequence of observations (Rabiner, 1989). A HMM can be visualized as a finite state machine composed of a finite set of states, $a_1, a_2, \ldots, a_n$, including a beginning and an ending states (G.Mirceva & Davcev, 2009). The HMM can generate a protein sequence by emitting symbols as it progresses through a series of states. Any sequence can be represented by a path through the model. Each state has probabilities associated to it: the transition and the emission probabilities. Although the states are hidden, there are several applications in which the states of the model can have physical meaning.
HMMs offer the advantages of having strong statistical foundations that are well-suited to natural language domains and they are computationally efficient (Seymore et al., 1999). Therefore, HMMs have been used for pattern recognition in many domains and, in special, in Bioinformatics (Durbin et al., 1998; Tavares et al., 2008).

6. Methodology

6.1 The data set and sequence encoding

A number of records of human globular proteins was selected and downloaded from the PDB. The original files were scanned and all annotated secondary structures were extracted by using a parser developed in Java programming language.
These primary sequences extracted from the files had a variable length from 2 to 29 amino acids, and they were divided into five classes, following the PDB nomenclature: *HELIX*, *SHEET*0, *SHEET*1 and *SHEET*2 and *TURN* with 1280, 284, 530, 498 and 119 sequences, respectively. *HELIX*, *SHEET*0 (*SHEET*1 and *SHEET*2) and *TURN* represent α-helices, β-sheets and turns, respectively. In order to properly train the classifiers it is also necessary a "negative" class. That is, a class dissimilar to the others that represent no secondary structure.

This was accomplished by creating a *NULL* class of 3000 variable-sized sequences. Therefore, the database created for this work had 5711 records, with six unbalanced classes.
The natural encoding of the protein sequence is a string of letters from the alphabet of letters representing the 20 proteinogenic amino acids. However, this encoding is not appropriate for some classification algorithms. Thus, another encoding was proposed, converting the string of amino acid symbols into a real-valued vector, by using a physico-chemical property of the amino acids, as suggested by (Weinert & Lopes, 2004). This was accomplished using the Kyte and Doolittle hydrophobicity scale. The real-valued vector was normalized in the range 1.00 - 10.00 (as shown in Table 1).

6.2 Computational methods

Three different computational approaches were used in this work. First, we applied several machine learning algorithms using the Weka workbench, as mentioned before. The algorithms in Weka were grouped into Bayesian, neural networks, meta-learners, trees, lazy-learners and rule-based. The input file for this algorithms was formatted to Attribute-Relation File Format (ARFF). This is an ASCII text file that describes a set of instances sharing a set of attributes.
An ARFF file has two sections: the header and data information. The header of the ARFF file contains the name of the relation, a list of attributes and their types. The convention used for data was: $p_1, p_2, \ldots, p_{29}$, corresponding to positions in the amino acid sequence from 1 to 29, followed by a nominal class attribute, that identifies the class of each instance (*HELIX*, *SHEET0*, *SHEET1*, *SHEET2*, *TURN* and *NULL*). Where p_i are real-valued, as explained in Section 6.1.
A second approach used was HMMs. To test this approach, we have used the HMM package for Weka[4]. Similar to other input files, the input file for this approach was also formatted as ARFF. The HMM classifiers only work on a sequence of data which in Weka is represented as a relational attribute. Data instances have a single nominal class attribute and a single relational sequence attribute. The instances in this relational attribute consist of single nominal data instances using the natural encoding of the protein sequence, i.e., the string of letters representing the amino acid sequence (for example, "GLY, TRP, LEU, ..., LEU").
Finally, Gene Expression Programming (GEP) was used for generating classification rules by means of the software GEPCLASS (Weinert & Lopes, 2006). The input file to GEP is similar to the ARFF file that was used before, but without header information, just with the real-valued data instances as detailed in the section 6.1.

7. Experiments and results

The main objective of this work is to compare the performance of ML algorithms and evolutionary computation approaches for protein secondary structure classification. The main motivation to test several algorithms is to identify the most suitable one for protein secondary structure classification.
The performance of the methods are measured according to their predictive accuracy and other statistical parameters drawn from confusion matrix. The processing time and memory load were not considered for the experimental analysis.
The training/testing methodology includes a 10-fold cross-validation (Kohavi, 1995) in which the data set is divided into 10 parts. In the first round, one part is used for testing and the

[4] Available at *http://www.doc.gold.ac.uk/ mas02mg/software/hmmweka/*

remaining, nine parts are used for training. This procedure is repeated until each partition has been used as the test set. The reported result is the weighted average of the 10 runs. The purpose of cross-validation is to avoid biased results when a small sample of data is used. The outcome of a classifier can lead to four different results, according to what was expected and what, in fact, was obtained. Therefore, when classifying unknown instances, four possible outcomes can be computed:

- tp: true positive: the number of instances (proteins) that are correctly classified, i.e., the algorithm predicts that the protein belongs to a given class and the protein really belongs to that class;
- fn: false positive: the number of instances that are wrongly classified, i.e., the algorithm predicts the protein that belongs to a given class but it does not belong to it;
- tn: true negative: the number of instances that are correctly classified as not belonging to a given class, i.e., the algorithm predicts that the protein does not belong to a given class, and indeed it does not belong to it.
- fp: false negative: the number of instances of a given class that are wrongly classified, i.e., the algorithm predicts that the protein does not belong to a given class but it does belong to it.

Combining the above-cited four outcomes obtained from a classifier, we have then calculated, for each class, metrics commonly used in ML: sensitivity (Se), specificity (Sp), the predictive accuracy and the Matthews Correlation Coefficient (MCC) (Matthews, 1975), defined in Equations 1, 2, 3 and 4, respectively. Then, the weighted average of these metrics was calculated. The results are shown in Table 3. The best results, according to these metrics, are shown in bold in the table.

$$Se = \frac{tp}{tp + fn} \tag{1}$$

$$Sp = \frac{tn}{tn + fp} \tag{2}$$

$$Accuracy = \frac{tp + tn}{tp + fp + fn + tn} \tag{3}$$

$$MCC = \frac{tp \times tn - fp \times fn}{\sqrt{(tp + fp) \times (tp + fn) \times (tn + fp) \times (tn + fn)}} \tag{4}$$

The visual comparison of performance of the classifiers was done using a ROC (Receiver Operating Characteristics) plot (Fawcett, 2006). The ROC plot is a useful technique for visualizing and comparing classifiers and is commonly used in decision making in ML, data mining and Bioinformatics (Sing et al., 2005; Tavares et al., 2008). It is constructed using the performance rate of the classifiers. The ROC analysis can be used for visualizing the behavior of diagnostic systems and medical decisions (Fawcett, 2006). In a ROC plot axes x and y are defined as (1-Sp) and Se, respectively. These axes can be interpreted as the relative trade-offs between the benefits and costs of a classifier. Thus, each classifier correspond to their (1-Sp, Se) pairs. The best prediction would be that lying as close as possible to the upper left corner, representing 100% of sensitivity (no fn) and specificity (no fp). Figure 5 shows the ROC plot for the classifiers evaluated in this work. It should be noted that some of the classifiers have achieved almost the same performance. Therefore, some points are superimposed in

Group	Method	*Se*	*Sp*	*MCC*	Accuracy rate (%)
Bayes	NaiveBayes	0.601	0.910	0.47	56.58
	AODE	0.688	0.292	0.57	68.78
Neural Net	MLP	0.599	0.774	0.39	59.87
	Logistic	0.559	0.604	0.21	55.90
	SMO	**0.792**	0.792	**0.67**	**79.16**
Meta	ADABoost	0.774	0.875	0.65	77.43
	logitBoost	0.769	0.879	0.65	76.89
	Bagging	0.741	0.854	0.60	74.07
Trees	J48	0.719	0.87	0.58	71.88
	RandomForest	0.774	0.874	0.65	77.41
	RepTree	0.717	0.868	0.58	71.72
Lazy	IB1	0.727	0.869	0.59	72.74
	IBk	0.683	0.86	0.53	68.33
	Kstar	0.730	0.874	0.59	72.96
Rules	Jrip	0.667	0.739	0.45	66.70
	PART	0.726	0.862	0.59	72.63
	Ridor	0.670	0.831	0.50	66.99
HMM		0.629	**0.919**	0.54	62.86
GEP		0.750	0.630	0.320	75.43

Table 3. Classifier performance

the graph. The top classifiers are identified in the ROC space: SMO (Sequential Minimal Optimization algorithm), ADABoost (ML meta-algorithm) and RandomForest (collection of tree-structured classifiers).

8. Conclusion and future works

Prediction of secondary structure protein has become an important research area in Bioinformatics. Since the beginning, similar sequences of proteins are used to predict the function of new proteins. In this work, several methods from ML and evolutionary computation for classifying protein secondary structures were compared. Considering sensitivity and specificity alone, SMO and NaiveBayes achieved the highest values, meaning that the first was good to detect secondary structures when they are present, and the latter was good to classify sequences that are not secondary structures. The highest sensitivity value was below 0.8, suggesting the presence of semantic noise in the data set given by the inherent variability of amino acid sequences in the secondary structure of proteins.

According to the results shown in Table 3 and considering the accuracy rate, we can conclude that SMO, Meta classifiers (ADABoost and logitBoost), RandomForest, and GEP methods achieved better performances, all above 75% of accuracy. These results can be considered very expressive, taking into account the difficulty of the classification problem, imposed, as mentioned before, mainly by biological variability of the secondary structure of proteins.

It is also possible to observe that the ROC plot shows the differences between methods more clearly than Table 3, when considering both, sensitivity and specificity. For instance, it is possible to observe that the HMM and NaiveBayes achieved the lowest number of false

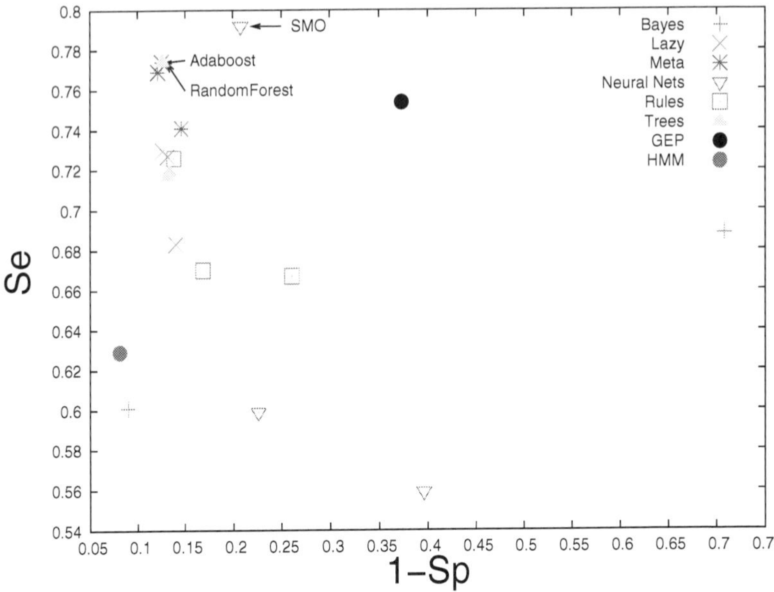

Fig. 5. ROC plot

negatives (highest specificity), however, they perform badly considering sensitivity. In the upper-left corner are the three best-performing methods. However, GEP is slightly distant from them, making clear that that considering only the accuracy rate may be misleading. The analysis of the classifiers using the MCC leads to similar results as the ROC curve. Therefore, based on our results, we can conclude that ROC and MCC are the best way to analyze the performance of methods in this classification problem.

It is important to mention that no effort was done to fine-tune parameters of any method. In all cases, the default parameters of the methods were used. However, it is a matter of fact that adjusting parameters of classifiers (for instance, in GEP and HMM methods), the overall performance can be significantly improved. On the other hand, such procedure could lead to a biased comparison of classifiers.

Although SMO and Meta-learners achieved the best classification performance, it should be highlighted the importance of rule-based methods (including GEP), since a set of classification rules are more comprehensive and expressive to humans than other type of classification method expressed by numbers.

Future work will include the use of hybrid techniques incorporating ML and evolutionary computation methods, as well as hierarchical classification methods.

9. Acknowledgements

This work was partially supported by the Brazilian National Research Council (CNPq) under grant no. 305669/2010-9 to H.S.Lopes; and a doctoral scholarship do C.M.V. Benítez from CAPES-DS.

10. References

Alberts, B., Johnson, A., Lewis, J., Raff, M., Roberts, K. & Walter, P. (2002). *Molecular Biology of The Cell*, Garland Science, New York, USA.

Benítez, C. & Lopes, H. (2010). Hierarchical parallel genetic algorithm applied to the three-dimensional hp side-chain protein folding problem, *Proc. of IEEE International Conference on Systems, Man and Cybernetics*, IEEE Computer Society, pp. 2669–2676.

Berman, H., Westbrook, J., Feng, Z., Gilliland, G., Bhat, T., Weissig, H., Shindyalov, I. & Bourne, P. (2000). The protein databank, *Nucleic Acids Research* 28(1): 235–242.

Bernstein, F., Koetzle, T., William, G., Meyer, D., Brice, M. & Rodgers, J. (1977). The protein databank: A computer-based archival file for macromolecular structures, *Journal of Molecular Biology* 112: 535–542.

Branden, C. & Tooze, J. (1999). *Introduction to Protein Structure*, Garland Publishing, New York, USA.

Breiman, L. (2001). Random forests, *Machine Learning* 45: 5–32.

Chang, B., Ratnaweera, A. & Watson, S. H. H. (2004). Particle swarm optimisation for protein motif discovery, *Genetic Programming and Evolvable Machines* 5: 203–214.

Cheng, J., Greiner, R., Kelly, J., Bell, D. & Liu, W. (2002). Learning bayesian networks from data: An information-theory based approach, *Artificial Intelligence* 137(1-2): 43–90.

Chiba, S., Sugawara, K. & Watanabe, T. (2001). Classification and function estimation of protein by using data compression and genetic algorithms, *Proc. of Congress on Evolutionary Computation*, IEEE Press, Piscataway, USA, pp. 839–844.

Cohen, W. W. (1995). Fast effective rule induction, *In Proceedings of the Twelfth International Conference on Machine Learning*, Morgan Kaufmann, pp. 115–123.

Cooper, G. (2000). *The Cell: A Molecular Approach*, Sinauer Associates, Sunderland, UK.

Davies, M. N., Secker, A., Freitas, A. A., Mendao, M., Timmis, J. & Flower, D. R. (2007). On the hierarchical classification of G protein-coupled receptors, *Bioinformatics* 23(23): 3113–3118.

Drenth, J. (1999). *Principles of Protein X-Ray Crystallography*, Springer-Verlag, New York, USA.

Durbin, R., Eddy, S., Krogh, A. & Mitchison, G. (1998). *Biological Sequence Analysis: Probabilistic Models of Proteins and Nucleic Acids*, Cambridge University Press, Cambridge, USA.

Fawcett, T. (2006). An introduction to ROC analysis, *Pattern Recognition Letters* 27: 861–874.

Ferreira, C. (2001). Gene expression programming: a new adaptative algorithm for solving problems, *Complex Systems* 13(2): 87–129.

Frank, E., Hall, M. & Pfahringer, B. (2003). Locally weighted naive bayes, *Proc. Conference on Uncertainty in Artificial Intelligence*, Morgan Kaufmann.

Frank, E. & Witten, I. H. (1998). Generating accurate rule sets without global optimization, *Proc. 15th International Conference on Machine Learning*, Morgan Kaufmann, pp. 144–151.

Freitas, A. (1998). *Data Mining and Knowledge Discovery with Evolutionary Algorithms*, Heidelberg, Germany.

Freund, Y. & Schapire, R. E. (1999). A short introduction to boosting, *Journal of Japanese Society for Artificial Intelligence* 14(5): 771–780.

Friedman, J., Hastie, T. & Tibshirani, R. (2000). Additive logistic regression: a statistical view of boosting, *Annals of Statistics* 28(2): 337–407.

Mirceva G. & Davcev, D. (2009). HMM-based approach for classifying protein structures, *International Journal of Bio-Science and Bio-Technology* 1(1): 37–46.

Griffiths, A., Miller, J., Suzuki, D., Lewontin, R. & Gelbart, W. (2000). *An Introduction to Genetic Analysis*, 7th edn, W.H. Freeman, New York, USA.

Hardin, C., Pogorelov, T. & Luthey-Schulten, Z. (2002). Ab initio protein structure prediction, *Current Opinion in Structural Biology* 12(2): 176–181.

Hunter, L. (1993). *Artificial Intelligence and Molecular Biology*, 1st edn, AAAI Press, Boston, USA.

Jaroniec, C., MacPhee, C., Bajaj, V., McMahon, M., Dobson, C. & Griffin, R. (2004). High resolution molecular structure of a peptide in an amyloid fibril determined by magic angle spinning NMR spectroscopy, *Proceedings of the National Academy of Sciences of the USA* 101(3): 711–716.

John, G. H. & Langley, P. (1995). Estimating continuous distributions in bayesian classifiers, *Proc. of the Eleventh Conference on Uncertainty in Artificial Intelligence*, Morgan Kaufmann, pp. 338–345.

Kalegari, D. & Lopes, H. (2010). A differential evolution approach for protein structure optimisation using a 2D off-lattice model, *International Journal of Bio-Inspired Computation* 2(3/4): 242–250.

Klabunde, T. & Hessler, G. (2002). Drug design strategies for targeting G-protein-coupled receptors, *Chembiochem* 3(10): 928–944.

Kohavi, R. (1995). A study of cross-validation and bootstrap for accuracy estimation and model selection, *Proc. of 14th International Joint Conference on Artificial Intelligence*, Morgan Kaufmann, San Francisco, USA, pp. 1137–1143.

Kohavi, R. (1996). Scaling up the accuracy of naive-bayes classifiers: a decision-tree hybrid, *Proc. of 2nd International Conference on Knowledge Discovery and Data Mining Conference*, pp. 202–207.

Kotsiantis, S. B. (2007). Supervised machine learning: A review of classification techniques, *Informatica* 31(3): 249–268.

Koza, J. (1996). Classifying protein segments as transmembrane domains using genetic programming and architecture-altering operations, *in* P. Angeline & K. Kinnear Jr. (eds), *Advances in Genetic Programming*, Vol. II, MIT Press, Cambridge, USA.

Kyte, J. & Doolittle, R. (1982). A simple method for displaying the hydropathic character of proteins, *Journal of Molecular Biology* 157: 105–132.

Landwehr, N., Hall, M. & Frank, E. (2005). Logistic model tree, *Machine Learning* 59(1): 161–205.

Lee, M., Duan, Y. & Kollman, P. (2001). State of the art in studying protein folding and protein structure prediction using molecular dynamics methods, *Journal of Molecular Graphics and Modelling* 19(1): 146–149.

Lewis, P., Momany, F. & Scheraga, H. (1973). Chain reversals in proteins, *Biochimica et Biophysica Acta* 303(2): 211–229.

Lodish, H., Berk, A., Matsudaira, P., Kaiser, C., Krieger, M., Scott, M., Zipursky, L. & Darnell, J. (2000). *Molecular Cell Biology*, 4th edn, W.H. Freeman, New York, USA.

Lopes, H. (2008). Evolutionary algorithms for the protein folding problem: A review and current trends, *Computational Intelligence in Biomedicine and Bioinformatics*, Vol. I, Springer-Verlag, Heidelberg, pp. 297–315.

Manning, A., Keane, J., Brass, A. & Goble, C. (1997). Clustering techniques in biological sequence analysis, *in* J. Komorowski & J. Zytkow (eds), *Principles of Data Mining and Knowledge Discovery*, Vol. 1263 of *Lecture Notes in Computer Science*, Springer-Verlag, Heidelberg, Germany, pp. 315–322.

Martin, B. (1995). *Instance-based learning: Nearest neighbor with generalisation*, Master's thesis, Department of Computer Science, University of Waikato, Waikato, New Zealand.

Matthews, B. (1975). Comparison of the predicted and observed secondary structure of T4 phage lysozyme, *Biochimica et Biophysica Acta* 405(2): 442–451.

Mitchell, T. M. (1997). *Machine Learning*, McGraw-Hill, New York, USA.

Nelson, D. & Cox, M. (2008). *Lehninger Principles of Biochemistry*, 5th edn, W.H. Freeman, New York, USA.

Nölting, B. (2006). *Protein Folding Kinetics*, 2nd edn, Springer-Verlag, Berlin, Germany.

Ohkawa, T., Namihira, D., Komoda, N., Kidera, A. & Nakamura, H. (1996). Protein structure classification by structural transformation, *Proc. of IEEE International Joint Symposia on Intelligence and Systems*, IEEE Computer Society Press, Piscataway, USA, pp. 23–29.

Pauling, L., Corey, R. & Branson, H. (1951a). Configurations of polypeptide chains with favored orientations of the polypeptide around single bonds: two pleated sheets, *Proceedings of the National Academy of Sciences of the USA* 37(11): 729–740.

Pauling, L., Corey, R. & Branson, H. (1951b). The structure of proteins: two hydrogen-bonded helicel configurations of the polypeptide chain, *Proceedings of the National Academy of Sciences of the USA* 37: 205–211.

Platt, J. C. (1998). Fast training of support vector machines using sequential minimal optimization, *in* C. B. B. Schoelkopf & A. Smola (eds), *Advances in Kernel Methods*, MIT Press, Cambridge, USA.

Quinlan, J. (1993). *C4.5: Programs for Machine Learning*, San Francisco, USA.

Rabiner, L. (1989). A tutorial on hidden Markov models and selected applications in speech recognition, 77(2): 257–286.

Scapin, M. & Lopes, H. (2007). A hybrid genetic algorithm for the protein folding problem using the 2D-HP lattice model., *in* A. Yang, Y. Shan & L. Bui (eds), *Success in Evolutionary Computation*, Vol. 92 of *Studies in Computational Intelligence*, Springer, Heidelberg, Germany, pp. 205–224.

Seymore, K., McCallum, A. & Rosenfeld, R. (1999). Learning hidden Markov model structure for information extraction, *AAAI 99 Workshop on Machine Learning for Information Extraction*, pp. 37–42.

Shmygelska, A. & Hoos, H. (2005). An ant colony optimisation algorithm for the 2D and 3D hydrophobic polar protein folding problem, *BMC Bioinformatics* 6: 30.

Sing, T., Sander, O., Beerenwinke, N. & Lengauer, T. (2005). ROCR: visualizing classifier performance in R, *Bioinformatics* 21: 3940–3941.

Sunde, M. & Blake, C. (1997). The structure of amyloid fibrils by electron microscopy and X-ray diffraction, *Advances in Protein Chemistry* 50: 123–159.

Tavares, L., Lopes, H. & Lima, C. (2008). A comparative study of machine learning methods for detecting promoters in bacterial DNA sequences, *in* D.-S. Huang, D. S. L. Donald C. Wunsch II & K.-H. Jo (eds), *Advanced Intelligent Computing Theories and Applications*, Vol. 5227 of *Lecture Notes in Computer Science*, Springer-Verlag, Heidelberg, pp. 959–966.

The UniProt Consortium (2010). The universal protein resource (UniProt) in 2010, *Nucleic Acids Research* 38: D142–D148.

Tsunoda, D. & Lopes, H. (2006). Automatic motif discovery in an enzyme database using a genetic algorithm-based approach, *Soft Computing* 10: 325–330.

Tsunoda, D., Lopes, H. & Freitas, A. (2011). A genetic programming method for protein motif discovery and protein classification, *Soft Computing* pp. 1–12. DOI 10.1007/s00500-010-0624-9.

Wang, T. L. J. & Ma, Q. (2000). Application of neural networks to biological data mining: A case study in protein sequence classification, *In Proc. of the Sixth ACM SIGKDD International Conference on Knowledge Discovery and Data Mining*, pp. 305–309.

Webb, G. I. (2000). Multiboosting: a technique for combining boosting and wagging, 40: 159–197.

Webb, G. I., Boughton, J. R. & Wang, Z. (2005). Not so naïve Bayes: aggregating one-dependence estimators, *Machine Learning* .

Weinert, W. & Lopes, H. (2004). Neural networks for protein classification, *Applied Bioinformatics* 3(1): 41–48.

Weinert, W. & Lopes, H. (2006). GEPCLASS: A classification rule discovery tool using gene expression programming, *in* X. Li, O. R. Zaïane & Z.-H. Li (eds), *Advanced Data Mining and Applications*, Vol. 4093 of *Lecture Notes in Computer Science*, Springer-Verlag, Heidelberg, Germany, pp. 871–880.

Witten, I., Frank, E. & Hall, M. (2011). *Data Mining: Practical Machine Learning Tools and Techniques*, 3rd edn, Morgan Kauffmann, San Francisco.

Wolstencroft, K., Lord, P., Tabernero, L., Brass, A. & Stevens, R. (2006). Protein classification using ontology classification, *Bioinformatics* 22(14): e530–e538.

Wüthrich, K. (1986). *NMR of Proteins and Nucleic Acids*, John Wiley & Sons, New York, USA.

Yanikoglu, B. & Erman, B. (2002). Minimum energy configurations of the 2-dimensional HP-model of proteins by self-organizing networks, *Journal of Computational Biology* 9(4): 613–620.

Zheng, Z. & Webb, G. I. (2000). Lazy learning of bayesian rules, *Machine Learning* 41: 53–87.

13

Functional Analysis of the Cervical Carcinoma Transcriptome: Networks and New Genes Associated to Cancer

Mauricio Salcedo et al.*
Laboratorio de Oncología Genómica, Unidad de Investigación en Enfermedades Oncológicas, Hospital de Oncología, CMN-SXXI, IMSS, México DF

1. Introduction

Cancer is one of the most important public health problem in Mexico and worldwide, especially for female population, breast and cervical cancer (CC) types are the most frequent. Incidence rates of CC are higher in developing countries 40/100,000 women per year vs. 10/100,000 in developed countries (1). In Mexico there are 12,000 new reports cases every year (2). The absence of the screening programs or comparatively ineffective screening programs lead to relatively late diagnosis of the disease and also in differences in the human papillomavirus (HPV) infection (3). Several types of HPV are associated with CC worldwide (4, 5), being the HPV16 the most frequent oncogenic type.

Epidemiological and experimental studies suggest that high risk HPV have an important role in cervical carcinogenesis. Persistent viral infection, genetic background in combination with constitutive expression of the viral oncogenes as E6 and E7, are decisive steps for malignant transformation, because these oncoproteins interact with the tumour suppressor proteins p53 and pRB, respectively for their degradation (6, 7). Finally, these interactions could induce cellular proliferation and genetic instability for example, which could promote the accumulation of mutations and aneuploidy (8). In conclusion, viral oncoproteins have a general impact in global profile of expressed genes, which could be analyzed by high-throughput methodologies. One of these techniques is DNA oligonucleotide-based microarray technology, which allows a rapid and high-throughput detection of thousands of transcripts simultaneously (9-11).

It has been published several studies about gene expression profiles in HPV infected cells. Mainly these reports are based on gene expression levels altered by E6 and E7 HPV oncoproteins (12-17). Regarding changes in gene expression profiles in cervical cancer

---

* Sergio Juarez-Mendez[1], Vanessa Villegas-Ruiz[1], Hugo Arreola[1], Oscar Perez[2], Guillermo Gómez[3], Edgar Roman-Bassaure[4], Pablo Romero[1], Raúl Peralta[1]
1 Laboratorio de Oncología Genómica, Unidad de Investigación en Enfermedades Oncológicas, Hospital de Oncología, CMN-SXXI, IMSS, México DF
2 Laboratorio de Oncología Experimental, Instituto Nacional de Pediatría, SS, México
3 Centro Nacional de Clínica de Displasias, Unidad de Oncología, Hospital General de México, SS.
4 Servicio de Oncología, Hospital General de México, SS.

samples, there are a few papers comparing normal cervical expressed genes versus tumors samples (12, 18), the major aim in those studies was to find potential tumor markers with clinical value. At present the list of the potential markers is short (p16, survivin).
We have published some works about alterations in gene expression in CC (19, 20). In those reports we observed that WNT pathway, calcium pathway and some cellular proteases (MMP11, cathepsin F) could be involved in cervical carcinogenesis. Thus, these findings are contributing to our knowledge about alterations in CC pathogenesis.
In the present work our microarray data obtained from microarray assay on CC samples (20) were newly managed and analyzed by using new bioinformatics suite programs and then, to get others genes altered in cervical cancer, as well as to define signaling pathways probably implicated in this type of cancer.

2. Material and methods

Biological samples. Eight squamous CC tissues stage IIB (according of International Federation of Obstetrics and Gynecology, FIGO) HPV16 positive were selected and two healthy "normal" cervical samples were originally studied. The normal samples were collected after hysterectomy by uterine myomatosis without HPV infection history. All DNAs from healthy and cervical cancer samples were subjected to PCR by using general oligonucleotides against to HPV; and to confirm the HPV type, the positive samples were then sequenced (data not shown). An important point to eliminate negative false, only the samples harboured at least 70% of tumour cells or normal epithelial cells were analyzed.
Total RNA was extracted from squamous CC and normal tissues using TRIzol reagent (GIBCO, BRL, USA). RNA was synthesized and labeled with CodeLink Express Assay Reagent Kit (Applied Microarrays, GE Healthcare, USA).

2.1 Microarray platform

CodeLink™ Human Whole Genome Microarrays offer comprehensive coverage of the Human genome, this array have ~57,000 probes and consider transcripts and ESTs (expression tagged sequences). In this system are included: 1,200 genes for oncogenesis process, 1,400 for cell cycle, 1,000 for cell-signaling, 3,000 for metabolism, 1,400 for developmental process, 2,700 of transcription and translation, 1,100 for immune and inflammation response, 800 for protein phosphorylation, 600 for apoptosis, 1,150 for ion transport, 400 for synaptic transmission, 200 for kinases, among other.
This array harbors 45,674 genes (based on unique UniGene IDs), 360 positive controls, 384 negative controls, 100 housekeeping genes, one specific and functionally validated probe and the oligonucleotide probe length of 30-mer.
This platform has been applied in different biological models (21-23). CodeLink Bioarray are recently introduced, single-color oligonucleotide microarrays, which differ from Affymetrix GeneChips in the following aspects: 1) this Bioarray use a single pre-synthesized, pre-validated 30-mer probe to detect each target transcript, whereas Gene-Chips use multiple in-situ synthesized, 25-mer probes; and 2) the surface of CodeLink Bioarrays is made of 3-dimensional aqueous gel matrix, whereas that of Affymetrix GeneChips is made of 2-dimensional glass matrix. These characteristics could suggest that CodeLink Bioarrays behave differently from GeneChips and may require different normalization strategies from the ones optimized for GeneChips (24)(Figure 1).

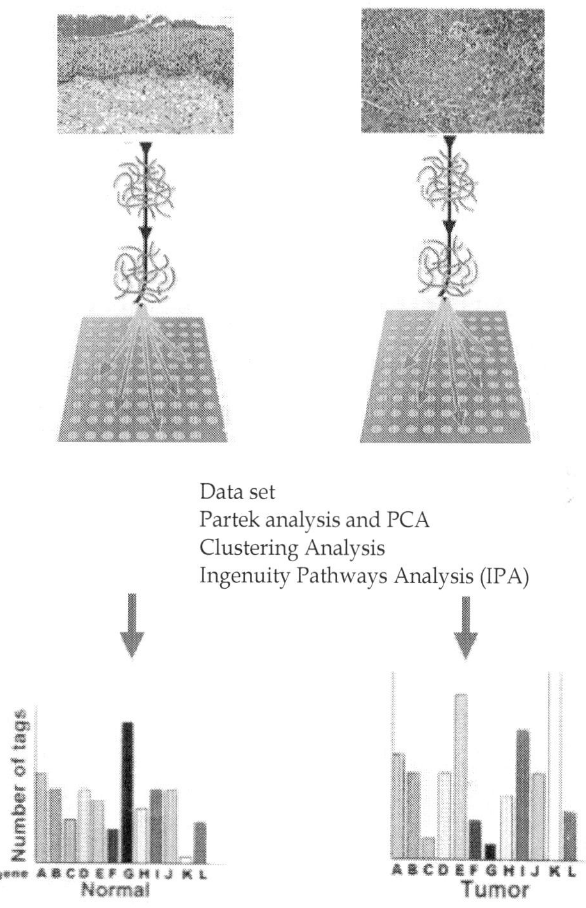

Fig. 1. General strategy of microarray analysis on cervical cancer samples. Originally RNA samples from normal epithelial cells or CC lesions were subjected to hybridization reaction on arrays. The data were analyzed by using different bioinformatics Tools to obtain differentially expressed genes.

3. Analysis of the data

Partek® Genomics Suite™ is a comprehensive suite of advanced statistics and interactive data visualization specifically designed to reliably extract biological signals from noisy data. The commercial software is unique in supporting all microarray and next generation sequencing technologies including gene expression and digital gene expression, exon/alternative splicing, RNA-Seq, copy number and association, ChIP-chip, ChIP-seq, and microRNAs in a single software package, allowing for analysis of multiple applications in one complete solution.

This kind of analysis will provide results with minimal noises generated from the internal controls. To perform this analysis is necessary to apply a software suite which is composed by three different statistical tests: Probe Statistical algorithm (MAS5), Probe Logarithmic Intensity Error (Plier) and Robust Multichip Analysis (RMA). The goal of these tests is to establish differences and similarities between internal controls and to get the most real data. In the present case, the normalized samples were analyzed RMA statistical tests eliminating the tags harboring variations.

In the global gene expression is difficult understand what happen with all of genes in different cellular process including the cancer, in this context a way of visualization data is a Principal Component Analysis or PCA (25). This method is a mathematical technique to reduction the effect of the gene expression sample in a small dimensional space, when there is less changes in the global gene expression the dimensional is smaller. Next for visualization of changes in gene expression in all samples we made a clustering analysis (26), in this method we used a K-means algorithm and let to classify to determine similitude and dissimilitude in all samples, to finish we applied methods of systems biology as Ingenuity Pathway Analysis) and gene classification to determine new list of candidates for subsequent lab verification and might help in the search for a cure for cancers.

3.1 Networks by Ingenuity Pathways Analysis (IPA)

This software (www.ingenuity.com) to help life science researchers explore, interpret, and analyze complex biological Systems, and is used to help researchers analyze 'omics data and model biological systems. This analysis was to identify Networks of interacting genes and other functional groups. A cut-off ratio of 2 was used to define genes.

4. Results

The best and most accurate method for identifying disease-causing genes is monitoring gene expression values in different samples using microarray technology. One of the shortcomings of microarray data is that they provide a small quantity of samples with respect to the number of genes. This problem reduces the classification accuracy of the methods, so gene selection is essential to improve the predictive accuracy and to identify potential marker genes for a disease. Among numerous existing methods for gene selection, PARTEK has become one of the leading methods, but its performance can be reduced because of the small sample size, noisy data and the fact that the methods remove redundant genes.

The original cervical dataset was already published and available in a previous report (20). This dataset was obtained by using CodeLink microarray platform and provides the expression levels of 57,000 probes for 2 normal tissues and 8 cervical cancers HPV16 positive. The data were pre-processed by carrying out a base 10 logarithmic transformation and normalized (see Figure 2). After first analysis of the data, using significance analysis of microarrays 3,248 genes well annotated were identified.

After that, the data already normalized were managed and analyzed using Partek genomics suite version 6.5 (Partek GS) obtaining a list of differentially expressed genes (Tumor versus Normal). Specific and new procedures are actually performed in analysis of microarray data. For instance, after internal control analysis, the raw data are normalized using the common software available on line or some other like Partek GS. This suite provides rigorous and easy-to-use statistical tests for differential expression of genes or exons, and a flexible and powerful statistical test to detect alternative splicing based on a powerful mixed model analysis of variance.

By one-side ANOVA statistical test a small group of 208 genes were selected with a false discovery ratio < 10%, from these, 111 overexpressed genes with a fold change >2, and 97 downregulated genes with a fold change of <-2 genes were observed (Tables 1 and 2).

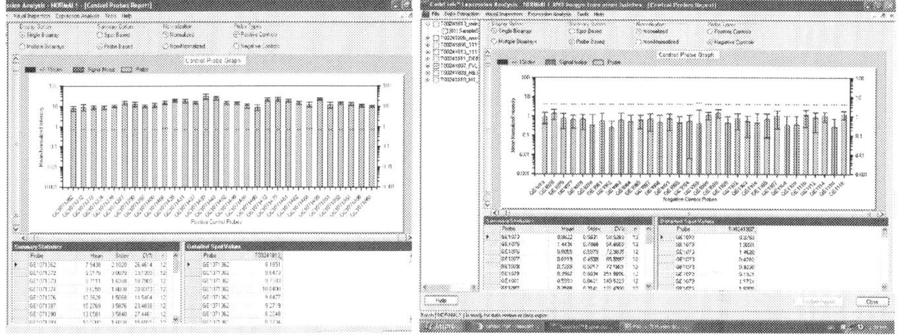

A)

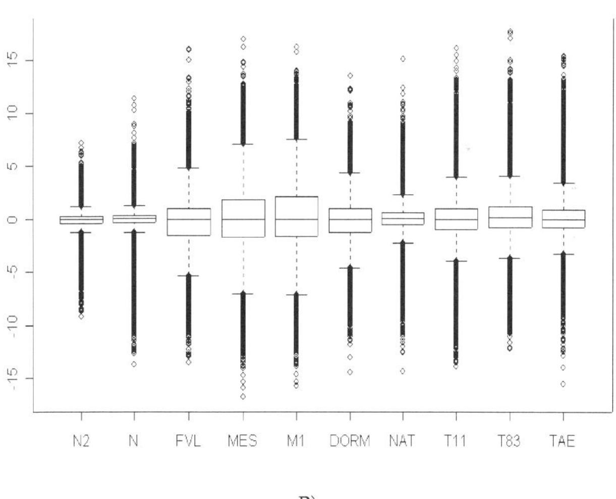

B)

Fig. 2. Normalization data set. The values of gene expression were normalyzed by medians. A) Left image is showing the normalization of positive internal controls harboured in the bioarray. The right image is showing the normalization of negative internal controls of the array. As example, these figures corresponds to a normal samples.

Id	Gene Name	Fold-Change
RASGRP3	RAS guanyl releasing protein 3 (calcium and DAG-regulated)	495.214
BG193781	-	263.988
AI738482	-	180.157
KNG1	kininogen 1	101.811
BM978180	-	98.8303
BG571599	-	86.5592
AHCYL1	adenosylhomocysteinase-like 1	85.9297
LOC196993	-	84.3881
BE869762	-	83.4775
BE217873	-	74.0974
PDC	phosducin	62.2193
BU180997	-	58.1258
AI911397	-	57.9307
T71115	-	51.9557
BI460109	-	51.0251
AI807757	-	50.3766
AA533894	-	49.0591
AI149692	-	48.6084
IL26	interleukin 26	48.4676
C6orf167	MMS22-like, DNA repair protein	41.5497
ZC3HC1	zinc finger, C3HC-type containing 1	40.224
KIAA0367	prune homolog 2 (Drosophila)	38.7778
BU618645	-	38.5742
AI681317	-	37.1748
C10orf6	family with sequence similarity 178, member A	36.6882
AI699935	-	36.6513
BX100783	-	36.5138
GABRB2	gamma-aminobutyric acid (GABA) A recepcor, beta 2	36.3921
AA179557	-	35.8407
CA773815	-	34.9108
CFTR	cystic fibrosis transmembrane conductance regulator (ATP-binding cassette sub-family C, member 7)	33.2819
HIST1H4B	histone cluster 1, H4b	31.9941
AI809950	-	31.7055
MAN1C1	mannosidase, alpha, class 1C, member 1	31.6513
ROR2	receptor tyrosine kinase-like orphan receptor 2	27.7424
BF432424	-	27.2275
AA676547	-	26.9504
BU852916	-	26.6901
AI028168	-	26.5172
BX111626	-	26.4598
BX103473	-	26.3868
PEG10	paternally expressed 10	26.2898
BF513637	-	26.279
BM702800	-	25.9976
AI802092	-	25.4967
BF130283	-	24.1707
CNGA3	cyclic nucleotide gated channel alpha 3	24.1082
LOC377064	-	24.0827
KIAA1045	KIAA1045	24.0353
ST7OT2	ST7 overlapping transcript 2 (non-protein coding)	23.2047
KLHDC1	kelch domain containing 1	23.0623
FLJ00012	ATG16 autophagy related 16-like 2 (S. cerevisiae)	22.9406
AI382167	-	22.7438
FLJ27365	-	22.5175
AA203713	-	20.951
INS	insulin	20.9348
MST1	macrophage stimulating 1 (hepatocyte growth factor-like)	20.9245
AI201147	-	20.8481
DEPDC2	phosphatidylinositol-3,4,5-trisphosphate-dependent Rac exchange factor 2	19.1111
BX104034	-	19.0765
AV652440	-	18.7252
SOAT	sterol O-acyltransferase 1	18.6437
LOC113386	-	18.306
CGI-69	solute carrier family 25, member 39	17.8068
FLJ32312	pseudoundylate synthase 10	17.7345
SMCY	lysine (K)-specific demethylase 5D	17.6437
MYOCD	myocardin	17.6379
BQ019477	-	17.3211
BCMO1	beta-carotene 15,15'-monooxygenase 1	17.1736
GABPB2	GA binding protein transcription factor, beta subunit 2	17.1521
TRY1	protease, serine, 1 (trypsin 1)	17.1496
AW194927	-	17.0466
FOLR3	folate receptor 3 (gamma)	16.8971
ZNF608	zinc finger protein 608	16.426
BM996042	-	16.35
KCNH3	potassium voltage-gated channel, subfamily H (eag-related), member 3	16.084
LOC389072	-	15.7376
LRRC7	leucine rich repeat containing 7	14.923
R39119	-	14.9203
AA227844	-	14.1789
SYMPK	symplekin	13.4518
BX119469	-	13.281
FLJ13941	MORN repeat containing 1	13.0872
AA977184	-	13.0121
AKAP5	A kinase (PRKA) ancher protein 5	12.861
BM993877	-	12.7869
AW140064	-	12.5985
AI809909	-	12.3174
MYOHD1	myosin XIX	11.958
AI821300	-	11.9086
AK123295	-	11.507
C14orf46	lin-52 homolog (C. elegans)	11.1136
AW265107	-	10.8928
BC041899	-	10.7334
AA609479	-	10.5942
AK055332	-	10.071
PARK2	parkinson protein 2, E3 ubiquitin protein ligase (parkin)	9.50627
MAPK1	mitogen-activated protein kinase 1	8.59309
AI792305	-	8.47742
CAMK2G	calcium/calmodulin-dependent protein kinase II gamma	8.30591
SNTB1	syntrophin, beta 1 (dystrophin-associated protein A1, 59kDa, basic component 1)	8.03086
ABLIM1	actin binding LIM protein 1	7.67597
AI139817	-	6.63686
PIP3-E	interaction protein for cytohesin exchange factors 1	6.27178
C22orf1	metallophosphoesterase domain containing 1	6.25449
ARHGAP22	Rho GTPase activating protein 22	5.97438
BC037923	-	5.32521
BI850933	-	5.24072
ADAMTS9	ADAM metallopeptidase with thrombospondin type 1 motif, 9	4.85455
GAS2L2	growth arrest-specific 2 like 2	4.74955
GRIA1	glutamate receptor, ionotropic, AMPA 1	2.5043

Table 1. List of genes and EST's (without gene name) up-regulated

Id	Gene Name	Fold-Change
RASGRP3	RAS guanyl releasing protein 3 (calcium and DAG-regulated)	495.214
KNG1	kininogen 1	101.811
AHCYL1	adenosylhomocysteinase-like 1	85.9297
PDC	phosducin	62.2193
IL26	interleukin 26	48.4676
C6orf167	MMS22-like, DNA repair protein	41.5497
ZC3HC1	zinc finger, C3HC-type containing 1	40.224
KIAA0367	prune homolog 2 (Drosophila)	38.7778
C10orf6	family with sequence similarity 178, member A	36.6882
GABRB2	gamma-aminobutyric acid (GABA) A receptor, beta 2	36.3921
CFTR	cystic fibrosis transmembrane conductance regulator (ATP-binding cassette sub-family C, member 7)	33.2819
HIST1H4B	histone cluster 1, H4b	31.9941
MAN1C1	mannosidase, alpha, class 1C, member 1	31.6513
ROR2	receptor tyrosine kinase-like orphan receptor 2	27.7424
PEG10	paternally expressed 10	26.2898
CNGA3	cyclic nucleotide gated channel alpha 3	24.1082
KIAA1045	KIAA1045	24.0353
ST7OT2	ST7 overlapping transcript 2 (non-protein coding)	23.2047
KLHDC1	kelch domain containing 1	23.0623
FLJ00012	ATG16 autophagy related 16-like 2 (S. cerevisiae)	22.9406
INS	insulin	20.9348
MST1	macrophage stimulating 1 (hepatocyte growth factor-like)	20.9245
DEPDC2	phosphatidylinositol-3,4,5-trisphosphate-dependent Rac exchange factor 2	19.1111
SOAT	sterol O-acyltransferase 1	18.6437
CGI-69	solute carrier family 25, member 39	17.8068
FLJ32312	pseudouridylate synthase 10	17.7345
SMCY	lysine (K)-specific demethylase 5D	17.6437
MYOCD	myocardin	17.6379
BCMO1	beta-carotene 15,15'-monooxygenase 1	17.1736
GABPB2	GA binding protein transcription factor, beta subunit 2	17.1521
TRY1	protease, serine, 1 (trypsin 1)	17.1496
FOLR3	folate receptor 3 (gamma)	16.8971
ZNF608	zinc finger protein 608	16.426
KCNH3	potassium voltage-gated channel, subfamily H (eag-related), member 3	16.084
LRRC7	leucine rich repeat containing 7	14.923
SYMPK	symplekin	13.4518
FLJ13941	MORN repeat containing 1	13.0872
AKAP5	A kinase (PRKA) anchor protein 5	12.861
MYOHD1	myosin XIX	11.958
C14orf46	lin-52 homolog (C. elegans)	11.1136
PARK2	parkinson protein 2, E3 ubiquitin protein ligase (parkin)	9.50627
MAPK1	mitogen-activated protein kinase 1	8.59309
CAMK2G	calcium/calmodulin-dependent protein kinase II gamma	8.30591
SNTB1	syntrophin, beta 1 (dystrophin-associated protein A1, 59kDa, basic component 1)	8.03086
ABLIM1	actin binding LIM protein 1	7.67597
PIP3-E	interaction protein for cytohesin exchange factors 1	6.27178
C22orf1	metallophosphoesterase domain containing 1	6.25449
ARHGAP22	Rho GTPase activating protein 22	5.97438
ADAMTS9	ADAM metallopeptidase with thrombospondin type 1 motif, 9	4.85455
GAS2L2	growth arrest-specific 2 like 2	4.74955
GRIA1	glutamate receptor, ionotropic, AMPA 1	2.5043

Table 1. List of genes up-regulated

Id	Gene Name	Fold-Change
SCGB1D2	secretoglobin, family 1D, member 2	-186.686
KRT4	keratin 4	-173.876
FOS	FBJ murine osteosarcoma viral oncogene homolog	-138.062
MAL	mal, T-cell differentiation protein	-133.173
WISP2	WNT1 inducible signaling pathway protein 2	-90.6242
UNQ698	suprabasin	-88.3632
FLJ22655	RERG/RAS-like	-88.2086
SPARCL1	SPARC-like 1 (hevin)	-86.9887
MGC45780	scavenger receptor class A, member 5 (putative)	-78.2205
RGS5	regulator of G-protein signaling 5	-73.238
MAMDC2	MAM domain containing 2	-69.9879
IGFBP6	insulin-like growth factor binding protein 6	-68.8765
C1orf10	cornulin	-67.0584
FHL1	four and a half LIM domains 1	-64.6338
TNA	C-type lectin domain family 3, member B	-59.3405
W57655	-	-57.5288
PLAC9	placenta-specific 9	-51.9896
DF	complement factor D (adipsin)	-50.8317
UNQ467	keratinocyte differentiation-associated protein	-50.2533
PTGDS	prostaglandin D2 synthase 21kDa (brain)	-50.1348
SCGB2A2	secretoglobin, family 2A, member 2	-49.633
DPT	dermatopontin	-47.7048
CRISP3	cysteine-rich secretory protein 3	-46.6048
BNC2	basonuclin 2	-46.1538
OGN	osteoglycin	-44.5308
EDN3	endothelin 3	-42.5161
DPT	dermatopontin	-38.5712
FXYD1	FXYD domain containing ion transport regulator 1	-36.7655
PPP1R3C	protein phosphatase 1, regulatory (inhibitor) subunit 3C	-35.3073
OSR2	odd-skipped related 2 (Drosophila)	-34.3512
ANKRD25	KN motif and ankyrin repeat domains 2	-32.6406
BF941677	-	-30.9503
MITF	microphthalmia-associated transcription factor	-30.7002
BQ007074	-	-30.1011
AL137566	-	-29.833
FBLN5	fibulin 5	-29.4494
PGM5	phosphoglucomutase 5	-28.8192
CRYAB	crystallin, alpha B	-28.3094
MFAP4	microfibrillar-associated protein 4	-27.6357
AA101632	-	-27.2784
COLEC12	collectin sub-family member 12	-26.3813
PCOLCE2	procollagen C-endopeptidase enhancer 2	-26.1851
COL14A1	collagen, type XIV, alpha 1	-25.1745
ARHGAP6	Rho GTPase activating protein 6	-24.9415
AEBP1	AE binding protein 1	-24.5738
TENC1	tensin like C1 domain containing phosphatase (tensin 2)	-22.8413
ANGPTL2	angiopoietin-like 2	-22.7692
COL4A1	collagen, type IV, alpha 1	-22.2278
LMO3	LIM domain only 3 (rhombotin-like 2)	-22.0822
NDRG2	NDRG family member 2	-21.5607
FLJ10970	transmembrane protein 100	-21.5054
CLCA4	chloride channel accessory 4	-21.3412
COX7A1	cytochrome c oxidase subunit VIIa polypeptide 1 (muscle)	-20.3295
MITF	microphthalmia-associated transcription factor	-20.1976
GNG11	guanine nucleotide binding protein (G protein), gamma 11	-20.1211
TACC1	transforming, acidic coiled-coil containing protein 1	-19.9851
ECG2	serine peptidase inhibitor, Kazal type 7 (putative)	-19.9095
AI734212	-	-19.527
AI765637	-	-19.1416
EBF	early B-cell factor 1	-18.9832
BAI3	brain-specific angiogenesis inhibitor 3	-18.8985
ARHGAP28	Rho GTPase activating protein 28	-18.5654
KCNAB1	:assium voltage-gated channel, shaker-related subfamily, beta membe	-18.2379
AA578982	-	-18.1596
CECR6	cat eye syndrome chromosome region, candidate 6	-18.0476
AF321976	-	-17.2499
EDN3	endothelin 3	-16.2896
MYLK	myosin light chain kinase	-16.0175
C7	complement component 7	-15.7473

Table 2. List of genes and EST's (without gene name) down-regulated

Id	Gene Name	Fold-Change
SCGB1D2	secretoglobin, family 1D, member 2	-186.686
KRT4	keratin 4	-173.876
FOS	FBJ murine osteosarcoma viral oncogene homolog	-138.062
MAL	mal, T-cell differentiation protein	-133.173
WISP2	WNT1 inducible signaling pathway protein 2	-90.6242
UNQ698	suprabasin	-88.3632
FLJ22655	RERG/RAS-like	-88.2086
SPARCL1	SPARC-like 1 (hevin)	-86.9887
MGC45780	scavenger receptor class A, member 5 (putative)	-78.2205
RGS5	regulator of G-protein signaling 5	-73.238
MAMDC2	MAM domain containing 2	-69.9879
IGFBP6	insulin-like growth factor binding protein 6	-68.8765
C1orf10	cornulin	-67.0584
FHL1	four and a half LIM domains 1	-64.6338
TNA	C-type lectin domain family 3, member B	-59.3405
PLAC9	placenta-specific 9	-51.9896
DF	complement factor D (adipsin)	-50.8317
UNQ467	keratinocyte differentiation-associated protein	-50.2533
PTGDS	prostaglandin D2 synthase 21kDa (brain)	-50.1348
SCGB2A2	secretoglobin, family 2A, member 2	-49.633
DPT	dermatopontin	-47.7048
CRISP3	cysteine-rich secretory protein 3	-46.6048
BNC2	basonuclin 2	-46.1538
OGN	osteoglycin	-44.5308
EDN3	endothelin 3	-42.5161
DPT	dermatopontin	-38.5712
FXYD1	FXYD domain containing ion transport regulator 1	-36.7655
PPP1R3C	protein phosphatase 1, regulatory (inhibitor) subunit 3C	-35.3073
OSR2	odd-skipped related 2 (Drosophila)	-34.3512
ANKRD25	KN motif and ankyrin repeat domains 2	-32.6406
MITF	microphthalmia-associated transcription factor	-30.7002
FBLN5	fibulin 5	-29.4494
PGM5	phosphoglucomutase 5	-28.8192
CRYAB	crystallin, alpha B	-28.3094
MFAP4	microfibrillar-associated protein 4	-27.6357
COLEC12	collectin sub-family member 12	-26.3813
PCOLCE2	procollagen C-endopeptidase enhancer 2	-26.1851
COL14A1	collagen, type XIV, alpha 1	-25.1745
ARHGAP6	Rho GTPase activating protein 6	-24.9415
AEBP1	AE binding protein 1	-24.5738
TENC1	tensin like C1 domain containing phosphatase (tensin 2)	-22.8413
ANGPTL2	angiopoietin-like 2	-22.7692
COL4A1	collagen, type IV, alpha 1	-22.2278
LMO3	LIM domain only 3 (rhombotin-like 2)	-22.0822
NDRG2	NDRG family member 2	-21.5607
FLJ10970	transmembrane protein 100	-21.5054
CLCA4	chloride channel accessory 4	-21.3412
COX7A1	cytochrome c oxidase subunit VIIa polypeptide 1 (muscle)	-20.3295
MITF	microphthalmia-associated transcription factor	-20.1976
GNG11	guanine nucleotide binding protein (G protein), gamma 11	-20.1211
TACC1	transforming, acidic coiled-coil containing protein 1	-19.9851
ECG2	serine peptidase inhibitor, Kazal type 7 (putative)	-19.9095
EBF	early B-cell factor 1	-18.9832
BAI3	brain-specific angiogenesis inhibitor 3	-18.8985
ARHGAP28	Rho GTPase activating protein 28	-18.5654
KCNAB1	potassium voltage-gated channel, shaker-related subfamily, beta member 1	-18.2379
CECR6	cat eye syndrome chromosome region, candidate 6	-18.0476
EDN3	endothelin 3	-16.2896
MYLK	myosin light chain kinase	-16.0175
C7	complement component 7	-15.7473
CCNG1	cyclin G1	-15.5806
SLC16A9	solute carrier family 16, member 9 (monocarboxylic acid transporter 9)	-15.5025
EMCN	endomucin	-15.2301
NOPE	immunoglobulin superfamily, DCC subclass, member 4	-15.0347
CLST11240	HIG1 hypoxia inducible domain family, member 1B	-14.9221
PRLR	prolactin receptor	-14.8317
EPHA3	EPH receptor A3	-14.6125
ZNF404	zinc finger protein 404	-14.4942
C21orf34	chromosome 21 open reading frame 34	-13.8307
NDN	necdin homolog (mouse)	-13.5665
A2M	alpha-2-macroglobulin	-13.2787
HPSE2	heparanase 2	-13.2025
ZNF471	zinc finger protein 471	-12.8793
C9orf13	sushi, von Willebrand factor type A, EGF and pentraxin domain containing 1	-12.4415
GYPC	glycophorin C (Gerbich blood group)	-12.1484
DSCR1L1	regulator of calcineurin 2	-11.7187
IL17D	interleukin 17D	-11.3345
RGS5	regulator of G-protein signaling 5	-10.2954
OSBPL1A	oxysterol binding protein-like 1A	-10.2234
C21orf34	chromosome 21 open reading frame 34	-9.41903
LIMS2	LIM and senescent cell antigen-like domains 2	-9.37894

Table 2. List of genes down-regulated

4.1 Analyzing microarray data with novel software suite

If we want to display the data in just two dimensions, we want as much of the variation in the data as possible captured in just two dimensions. Principal component analysis or PCA has been developed for this purpose. Applying this PCA method in the cervical data we observed some expected differences (Figure 3).

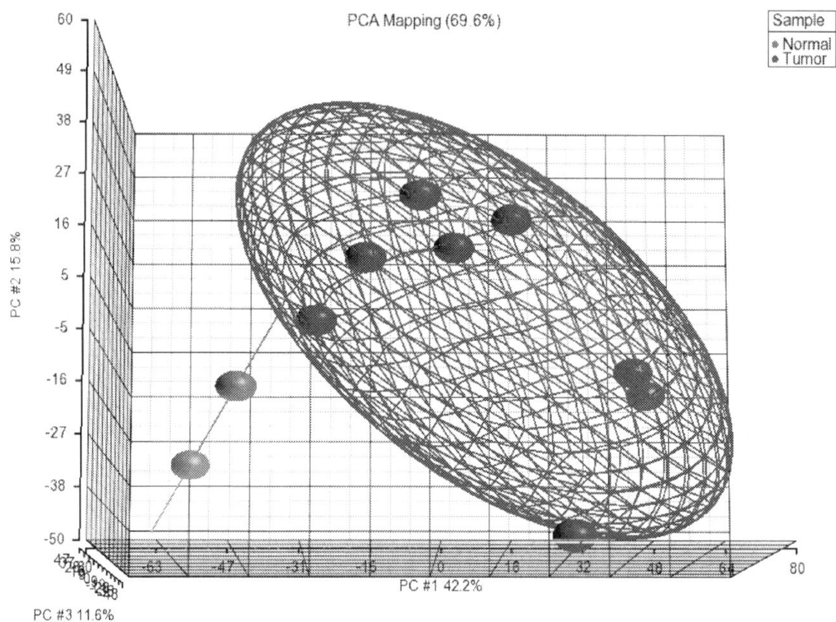

Fig. 3. Principal component analysis of cervical cancer samples. As expected, after perform PCA analysis, the normal samples (red balls) were grouped out of CC Group (blue balls) in a 3-D image.

In order to obtain a graphical representation of the differences between normal and tumor tissues, hierarchical cluster analysis was performed on all samples with a pseudo-color visualization matrix of the 208 selected genes grouping with greater intra-group similarity and differences between groups. The phylogenetic tree resulting from the hierarchical complete linkage-clustering algorithm is shown in figure 4. The figure shows those genes that are changing respect to health cervical tissue. In this method of clustering, allows to do relationships among objects (genes or samples) and are represented by a tree whose branch lengths reflect the degree of similarity between the objects, as assessed by a pairwise similarity function. The computed trees can be used to arrange individual samples in the original data table; this allows the samples or groups of samples with similar expression patterns to be shown adjacent to each other.

In general, tumor samples showed heterogeneity among them compared with normal samples, which had a more homogeneous gene expression profile. So, clustering analysis in CC failed to show significant segregation of patients based on expression profiling possibly

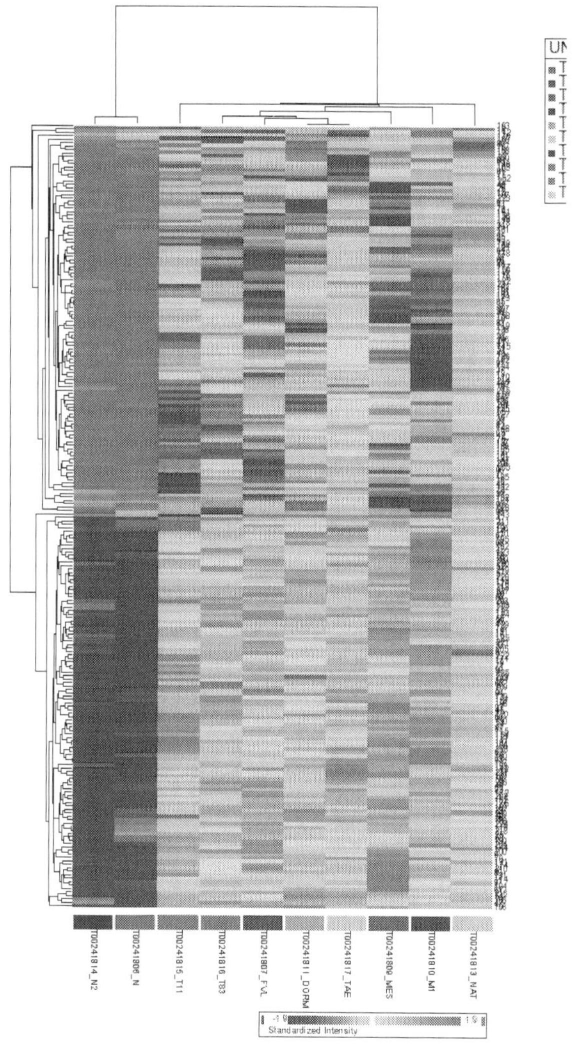

Fig. 4. Hierarchical clustering of the gene expression data for cervical tissues. Clustering analysis were performed for all tumours and normal samples. The data were clustered using the standard hierarchical method with ward linkage and using the Perason correlation to determine distance tumour. Before clustering, the data was filtered to remove genes that were scored absent in 75% or more of the samples as they are likely to be measuring noise in the system. The cluster of normal samples exhibit nearly identical patterns of gene expression changes, on contrary, as expected the invasive samples grouped in a different branch showing a heterogeneous gene expression. Blue color indentify downregulation and red color an overexpression status.

due to the heterogeneous nature of the samples as well as the relatively small numbers of samples in this study. Even when the samples were subjected to rigorous procedures of analysis, the special selection of the patients, including age, clinical stage, HPV16 positive and contraceptive oral status avoiding any bias, in this stage of the carcinogenesis process (stage IIb) the pattern of gene expression is quite different between samples.

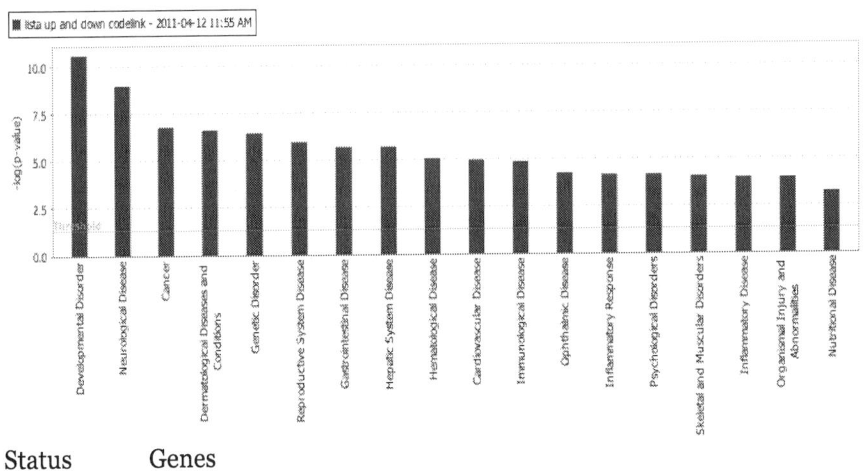

Status	Genes
Developmental disorder	
Up	*ALDH1A1, ALDOC, APAF1, BMP4, COMP, CREB1, CRHR1,COK1, DSP, FGF2, GHSR, GLI2, HPRT1, HRH, IL11, PTEN, ROR2, SLC16A2*
Down	*AGER, AHR, AGR1, ASCL1, ASPH, ATM, CAV3, DCC, DRD3, FANCD2, FGF8, FSL2, GATA4, GPX3, H2FZ, IKBKB, PTK2B, RMRP, TEAD1, TNNT2*
Neurological disease	
Up	*ALDH1A1, CD5, CENTFR, CREB1, CXCL10, E3F5, FGF2, GFI1, HLA-B, HPRT1, HRH1, HSPA5, IL1RN, IMPA2, KCNC3, MED12, PTEN, SCLC16A2,* ***PARK2, PEG10***
Down	*ADCYAP1, ADCYAP1R, AGER, ATM, CHI3L1, CHRNB2, DRD3, GJB2, IKBK, LPAR1, NDP, S1PR1*
Cancer tumorigenesis	
Up	*AKR1C1/2, ALDH1A1, ALDOC, APAF1, BMP4, BNC1, CALB2, CASP7, CD36, CD3G, CD247, CRABP1 CREB1, CTSL1, CXCL2, DAB2, DNASE1, DOK1, DSP, ENPEP, EPS15, FDXR, FGF2, GFI1, GLI2, HLA-B, HLA-E. HMX2,* ***HOXA9****, HPRT1, HR, HRH1, HSPA5, IL1RN, IL7R, INPP5D, ITGAE, KLF9, PTEN, RHOG, TXL1*
Down	*AGER, AHR, AMFR, AMH, ASCL1, ASPH, ATM, ATP2A2, BMP3, CFH, CFHR1, CHI3L1, CHRNA3, CHRBN2, CHRNE, CHRNG, CST6, DCC, DRDE, EIF4A1, EPJB6, FANCD2, FGF8, FHL2, FOLR1, GATA2, GJB2, GPX3, IGLL1, IKBK, LPAR, LTB4R, S1PR1, ADCYAPP1R1*

Table 3. Disease and Disorders functions in cervical cancer

We used IPA to investigate the biological relevant of the observed genome-wide expressed gene changes by categorizing our data set into biological functions and/or diseases Ingenuity Pathway analysis was applied. The 208 genes annotated list by PARTEK analysis was submitted to the visualization IPA tool. This bioinformatics tool is employed for visualizing expression data in the context of KEGG biological pathways; the importance IPA is that retrieves an impact factor (IF) of genes that entire pathway involved, which can help to obtain a clearer notion of the alteration level in each biological pathway, and understand the complexity of these different process of the cancer cell. We imported a list of significantly up and down regulated genes (with extension .txt) into the program to convert the expression data into illustrations in an attempt to explore altered mechanisms in CC. To overcome any possible incorrect IF in altered pathways due to different size of samples, we submitted a similar quantity of up and down regulated genes. This allowed confirming that genes involved in several metabolic pathways were altered in CC (see networks).
We were able to associate biological functions and diseases to the experimental results. Fifteen pathways were obtained with a high score. Table 3 is showing the genes and the top three disorders/disease of "small networks" based in the analysis of the data. As can be seen, a clear route in cancer as it is known was not observed but some genes have been previously associated; however, these data give important information involving *"non canonical"* pathways in cancer.
Finally, in the Figure 5 is showed a "hypothetical network in CC" based from the 15 small networks. In addition to gene expression values, the proposed method uses Gene Ontology, which is a reliable source of information on genes. The use of Gene Ontology can compensate, in part, for the limitations of microarrays, such as having a small number of samples and erroneous measurement results.

5. Discussion

In our results, non classical "cancer genes" were conserved, respect to expected genes as MYC, FOS, RB, P53, HIF, etc. However, in the "strict sense of the word" when is considered a cancer gene? By instance, over-expression, down-regulation, point mutation, amplification, loss of heterozygosity, polymorphisms, epigenetic changes, etc. Thus, any gene could be considered like cancer gene, if they are following special criteria as recently was reported (27).
In this context, we decided to explore two non-related genes in cervical cancer *PARK2* gene. Interestingly, *PARK2* gene mutations (point mutations and exonic deletions) were first identified in autosomal recessive juvenile-onset parkinsonism. This gene is mapped to 6q25.2-q27 containing 12 small exons, and encodes parkin protein which functions as an E3 ligase, ubiquitinating proteins for destruction by the proteosome. Several substrates for parkin have been identified, including a 22kD glycosolated form of synuclein, parkin-associated endothelin receptor-like receptor (Pael-R), and CDCrel-1. Over-expression of Pael-R causes it to become ubiquinated, insoluble, and unfolded, and lead to endoplasmic reticulum stress and cell death (for review see 28). The location of Parkin is in a chromosomal region that is frequently deleted in multiple tumor types, including hepatocellular carcinoma (HCC), ovarian cancer, and breast cancer. The Parkin gene is within FRA6E, the third most active common fragile site (29,30). Interestingly, all three fragile sites regions were found consistently deleted in HCC (31) as well as in ovarian, breast, and prostate cancers. Further PARKIN protein overexpression did not lead to

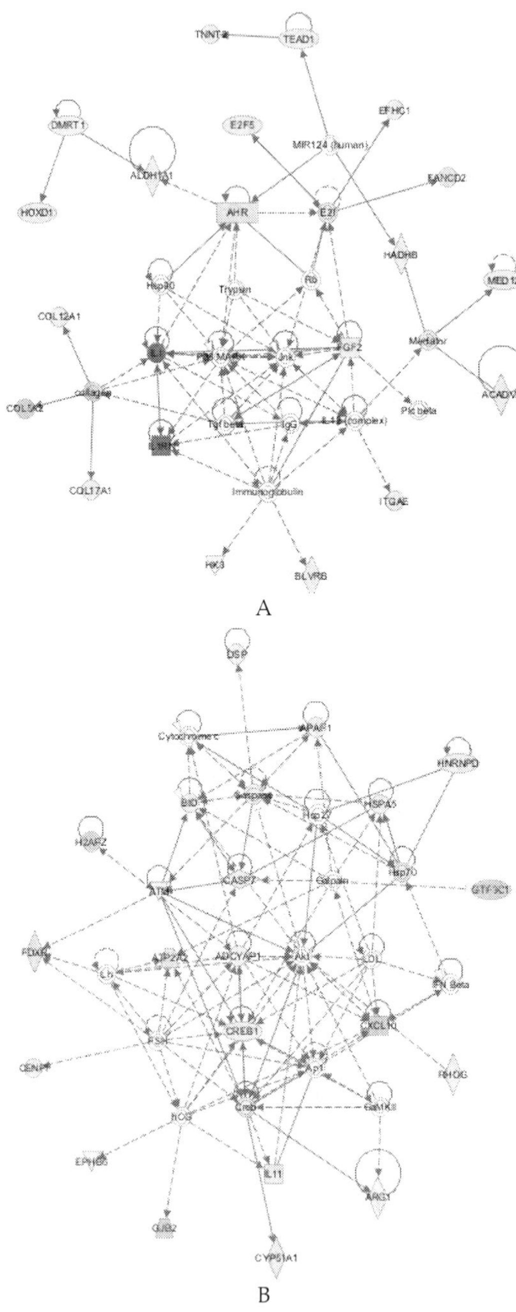
TEAD1
DMRT1
E2F5
EFHC1
MIR124 (human)
FANCD2
HOXD1
AHR
HADHB
MED12
Hsp90
Trypsin
Rb
COL12A1
Mediator
Jnk
ACADVL
Tgf beta
IgG
IL12 (complex)
COL17A1
Immunoglobulin
ITGAE
HK3
BLVRB
A
DSP
Cytochrome c
APAF1
HNRNPD
BID
Hsp27
HSPA5
H2AFZ
Hsp70
CASP7
Calpain
GTF3C1
Lh
ADCYAP1
Akt
LDL
IFN Beta
CREB1
FSH
CENPF
Ap1
RHOG
Creb
hCG
IL11
ARG1
GJB2
CYP51A1
B

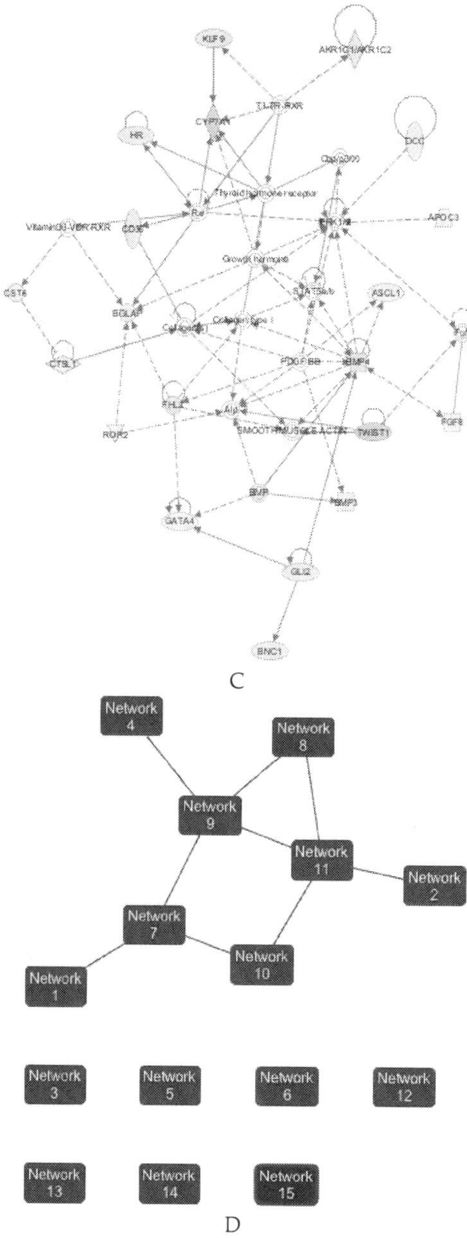

Fig. 5. Networks in cervical cancer. Top three networks diagrams generated as graphical representations of the molecular relationship between genes and gene product. The gene

products are represented as nodes (shapes) and the biological relationship between two nodes is represented as an edge (line). A) Network 1. skeletal and muscular system development and function, embryonic development, tissue development, B) network 2; dermatological diseases and conditions, cardiovascular disease, organismal injury and abnormalities, C) network 3; cell cycle, organismal functions, organismal injury and abnormalities, D) Ingenuity Pathways Analysis network of genes associated with CC. This network diagram shows the biological association of 8 focus networks associated with cervical cancer as graphical representation of the molecular relationship between genes/gene produtcs.

increased sensitivity to all pro-apoptotic induction but may show specificity for a certain type of cellular stress. At present, *Parkin* gene could be considered as new tumor suppressor gene (32). In our case, *per se* Parkin expression in CC is interesting. It is widely described that p53 master gene is constitutively expressed but under stress conditions as HPV infection, DNA damage or point mutations its half-time life of the protein is increased. Similar situation might be observed for *Parkin* gene due to increased expression in CC. This could be supported because the 6q25 cytogenetic region in CC is not altered as happen for *TP53* gene (33). To this respect, we could hypothesize that a new overexpression of *Parkin* gene could be involved in invasion cervical carcinogenesis. These findings could demonstrate that the genetic context in which a mutation occurs can play a significant role in determining the type of illness produced or associated.

It has been established that although we inherit two copies of all genes (except those that reside on the sex chromosomes), there is a subset of these genes in which only the paternal or maternal copy is functional. This phenomenon of monoallelic, parent-of-origin expression of genes is termed genomic imprinting. Imprinted genes are normally involved in embryonic growth and behavioral development, but occasionally they also function inappropriately as oncogenes and tumor suppressor genes (34). Furthermore, it is well know that a variety of genetic changes influence the development and progression of cancer. These changes may result from inherited or spontaneous mutations that are not corrected by repair mechanisms prior to DNA replication. It is increasingly clear that so called epigenetic effects that do not affect the primary sequence of the genome also play an important role in tumorigenesis (35).

Other gene overexpressed seen in this analysis was *PEG10* gene. This gene is mapped in chromosme 7q21. PEG10 protein prevents apoptosis in hepatocellular carcinoma cells through interaction with SIAH1, a mediator of apoptosis. May also have a role in cell growth promotion and hepatoma formation. Inhibits the TGF-beta signaling by interacting with the TGF-beta receptor ALK1. This is a paternally expressed imprinted gene that encodes transcripts containing two overlapping open reading frames (ORFs), RF1 and RF1/RF2, as well as retroviral-like slippage and pseudoknot elements, which can induce a -1 nucleotide frame-shift. Increased expression of this gene is associated with hepatocellular carcinomas. These findings link to cancer genetics and epigenetic by showing that a classic proto-oncogene, *MYC*, acts directly upstream of a proliferation-positive imprinted gene, PEG10 (36,37).

The *HOX* genes are a family of transcription factors that bind to specific sequences of DNA in target genes regulating their expression. The role of *HOX* genes in adult cell differentiation is still obscure, but growing evidence suggests that they may play an important role in the development of cancer. We have previously reported that some *HOX* genes could be related to CC. Specifically, *HOXA9* was observed expressed in cervical cancer by RT-PCR end point. In the present work, the data are showing that statistically significative *HOXA9* gene is differentially expressed in CC. Together to *HOXB13, D9, D10,* and *HOXC* cluster (*HOXC9, C11-C13*) genes this family of genes might be an important factor involved in CC (35).

It is clear that the most altered genes in CC are not commonly associated to cancer process. This fact could suggest: 1) the "classic genes of cancer" are statistically significant altered with tiny values, but there are some exceptions and specific tumor types as neuroblastomas and *N-MYC* gene, *Her2/neu* in breast cancer. 2) At least in stage IIb of cervical carcinogenesis could be involved genes related to "cellular economy" but not belonging to genes of cancer. This is supported by recent reports showing molecular alterations in genes not previously related to cancer (27). 3) The extreme values (high or low) in microarray analysis not always represent strong candidates of markers in the models performed. What about the most frequent?. (4) Integrative genomics, multicentre protocols in well selected samples, stratified stages and clinical follow-up, will be the clue to get cancer hallmarks. In addition, the study of the molecular function of selected genes strengthened the hypothesis that these genes are involved in the process of cancer growth.

The data information obtained from microarray analysis should be validated because can appear errors in positive and negative false due to nature of the massive assays. In this context, in order to confirm the microarray data, additional molecular tool as end point PCR, real time PCR, northern blot, immunohistochemistry should be performed and to obtain results. (Mendez S. An Integrative microarray gene expression analysis, approach identifies candidates' array multi-experiments in Ovary Tumours, submitted to publication 2011).

6. Acknowledgements

This work was partially supported by CONACYT (México) grants 69719 AND 87244 from FONDOS SECTORIALES. We appreciate the technical assistance of Laboratorio de Oncología Genómica, CIS, HO-IMSS. Sergio JUAREZ and Mauricio SALCEDO made similar efforts in the present work.

7. References

[1] Parkin DM, Bray F, Ferlay J, Pisani P. Global cancer statistics, 2002. CA Cancer J Clin. 2005, 55(2):74-108

[2] Secretaría de Salud (Méx). Registro Histopatológico de Neoplasias en México. México: Secretaría de Salud, México D.F; 1999.

[3] Munoz N, Xavier Bosch F. Cervical cancer and human papillomavirus: epidemiological evidence and perspectives for prevention. Salud Publica Mex 1997; 39:274-282.

[4] Boffetta P, Parkin DM. Cancer in developing countries. CA Cancer J Clin 1994 44:81-90

[5] Walboomers JM, Jacobs MV, Manos MM, Bosch FX, Kummer JA, Shah K, et al. Human papillomavirus is a necessary cause of invasive cervical cancer worldwide. J Pathol 1999, 189: 12-19.

[6] Scheffner, M., B. A. Werness, J. M. Huibregtse, A. J. Levine, and P. M. Howley. The E6 oncoprotein encoded by human papillomavirus types 16 and 18 promotes the degradation of p53. Cell 1990, 63:1129–1136.

[7] Boyer SN, Wazer DE, Band V. E7 protein of human papilloma virus-16 induces degradation of retinoblastoma protein through the ubiquitin-proteasome pathway. Cancer Res. 1996, 56(20):4620-4624.

[8] Duensing S, Munger K. The human papillomavirus type 16 E6 and E7 oncoproteins independently induce numerical and structural chromosome instability. Cancer Res. 2002, 62(23):7075-7082.

[9] Nees M, van Wijngaarden E, Bakos E, Schneider A, Dürst M. Identifcation of novel molecular markers which correlate with HPVinduced tumor progression. Oncogene 1998, 16:2447–2458.

[10] Brown P, Botstein R. Exploring the new world of the genome with DNA microarrays. Nature Genet 1999, 21 (suppl):33-37.

[11] Lockhart DJ, Winzeler EA. Genomics, gene expression and DNA arrays. Nature, 2000, 405:827-836.

[12] Ruutu M, Peitsaro P, Johansson B, Syrjanen S: Transcriptional profiling of a human papillomavirus 33-positive squamous epithelial cell line which acquired a selective growth advantage after viral integration. Int J Cancer 2002, 100: 318-326.

[13] Duffy CL, Phillips SL, Klingelhutz AJ: Microarray analysis identifies differentiation-associated genes regulated by human papillomavirus type 16 E6. Virol 2003, 314: 196-205.

[14] Thomas JT, Oh ST, Terhune SS, Laimins LA: Cellular changes induced by low-risk human papillomavirus type 11 in keratinocytes that stably maintain viral episomes. J Virol 2001, 75: 7564-7571.

[15] Garner-Hamrick PA, Fostel JM, Chien WM, Banerjee NS, Chow LT, Broker TR, Fisher C: Global effects of human papillomavirus type 18 E6/E7 in an organotypic keratinocyte culture system. J. Virol 2004, 78: 9041-9050.

[16] Toussaint-Smith E, Donner DB, Roman A: Expression of human papillomavirus type 16 E6 , E7 oncoproteins in primary foreskin keratinocytes is sufficient to alter the expression of angiogenic factors. Oncogene 2004, 23: 2988-2995.

[17] Nees M, Geoghegan JM, Hyman T, Frank S, Miller L, Woodworth CD: Papillomavirus type 16 oncogenes downregulate expression of interferon-responsive genes and upregulate proliferation-associated, NF-kappaB-responsive genes in cervical keratinocytes. J Virol 2001, 75: 4283-4296.

[18] Shim C, Zhang, Hun C, Lee. Profiling of differentially expressed genes in human primary cervical cancer by complementary DNA expression array. Clin Cancer Res 1998, 4:3045-3050.

[19] Vazquez-Ortiz G, García JA, Ciudad CJ, Noé V, Peñuelas S, López-Romero R, Mendoza-Lorenzo P, Piña-Sánchez P, Salcedo MDifferentially expressed genes between high-risk human papillomavirus types in human cervical cancer cells. Int J Gynecol Cancer. 2007, 17:484-491

[20] Pérez-Plasencia C, Vázquez-Ortiz G, López-Romero R, Piña-Sanchez P, Moreno J, Salcedo M. Genome wide expression analysis in HPV16 cervical cancer: identification of altered metabolic pathways. Infect Agent Cancer. 2007, 6;2:16

[21] Davidson LA, Nguyen DV, Hokanson RM, Callaway ES, Isett RB, Turner ND, Dougherty ER, Wang N, Lupton JR, Carroll RJ, Chapkin RS. Chemopreventive n-3 polyunsaturated fatty acids reprogram genetic signatures during colon cancer initiation and progression in the rat. Cancer Res. 2004, 64:6797-6804.

[22] Ramakrishnan R, Dorris D, Lublinsky A, Nguyen A, Domanus M, Prokhorova A, Gieser L, Touma E, Lockner R, Tata M, Zhu X, Patterson M, Shippy R, Sendera TJ, Mazumder A. An assessment of Motorola CodeLink microarray performance for gene expression profiling applications. Nucleic Acids Res. 2002, 30:e30.

[23] Nie AY, McMillian M, Parker JB, Leone A, Bryant S, Yieh L, Bittner A, Nelson J, Carmen A, Wan J, Lord PG. Predictive toxicogenomics approaches reveal underlying molecular mechanisms of nongenotoxic carcinogenicity. Mol Carcinog. 2006, 45:914-933.

[24] Wu W, Dave N, Tseng G, Richards T, Xing EP, Kaminski N. Comparison of normalization methods for CodeLink Bioarray data. BMC Bioinformatics 2005, 6:309

[25] Reich D, Price Al, Pattterson N. Principal component analysis of genetic data. Nature Genetics. 2008, 40(5): 491-492

[26] Eisen MB, Spellman PT, Brown PO, Botstein D. Cluster analysis and display of genome-wide expression patterns. Proc Natl Acad Sci U S A 1998, 95: 14863–14868

[27] Seltzer MJ, Bennett BD, Joshi AD, Gao P, Thomas AG, Ferraris DV, Tsukamoto T, Rojas CJ, Slusher BS, Rabinowitz JD, Dang CV, Riggins GJ. Inhibition of glutaminase preferentially slows growth of glioma cells with mutant IDH1. Cancer Res. 2010, 70:8981-8987.

[28] Schapira A. Etiology of Parkinson's disease Neurol 2006, 4:S10-S23

[29] Kahkonen M. Population cytogenetics of folate-sensitive fragile sites. Common fragile sites. Hum Genet 1988, 80:344–348.

[30] Denison SR, Wang F, Mueller B, Kock N, Phillips LA, Klein C, Smith DI. Genetic alteration in the common fragile site gene PARK2 in ovarian and other cancers. Oncogene 2003, 22:8370– 8378.

[31] Zhao P, Song X, Nin YY, Lu YL, Li XH. Loss of fragile histidine triad protein in human hepatocellular carcinoma. World J Gastroenterol 2003, 9:1216–1219.

[32] Wang F, Denison S, Lai JP, Philips LA, Montoya D, Kock N, Schule B, Klein C, Shridhar V, Roberts LR, Smith D. Parkin Gene Alterations in Hepatocellular Carcinoma. Genes Chromosome & cancer 2004, 40:85–96

[33] Hidalgo A, Baudis M, Petersen I, Arreola H, Piña P, Vazquez G, Hernandez D, Gonzalez JL, Lazos M, Lopez R, Salcedo M. Microarray comparative genomic hybridization detection of chromosomal imbalances in uterine cervix carcinoma. BMC Cancer 2005, 5, 77

[34] Jirtle RL. Genomic imprinting and cancer. Exp Cell Res. 1999, 248:18-24.

[35] Plass C, Soloway PD. DNA methylation, imprinting and cancer. Eur J Hum Genet 2002, 10:6-16

[36] Li CM, Margolin AA, Salas M, Memeo L, Mansukhani M, Hibshoosh H, Szabolcs M, Klinakis A, Tycko B. PEG10 Is a c-MYC Target Gene in Cancer Cells. Cancer Res 2006, 66: 665-672
[37] Tsuji K, et al. PEG10 is a probable target for the amplification at 7q21 detected in hepatocellular carcinoma. Cancer Genet Cytogenet, 2010, In press.
[38] López R, Garrido E, Vázquez G, Piña P, Pérez C, Alvarado I, Salcedo M. A subgroup of HOX Abd-B gene is differentially expressed in cervical cancer. Int J Gynecol Cancer. 2006, 16:1289-96

14

Number Distribution of Transmembrane Helices in Prokaryote Genomes

Ryusuke Sawada and Shigeki Mitaku
Nagoya University
Japan

1. Introduction

Number distribution of transmembrane helices represents genetic feature of survival strategy, because the number of transmembrane helices is closely related to the functional group of membrane proteins: for example, most of membrane proteins that have six transmembrane helices belong to transporter functional group. Survival strategies were obtained by evolutionary mechanism that changes the genome sequences. Comparisons of number distributions of transmembrane helices among species that have different survival strategies help us to understand the evolutionary mechanism that has increased the categories of membrane proteins.

Some studies about how the categories of protein functions have been increased during evolution were performed using protein database (Chothia et al., 2003; Huynen & van Nimwegen, 1998; Koonin et al., 2002; Qian et al., 2001; Vogel et al., 2005). However, these studies were carried out by the analysis almost for soluble proteins. Classification of protein function groups are often carried out by the empirical methods such as sequence homology that use sequence information of three-dimensional structure resolved proteins as template sequences for each functional group. However three-dimensional structure resolved membrane proteins were much less than that for the soluble proteins because of experimental difficulty of membrane proteins.

In the previous study, we developed membrane protein prediction system SOSUI and signal peptide prediction system SOSUIsignal (Gomi et al., 2004; Hirokawa et al., 1998). By combination of those systems, number of transmembrane helices can be predicted based not on empirical but on physicochemical parameters. Therefore, it is possible to investigate the number distribution of transmembrane regions in membrane proteins comprehensively among various genomes by using SOSUI and SOSUIsignal.

2. Membrane protein prediction systems

SOSUI prediction software (Hirokawa et al., 1998; Mitaku et al., 2002) for transmembrane helix regions uses physicochemical features of transmembrane helix segments. Transmembrane helix regions have three common features: (1) a hydrophobic core at the center of the helix segment; (2) amphiphilicity at both termini of each helix region; and (3) length of transmembrane helix regions. These features are essential factors for the transmembrane segment to stably present at the cell membrane. The SOSUI system first enumerates candidates of transmembrane regions by the average hydrophobicity of segments which are then discriminated by the distributions of the hydrophobicity and the amphiphilicity around the candidate segments.

SOSUIsignal (Gomi et al., 2004) predicts signal peptides that are removed from proteins that are secreted to the extracellular space via the secretary process. Signal peptides are present at the amino terminal segment of their respective proteins; the physicochemical features N-terminal structure is recognized by molecular modules during the cleavage process. The SOSUIsignal system is similar to the SOSUI system in that candidates are first enumerated by the average hydrophobicity at the amino terminal region and then real signal peptides are discriminated by several parameters.

By focusing on these physicochemical features, accuracy of the prediction systems is very good: approximately 95%for SOSUI and 90% for SOSUIsignal. By using these softwares, we can estimate not only function unknown protein sequence but also simulated ones.

3. Typical number distribution of transmembrane regions in membrane proteins

We investigated the population for number groups of transmembrane helices for 557 prokaryote genomes using SOSUI and SOSUIsignal. Figure 1 shows the results of the analysis of the membrane protein encoded in the *E. coli* genome as a typical example; the

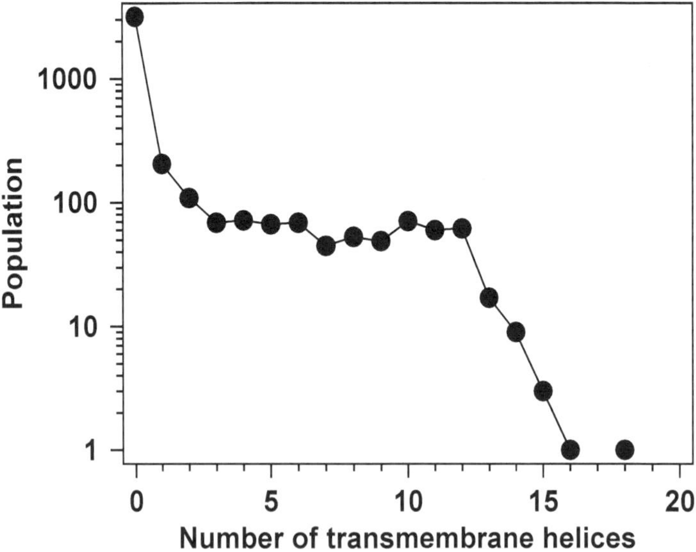

Fig. 1. Number distribution of transmembrane helices for *E. coli*. Estimation of the number of transmembrane helices for membrane proteins was performed using SOSUI and SOSUIsignal. Transmembrane helices number zero means soluble proteins.

largest category of membrane proteins comprised proteins with only one transmembrane-spanning helix, and the second largest category comprised proteins with two transmembrane-spanning helices. The populations within each category decreased gradually up to 4 transmembrane helices and then there is a plateau from 4-13 helices. The population within each category decreased rapidly for categories comprising proteins with more than 13 transmembrane helices and there were apparently no proteins with more than 16 transmembrane helices. These results indicated that membrane proteins that have a particular number of transmembrane helices, such as 12, are important for *E. coli*.

4. Variety of number distribution of transmembrane regions among organisms

The general trend of the number distribution of transmembrane helices were very similar among 557 prokaryotic genomes, but the fine structures of the number distribution of transmembrane helices can change during the evolution. Four graphs in Fig.2 show the results for the analysis for four prokaryotic genomes: A, *Pyrobaculum calidifontis*; B, *Thermoplasma volcanium*; C, *Pseudomonas putida* and D, *Thermotoga petrophila*. We selected these four kinds of organisms for showing how the number distributions of transmembrane helices are different among organisms. The number distribution of transmembrane helices for *P. calidifontis* did not show significant shoulder at 13 helices as *E. Coli*. The shape of the distribution for *T. volcanium* was very similar to that for *E. Coli*, although the population was much smaller. A significant peak was observed at 12 helices for *P. putida*. A peak at 6 helices was observable for *T. petrophila*. Despite of the difference in the number distribution of transmembrane helices among organisms, the general trend of the distribution suggests the existence of a target distribution.

5. Number distribution of transmembrane helices in proteins in organisms grouped by GC contents

If the difference in the number distribution is due to the fluctuation around a target distribution, the difference would decrease by averaging of the distribution of many organisms. In contrast if the difference is due to some systematic change among organisms, the difference would not disappear by a simple procedure of the averaging. The GC content of genomes differs widely among species, from 0.3 to 0.7, and it is well known that various characteristics of prokaryotic cells systematically change according to GC content. Therefore, we investigated whether the distribution in the number of transmembrane helices per protein changed according to the GC content. Genomes for 557 prokaryotes were classified into nine groups with different GC content. In Fig. 3, the average number distributions of transmembrane helices in the nine groups indicated that the distributions were unchanged despite differences in GC content. The membrane-protein profile of the nine groups shared a common feature in that the general shapes of the curves were the same; the curves gradually decreased in the population of each category of membrane protein and there was a shoulder at the categories with 12 transmembrane helices. This result strongly suggests that the difference in the fine structures of the number distribution is due to the fluctuation around a general curve of Fig. 3. Then, aquestion arises about the natural selections: Is the general curve formed by the pressure of natural selection?

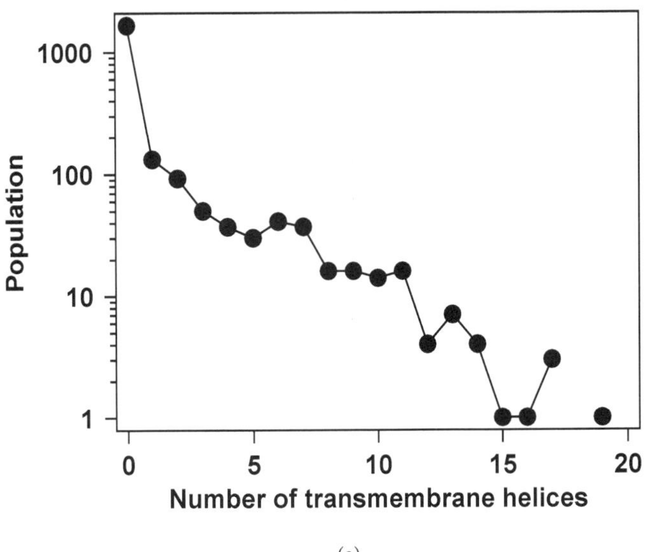
1000
100
10
1
Population
0
5
10
15
20
Number of transmembrane helices

(a)

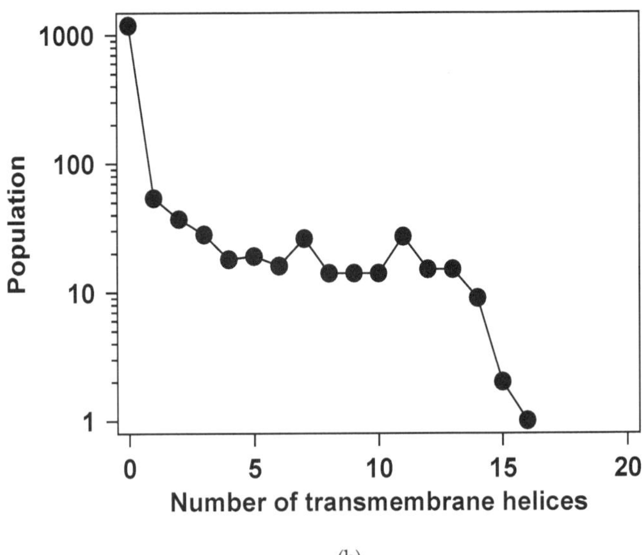
1000
100
10
1
Population
0
5
10
15
20
Number of transmembrane helices

(b)

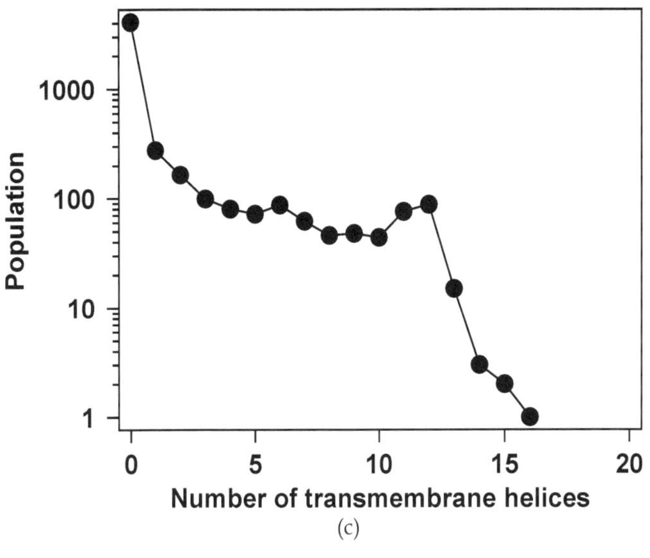

(c)

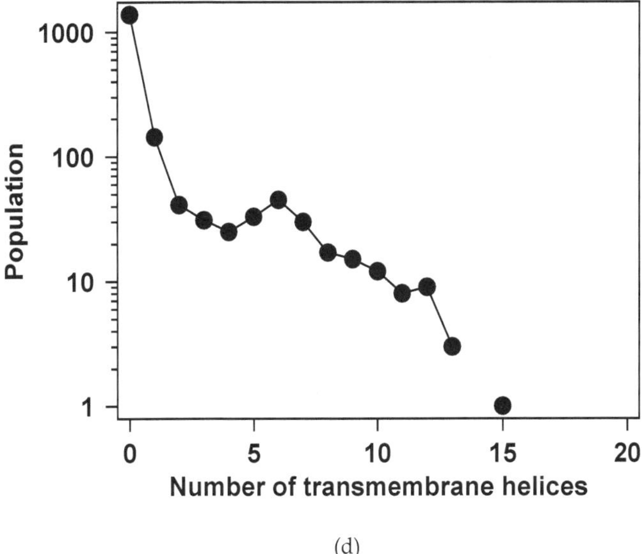

(d)

Fig. 2. Number distributions of transmembrane helices for four prokaryotes *P. calidifontis* (A), *T. volcanium* (B), *P. putida* (C) and *T. petrophila* (D).

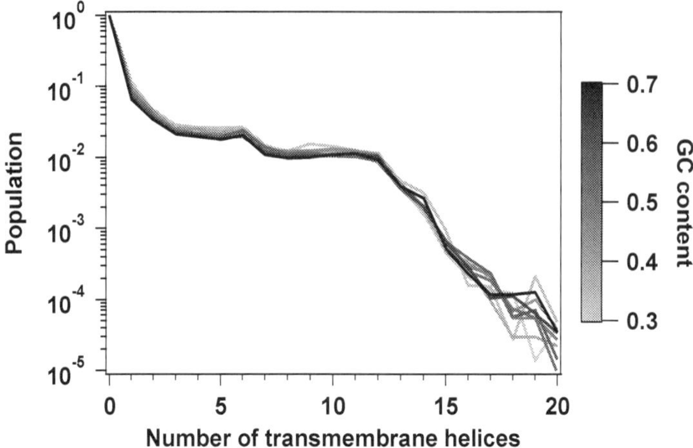

Fig. 3. Number distributions in nine groups of organisms classified by GC content. 557 prokaryotes were divided into nine groups according to GC content (0.3, 0.35, 0.4, 0.45, 0.5, 0.55, 0.6, 0.65 and 0.7).

6. Random sequence simulation

Presumably, functionally important proteins are maintained in biological genomes by natural selection. Therefore, the general curve of the number distribution of transmembrane helices must reflect natural selection that occurs during biological evolution. The prediction systems, SOSUI and SOSUIsignal, have the great advantage that they are applicable to any amino acid sequences independent of empirical information because they are based mainly on the physicochemical parameters of amino acids. So, we planned to use the prediction system for comparing the number distributions of transmembrane helices between the real genomes and the simulated genomes in which comprehensive mutations are introduced with any pressure of the natural selection. Therefore, we investigated the effect of random mutation uncoupled from natural selection on the number distribution of transmembrane helices using random sequence simulations. The *E. coli* genome was used for the random sequence simulation. At each simulation step, one in every 100 amino acids in all protein sequences was mutated randomly. When the amino acids were mutated, the new amino acids were determined according to the genomic amino acid composition of the *E. coli* genome. Distributions of number of transmembrane helices were estimated by using membrane protein prediction systems SOSUI and SOSUIsignal after each simulation step. Simulations were reiterated until 500 simulation steps.

Distributions of transmembrane helices in membrane proteins for simulated genomes are shown in Fig. 4. As the simulation steps proceed, the number of membrane proteins with more than six transmembrane helices decreased monotonously and the shoulder in the distribution around 12 transmembrane helices disappeared. Beyond 300 simulation steps at which the sequences were completely randomized, the distribution became very similar to a single exponential decay. A broken line in Fig.4 represents the single exponential decay curve, $y = 2090e^{-0.87x}$, which was obtained least square deviation analysis for the averaged distribution between 300 and 500 steps.

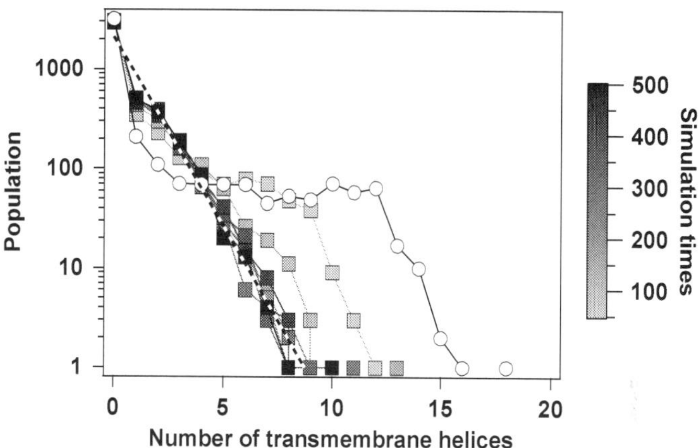

Fig. 4. Number distribution of transmembrane helices in *E. coli* genomes subjected to random mutation of amino acid sequences. Distributions of simulated genomes are represented as gray scaled rectangles. Open circle indicates the distribution for the original (simulation step zero) genome of *E. coli*. Dotted line represents the fitting result of $y = 2090e^{-0.87x}$ for the averaged distribution between 300 and 500 simulation steps.

After 300 simulation steps, shapes of the distributions of number of transmembrane helices for simulated genomes were almost unchanged in spite of additional mutations. A single exponential distribution for simulated genome can be explained by a kind of reaction in the evolutionary time scale changing the number of membrane proteins due to extensive mutations.

$$TM_i \leftrightarrow TM_{i+1}$$

in which TM_i represents a membrane protein with i transmembrane helices. If the equilibrium constant is the same among the distinct equilibrium state, the shape would become the exponential, as follow:

$$k^- \langle TM_n \rangle = k^+ \langle TM_{n-1} \rangle$$

$$\langle TM_n \rangle = \frac{k^+}{k^-} \langle TM_{n-1} \rangle = \left(\frac{k^+}{k^-} \right)^n \langle TM_0 \rangle$$

where $<TM_n>$, $<TM_{n-1}>$ and $<TM_0>$ represent the population of the membrane protein, with n, n-1 and 0 helices, respectively, and k^+/k^- means the equilibrium constant. In the simulation, the equilibrium constants for each transmembrane helices number group are the same from the algorithm of the prediction systems, and a single exponential decay in the computer experiment is well interpreted by this model. However, in the real genome, the shape of distribution is not exponential, showing a significant plateau and shoulder. This indicates that there equilibrium constants for each transmembrane helices number groups are not same. This may be due to the difference of the functional importance among membrane protein groups.

7. References

Chothia, C., Gough, J., Vogel, C. & Teichmann, S.A. (2003). Evolution of the protein repertoire. *Science*, Vol. 300, No. 5626, 1701-1703

Gomi, M., Sonoyama, M. & Mitaku, S. (2004). High performance system for signal peptide prediction: SOSUIsignal. *Chem-Bio Informatics Journal*, Vol. 4, No. 4, 142-147

Hirokawa, T., Boon-Chieng, S. & Mitaku, S. (1998). SOSUI: classification and secondary structure prediction system for membrane proteins. *Bioinformatics*, Vol. 14, No. 4, 378-379

Huynen, M.A. & van Nimwegen, E. (1998). The frequency distribution of gene family sizes in complete genomes. *Mol Biol Evol*, Vol. 15, No. 5, 583-589

Koonin, E.V., Wolf, Y.I. & Karev, G.P. (2002). The structure of the protein universe and genome evolution. *Nature*, Vol. 420, No. 6912, 218-223

Mitaku, S., Hirokawa, T. & Tsuji, T. (2002). Amphiphilicity index of polar amino acids as an aid in the characterization of amino acid preference at membrane-water interfaces. *Bioinformatics*, Vol. 18, No. 4, 608-616

Qian, J., Luscombe, N.M. & Gerstein, M. (2001). Protein family and fold occurrence in genomes: power-law behaviour and evolutionary model. *J Mol Biol*, Vol. 313, No. 4, 673-681

Vogel, C., Teichmann, S.A. & Pereira-Leal, J. (2005). The relationship between domain duplication and recombination. *J Mol Biol*, Vol. 346, No. 1, 355-365

15

Classifying TIM Barrel Protein Domain Structure by an Alignment Approach Using Best Hit Strategy and PSI-BLAST

Chia-Han Chu[1], Chun Yuan Lin[2], Cheng-Wen Chang[1], Chihan Lee[3] and Chuan Yi Tang[4]
[1]*National Tsing Hua University,*
[2]*Chang Gung University,*
[3]*Chipboud Technology Corporation,*
[4]*Providence University*
Taiwan

1. Introduction

High-tech large-scale sequencing projects have identified a massive number of amino acid sequences for both known and putative proteins, but information on the three-dimensional (3D) structures of these proteins is limited. Several structure databases, such as the Structural Classification of Proteins (SCOP (Andreeva et al., 2008), release version 1.73) and the Class, Architecture, Topology, and Homologous superfamily (CATH (Cuff et al., 2009), release version 3.2.0), contain fewer than 60,000 entries in the Protein Data Bank (PDB (Berman et al., 2000), released on 12 May, 2009). This number of entries constitutes only about 15% of entries in Swiss-Prot (Bairoch et al., 2004), release version 57.2, with more than 400,000 entries). Either X-ray diffraction or NMR can be used to determine the 3D structure of a protein, but each method has its limitation (Dubchak et al., 1995). As such, extracting structural information from sequence databases is an important and complementary alternative to these experimental methods, especially when swiftly determining protein functions or discovering new compounds for medical or therapeutic purposes.
From ASTRAL SCOP 1.73, it has been estimated that ~10% of known enzymes have triosephosphate isomerase (TIM) barrel domains. Moreover, TIM barrel proteins have been identified in five of six enzyme classes, oxidoreductases, transferases, hydrolases, lyases and isomerases, in the Enzyme nomenclature (ENZYME (Bairoch, 2000), released on 5 May, 2009) database; the ligases class does not contain TIM barrel protein. TIM barrel proteins are diverse in sequence and functionality and thus represent attractive targets for protein engineering and evolutionary studies. It is therefore important to examine TIM barrel protein domain structure classification in SCOP and ENZYME.
In SCOP, there are six levels of hierarchy: class, fold, superfamily, family, protein domain and species. The classification of protein structures has, more recently, been facilitated by computer-aided algorithms. Previous research (Chou & Zhang, 1995; Dubchak et al., 1995; Lin et al., 2005, 2007) has shown that an overall prediction accuracy rate of 70-90% can be

easily achieved by using only amino acid sequence information to classify most of proteins into four major classes in SCOP (all-alpha (α), all-beta (β), alpha/beta (α/β) and alpha+beta ($\alpha+\beta$)) (Murzin, 1995). For the α/β class (constituting TIM barrel proteins), the overall prediction accuracy rate achieved 97.9% (Lin et al., 2005, 2007). However, less optimal results were obtained if a more complicated category was used, such as protein folding patterns. The overall prediction accuracy rate for classifying 27 fold categories in SCOP only achieved only 50-70% using amino acid sequence information (Ding & Dubchak, 2001; Huang et al., 2003; Lin et al., 2005, 2007; Shen & Chou, 2006; Vapnik, 1995; Yu et al., 2003). Although the classification for the SCOP fold category is still a challenge, the overall prediction accuracy rate for the TIM barrel fold is 93.8% (Yu et al., 2003). Based on the above results, it is possible to further classify TIM barrel proteins into the SCOP superfamily and family categories. Four projection methods, PRIDE (Carugo & Pongor, 2002; Gáspári et al., 2005), SGM (Rogen & Fain, 2003), LFF (Choi et al., 2004) and SSEF (Zotenko et al., 2006, 2007), have been proposed for protein structure comparisons. Zotenko *et al.* (Zotenko et al., 2006) compared these four methods for classifying proteins into the SCOP fold, superfamily and family categories and showed that the SSEF method had the best overall prediction accuracy rate. The SSEF method utilizes 3D structure information to generate the triplet of secondary structure elements as the footprints in the comparisons.

Hence, in this chapter, an alignment approach using the pure best hit strategy, denoted PBH, is proposed to classify the TIM barrel protein domain structures in terms of the superfamily and family categories in SCOP. This approach requires only amino acid sequence information to generate alignment information, but secondary and 3D structure information is also applied in this approach, respectively, to compare the performances with each other. This work is also used to perform the classification for the class category in ENZYME. Two testing data sets, TIM40D and TIM95D from ASTRAL SCOP 1.71 (Chandonia et al., 2004), were tested to evaluate this alignment approach. First, for any two proteins, we adopt the tools CLUSTALW (Thompson et al., 1994), SSEA (Fontana et al., 2005) and CE (Shindyalov & Bourne, 1998) to align the amino acid sequences, secondary and 3D structures, respectively, to obtain the scores of sequence identity, secondary structure identity and RMSD. These scores are then used to build an alignment-based protein-protein identity score network. Finally, a PBH strategy is used to determine the prediction result of a target protein by selecting the protein having the best score for the target protein according to this network. This score can be calculated by a single parameter, such as sequence identity, or mixed parameters by combing two or three single parameters, such as combining sequence identity and secondary structure identity. In this chapter, we only consider the single parameter. To verify the stability of the proposed alignment approach, we also use the novel TIM barrel proteins in TIM40D and TIM95D from ASTRAL SCOP 1.73 that do not existed in ASTRAL SCOP 1.71. For this test, the alignment-based protein-protein identity score network constructed by the TIM barrel proteins from ASTRAL SCOP 1.71 and the PBH strategy are used to predict the classification result for each novel TIM barrel protein. In addition, we further adopt the PSI-BLAST method as a filter for the PBH strategy, denoted the BHPB strategy, to reduce the number of false positives. The experimental results demonstrated that the alignment approach with the PBH strategy or BHPB strategy is a simple and stable method for TIM barrel protein domain structure classification, even when only the amino acid sequence information is available.

2. Materials

2.1 TIM barrel proteins from ASTRAL SCOP 1.71

Two data sets, TIM40D and TIM95D, were used to evaluate the proposed PBH and BHPB alignment strategies. TIM40D contains 272 TIM barrel protein domain sequences (abbreviated to TIM sequences) extracted from the 40D set in ASTRAL SCOP 1.71, in which any two proteins must have ≤40% sequence identity based on PDB SEQRES records. TIM95D contains 439 TIM sequences extracted from the 95D set in ASTRAL SCOP 1.71, in which any two proteins must have ≤95% sequence identity based on PDB SEQRES records. For TIM40D and TIM95D, we directly retrieved amino acid sequences and 3D structures from ASTRAL SCOP 1.71 but excluded redundant and possible mutant data. Secondary structure information for each TIM barrel protein with eight states (H, I, G, E, B, S, T and _) was first derived from the digital shape sampling and processing (DSSP (Kabsch & Sander, 1983)) program. Then the eight states for each TIM barrel protein were then reduced to three states (H, E and C)

Superfamily categories	Index	N_{40D}*	N_{95D}*
Triosephosphate isomerise (TIM)	1	3	16
Ribulose-phosphate binding barrel	2	19	30
Thiamin phosphate synthase	3	2	2
FMN-linked oxidoreductases	4	15	22
Inosine monophosphate dehydrogenase (IMPDH)	5	3	5
PLP-binding barrel	6	8	10
NAD(P)-linked oxidoreductase	7	8	21
(Trans)glycosidases	8	82	134
Metallo-dependent hydrolases	9	18	22
Aldolase	10	31	48
Enolase C-terminal domain-like	11	12	24
Phosphoenolpyruvate/pyruvate domain	12	12	22
Malate synthase G	13	1	2
RuBisCo, C-terminal domain	14	4	10
Xylose isomerise-like	15	7	15
Bacterial luciferase-like	16	7	9
Nicotinate/Quinolinate PRTase C-terminal domain-like	17	4	5
PLC-like phosphodiesterases	18	5	5
Cobalamin (vitamin B12)-dependent enzymes	19	5	6
tRNA-guanine transglycosylase	20	2	2
Dihydropteroate synthetase-like	21	4	6
UROD/MetE-like	22	4	4
FAD-linked oxidoreductase	23	3	3
Pyridoxine 5′-phosphate synthase	24	1	1
Monomethylamine methyltransferase MtmB	25	1	1
Homocysteine S-methyltransferase	26	2	3
(2r)-phospho-3-sulfolactate synthase ComA	27	1	2
Radical SAM enzymes	28	3	3
GlpP-like	29	1	1
CutC-like	30	1	1
ThiG-like	31	1	2
TM1631-like	32	2	2

*N_{40D}: the number of TIM sequences in TIM40D
*N_{95D}: the number of TIM sequences in TIM95D

Table 1. Non-redundant data sets, TIM40D and TIM95D, of superfamily categories in SCOP

Index	N_{40D}*	N_{95D}*	Index	N_{40D}*	N_{95D}*	Index	N_{40D}*	N_{95D}*
1.1	3	16	9.6	4	5	16.3	2	3
2.1	2	5	9.7	1	1	16.4	2	2
2.2	2	4	9.8	1	1	17.1	2	3
2.3	4	7	9.9	1	1	17.2	2	2
2.4	10	13	9.11	1	1	18.1	1	1
2.5	1	1	9.12	1	1	18.2	2	2
3.1	2	2	9.13	1	1	18.3	2	2
4.1	15	22	10.1	18	29	19.1	2	2
5.1	3	5	10.2	2	3	19.2	1	1
6.1	7	9	10.3	3	5	19.3	1	2
6.2	1	1	10.4	3	6	19.4	1	1
7.1	8	21	10.5	3	3	20.1	2	2
8.1	25	48	10.6	2	2	21.1	2	4
8.3	26	41	11.1	1	6	21.2	2	2
8.4	4	12	11.2	11	18	22.1	2	2
8.5	13	18	12.1	1	5	22.2	2	2
8.6	3	3	12.2	1	2	23.1	2	2
8.7	2	2	12.3	1	2	23.2	1	1
8.8	3	3	12.5	4	4	24.1	1	1
8.9	1	1	12.7	4	6	25.1	1	1
8.10	1	2	12.8	1	3	26.1	2	3
8.11	1	1	13.1	1	2	27.1	1	2
8.12	1	1	14.1	4	10	28.1	1	1
8.13	1	1	15.1	1	1	28.2	1	1
8.14	1	1	15.2	1	1	28.3	1	1
9.1	1	2	15.3	2	10	29.1	1	1
9.1	2	2	15.4	1	1	30.1	1	1
9.2	1	3	15.5	1	1	31.1	1	2
9.3	2	2	15.6	1	1	32.1	2	2
9.4	1	1	16.1	2	2	-	-	-
9.5	1	1	16.2	1	2	-	-	-

*N_{40D}: the number of TIM sequences in TIM40D
*N_{95D}: the number of TIM sequences in TIM95D

Table 2. Non-redundant data sets, TIM40D and TIM95D, of family categories in SCOP

Class categories	Index	N_{40D}*	N_{95D}*
Oxidoreductases	1	27	46
Transferases	2	31	53
Hydrolases	3	68	106
Lyases	4	58	97
Isomerases	5	23	49
undefined	-	67	91

*N_{40D}: the number of TIM sequences in TIM40D
*N_{95D}: the number of TIM sequences in TIM95D
The sum of N_{40D} and N_{95D} are 274 and 442, respectively. TIM sequences:
"d1pii_2" and "d1pii_1" in TIM40D and TIM95D have multiple EC numbers for class categories;
"d1b9ba_" and "d1jvna1" in TIM95D have multiple EC numbers for class categories

Table 3. Non-redundant data sets, TIM40D and TIM95D, of class categories in ENZYME

according to the scheme outlined by Jones (Jones, 1999). The TIM sequence "d1cwn_" (SCOP id) in TIM95D was excluded because of lack of secondary structure information (only 438 TIM sequences in TIM95D were tested). The TIM barrel proteins (from ASTRAL SCOP 1.71 and the Universal Protein Resource (UniProt (Bairoch et al., 2005))) for each of TIM40D and TIM95D were classified into 32 superfamily categories, 91 family categories and 5 class categories (Tables 1, 2 and 3; supplemental Table S1(Chu, 2011)).

2.2 Novel TIM barrel proteins from ASTRAL SCOP 1.73

Novel TIM barrel proteins from ASTRAL SCOP 1.73 that do not exist in ASTRAL SCOP 1.71 were also tested. The intersection among the TIM barrel proteins from ASTRAL SCOP 1.71 and 1.73 for TIM40D (Figure 1(A)) and TIM95D (Figure 1(B)) are shown. The number of TIM sequences are represented in green (ASTRAL SCOP 1.71), light blue (ASTRAL SCOP 1.73) and orange (ASTRAL SCOP 1.73 that are not presented in 1.71). In TIM40D (Figure 1(A)), we identified 258 TIM sequences (ASTRAL SCOP 1.71 and 1.73), 14 TIM sequences (exclusively ASTRAL SCOP 1.71) and 64 novel TIM sequences (exclusively ASTRAL SCOP 1.73: 12 of 64 were categorized as new). In TIM95D (Figure 1(B)), we identified 439 TIM sequences (ASTRAL SCOP 1.71 and 1.73) and 79 novel TIM sequences (exclusively ASTRAL SCOP 1.73: 12 of 79 were categorized as new). These 12 novel TIM sequences within the new categories were identical and thus were excluded in the alignment approach. Hence, 52 (TIM40D) and 67 (TIM95D) novel TIM sequences from ASTRAL SCOP 1.73 were used to evaluate the stability of the proposed PBH alignment strategy, respectively. (see supplemental Table S2 (Chu, 2011)).

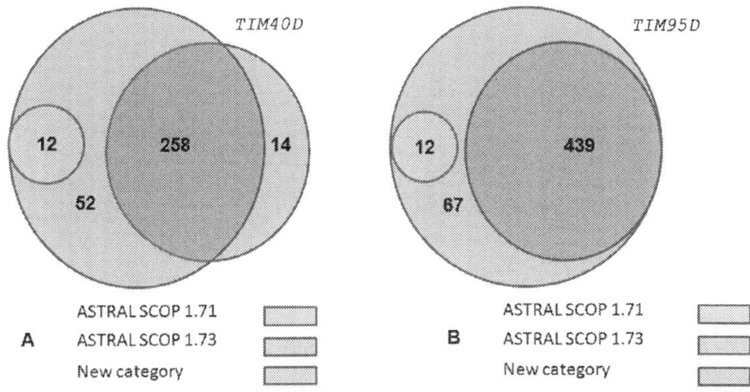

Fig. 1. Intersection among TIM sequences for TIM40D and TIM95D between ASTRAL SCOP 1.71 and 1.73. (A) In TIM40D, there are 272 (ASTRAL SCOP 1.71) and 322 (ASTRAL SCOP 1.73) TIM sequences. (B) In TIM95D, there are 439 (ASTRAL SCOP 1.71) and 518 (ASTRAL SCOP 1.73) TIM sequences.

3. Results and discussion

3.1 Performance analysis

The standard percentage prediction accuracy rate Q_i (Rost & Sander, 1993) was used to evaluate the proposed alignment approach and Q_i is defined as

$$Q_i = \frac{p_i}{n_i} \times 100, \quad (1)$$

where n_i is the number of test proteins in the ith superfamily/family/class category and p_i is the number of test proteins being correctly predicted in the ith superfamily/family/class. The overall prediction accuracy rate Q is given by

$$Q = \sum_{i=1}^{k} q_i Q_i , \quad (2)$$

where $q_i = n_i/K$, where K is the total number of test proteins. Q_i is equivalent to Recall (Gardy et al., 2003), which is defined as

$$\text{Recall}_i = \frac{TP_i}{(TP_i + FN_i)}, \quad (3)$$

where TP_i (true positives) is the number of correctly predicted proteins in the ith superfamily/family/class category, and FN_i (false negatives) is the number of missed proteins in the ith superfamily/family/class category. Precision (Gardy et al., 2003) was also used to evaluate the proposed alignment approach. Precision is defined as

$$Precision_i = \frac{TP_i}{(TP_i + FP_i)}, \quad (4)$$

where FP_i (false positives) is the number of pseudo proteins predicted in the ith superfamily/family/class category. In addition, the Matthews Correlation Coefficient (*MCC* for short) (Matthews, 1975) was used to measure the prediction quality of classifications by utilizing the proposed PBH and BHPB alignment strategies. *MCC* accounts for TP_i, FP_i, TN_i and FN_i as a balanced measure, which can be used for categories with varying sizes. *MCC* returns a value +1 for the perfect prediction quality, 0 for the average random prediction quality, or -1 for an inverse prediction quality. The formula of *MCC* is defined as

$$MCC = \frac{TP_i \times TN_i - FP_i \times FN_i}{\sqrt{(TP_i + FP_i) \times (TP_i + FN_i) \times (TN_i + FP_i) \times (TN_i + FN_i)}} \quad (5)$$

3.2 Alignment approach with the PBH strategy

Zotenko *et al.* (Zotenko et al., 2006) compared four projection methods, PRIDE, SGM, LFF and SSEF, to classify the 40D data set (ASTRAL SCOP 1.69) into superfamily and family categories in SCOP. There are 246 TIM barrel proteins classified into 24 superfamily categories and 210 TIM barrel proteins classified into 42 family categories. Based on the overall Q values, SSEF outperformed LFF, SGM and PRIDE for TIM barrel protein structure classification (Table 4).

For the proposed PBH alignment strategy, the overall Q values for TIM40D and TIM95D from ASTRAL SCOP 1.71 are shown in Table 5. In Table 5, the threshold (see Methods) is determined without decreasing the Q value, which is achieved without a threshold. The Q

Method	SSEF Q (%)	LFF Q (%)	SGM Q (%)	PRIDE Q (%)
Superfamily*	78.0	73.6	63.0	41.5
Family*	74.3	72.9	57.1	45.2

*: the performances of SSEF, LFF, SGM and PRIDE are extracted from the additional file (Zotenko et al., 2006)

Table 4. Overall *Q* values for SSEF, LFF, SGM and PRIDE in TIM40D (ASTRAL SCOP 1.69)

value will decrease when a score, which is higher or lower than the threshold given in Table 5, is assigned as the threshold. For TIM40D, the best *Q* value (84.2%) for the superfamily classification is derived according to secondary structure identity, 76.1% for the family classification is derived according to sequence identity and 48.2% for the class classification is derived according to sequence identity (Table 5). The *Q* value of 48.2% for the class classification is not valid. In TIM40D, 67 of 274 TIM sequences with undefined class categories (derived from UniProt) were initially assumed to be false negatives before the test (see Discussion). Using amino acid sequence or secondary structure information, the PBH alignment strategy yields results as good as SSEF (footprint information). This alignment approach will be useful for TIM barrel proteins lacking 3D structure information. Moreover, for the class classification, the *Q* value of 48.2% is better than the *Q* value of 35% (under Rank 1 condition) by a non-alignment method proposed by Dobson and Doig (Dobson & Doig, 2005). For TIM95D, the best *Q* value (93.2%) for the superfamily classification is derived according to secondary structure identity, 90.0% for the family classification is derived according to sequence identity and 65.2% for the class classification is derived according to secondary structure identity. Similarly, for the class classification, 91 of 442 TIM sequences with undefined class categories in TIM95D were initially assumed to be false negatives before the test.

	Method	Sequence identity		Secondary structure identity		RMSD	
		Q (%)	Threshold	*Q* (%)	Threshold	*Q* (%)	Threshold
TIM40D	Superfamily	83.1	<14	84.2	<67	40.4	>1.9
	Family	76.1	<14	75.0	<67	37.1	>1.9
	Class (ENZYME)	48.2	<13	47.4	<67	21.2	>1.8
TIM95D	Superfamily	92.5	<14	93.2	<68	68.0	>2.0
	Family	90.0	<14	89.3	<68	66.4	>2.0
	Class (ENZYME)	64.0	<16	65.2	<72	48.0	>1.8

Table 5. Overall *Q* values for the PBH alignment strategy in TIM40D and TIM95D (ASTRAL SCOP 1.71)

Overall, the *Q* values of the PBH alignment strategy using secondary structure information are similar to those using amino acid sequence information in TIM40D and TIM95D. For practical purposes, however, it may be best to use only amino acid sequence information. In addition, the *Q* values of the PBH alignment strategy using the RMSD do not yield valid results. The RMSD (global alignment result in this chapter) may not be a valid feature for

the alignment approach to perform TIM barrel protein domain structure classification. In Table 5, the threshold is too low for sequence identity, suggesting that the sequence identity of the target and its selected proteins (within the same category) is low. When the threshold is set higher than the above sequence identity, the target protein becomes a false negative and then the *Q* value under the "no threshold" condition will decrease. The threshold is high for secondary structure identity, suggesting that the secondary structure identity of the target and its selected proteins (within the same category) is high. These results imply that although TIM barrel proteins have diverse sequences they have very similar secondary structures. This inference matches the recent observation for the TIM barrel proteins.
Tables 6 and 7 show overall Q and Precision values for various categories in TIM40D and TIM95D from ASTRAL SCOP 1.71, respectively. Only the categories with more than ten TIM sequences are listed; tests for RMSD were omitted because the results were invalid. In Tables 6 and 7, the threshold is determined with the best Precision value without decreasing the Q value, which is achieved without a threshold. Precision values with the threshold outperform or are equals to those without the threshold. However, it is very difficult to determine the appropriate threshold to obtain the best Precision value for routine alignment practices. This problem may be omitted by the BHPB strategy to reduce the number of false positives.

	Method	**Sequence identity**			**Secondary structure identity**		
	Index	*Q* (%)	**Precision[1]** (%)	**Precision[2]** (%)	*Q* (%)	**Precision[1]** (%)	**Precision[2]** (%)
Superfamily	2	94.7	64.3	78.3(17)	84.2	64.0	84.2(78)
	4	73.3	78.6	100.0(18-22)	73.3	100.0	100.0(<77)
	8	87.8	92.3	93.5(13)	86.6	95.9	95.9(<67)
	9	83.3	78.9	78.9(<15)	72.2	86.7	92.9(68-70)
	10	83.9	78.8	81.3(16)	96.8	78.9	81.0(74)
	11	83.3	76.9	100.0(17-18)	91.7	91.7	100.0(73-75)
	12	75.0	90.0	100.0(15)	83.3	100.0	100.0(<77)
Family	2.4	100.0	66.7	100.0(19-29)	100.0	71.4	100.0(80-86)
	4.1	73.3	78.6	100.0(18-22)	73.3	100.0	100.0(<77)
	8.1	88.0	84.6	88.0(13)	92.0	95.8	95.8(<67)
	8.3	84.6	95.7	95.7(<17)	84.6	88.0	88.0(<73)
	8.5	100.0	86.7	100.0(17)	92.3	100.0	100.0(<75)
	10.1	83.3	78.9	100.0(18-19)	88.9	72.7	84.2(76)
	11.2	90.9	76.9	100.0(17-18)	90.9	83.3	100.0(76-79)
Class (ENZYME)	1	66.7	64.3	85.7(21-22)	74.1	71.4	71.4(<72)
	2	45.2	58.3	58.3(<13)	45.2	66.7	70.0(64-66)
	3	60.3	60.3	62.1(14)	54.4	61.7	63.8(68-70)
	4	77.6	65.2	71.4(16)	75.9	62.0	69.8(74)
	5	61.0	53.8	61.0(15-16)	65.2	45.5	48.4(70-71)

Table 6. Overall *Q* and Precision values for the PBH alignment strategy in TIM40D (ASTRAL SCOP 1.71)

	Method	Sequence identity			Secondary structure identity		
	Index	Q (%)	Precision[1] (%)	Precision[2] (%)	Q (%)	Precision[1] (%)	Precision[2] (%)
Superfamily	1	100.0	94.1	100.0(17-44)	100.0	88.9	94.1(80-85)
	2	96.7	78.4	82.9(17)	90.0	84.4	96.4(79)
	4	90.9	87.0	90.9(15-16)	86.4	95.0	100.0(70-81)
	6	90.0	100.0	100.0(<22)	100.0	100.0	100.0(<84)
	7	100.0	95.2	100.0(16-23)	100.0	100.0	100.0(<82)
	8	91.8	98.4	98.4(<14)	95.5	98.5	98.5(<68)
	9	81.8	94.7	94.7(<16)	77.3	89.5	94.4(68-74)
	10	95.8	86.8	88.5(16)	97.9	87.0	94.0(76-77)
	11	100.0	96.0	100.0(17-21)	100.0	96.0	100.0(74-80)
	12	100.0	100.0	100.0(<26)	100.0	100.0	100.0(<79)
	14	100.0	100.0	100.0(<31)	100.0	83.3	100.0(74-85)
	15	93.3	82.4	82.4(<16)	93.3	93.3	93.3(<76)
Family	1.1	100.0	94.1	100.0(17-44)	100.0	88.9	94.1(80-85)
	2.4	100.0	76.5	100.0(20-29)	100.0	86.7	100.0(77-86)
	4.1	90.9	87.0	90.9(15-16)	86.4	95.0	100.0(70-81)
	7.1	100.0	95.2	100.0(16-23)	100.0	100.0	100.0(<82)
	8.1	95.8	97.9	97.9(<14)	97.9	97.9	97.9(<68)
	8.3	92.7	100.0	100.0(<17)	92.7	92.7	95.0(73-74)
	8.4	100.0	92.3	100.0(16-35)	100.0	100.0	100.0(<85)
	8.5	100.0	90.0	100.0(17)	94.4	94.4	100.0(72-74)
	10.1	96.6	87.5	100.0(18-19)	96.6	84.8	93.3(76-77)
	11.2	100.0	94.7	100.0(17-21)	100.0	94.7	100.0(74-80)
	14.1	100.0	100.0	100.0(<31)	100.0	83.3	100.0(74-85)
	15.3	100.0	66.7	100.0(18-67)	100.0	90.9	100.0(76-95)
Class (ENZYME)	1	89.1	80.4	91.1(21-23)	89.1	80.4	82.0(70-71)
	2	77.4	75.9	75.9(<16)	79.2	87.5	87.5(<73)
	3	67.0	71.7	72.4(14-15)	69.8	71.2	73.3(72-73)
	4	91.8	80.2	84.0(17)	90.7	79.3	80.7(73)
	5	83.7	78.8	82.0(15)	87.8	79.6	79.6(<72)

Table 7. Overall Q and Precision values for the PBH alignment strategy in TIM95D (ASTRAL SCOP 1.71)

The Q values for Ribulose-phosphate binding barrel, (Trans)glycosidases, Aldolase and Enolase C-terminal domain-like in the superfamily categories were above 84.2%, whereas FMN-linked oxidoreductases, Metallo-dependent hydrolases, and Phosphoenolpyruvate/pyruvate domain had lower Q values (Table 6). All of family categories except for FMN-linked oxidoreductases had Q values above 76.1%. Only one of the class categories, Transferases, had a Q value below 48.2%. For the superfamily categories, Ribulose-phosphate binding barrel, (Trans)glycosidases and Metallo-dependent hydrolases, the Q values derived according to sequence identity were better than those derived according to secondary structure identity. In contrast, the Q values of Aldolase, Enolase C-terminal domain-like and Phosphoenolpyruvate/pyruvate domain derived according to secondary structure identity were better than those derived according to sequence identity. FMN-linked oxidoreductases yielded the same *Q* values based on sequence identity and secondary structure identity. For the family category, Type II chitinase had a better *Q* value derived according to sequence identity than derived according to secondary structure identity. Amylase, catalytic domain and Class I aldolase produced better *Q* values derived according to secondary structure identity than derived according to sequence identity. Tryptophan biosynthesis enzymes, FMN-linked oxidoreductases, beta-glycanases and D-glucarate dehydratase-like had the same *Q* values derived according to sequence identity and secondary structure identity. For the class categories, the *Q* values of Hydrolases and Lyases derived according to sequence identity were better than those derived according to secondary structure. Oxidoreductases and Isomerases yielded better *Q* values derived according to secondary structure identity than derived according to sequence identity. The remaining category, Transferases, had the same *Q* values derived according to sequence identity and secondary structure identity. These results demonstrated that the proposed PBH alignment strategy is more useful for certain TIM barrel proteins than others.

In TIM95D superfamily categories, FMN-linked oxidoreductases and Metallo-dependent hydrolases yielded Q values less than 93.2% (Table 7); others all yielded Q values above 93.2%. All of family categories obtained Q values above 90.0% and no class category obtained Q values below 65.2%. For the superfamily categories, Ribulose-phosphate binding barrel, FMN-linked oxidoreductases and Metallo-dependent hydrolases, and for the family categories, FMN-linked oxidoreductases and Type II chitinase, the Q values derived according to sequence identity were better than those derived according to secondary structure identity. For other superfamily and family categories, the same Q values were obtained using sequence identity and secondary structure identity. For the class categories, Transferases, Hydrolases and Isomerases had better Q values derived according to secondary structure identity than those derived according to sequence identity; Lyases had a better Q value derived according to sequence identity than that derived according to secondary structure identity. The last category, Oxidoreductases, produced the same Q values derived according to sequence identity and secondary structure identity.

3.3 Estimating stability using the PBH alignment strategy

Novel TIM sequences in TIM40D (n=52) and TIM95D (n=67) from ASTRAL SCOP 1.73 were used to estimate the stability of the proposed PBH alignment strategy. Table 8 presents the overall *Q* values for novel TIM sequences. The definition and observation of the threshold in Table 8 is the same as that in Table 5. In Table 8, the best *Q* values for the superfamily, family and class classifications in TIM40D and TIM95D from ASTRAL SCOP 1.73 were derived according to sequence identity. For TIM40D, the best *Q* value was 94.2% for superfamily, 90.4% for family and 40.4% for class; for TIM95D, the best *Q* value was 91.0% for superfamily, 88.1% for family and 47.8% for class. Similarly, for the class classification, 20 of 52 (TIM40D) and 25 of 67 (TIM95D) novel TIM sequences with undefined class categories were initially assumed to be false negatives before the test (see supplemental Table S3 (Chu, 2011)). These results suggest that the proposed PBH alignment strategy is stable and suitable for TIM barrel protein domain structure classification.

	Method	Sequence identity		Secondary structure identity		RMSD	
		Q (%)	Threshold	*Q* (%)	Threshold	*Q* (%)	Threshold
TIM40D	Superfamily	94.2	<16	90.4	<72	57.7	>1.7
	Family	90.4	<16	84.6	<74	55.8	>1.7
	Class (ENZYME)	40.4	<17	40.4	<72	25.0	>1.7
TIM95D	Superfamily	91.0	<16	86.6	<72	59.7	>1.8
	Family	88.1	<16	80.6	<78	58.2	>1.8
	Class (ENZYME)	47.8	<14	44.8	<72	29.9	>1.7

Table 8. Overall *Q* values for novel TIM sequences in TIM40D and TIM95D (ASTRAL SCOP 1.73)

3.4 Alignment strategy with the BHPB strategy

The high *Q* value derived according to sequence identity using the PBH alignment strategy can decrease the false positives via the homologous finding method. PSI-BLAST is an established method that detects subtle relationships between proteins that are structurally distant or functionally homologous owing to a position-specific scoring matrix generated from multiple alignments of the top-scoring BLAST responses to a given query sequence. The PSI-BLAST package was integrated into the NCBI standalone BLAST package (Altschul et al., 1997). All of our tests were implemented using Perl combined with the CPAN bioperl package (http://www.cpan.org/).

Table 9 presents the overall *Q* values for TIM sequences in TIM40D and TIM95D from ASTRAL SCOP 1.71 using PSI-BLAST as a filter. The definition and observation of the threshold in Table 9 is the same as that in Table 5. For TIM40D (Table 9), the best *Q* values acquired according to sequence identity were 76.1% for superfamily, 73.9% for family, and 41.6% for class. For TIM95D, the secondary structure identity was used obtain the best *Q* value of 62.2% for class, whereas sequence identity was used to obtain the best *Q* values for superfamily (88.8%) and family (88.4%). Based on Tables 5 and 9, the *Q* values obtained using the BHPB alignment strategy were slightly lower than those obtained using the PBH alignment strategy. The lower *Q* values may be a consequence of proteins for which no homolog was found using PSI-BLAST method; such proteins were thus false negatives. Although the overall *Q* values using the PBH alignment strategy were higher than those using the BHPB alignment strategy, Precision values obtained using the BHPB alignment strategy were higher than those using the PBH alignment strategy. Tables 10 (TIM40D from ASTRAL SCOP 1.71) and 11 (TIM95D) show the overall *Q* and Precision values for TIM sequences within various categories. The definitions of the threshold in Tables 10 and 11 are the same as that in Tables 6 and 7, respectively.

3.4.1 Q analysis

In Table 10, for the superfamily categories, the same categories as those observed in Table 6 obtained *Q* values above 76.1%; however, all of the categories obtained better or equal *Q* values derived according to sequence identity than derived according to secondary

	Method	Sequence identity		Secondary structure identity		RMSD	
		Q (%)	Threshold	*Q* (%)	Threshold	*Q* (%)	Threshold
TIM40D	Superfamily	76.1	<14	73.5	<67	39.0	>1.9
	Family	73.9	<14	70.6	<67	37.1	>1.9
	Class (ENZYME)	41.6	<17	40.9	<73	20.1	>1.8
TIM95D	Superfamily	88.8	<14	87.2	<68	67.4	>2.0
	Family	88.4	<14	86.3	<68	66.4	>2.0
	Class (ENZYME)	61.3	<18	62.2	<73	47.7	>1.8

Table 9. Overall *Q* values for the BHPB alignment strategy in TIM40D and TIM95D (ASTRAL SCOP 1.71)

Method		Sequence identity			Secondary structure identity		
	Index	*Q* (%)	Precision[1] (%)	Precision[2] (%)	*Q* (%)	Precision[1] (%)	Precision[2] (%)
Superfamily	2	89.5	100.0	100.0(<20)	84.2	80.0	88.9(79)
	4	73.3	100.0	100.0(<23)	73.3	100.0	100.0(<77)
	8	78.0	100.0	100.0(<14)	74.4	98.4	98.4(<67)
	9	44.4	100.0	100.0(<26)	44.4	88.9	100.0(78-79)
	10	83.9	96.3	96.3(<17)	83.9	92.9	92.9(<75)
	11	83.3	90.9	100.0(17-18)	83.3	100.0	100.0(<80)
	12	75.0	90.0	100.0(15)	75.0	100.0	100.0(<77)
Family	2.4	100.0	100.0	100.0(<30)	100.0	90.9	100.0(77-86)
	4.1	73.3	100.0	100.0(<23)	73.3	100.0	100.0(<77)
	8.1	88.0	100.0	100.0(<14)	92.0	100.0	100.0(<67)
	8.3	84.6	100.0	100.0(<17)	76.9	95.2	95.2(<75)
	8.5	92.3	100.0	100.0(<18)	84.6	100.0	100.0(< 81)
	10.1	83.3	93.8	100.0(18-19)	83.3	88.2	93.8(75-76)
	11.2	90.9	90.9	100.0(17-18)	90.9	100.0	100.0(<80)
Class (ENZYME)	1	66.7	78.3	85.7(21-22)	70.4	73.1	76.0(75-76)
	2	38.7	60.0	63.2(17)	32.3	71.4	71.4(<73)
	3	42.6	61.7	69.0(18)	44.1	60.0	62.5(72-73)
	4	70.7	78.8	85.4(18-19)	69.0	74.1	76.9(75-76)
	5	60.9	63.6	63.6(<17)	56.5	56.5	59.1(78-79)

Table 10. Overall *Q* and Precision values for the BHPB alignment strategy in TIM40D (ASTRAL SCOP 1.71)

structure identity. For the family categories, only FMN-linked oxidoreductases had a *Q* value less than 73.9%. Amylase, catalytic domain was the only category that had a lower *Q* value when using sequence identity instead of secondary structure identity. For the class categories, only Transferases had a *Q* value less than 41.6%. For Transferases, Lyases and Isomerases, the *Q* values derived according to sequence identity were higher than those derived according to secondary structure identity. In Table 11, for the superfamily categories, (Trans)glycosidases, Metallo-dependent hydrolases and Xylose isomerase-like, had Q values less than 88.8%. For the family categories, only beta-glycanases had a Q value less than 88.4%. For all superfamily and family categories, the Q values derived according to sequence identity were higher than or equal to those derived according to secondary structure identity. All of the class categories had Q values higher than 62.2%. For Hydrolases and Isomerases, the Q values derived according to secondary structure identity were higher than those derived according to sequence identity.

	Method	Sequence identity			Secondary structure identity		
	Index	Q (%)	Precision[1] (%)	Precision[2] (%)	Q (%)	Precision[1] (%)	Precision[2] (%)
Superfamily	1	100.0	100.0	100.0(<45)	100.0	94.1	94.1(<86)
	2	96.7	100.0	100.0(<18)	90.0	93.1	96.4(77-79)
	4	90.9	100.0	100.0(<17)	86.4	100.0	100.0(<82)
	6	90.0	100.0	100.0(<22)	90.0	100.0	100.0(<88)
	7	100.0	100.0	100.0(<24)	100.0	100.0	100.0(<82)
	8	86.6	100.0	100.0(<14)	85.1	99.1	99.1(<72)
	9	63.6	100.0	100.0(<26)	63.6	93.3	100.0(78-79)
	10	93.8	97.8	97.8(<17)	93.8	93.8	95.7(75-77)
	11	100.0	100.0	100.0(<22)	100.0	100.0	100.0(<81)
	12	100.0	100.0	100.0(<26)	100.0	100.0	100.0(<79)
	14	100.0	100.0	100.0(<31)	100.0	100.0	100.0(<86)
	15	80.0	100.0	100.0(<18)	80.0	100.0	100.0(<80)
Family	1.1	100.0	100.0	100.0(<45)	100.0	94.1	94.1(<86)
	2.4	100.0	100.0	100.0(<30)	100.0	92.9	100.0(77-86)
	4.1	90.0	100.0	100.0(<17)	86.4	100.0	100.0(<82)
	7.1	100.0	100.0	100.0(<24)	100.0	100.0	100.0(<82)
	8.1	95.8	100.0	100.0(<14)	95.8	100.0	100.0(<68)
	8.3	87.8	100.0	100.0(<17)	85.4	97.2	97.2(<75)
	8.4	100.0	100.0	100.0(<36)	100.0	100.0	100.0(<85)
	8.5	94.4	100.0	100.0(<18)	88.9	100.0	100.0(<83)
	10.1	93.1	96.4	100.0(18-25)	93.1	93.1	96.4(75-86)
	11.2	100.0	100.0	100.0(<22)	100.0	100.0	100.0(<81)
	14.1	100.0	100.0	100.0(<31)	100.0	100.0	100.0(<86)
	15.3	100.0	100.0	100.0(<68)	100.0	100.0	100.0(<96)
Class (ENZYME)	1	89.1	87.2	91.1(21-23)	87.0	83.3	85.1(78-79)
	2	75.4	81.6	83.3(17)	75.4	88.9	88.9(<73)
	3	59.4	72.4	74.1(18)	64.2	73.9	74.7(72-73)
	4	89.7	87.9	88.8(17)	88.7	86.9	88.7(77)
	5	81.6	93.0	93.0(<19)	83.7	85.4	85.4(<80)

Table 11. Overall Q and Precision values for the BHPB alignment strategy in TIM95D (ASTRAL SCOP 1.71)

3.4.2 Precision analysis

In Tables 10 and 11, Precision values with the threshold were higher or equal to those without the threshold, thus making it difficult to determine the feasible threshold to obtain the best Precision value for routine alignment practices. However, the differences between Precision values with and without the threshold were greatly reduced by using the BHPB alignment strategy with the exception of Hydrolases of TIM40D. Using the BHPB alignment strategy in TIM40D, the average Precision values without the threshold were 96.7% for superfamily, 97.8% for family, and 68.5% for class (Table 10). Using the BHPB alignment strategy in TIM95D, the average Precision values without the threshold were 99.8% for superfamily, 99.7% for family, and 84.4% for class (Table 11). The best average Precision values were derived according to sequence identity. The PSI-BLAST method in the BHPB alignment strategy can filter out some of the false positives. Figures 2 and 3 indicate the

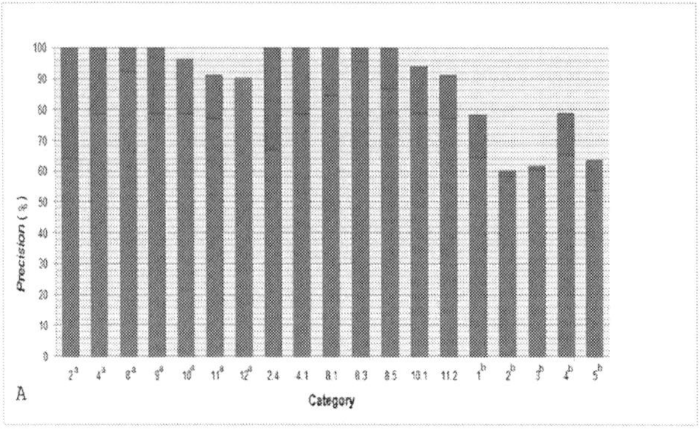

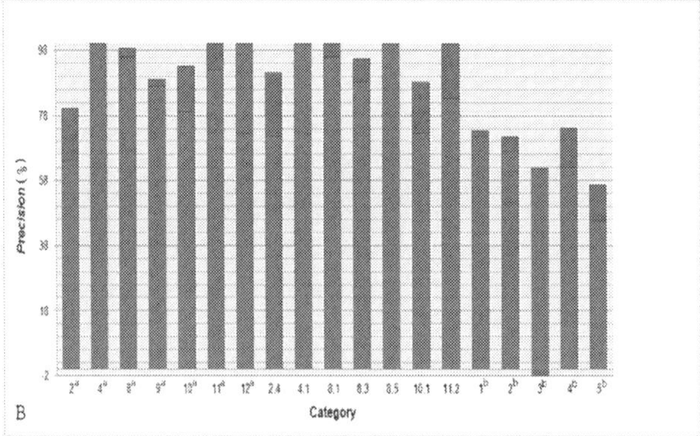

Fig. 2. The increase in Precision values for TIM40D (ASTRAL SCOP 1.71) using the BHPB alignment strategy. (A) Increase in Precision values for TIM40D derived according to sequence identity. (B) Increase in Precision values for TIM40D derived according to secondary structure identity. Superscript 'a' or 'b' indicates the superfamily categories or the class categories, respecively. Categories without a superscript indicated the family categories. The blue bar indicates Precision values using the PBH alignment strategy and the red bar indicates the increase in Precision values using the BHPB alignment strategy.

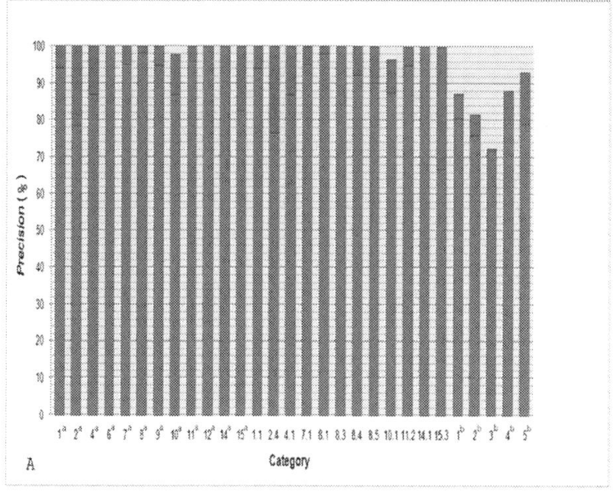

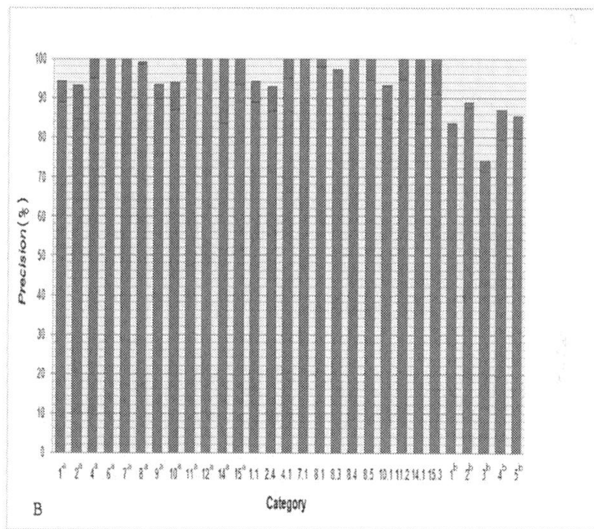

Fig. 3. The increase in Precision values for TIM95D (ASTRAL SCOP 1.71) using the BHPB alignment strategy. (A) Increase in Precision values for TIM95D derived according to sequence identity. (B) Increase in Precision values for TIM95D derived according to secondary structure identity. Superscript 'a' or 'b' indicates the superfamily categories or the class categories, respecively. Categories without a superscript indicated the family categories. The blue bar indicates Precision values using the PBH alignment strategy and the red bar indicates the increase in Precision values using the BHPB alignment strategy.

increase in Precision values for TIM40D and TIM95D from ASTRAL SCOP 1.71 using the BHPB alignment strategy as compared to the PBH alignment strategy, respectively. The increase in Precision values was computed by comparing the results shown in Tables 6, 7, 10 and 11 (see supplemental Table S4 (Chu, 2011)). Based on Figures 2 and 3, Precision values for almost all categories improved when using the BHPB alignment strategy. The average increases in Precision values for TIM40D using sequence identity were 16.8% for superfamily, 16.7% for family and 8.1% for class. The average increases in Precision values using secondary structure identity were 7.1% for superfamily, 9.5% for family and 7.0% for class. The average increases in Precision values derived according to sequence identity were higher than those derived according to secondary structure identity for TIM40D and TIM95D. Thus, the BHPB alignment strategy yields higher Precision values than the PBH alignment strategy.

3.4.3 MCC analysis

Figure 4 presents the *MCC* measures of (1) the PBH alignment strategy derived according to sequence identity (PBH(1D) for short), (2) the PBH alignment strategy derived according to secondary structure identity (PBH(2D) for short), (3) the BHPB alignment strategy derived according to sequence identity (BHPB(1D) for short) and (4) the BHPB alignment strategy derived according to secondary structure identity (BHPB(2D) for short) for TIM40D and TIM95D from ASTRAL SCOP 1.71, respectively. (see supplemental Table S5 (Chu, 2011))

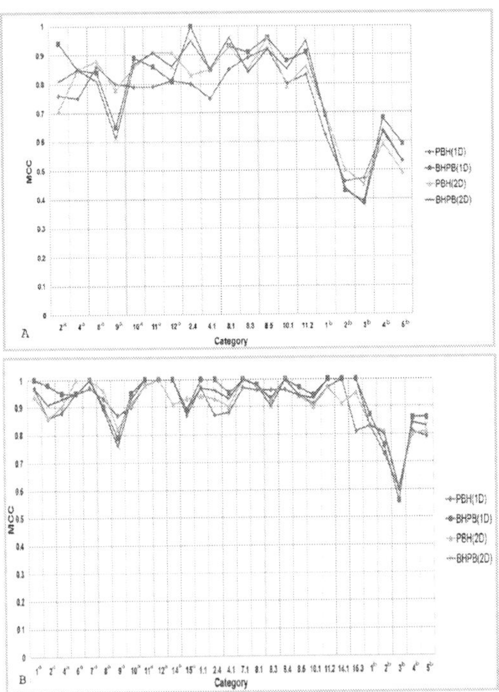

Fig. 4. *MCC* scores of PBH(1D), PBH(2D), BHPB(1D) and BHPB(2D) for TIM40D and TIM95D (ASTRAL SCOP 1.71). (A) *MCC* scores for TIM40D. (B) *MCC* scores for TIM95D. Superscript 'a' or 'b' indicates the superfamily categories or the class categories. Categories without a superscript indicate the family categories.

Using the PBH and BHPB alignment strategies, all of the superfamily categories had *MCC* scores greater than 0.7 except Metallo-dependent hydrolases when using the BHPB alignment strategy (Figure 4(A)); Ribulose-phosphate binding barrel, Enolase C-terminal domain-like, and Phosphoenolpyruvate/pyruvate domain had *MCC* scores greater than 0.9. All of the family categories had *MCC* scores greater than 0.7; Tryptophan biosynthesis enzymes, Amylase, catalytic domain, beta-glycanases, Type II chitinase, and D-glucarate dehydratase-like had *MCC* scores greater than 0.9. All of the class categories had *MCC* scores between 0.3~0.7, which is not an optimal score. From Figure 4(B), all of the superfamily and family categories had *MCC* scores greater than 0.7; 13 categories had the optimal *MCC* score (+1), indicating perfect prediction quality. All of the class categories had *MCC* scores between 0.5~0.9. The above results demonstrate that the proposed PBH or BHPB alignment strategy yielded high prediction quality for TIM barrel protein domain structure classification.

3.5 Discussion

Here we further investigate why the alignment approach with the PBH or BHPB strategy is not sufficient to classify the class category. For the above experiments, all of the EC annotations for TIM sequences in TIM40D and TIM95D from ASTRAL SCOP 1.71 and 1.73 were derived from UniProt. There are 24.5% TIM40D (67 of 274) and 20.6% TIM95D (91 of 442) TIM sequences listed as undefined in class from ASTRAL SCOP 1.71; there are 38.5% TIM40D (20 of 52) and 37.3% TIM95D (25 of 67) novel TIM sequences listed as undefined in class from ASTRAL SCOP 1.73. These TIM sequences with undefined class categories were initially assumed to be false negatives before the test. Therefore, the *Q* values for class obtained by the PBH or the BHPB alignment strategy derived according to sequence identity or secondary structure identity is poor. However, the ENZYME functions of some of these TIM sequences with undefined class categories derived from UniProt have been described in PDB. Thus, the EC annotations derived from PDB were integrated into TIM40D and TIM95D from ASTRAL SCOP 1.71 and 1.73 (see supplemental Table S3 (Chu, 2011)), and the above experiments for the class classification were repeated. After the PDB integrations, 13.6% TIM40D (38 of 279) and 11.1% TIM95D (50 of 450) TIM sequences remained undefined from ASTRAL SCOP 1.71; further, 11.5% TIM40D (6 of 52) and 9.0% TIM95D (6 of 67) novel TIM sequences remained undefined from ASTRAL SCOP 1.73. These six novel TIM sequences were identical in TIM40D and TIM95D from ASTRAL SCOP 1.73.

3.5.1 Improvement in Q

Figure 5 compares the *Q* values for TIM40D and TIM95D from ASTRAL SCOP 1.71 with UniProt and PDB EC annotations using the PBH and BHPB alignment strategies. (see supplemental Table S6 (Chu, 2011)) The *Q* values for the class classification using the PBH and BHPB alignment strategies improved after integrating the PDB EC annotations. By integrating the PDB EC annotations, some of the false negatives from UniProt were eliminated. The alignment approach using either the PBH or BHPB strategy was useful for the class classification. For TIM40D, the best *Q* value of 62.0% (an increase from 48.2%) for class was derived according to sequence identity or secondary structure identity using the PBH alignment strategy; the best *Q* value of 53.4% (an increase from 41.6%) for class was derived according to sequence identity using the BHPB alignment strategy. For TIM95D, the best *Q* value of 78.2% (an increase from 65.2%) for class was derived according to secondary structure identity using the PBH alignment strategy; the best Q value of 72.9% (an increase from 62.2%) for class was derived according to sequence identity or secondary structure identity using the BHPB alignment strategy. For the novel TIM sequences in TIM40D and TIM95D from ASTRAL SCOP 1.73, the best Q values were 73.1% (TIM40D) and 79.1% (TIM95D) using the PBH alignment strategy (see supplemental Table S6 (Chu, 2011)).

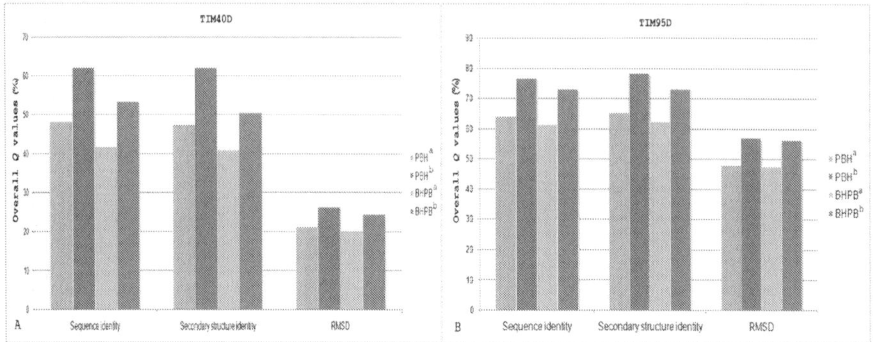

Fig. 5. Comparisons for TIM40D and TIM95D (ASTRAL SCOP 1.71) with UniProt and PDB EC annotations. (A) The Q values for TIM40D. (B) The Q values for TIM95D. Superscript 'a' or 'b' indicates TIM sequences available using only UniProt EC annotations or TIM sequences available using UniProt and PDB EC annotations.

3.5.2 Improvement in MCC

Table 12 presents *MCC* scores for TIM sequences in TIM40D and TIM95D from ASTRAL SCOP 1.71 with UniProt and PDB EC annotations using the PBH and BHPB alignment strategies. All of the class categories had *MCC* scores between 0.4~0.8 in TIM40D; greater than 0.7 in TIM95D; Oxidoreductases and Lyases also had *MCC* scores greater than 0.9 using the BHPB alignment strategy. Hence, the proposed PBH or BHPB alignment strategy also yielded high prediction quality for class.

	Category	Index	Sequence identity *MCC*		Secondary structure identity *MCC*	
			PBH	BHPB	PBH	BHPB
TIM40D	Class (ENZYME)	1	0.72	0.80	0.75	0.76
		2	0.49	0.47	0.50	0.42
		3	0.68	0.61	0.67	0.57
		4	0.67	0.69	0.67	0.65
		5	0.50	0.53	0.52	0.49
TIM95D	Class (ENZYME)	1	0.89	0.93	0.87	0.89
		2	0.76	0.78	0.81	0.80
		3	0.79	0.74	0.81	0.73
		4	0.87	0.91	0.86	0.90
		5	0.74	0.78	0.81	0.79

Table 12. *MCC* scores for TIM40D and TIM95D (ASTRAL SCOP 1.71) with UniProt and PDB EC annotations

3.5.3 Inferring ENZYME function for TIM barrel proteins with undefined class categories

After integrating the PDB EC annotations into the above tests, there remained 38 (TIM40D) and 50 (TIM95D) TIM sequences with undefined class categories from ASTRAL SCOP 1.71; 6 novel TIM sequences had undefined class categories from ASTRAL SCOP 1.73. Therefore, we used the proposed alignment approach to infer the ENZYME functions for TIM barrel proteins with undefined class.

We first assessed the classification results of the class categories by the PBH alignment strategy for TIM sequences in TIM40D and TIM95D from ASTRAL SCOP 1.71 with UniProt and PDB EC annotations. We found that the target protein and its selected protein belong to the same superfamily category for most of the true positives identified in the alignment. Table 13 presents statistics for true positives and false negatives for class using the PBH alignment strategy. For true positives, 94% (162 of 173) and 99% (342 of 344) of the target and its selected proteins belonged to the same superfamily category derived according to PBH(1D) in TIM40D and TIM95D, respectively. For false negatives, however, 38% (40 of 106) and 31% (33 of 106) of the target and its selected proteins belonged to the same superfamily category derived according to PBH(1D) in TIM40D and TIM95D, respectively.

Statistic	TIM40D				TIM95D			
	PBH(1D)		PBH(2D)		PBH(1D)		PBH(2D)	
	TP_i	FP_i	TP_i	FP_i	TP_i	FP_i	TP_i	FP_i
s,f	154.0	32.0	146.0	37.0	335.0	31.0	332.0	31.0
$s,\bar{f}$	8.0	8.0	13.0	9.0	7.0	2.0	11.0	2.0
$\bar{s}$	11.0	28.0	14.0	22.0	2.0	23.0	9.0	15.0
sum	173.0	106.0	173.0	106.0	344.0	106.0	352.0	98.0

s : Target and its selected proteins belong to the same superfamily category

$\bar{s}$: Target and its selected proteins belong to the different superfamily categories

f : Target and its selected proteins belong to the same family category

$\bar{f}$: Target and its selected proteins belong to the different family categories

Table 13. Statistical results for true positives and false negatives for class using the PBH alignment strategy

Overall, 58% (of 279) and 76% (of 450) of the target and its selected proteins belonged to the same superfamily and class categories derived according to PBH(1D) in TIM40D and TIM95D, respectively. Similar observations were made based on PBH(2D) in TIM40D and TIM95D. We observed 19 (PBH(1D)) and 23 (PBH(2D)) TIM sequences with undefined class categories in TIM40D with the same superfamily category, respectively. We observed 19 (PBH(1D)) and 26 (PBH(2D)) TIM sequences with undefined class categories in TIM95D with the same superfamily category. Therefore, it may be possible to infer the ENZYME functions for TIM barrel proteins with undefined class categories, especially for TIM95D, according to the classification results predicted by the proposed alignment approach. Table 13 also shows that 14% of 279 and 7% of 450 target and selected proteins belong to the same superfamily category, but they belong to different class categories derived according to PBH(1D) in TIM40D and TIM95D, respectively. Hence, all of TIM sequences of undefined class may not be correctly inferred by the proposed alignment approach with the PBH or the BHPB strategy. In the future, information regarding the active sites will be used in the proposed alignment approach to remedy discrepancies in undefined class. In the following test cases, all of the alignment results were displayed by DS Visualizer (Accelrys). The split structure superposition was displayed utilizing PyMol Molecular Viewer (DeLano, 2002).

4. Methods

4.1 The alignment approach with the PBH strategy

An alignment approach with the PBH strategy was proposed to perform TIM barrel protein domain structure classification (Figure 6). TIM40D and TIM95D can be used as the input for this alignment approach. In the alignment methods block, three alignment tools, CLUSTALW, SSEA and CE, were adopted to align any two of proteins by the amino acid sequences, secondary structures and 3D structures, respectively, to obtain the scores of sequence identity, secondary structure identity and RMSD. CLUSTALW is an established multiple sequence alignment tool (global alignment) for DNA/RNA or protein sequences based on a progressive pair-wise alignment method by considering sequence weighting, variations in amino acid substitution matrices and residue-specific gap penalty scores. It is widely used by biologists to investigate evolutional relationships among multiple protein sequences. CLUSTALW may not be the best choice for the sequence alignment because of recent advancements in programming, but it is still suitable for this alignment approach for two reasons. First, we simply want to obtain the score of sequence identity for any two proteins rather than the actual alignment information. Hence, the sequence identity score obtained by CLUSTALW is not significantly different from that obtained by other tools. Second, the design of most of other tools is focused on revising the multiple sequence alignment results, not improving the pair-wise alignment results, even using the pair-wise alignment results by CLUSTALW. SSEA is a multiple protein secondary structure alignment tool (either global or local alignment) that aligns entire elements (rather than residue-based elements [20]) of multiple proteins based on the H, C, and E states of SSEs. CE is a popular and accurate pair-wise protein 3D structural alignment tool that aligns residues in sequential order in space. If a protein domain sequence is not continuous, however, each continuous fragment in the domain will be aligned against the other protein using the CE alignment tool. Two criteria were adopted to resolve this problem. First, the sequence length of the continuous fragment must be at least 30 residues, and second the minimal RMSD of any two aligned fragments must be chosen. The default parameters of CLUSTALW (accurate, but slow mode in setting your pairwise alignment options) and SSEA (global alignment version) were used to align any two proteins in TIM40D and TIM95D to obtain scores for sequence and secondary structure identities with normalized values ranging from 0-100. The default parameters of CE were used to align any two proteins in TIM40D and TIM95D to obtain RMSD scores. After using CLUSTALW, SSEA and CE, these scores were used to build an alignment-based protein-protein identity score network.

In the best hit strategy block, each protein in the network was first considered as a target protein. Each target protein was then used to map the remaining proteins in the network. Finally, the prediction result of each target protein was determined by selecting the remaining proteins in the network according to certain parameters, which are critical for classification of the target protein. In our method, a PBH strategy is used to determine the prediction result of a target protein by selecting the protein that has the best score for the target protein according to this network. This score is calculated by a single parameter (sequence identity, secondary structure identity, or RMSD). For the sequence or secondary structure identity, the remaining protein with the highest score for the target protein is selected; for the RMSD, the remaining protein with the lowest score for the target protein is selected. For n proteins in the network, the time complexity is $O(n^2)$ for n target proteins to find all selected proteins in this network using the PBH strategy owing to the bidirectional aspect of the network. We used Perl to implement the PBH finding program because it supports powerful data structures.

The single parameter threshold was applied in this classification model. When a threshold is given for this approach, a target protein is assigned to a null situation as a false negative if the highest score of sequence identity or secondary structure identity (or lowest score of RMSD) for the target protein among all remaining proteins is less than (or larger than, for RMSD) this threshold. Although the overall prediction accuracy cannot be improved by the threshold concept, it may be decreased when an unfavorable threshold is given; however, the number of false positives may be reduced when an appropriate threshold is used. In other words, Precision values may be improved by the threshold concept. Nevertheless, an appropriate threshold is very difficult to attain for the classification problem in a practical setting. Therefore, in the experimental tests, an appropriate threshold was chosen after processing was complete. Using the threshold concept, we observed the best possible Precision values by this alignment approach and the properties of TIM barrel proteins.

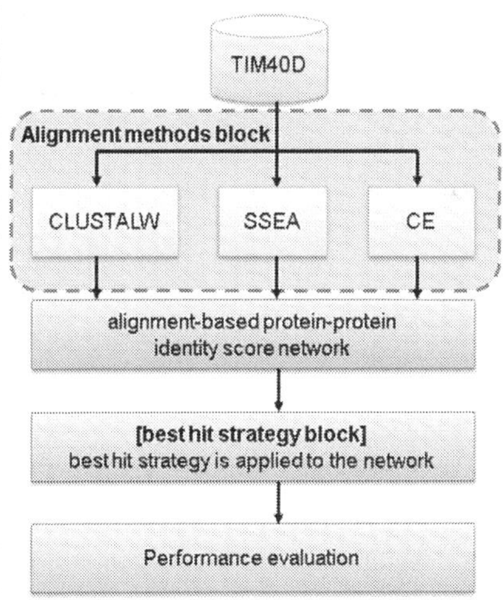

Fig. 6. Flow chart of the alignment approach with the PBH strategy

To experimentally test the novel TIM sequences from ASTRAL SCOP 1.73, the flow chart of the alignment approach with the PBH strategy is slightly different than that shown in Figure 6. For this test, the input is the novel TIM sequences from ASTRAL SCOP 1.73; however, the alignment-based protein-protein identity score network is built by TIM sequences from

ASTRAL SCOP 1.71. Therefore, the target protein is a novel TIM sequence from ASTRAL SCOP 1.73, and the remaining proteins are obtained from the TIM sequences (ASTRAL SCOP 1.71). All tools and materials used for this research are accessible from (Chu, 2011).

4.2 The alignment approach with the BHPB strategy

PSI-BLAST is a position-specific iterative BLAST that results from refinement of the position-specific scoring matrix (PSSM) and the next iterative PSSM. The position-specific scoring matrix is automatically constructed from a multiple alignment with the highest scoring hits in the BLAST search. The next iterative PSSM is generated by calculating position-specific scores for each position in the previous iteration. PSI-BLAST is typically used instead of BLAST to detect subtle relationships between proteins that are structurally distant or functionally homologous. Therefore, it is possible to utilize PSI-BLAST as a filter prior to the PBH strategy, denoted the BHPB strategy. The BHPB strategy can filter out potential false positives, which may improve Precision values. The flow chart of the alignment approach with the BHPB strategy is also slightly different than that shown in Figure 6. In the best hit strategy block, each target protein in the network is used to map a subset, but not all, of the remaining proteins in the network. This subset of remaining proteins is grouped from the network using PSI-BLAST method for the target protein. Hence, the selected protein with the best score for any target protein by the BHPB strategy may not be the same as that by the PBH strategy.

5. Conclusion

At the amino acid sequence level, TIM barrel proteins are very diverse; however, these proteins contain very similar secondary structures. Our results demonstrate that the alignment approach with the PBH strategy or BHPB strategy is a simple and stable method for TIM barrel protein domain structure classification, even when only amino acid sequence information is available.

6. Acknowledgment

Part of this work was supported by National Science Council (NSC) under contract NSC95-2627-B-007-002. The authors would like to thank Shu Hao Chang to help us to collect the TIM barrel proteins from the SCOP 1.71 version.

7. References

Altschul, S.F.; Madden, T.L.; Schäffer, A.A.; Zhang, J.; Zhang, Z.; Miller, W. & Lipman, D.J. (1997). Gapped BLAST and PSI-BLAST: a new generation of protein database search programs. *Nucleic Acids Research*, 12.06.2008, Available from ftp://ncbi.nlm.nih.gov/blast/.

Andreeva, A.; Howorth, D.; Chandonia, J.-M.; Brenner, S.E.; Hubbard, T.J.P.; Chothia, C. & Murzin, A.G. (2008). Data growth and its impact on the SCOP database: new developments. *Nucleic Acids Research*, Vol.36, pp. D419-D425.

Bairoch, A. (2000). The ENZYME database in 2000. *Nucleic Acids Research*, Vol.28, pp. 304-305.

Bairoch, A.I.; Apweiler, R.; Wu, C.H.; Barker, W.C.; Boeckmann, B.; Ferro, S.; Gasteiger, E.; Huang, H.; Lopez, R.; Magrane, M.; Martin, M.J.; Natale, D.A.; O'Donovan, C.; Redaschi, N. & Yeh, L.S. (2005). The Universal Protein Resource (UniProt). *Nucleic Acids Research*, Vol.33, pp. D154-D159.

Bairoch, A.; Boeckmann, B.; Ferro, S. & Gasteiger, E. (2004). Swiss-Prot: Juggling between evolution and stability. *Briefings in Bioinformatics*, Vol.5, pp. 39-55.

Berman, H.M.; Westbrook, J.; Feng, Z.; Gilliland, G.; Bhat, T.N.; Weissig, H.; Shindyalov, I.N. & Bourne, P.E. (2000). The Protein Data Bank. *Nucleic Acids Research*, Vol.28, pp. 235-242.

Carugo, O. & Pongor, S. (2002). Protein fold similarity estimated by a probabilistic approach based on C_{α}-C_{α} distance comparison. *Journal of Molecular Biology*, Vol.315, pp. 887-898.

Chandonia, J.M.; Hon, G.; Walker, N.S.; Lo, C.L.; Koehl, P.; Levitt, M. & Brenner, S.E. (2004). The ASTRAL compendium in 2004. *Nucleic Acids Research*, Vol.32, pp. D189-D192.

Choi, I.; Kwon, J. & Kim, S. (2004). Local feature frequency profile: a method to measure structural similarity in proteins. *Proceeding of the National Academy of Sciences of the United States of America*, Vol.101, pp. 3797-3802.

Chou, K.C. & Zhang, C.T. (1995). Prediction of protein structural classes. *Critical Reviews in Biochemistry and Molecular Biology*, Vol.30, pp. 275-349.

Chu, C.-H. (2011). TIM barrel supplemental data for the InTech bookchapter. *National Tsing Hua Uinversity, Computational Systems Biology & Bio-Medicine Laboratory*, 03.01.2011, Available from http://oz.nthu.edu.tw/~d938301/InTech/bookchapter/

Cuff, A.L.; Sillitoe, I.; Lewis, T.; Redfern, O.C.; Garratt, R.; Thornton, J. & Orengo, C.A. (2009). The CATH classification revisited—architectures reviewed and new ways to characterize structural divergence in superfamilies. *Nucleic Acids Research*, Vol.37, pp. D310-D314.

DeLano, W.L. (2002).The PyMOL Molecular Graphics System, DeLano Scientific, Palo Alto, CA, USA. http://www.pymol.org.

Ding, C.H.Q. & Dubchak, I. (2001). Multi-class protein fold recognition using support vector machines and neural networks. *Bioinformatics*, Vol.17, pp. 349-358.

Dobson, P.D. & Doig, A.J. (2005). Predicting enzyme class from protein structure without alignments. *Journal of Molecular Biology*, Vol.345, pp. 187-199.

Dubchak, I.; Muchnik, I.; Holbrook, S.R. & Kim, S.H.(1995). Prediction of protein folding class using global description of amino acid sequence. *Proceeding of the National Academy of Sciences of the United States of America*, Vol.92, pp. 8700-8704.

Fontana, P.; Bindewald, E.; Toppo, S.; Velasco, R.; Valle, G. & Tosatto, S.C.E. (2005). The SSEA server for protein secondary structure alignment. *Bioinformatics*, Vol.21, pp. 393-395.

Gardy, J.L.; Spencer, C.; Wang, K.; Ester, M.; Tusnday, G.E.; Simon, I.; Hua, S.; deFays, K.; Lambert, C.; Nakai, K. & Brinkman, F.S.L. (2003). PSORT-B: improving protein subcellular localization prediction for gram-negative bacteria. *Nucleic Acids Research*, Vol.31, pp. 3613-3617.

Gáspári, Z.; Vlahovicek, K. & Pongor, S. (2005). Efficient recognition of folds in protein 3D structures by the improved PRIDE algorithm. *Bioinformatics*, Vol.21, pp. 3322-3323.

Gloster, T.M.; Roberts, S.; Ducros, V.M.-A.; Perugino, G.; Rossi, M.; Hoos, R.; Moracci, M.; Vasella, A. & Davies, G.J. (2004). Structural studies of the β-Glycosidase from

Sulfolobus solfataricus in complex with covalently and noncovalently bound inhibitors. *Biochemistry*, Vol.43, pp. 6101-6109.

Huang, C.D.; Lin, C.T. & Pal, N.R. (2003). Hierarchical learning architecture with automatic feature selection for multi-class protein fold classification. *IEEE Transactions on NanoBioscience*, Vol.2, pp. 503–517.

Jones, D.T. (1999). Protein secondary structure prediction based on position-specific scoring matrices. *Journal of Molecular Biology*, Vol.292, pp. 195-202.

Kabsch, W. & Sander, C. (1983). Dictionary of protein secondary structure: pattern recognition of hydrogen-bonded and geometrical features. *Biopolymers*, Vol.22, pp. 2577-2637.

Lin, K.L.; Lin, C.Y.; Huang, C.D.; Chang, H.M.; Yang, C.Y.; Lin, C.T.; Tang, C.Y. & Hsu, D.F. (2005). Methods of improving protein structure prediction based on HLA neural network and combinatorial fusion analysis. *WSEAS Transations on Information Science and Applications*, Vol.2, pp. 2146-2153.

Lin, K.L.; Lin, C.Y.; Huang, C.D.; Chang, H.M.; Yang, C.Y.; Lin, C.T.; Tang, C.Y. & Hsu, D.F. (2007). Feature combination criteria for improving accuracy in protein structure prediction. *IEEE Transactions on NanoBioscience*, Vol.6, pp. 186-196.

Matthews, B.W. (1975). Comparison of the predicted and observed secondary structure of T4 phage lysozyme. *Biochimica et Biophysica Acta*, Vol.405, pp. 442-451.

Murzin, A.G.; Brenner, S.E.; Hubbard, T. & Chothia, C. (1995). SCOP: A structural classification of proteins database for the investigation of sequence and structures. *Journal of Molecular Biology*, Vol.247, pp. 536-540.

Rogen, P. & Fain, B. (2003). Automatic classification of protein structure by using gauss integrals. *Proceeding of the National Academy of Sciences of the United States of America*, Vol.100, pp. 119-124.

Rost, B. & Sander, C. (1993). Prediction of protein secondary structure at better than 70% accuracy. *Journal of Molecular Biology*, Vol.232, pp. 584-599.

Shen, H.B. & Chou, K.C. (2006). Ensemble classifier for protein fold pattern recognition. *Bioinformatics*, Vol.22, pp. 1717-1722.

Shindyalov, I.N. & Bourne, P.E. (1998). Protein structure alignment by incremental combinatorial extension (CE) of the optimal path. *Protein Engineering*, Vol.11, pp. 739-747.

Thompson, J.D.; Higgins, D.G. & Gibson, T.J. (1994). CLUSTAL W: improving the sensitivity of progressive multiple sequence alignment through sequence weighting, positions-specific gap penalties and weight matrix choice. *Nucleic Acids Research*, Vol.22, pp. 4673-4680.

Vapnik, V.N. (1995). *The Nature of Statistical Learning Theory*, New York: Springer-Verlag.

Yu, C.S.; Wang, J.Y.; Yang, J.M.; Lyu, P.C.; Lin, C.J. & Hwang, J.K. (2003). Fine-grained protein fold assignment by support vector machines using generalized *n*Peptide coding schemes and jury voting from multiple-parameter sets. *Proteins*, Vol.50, pp. 531-536.

Zotenko, E.; Dogan, R.I.; Wilbur, W.J.; O'Leary, D.P. & Przytycka, T.M. (2007). Structural footprinting in protein structure comparison: The impact of structural fragments. *BMC Structural Biology*, Vol.7, pp. 53.

Zotenko, E.; O'Leary, D.P. & Przytycka, T.M. (2006). Secondary structure spatial conformation footprint: a novel method for fast protein structure comparison and classification. *BMC Structural Biology*, Vol.6, pp. 12.

16

Identification of Functional Diversity in the Enolase Superfamily Proteins

Kaiser Jamil[1] and M. Sabeena[2]
[1]*Head, Genetics Department, Bhagwan Mahavir Medical Research Centre, Mahavir Marg, Hyderabad, & Dean, School of Life Sciences (SLS), Jawaharlal Nehru Institute of Advanced Studies, (JNIAS)*
[2]*Research Associate, Centre for Biotechnology and Bioinformatics- (CBB), Jawaharlal Nehru Institute of Advanced Studies, (JNIAS), Secunderabad*
India

1. Introduction

The Escherichia coli K12 genome is a widely studied model system. The members of the Enolase superfamily encoded by *E.coli* catalyze mechanistically diverse reactions that are initiated by base-assisted abstraction of the α-proton of a carboxylate anion substrate to form an enodiolate intermediate (Patricia C ,1996). Six of the eight members of the Enolase superfamily encoded by the Escherichia coli K-12 genome have known functions (John F, 2008). The members share a conserved tertiary structure with a two-domain architecture, in which three carboxylate ligands for the Mg^{2+} ion as well as the acid/base catalysts are located at the C-terminal ends of the β-strands in a $(\beta/\alpha)_7\beta$-barrel [modified $(\beta/\alpha)_8$- or TIM-barrel] domain and the specificity-determining residues are located in an N-terminal α+β capping domain.

The rapid accumulation of data has led to an extraordinary problem of redundancy, which must be confronted in almost any type of statistical analysis. An important goal of bioinformatics is to use the vast and heterogeneous biological data to extract patterns and make discoveries that bring to light the "unifying" principles in biology. (Kaiser Jamil, 2008)Because these patterns can be obscured by bias in the data, we approach the problem of redundancy by appealing to a well known unifying principle in biology, evolution. Bioinformatics has developed as a data-driven science with a primary focus on storing and accessing the vast and exponentially growing amount of sequence and structure data (Gerlt JA, 2005)

Protein sequences and their three-dimensional structures are successful descendants of evolutionary process. Proteins might have considerable structural similarities even when no evolutionary relationship of their sequences can be detected (Anurag Sethi, 2005). This property is often referred to as the proteins sharing only a "fold". Of course, there are also sequences of common origin in each fold, called a "superfamily", and in them groups of sequences with clear similarities, are designated as "family".

The concept of protein superfamily was introduced by Margaret Dayholff in the 1970 and was used to partition the protein sequence databases based on evolutionary consideration

(Lindahl E, 2000). The objective of this study was to analyse the functional diversity of the enolase gene superfamily. The gene superfamily consisting of twelve genes possess enzymatic functions such as L-Ala-D/L-Glu epimerase, Glucarate dehydratase, D-galactarate dehydratase, 2-hydroxy-3-oxopropionate reductase,l-o-succinylbenzoate synthase, D-galactonate dehydratase,[12]. 5-keto-4-deoxy-D-glucarate aldolase, L-rhamnonate dehydratase, 2-keto-3-deoxy-L-rhamnonate aldolase, Probable galactarate transporter, and Probable glucarate transporter (Steve EB ,1998)
This study was carried out to determine the Probable glucarate transporter (D-glucarate permease) features relating enolase superfamily sequences to structural hinges, which is important for identifying domain boundaries, and designing flexibility into proteins functions also helps in understanding structure-function relationships.

2. Methodology

Enolase Superfamily Study/Analysis

Enolase Sequence Retrieval from Biological Databases

Sequence Analysis and Alignment (Using BLAST Program)

Multiple Sequence Alignment (Clustal W algorithm)

Sequence Alignment retrieval and improving of alignment using Jalview Program

SCI -PHY server for superfamily and subfamily prediction

ConSurf Server for residue Conservation analysis

Pattern Recognition Using ScanProsite

Visualization of the key residues represents superfamily in visualization program Rasmol

Flowchart represents the materials and methods

2.1 UniProt KB for genomic sequence analysis

Enolase sequence from *E.coli* formed the basis for this study. The protein sequences were derived from UniProt KB, we found twelve sequences (Table 1). Most of the sequences in UniProt KB were derived from the conceptual translation of nucleotide sequences. The advantage of using UniProt KB was that it provides a stable, comprehensive, freely accessible central resource on protein sequences and functional annotation. UniProt comprises of four major components, each optimized for different uses: the UniProt Archive, the UniProt Knowledgebase, the UniProt Reference Clusters and the UniProt Metagenomic and Environmental Sequence Database. We used this knowledge based computational analysis which helps for the functional annotation for the gene sequences shown below:

S.No	Accession Id	Sequence Name	Sequence
1.	P0A6P9	ENO_ECOLI Enolase OS=*Escherichia coli* (strain K12) GN=eno PE=1 SV=2	MSKIVKIIGREIIDSRGNPTVEAEVHLEGGFVGMA AAPSGASTGSREALELRDGDKSRFLGKGVTKAVA AVNGPIAQALIGKDAKDQAGIDKIMIDLDGTENK SKFGANAILAVSLANAKAAAAAKGMPLYEHIAE LNGTPGKYSMPVPMMNIINGGEHADNNVDIQEF MIQPVGAKTVKEAIRMGSEVFHHLAKVLKAKGM NTAVGDEGGYAPNLGSNAEALAVIAEAVKAAG YELGKDITLAMDCAASEFYKDGKYVLAGEGNKA FTSEEFTHFLEELTKQYPIVSIEDGLDESDWDGFAY QTKVLGDKIQLVGDDLFVTNTKILKEGIEKGIANS ILIKFNQIGSLTETLAAIKMAKDAGYTAVISHRSGE TEDATIADLAVGTAAGQIKTGSMSRSDRVAKYN QLIRIEEALGEKAPYNGRKEIKGQA
2.	P51981	AEEP_ECOLI L-Ala-D/L-Glu epimerase OS=*Escherichia coli* (strain K12) GN=ycjG PE=1 SV=2	MRTVKVFEEAWPLHTPFVIARGSRSEARVVVVEL EEEGIKGTGECTPYPRYGESDASVMAQIMSVVPQL EKGLTREELQKILPAGAARNALDCALWDLAARR QQQSLADLIGITLPETVITAQTVVIGTPDQMANSA STLWQAGAKLLKVKLDNHLISERMVAIRTAVPD ATLIVDANESWRAEGLAARCQLLADLGVAMLEQ PLPAQDDAALENFIHPLPICADESCHTRSNLKAL KGRYEMVNIKLDKTGGLTEALALATEARAQGFSL MLGCMLCTSRAISAALPLVPQVSFADLDGPTWLA VDVEPALQFTTGELHL
3.	P0AES2	GUDH_ECOLI Glucarate dehydratase OS=*Escherichia coli* (strainK12) GN=gudD PE=1 SV=2	MSSQFTTPVVTEMQVIPVAGHDSMLMNLSGAHA PFFTRNIVIIKDNSGHTGVGEIPGGEKIRKTLEDAIP LVVGKTLGEYKNVLTLVRNTFADRDAGGRGLQT FDLRTTIHVVTGIEAMLDLLGQHLGVNVASLLGD GQQRSEVEMLGYLFFVGNRKATPLPYQSQPDDSC DWYRRHEEAMTPDAVVRLAEAAYEKYGFNDFK LKGGVLAGEEEAESIVALAQRFPQARITLDPNGA WSLNEAIKIGKYLKGSLAYAEDPCGAEQGFSGRE VMAEFRRATGLPTATNMIATDWRQMGHTLSLQS VDIPLADPHFWTMQGSVRVAQMCHEFGLTWGS HSNNHFDISLAMFTHVAAAAPGKITAIDTHWIW QEGNQRLTKEPFEIKGGLVQVPEKPGLGVEIDMD QVMKAHELYQKHGLGARDDAMGMQYLIPGWT FDNKRPCMVR

S.No	Accession Id	Sequence Name	Sequence
4.	P39829	GARD_ECOLI D-galactarate dehydratase OS=*Escherichia coli* (strain K12) GN=garD PE=1 SV=2	MANIEIRQETPTAFYIKVHDTDNVAIIVNDNGLK AGTRFPDGLELIEHIPQGHKVALLDIPANGEIIRYG EVIGYAVRAIPRGSWIDESMVVLPEAPPLHTLPLA TKVPEPLPPLEGYTFEGYRNADGSVGTKNLLGITT SVHCVAGVVDYVVKIIERDLLPKYPNVDGVVGLN HLYGCVAINAPAAVVPIRTIHNISLNPNFGGEVM VIGLGCEKLQPERLLTGTDDVQAIPVESASIVSLQD EKHVGFQSMVEDILQIAERHLQKLNQRQRETCPA SELVVGMQCGGSDAFSGVTANPAVGYASDLLVR CGATVMFSEVTEVRDAIHLLTPRAVNEEVGKRLL EEMEWYDNYLNMGKTDRSANPSPGNKKGGLAN VVEKALGSIAKSGKSAIVEVLSPGQRPTKRGLIYA ATPASDFVCGTQQVASGITVQVFTTGRGTPYGLM AVPVIKMATRTELANRWFDLMDINAGTIATGEET IEEVGWKLFHFILDVASGKKKTFSDQWGLHNQL AVFNPAPVT
5.	P29208	MENC_ECOLI o-succinylbenzoate synthase OS=*Escherichia coli* (strain K12) GN=menC PE=1 SV=2	MRSAQVYRWQIPMDAGVVLRDRRLKTRDGLYV CLREGEREGWGEISPLPGFSQETWEEAQSVLLAW VNNWLAGDCELPQMPSVAFGVSCALAELTDTLP QAANYRAAPLCNGDPDDLILKLADMPGEKVAK VKVGLYEAVRDGMVVNLLLEAIPDLHLRLDANR AWTPLKGQQFAKYVNPDYRDRIAFLEEPCKTRD DSRAFARETGIAIAWDESLREPDFAFVAEEGVRAV VIKPTLTGSLEKVREQVQAAHALGLTAVISSSIESS LGLTQLARIAAWLTPDTIPGLDTLDLMQAQQVRR WPGSTLPVVEVDALERLL
6.	Q6BF17	DGOD_ECOLI D-galactonate dehydratase OS=*Escherichia coli* (strain K12) GN=dgoD PE=1 SV=1	MKITKITTYRLPPRWMFLKIETDEGVVGWGEPVIE GRARTVEAAVHELGDYLIGQDPSRINDLWQVMY RAGFYRGGPILMSAIAGIDQALWDIKGKVLNAPV WQLMGGLVRDKIKAYSWVGGDRPADVIDGIKTL REIGFDTFKLNGCEELGLIDNSRAVDAAVNTVAQ IREAFGNQIEFGLDFHGRVSAPMAKVLIKELEPYR PLFIEEPVLAEQAEYYPKLAAQTHIPLAAGERMFS RFDFKRVLEAGGISILQPDLSHAGGITECYKIAGM AEAYDVTLAPHCPLGPIALAACLHIDFVSYNAVL QEQSMGIHYNKGAELLDFVKNKEDFSMVGGFFK PLTKPGLGVEIDEAKVIEFSKNAPDWRNPLWRHE DNSVAEW
7.	P23522	GARL_ECOLI 5-keto-4-deoxy-D-glucarate aldolase OS=*Escherichia coli* (strain K12) GN=garL PE=1 SV=2	MNNDVFPNKFKAALAAKQVQIGCWSALSNPIST EVLGLAGFDWLVLDGEHAPNDISTFIPQLMALKG SASAPVVRVPTNEPVIIKRLLDIGFYNFLIPFVETKE EAELAVASTRYPPEGIRGVSVSHRANMFGTVADY FAQSNKNITILVQIESQQGVDNVDAIAATEGVDGI FVGPSDLAAALGHLGNASHPDVQKAIQHIFNRA SAHGKPSGILAPVEADARRYLEWGATFVAVGSDL GVFRSATQKLADTFKK

S.No	Accession Id	Sequence Name	Sequence
8.	P77215	RHAMD_ECOLI L-rhamnonate dehydratase OS=*Escherichia coli* (strain K12) GN=yfaW PE=1 SV=2	MTLPKIKQVRAWFTGGATAEKGAGGGDYHDQG ANHWIDDHIATPMSKYRDYEQSRQSFGINVLGTL VVEVEAENGQTGFAVSTAGEMGCFIVEKHLNRFI EGKCVSDIKLIHDQMLSATLYYSGSGGLVMNTISC VDLALWDLFGKVVGLPVYKLLGGAVRDEIQFYA TGARPDLAKEMGFIGGKMPTHWGPHDGDAGIR KDAAMVADMREKCGEDFWLMLDCWMSQDVN YATKLAHACAPYNLKWIEECLPPQQYESYRELKR NAPVGMMVTSGEHHGTLQSFRTLSETGIDIMQPD VGWCGGLTTLVEIAAIAKSRGQLVVPHGSSVYSH HAVITFTNTPFSEFLMTSPDCSTMRPQFDPILLNEP VPVNGRIHKSVLDKPGFGVELNRDCNLKRPYSH
9.	P76469	KDRA_ECOLI 2-keto-3-deoxy-L-rhamnonate aldolase OS=*Escherichia coli* (strain K12) GN=yfaU PE=1 SV=1	MNALLSNPFKERLRKGEVQIGLWLSSTTAYMAEI AATSGYDWLLIDGEHAPNTIQDLYHQLQAVAPY ASQPVIRPVEGSKPLIKQVLDIGAQTLLIPMVDTAE QARQVVSATRYPPYGERGVGASVARAARWGRIE NYMAQVNDSLCLLVQVESKTALDNLDEILDVEGI DGVFIGPADLSASLGYPDNAGHPEVQRIIETSIRRI RAAGKAAGFLAVAPDMAQQCLAWGANFVAVG VDTMLYSDALDQRLAMFKSGKNGPRIKGSY
10.	P0AA80	GARP_ECOLI Probable galactarate transporter OS=*Escherichia coli* (strain K12) GN=garP PE=1 SV=1	MILDTVDEKKKGVHTRYLILLIIFIVTAVNYADRA TLSIAGTEVAKELQLSAVSMGYIFSAFGWAYLLM QIPGGWLLDKFGSKKVYTYSLFFWSLFTFLQGFVD MFPLAWAGISMFFMRFMLGFSEAPSFPANARIVA AWFPTKERGTASAIFNSAQYFSLALFSPLLGWLTF AWGWEHVFTVMGVIGFVLTALWIKLIHNPTDHP RMSAEELKFISENGAVVDMDHKKPGSAAASGPK LHYIKQLLSNRMMLGVFFGQYFINTITWFFLTWFP IYLVQEKGMSILKVGLVASIPALCGFAGGVLGGVF SDYLIKRGLSLTLARKLPIVLGMLLASTIILCNYTN NTTLVVMLMALAFFGKGFGALGWPVISDTAPKEI VGLCGGVFNVFGNVASIVTPLVIGYLVSELHSFNA ALVFVGCSALMAMVCYLFVVGDIKRMELQK
11.	P0ABQ2	GARR_ECOLI 2-hydroxy-3-oxopropionate reductase OS=*Escherichia coli* (strain K12) GN=garR PE=1 SV=1	MKVGFIGLGIMGKPMSKNLLKAGYSLVVADRNP EAIADVIAAGAETASTAKAIAEQCDVIITMLPNSP HVKEVALGENGIIEGAKPGTVLIDMSSIAPLASREI SEALKAKGIDMLDAPVSGGEPKAIDGTLSVMVGG DKAIFDKYYDLMKAMAGSVVHTGEIGAGNVTKL ANQVIVALNIAAMSEALTLATKAGVNPDLVYQA IRGGLAGSTVLDAKAPMVMDRNFKPGFRIDLHIK DLANALDTSHGVGAQLPLTAAVMEMMQALRA DGLGTADHSALACYYEKLAKVEVTR

S.No	Accession Id	Sequence Name	Sequence
12.	Q46916	GUDP_ECOLI Probable glucarate transporter OS=*Escherichia coli* (strain K12) GN=gudP PE=1 SV=1	MSSLSQAASSVEKRTNARYWIVVMLFIVTSFNYG DRATLSIAGSEMAKDIGLDPVGMGYVFSAFSWAY VIGQIPGGWLLDRFGSKRVYFWSIFIWSMFTLLQG FVDIFSGFGIIVALFTLRFLVGLAEAPSFPGNSRIVA AWFPAQERGTAVSIFNSAQYFATVIFAPIMGWLT HEVWSHVFFFMGGLGIVISFIWLKVIHEPNQHPG VNKKELEYIAAGGALINMDQQNTKVKVPFSVKW GQIKQLLGSRMMIGVYIGQYCINALTYFFITWFPV YLVQARGMSILKAGFVASVPAVCGFIGGVLGGIIS DWLMRRTGSLNIARKTPIVMGMLLSMVMVFCNY VNVEWMIIGFMALAFFGKGIGALGWAVMADTA PKEISGLSGGLFNMFGNISGIVTPIAIGYIVGTTGSF NGALIYVGVHALIAVLSYLVLVGDIKRIELKPVAG Q

Table 1. Enolase Sequences from *E.coli* -K12 Strain (from UNIPROT-KB in Fasta format)

2.2 BLAST program for sequence analysis and alignment

Basic Local Alignment Search Tool (BLAST) is one of the most heavily used sequence analysis tools we have used to perform Sequence Analysis and Alignment. BLAST is a heuristic that finds short matches between two sequences and attempts to start alignments. In addition to performing alignments, BLAST provides statistical information to help decipher the biological significance of the alignment as 'expect' value. (Scott McGinnis, 2004). Using this BLAST program the twelve gene sequences were aligned against archaea and bacteria. The sequences were sorted out according to the existing gene names with similarity and the fused genes were removed.

2.3 Clustal W program for multiple sequence alignment

Multiple sequence alignments are widely acknowledged to be powerful tools in the analysis of sequence data.(Sabitha Kotra et al 2008) Crucial residues for activity and for maintaining protein secondary and tertiary structures are often conserved in sequence alignments. Hence, multiple sequence alignment was done for all the enolase gene sequences based on the ClustalW algorithm using the tool BioEdit software program. We determined the alignments which is the starting points for evolutionary studies. Similarity is a percentage sequence match between nucleotide or protein sequences. The basic hypothesis involved here was that similarity relates to functionality, if two sequences are similar, they will have related functionalities.

Realigned the obtained Multiple Sequence Alignments (MSA) using ClustalW (Muhummad Khan and Kaiser Jamil, 2010). Using MSA we could obtain high score for the conserved regions, compared to the reported query sequences. So we viewed the multiple alignment result using a program 'Jalview' which improved the multiple alignment. With this program we could extract and get the complete alignment of all sequences for realigning to the query sequence to get better results (Fig. 1). Jalview is a multiple alignment editor written in Java. It is used widely in a variety of web pages which is available as a general purpose alignment editor. The image below shows the result when Jalview has taken the full length sequences and realigned them (using Clustalw) to the query sequence. The alignment has far fewer gaps and more similarities to the entire portion of the query sequences.

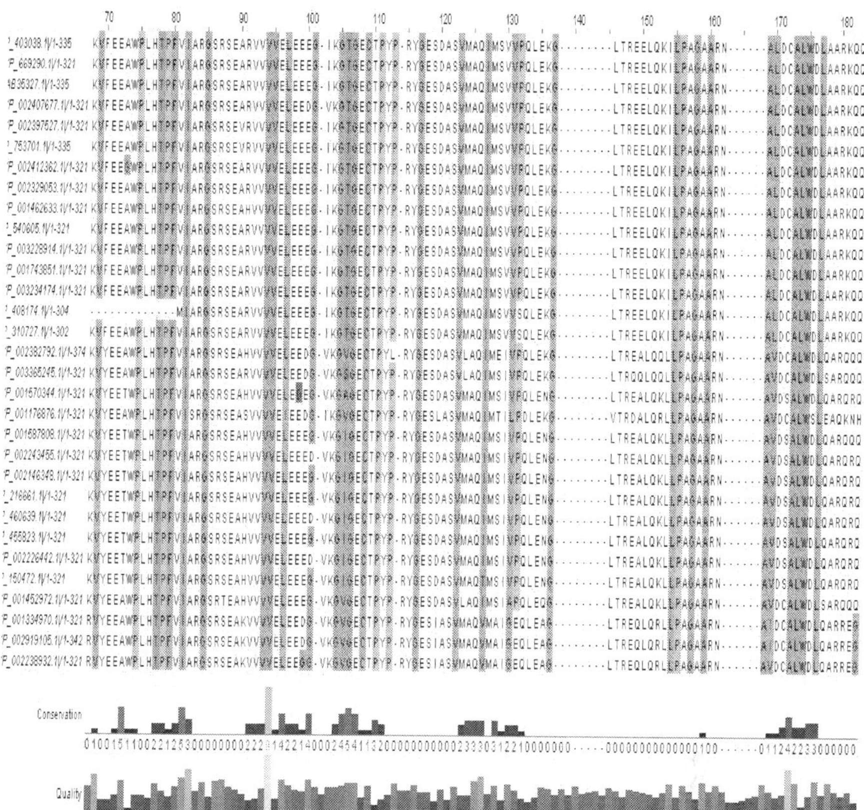

Fig. 1. Multiple Sequence Alignment as shown in Jalview

2.4 SCI –PHY server for superfamily and subfamily prediction

Using SCI-PHY server we found subfamilies/subclasses present in the aligned sequences, which merged into five groups. The corresponding pattern for each group of subfamily sequences was found by using ScanProsite and PRATT. A low-level simple pattern-matching application can prove to be a useful tool in many research settings (Doron Betel, 2000). Many of these applications are geared toward heuristic searches where the program finds sequences that may be closely related to the query nucleotide/protein sequences.

2.5 ConSurf server for conservation analysis

For each subfamily sequences the corresponding PDB ID using ConSurf Server was determined. ConSurf-DB is a repository of ConSurf Server which used for evolutionary conservation analysis of the proteins of known structures in the PDB. Sequence homologues of each of the PDB entries were collected and aligned using standard methods. The algorithm behind the server takes into account the phylogenetic relations between the aligned proteins and the stochastic nature of the evolutionary process explicitly. The server assigned the conservation level for each position in the multiple sequence alignment (Ofir Goldenberg, 2002). Identified specific pattern for each of the FASTA format sequence from PDB files using ScanProsite and some of the key residues that comprise the functionally important regions of the protein (Ofir Goldenberg, 2002). We determined the residues present in each of PDB files denoting subfamilies using Swiss PDB Viewer. Mapped out all the residues in color with the help of Rasmol by finding the specific pattern.

3. Results and discussion

This study is an attempt to determine the functional diversity in enolase superfamily protein. The approach we used is a all pairwise alignment of the sequences followed by a clustering of statistically significant pairs into groups or subfamilies by making sure that there is a common motif holding all the members together. Multiple sequence alignment and pattern recognition methods were included in this. The study analyzed the possible subfamilies in Enolase protein superfamily which shares in organisms such as archaea, bacteria with respect to *E.coli* and finally predicted five superfamilies which may play a role in functional diversity in Enolase superfamily protein.

Generally a protein's function is encoded within putatively functional signatures or motifs that represent residues involved in both functional conservation and functional divergence within a set of homologous proteins at various levels of hierarchy that is, super-families, families and sub-families. Protein function divergence is according to local structural variation around the active sites (Changwon K, 2006). Even when proteins have similar overall structure, the function could be different from each other. Accurate prediction of residue depth would provide valuable information for fold recognition, prediction of functional sites, and protein design. Proteins might have considerable structural similarities even when no evolutionary relationship of their sequences can be detected. This property is often referred to as the proteins sharing ie; a "fold". Of course, there are also sequences of common origin in each fold, called a "superfamily", and in them there are groups of sequences with clear similarities designated as "family". These sequence-level superfamilies can be categorized with many Bioinformatics approaches (LevelErik L , 2002)

3.1 Functional/ structural validation

The functions of the five identified protein family include:

3.1.1 Group 1

Mandelate racemase / muconate lactonizing enzyme family signature-1: which is an independent inducible enzyme cofactor. Mandelate racemase (MR) and muconate lactonizing enzyme (MLE) catalyses separate and mechanistically distinct reactions necessary for the catabolism of aromatic acids Immobilization of this enzyme leads to an enhanced activity and facilitates its recovery

MR_MLE_1 Mandelate racemase / muconate lactonizing enzyme family signature 1: (Fig.2)

Polymer: 1

Type: polypeptide(L)

Length: 405 Chains:A, B, C, D, E, F, G, H

Functional Protein: PDB ID: 3D46 chain A in E-val 0.0.

Possible amino acid pattern found in chain A

I-x(1,3)-Q-P-D-[ALV]-[ST]-H-[AV]-G-G-I-[ST]-E-x(2)-K-[IV]-A-[AGST]-[LM]-A-E-[AS]-[FY]-D-V-[AGT]-[FLV]-[AV]-[LP]-H-C-P-L-G-P-[IV]-A-[FL]-A-[AS]-[CS]-L-x-[ILV]-[DG]
Key Residues

THR 136, SER 138, CYS 139,VAL 140, Asp 141, ALA 143, LEU 144, ASP 146, LEU 147, GLY 149, LYS 150, PRO 155, VAL 156, LEU 159, LEU 160, GLY 161

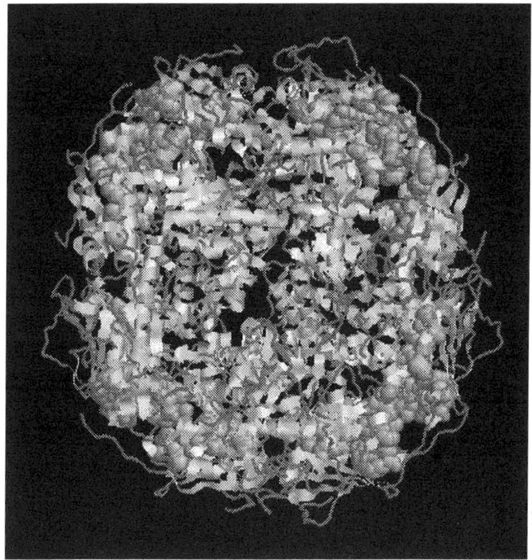

Fig. 2. Functional Protein Information (PDB Id: 3D46) The residues in yellow colour represents the identified functional residues in Group 1

3.1.2 Group 2

TonB-dependent receptor proteins signature-1 : TonB-dependent receptors is a family of beta-barrel proteins from the outer membrane of Gram-negative bacteria. The TonB complex senses signals from outside the bacterial cell and transmits them via two membranes into the cytoplasm, leading to transcriptional activation of target genes
TONB_DEPENDENT_REC_1 TonB-dependent receptor proteins signature 1 : (Fig.3)

Polymer:1

Type:polypeptide(L)

Length:99

Chains:A, B

Functional Protein: PDB ID: 3LAZ

Possible amino acid pattern found in 3LAZ

T-K-R-G-L-I-Y-A-A-T-P-A-S-D-F-V-C-G-T-Q-Q-V-A-S-G-I-T-V-Q-V-F-T-T-G-R-G-T-P-Y-G-L-M-A-V-P-V-I-K-M-A

Key Residues

GLU 88, SER89, VAL91, VAL92, PRO94, GLU95

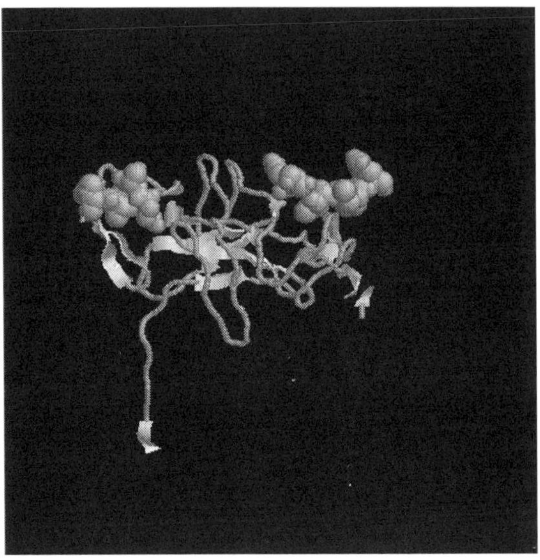

Fig. 3. Functional Protein Information (PDB Id: 3LAZ). The residues in yellow colour represents the identified functional residues in Group 2

3.1.3 Group 3

3-hydroxyisobutyrate dehydrogenase signature : This enzyme is also called beta-hydroxyisobutyrate dehydrogenase. This enzyme participates in valine, leucine and isoleucine degradation.

3_HYDROXYISOBUT_DH 3-hydroxyisobutyrate dehydrogenase signature : (Fig.4. a and Fig.4. b)

Polymer:1

Type:polypeptide(L)

Length:295

Chains:A, B

Functional Protein: PDB ID: 1YB4

Possible amino acid pattern found in 1YB4

G-[IMV]-[EK]-F-L-D-A-P-V-T-G-G-[DQ]-K-[AG]-A-x-E-G-[AT]-L-T-[IV]-M-V-G-G-x(2)-[ADEN]-[ILV]-F-x(2)-[LV]-x-P-[IV]-F-x-A-[FM]-G-[KR]-x-[IV]-[IV]-[HY]-x-G

Key Residues

PHE5, ILE6, GLY7, LEU8, GLY 9, GLY 12, ALA 16, ASN 18

Polymer:1

Type:polypeptide(L)

Length:299

Chains:A

Alternate: 3_HYDROXYISOBUT_DH 3-hydroxyisobutyrate dehydrogenase signature :

Functional Protein: PDB ID: 1VPD

Possible amino acid pattern found in1VPD

G-[ADET]-x-G-[AS]-G-x(1,2)-T-x(0,1)-K-L-[AT]-N-Q-[IV]-[IMV]-V-[AN]-x-[NT]-I-A-A-[MV]-[GS]-E-A-[FLM]-x-L-A-[AT]-[KR]-[AS]-[GV]-x-[ADNS]-[IP]

OR

K-L-A-N-Q-x(0,1)-I-x(0,1)-V-[AN]-x-N-I-[AQ]-A-[MV]-S-E-[AS]-[FL]-x-L-A-x-K-A-G-[AIV]-[DENS]-[PV]-[DE]-x-[MV]-[FY]-x-A-I-[KR]-G-G-L-A-G-S-[AT]-V-[LM]-[DN]-A-K

Key Residues

PHE7, ILE8, GLY9, LEU10, GLY11, GLY14, SER18, ASN20

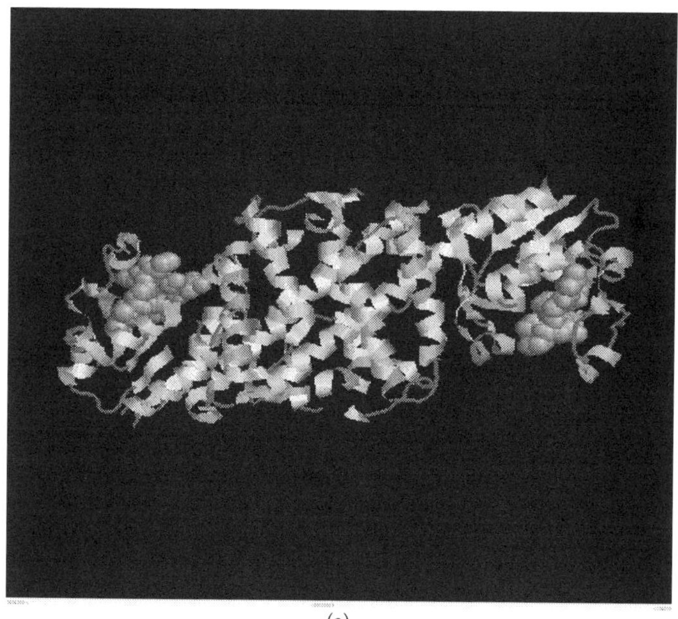

(a)

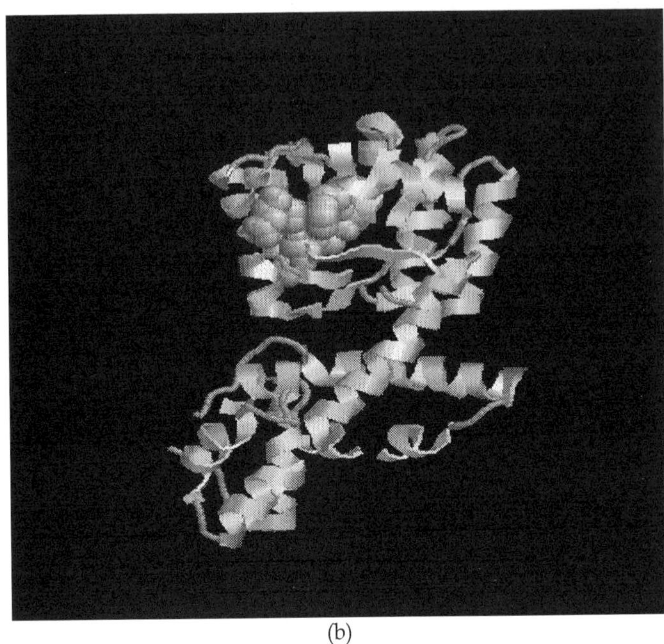

(b)

Fig. 4. a. Functional Protein Information (PDB Id: 1YB4) The residues in yellow colour represents the identified functional residues in Group 3. Also. b Functional Protein Information (PDB Id: 1VPD). The residues in yellow colour represents the identified functional residues in Group 3

3.1.4 Group 4

Enolase signature : Enolase, also known as phosphopyruvate dehydratase, is a metalloenzyme responsible for the catalysis of the conversion of 2-phosphoglycerate (2-PG) to phosphoenolpyruvate (PEP), the ninenth and penultimate step of glycolysis. Enolase can also catalyze the reverse reaction, depending on environmental concentrations of substrates.

Polymer:1

Type:polypeptide(L)

Length:431

Chains:A, B, C, D

Functional Protein: PDB Id: 1E9I

ENOLASE Enolase signature: (Fig.5. a and Fig.5. b)

Possible amino acid pattern found in 1E9I

G-x(0,1)-D-D-[IL]-F-V-T-[NQ]-[PTV]-[DEKR]-x-[IL]-x(2)-G-[IL]-x(4)-[AGV]-N-[ACS]-[ILV]-L-[IL]-K-x-N-Q-[IV]-G-[ST]-[LV]-x-[DE]-[AST]-[FILM]-[ADES]-A-[AIV]-x(2)-[AS]-x(3)-[GN]

Key Residues

ILE 338, LEU339, ILE340, LYS341, ASN343, GLN344, ILE 345, GLY346, SER347, LEU348, THR349, GLU350, THR351

Alternate : ENOLASE Enolase signature

Polymer:1

Type:polypeptide(L)

Length:427

Chains:A, B

Functional Protein: PDB ID: 2PA6

Possible amino acid pattern found in 2PA6

S-x(1,2)-S-G-[DE]-[ST]-E-[DG]-[APST]-x-I-A-D-[IL]-[AS]-V-[AG]-x-[AGNS]-[ACS]-G-x-I-K-T-G-[AS]-x-[AS]-R-[GS]-[DES]-R-[NTV]-A-K-Y-N-[QR]-L-[ILM]-[ER]-I-E-[EQ]-[ADE]-L-[AEGQ]

Key Residues

LEU 336, LEU337, LEU338, LYS339, ASN341, GLN342, ILE343, GLY344,THR345, LEU 346, SER347, GLU348, ALA 349

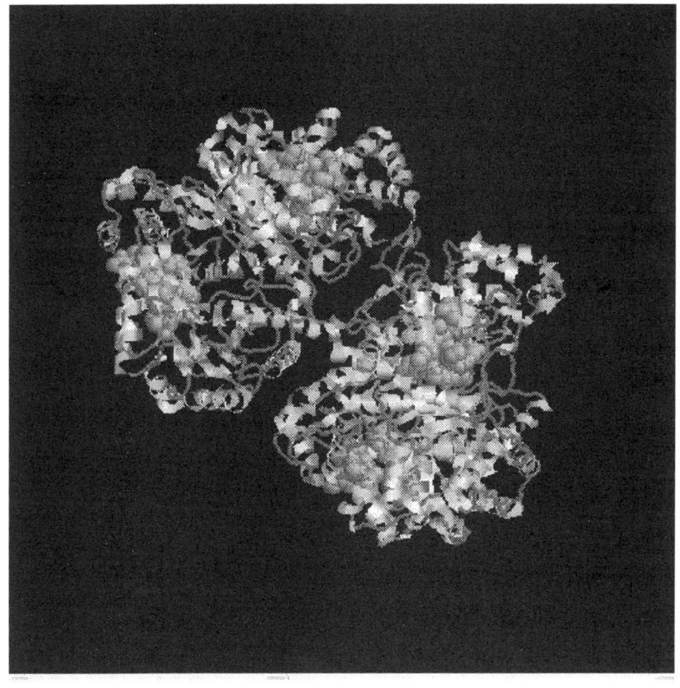

(a)

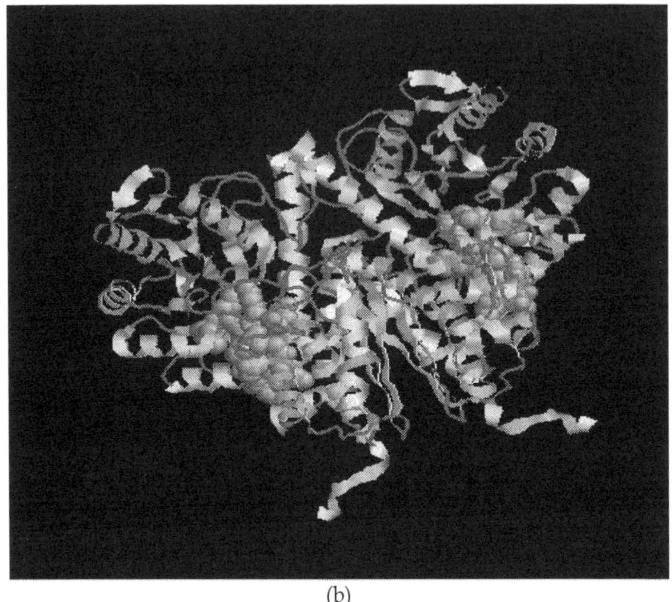

(b)

Fig. 5. a Functional Protein Information (PDB Id: 1E91) The residues in yellow colour represents the identified functional residues in Group 4. Also. b Functional Protein Information (PDB Id: 2PA6). The residues in yellow colour represents the identified functional residues in Group 4

3.1.5 Group 5

Glycerol-3-phosphate transporter (glpT) family of transporters signature :(Fig.6)
The major facilitator superfamily represents the largest group of secondary membrane transporters in the cell.

Molecule:Glycerol-3-phosphate transporter

Polymer:1

Type:polypeptide(L)

Length:451

Chains:A

Functional Protein: PDB ID: 1PW4

Possible amino acid pattern found in 1PW4

P-x(2,3)-R-x(0,1)-G-x-A-x-[AGS]-[FILV]-x(3)-[AGS]-x(3)-[AGS]-x(2)-[AILV]-x-[APST]-[IPV]-x(2)-[AG]-x-[ILV]-[ASTV]-x(3)-G-x(3)-[ILMV]-[FY]-x(3)-[AGV]-[AGILPV]-x-[GS]-[FILMV]

Key Residues

GLU153, ARG154, GLY155, SER159, VAL160, TRP161, ASN162, ALA164, ASN166, VAL167, GLY168, GLY169

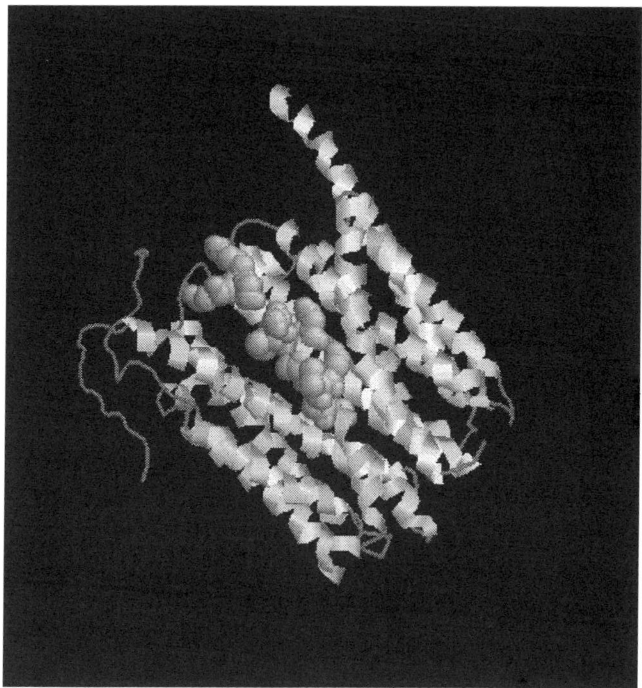

Fig. 6. Functional Protein Information (PDB Id: 1PW4). The residues in yellow colour represents the identified functional residues in Group 5

4. Conclusion

Identification of the specificity-determining residues in the various protein family studies has an important role in bioinformatics because it provides insight into the mechanisms by which nature achieves its astonishing functional diversity, but also because it enables the assignment of specific functions to uncharacterized proteins and family prediction. Genomics has posed the challenge of determination of protein function from sequence or 3-

D structure. Functional assignment from sequence relationships can be misleading, and structural similarity does not necessarily imply functional similarity. Our studies on the analysis of the superfamily revealed, for the first time, that in these species (archaea and bacteria) using *E. coli.* as a genomic model, we can contribute important insights for understanding their structural as well as functional relationships. The computational prediction of these functional sites for protein structures provides valuable clues for functional classification.

5. Acknowledgement

The authors gratefully acknowledge the support from JNIAS for the successful completion of this project.

6. References

Muhummad Khan and Kaiser Jamil (2008) Genomic distribution, expression and pathways of cancer metasignature genes through knowledge based data mining. International Journal of Cancer Research 1 (1), PP1-9, ISSN 1811-9727

Muhummad Khan and Kaiser Jamil (2008), Study on the conserved and polymorphic sites of MTHFR using bioinformatic approaches. Trends in Bioinformatics 1 (1) 7-17.

Sabitha Kotra, Kishore Kumar Madala and Kaiser Jamil (2008), Homology Models of the Mutated EGFR and their Response towards Quinazolin Analogues; J. Molecular Graphics and modeling , Vol-27, pp244-254.

Muhummadh Khan and Kaiser Jamil (2010) Phylogeny reconstruction of ubiquitin conjugating (E2) enzymes. Biology and Medicine Vol 2 (2), 10-19.

Patricia C, Babbitt, Miriam S. Hasson, Joseph E. Wedekind, David R. J. Palmer, William C. Barrett, George H. Reed, Ivan Rayment, Dagmar Ringe, George L. Kenyon, and John A. Gerlt (1996); The Enolase Superfamily: A General Strategy for Enzyme-Catalyzed Abstraction of the α-Protons of Carboxylic Acid *Biochemistry* 35 (51), pp 16489–16501

Babbitt PC Hasson MS, Wedekind JE, Palmer DR, Barrett WC, Reed GH, Rayment I, Ringe D, Kenyon GL, Gerlt JA (1996) The enolase superfamily: a general strategy for enzyme-catalyzed abstraction of the alpha-protons of carboxylic acids, Biochemistry. 35(51):16489-50

John F. Rakus, Alexander A. Fedorov, Elena V. Fedorov, Margaret E. Glasner, Brian K. Hubbard, Joseph D. Delli, Patricia C. Babbitt, Steven C. Almo and John A. Gerlt, (2008) Evolution of Enzymatic Activities in the Enolase Superfamily: l-Rhamnonate Dehydratase, *Biochemistry* 47 (38), pp 9944–9954

Gerlt JA, Babbitt PC, Rayment I.(2005). Divergent evolution in the enolase superfamily: the interplay of mechanism and specificity. Arch Biochem Biophys. 1;433(1):59-7

Anurag Sethi, Patrick O'Donoghue, and Zaida Luthey-Schulten (2005) Evolutionary profiles from the QR factorization of multiple sequence alignments, PNAS vol. 102 no. 11 4045-4050

Lindahl E, Elofsson A, (2000) Identification of related proteins on family, superfamily and fold level. Journal of Molecular Biology 295: 3, 613-625

Steven E. Brenner, Cyrus Chothia, and Tim J. P. Hubbard (*1998*) Assessing sequence comparison methods with reliable structurally identified distant evolutionary relationships PNAS May 26: 95 6073-6078

Dayhoff, M.O. (1974) Computer analysis of protein sequences, Fed. Proc. 33, 2314-2316,.

Scott McGinnis (2004) BLAST: at the core of a powerful and diverse set of sequence analysis tools, Nucleic Acids Research, Vol. 32

Hubbard BK, Koch M, Palmer DR, Babbitt PC, Gerlt JA. (1998) Evolution of enzymatic activities in the enolase superfamily: characterization of the (D)-glucarate/galactarate catabolic pathway in Escherichia coli. Biochemistry. 13;37(41):14369-75.

David R. J. Palmer, James B. Garrett,V. Sharma, R. Meganathan, Patricia C. Babbitt, and John A. Gerlt, (1999) Unexpected Divergence of Enzyme Function and Sequence: "N-Acylamino Acid Racemase" Is o-Succinylbenzoate Synthase, Biochemistry, 38 (14), pp 4252-4258

Satu Kuorelahti Paula Jouhten, Hannu Maaheimo, Merja Penttila and Peter Richard (2006) l-galactonate dehydratase is part of the fungal path for d-galacturonic acid catabolism Molecular Microbiology 61:4 1060 - 1068

Brian K. Hubbard, Marjan Koch, David R. J. Palmer, Patricia C. Babbitt, and John A. Gerlt (1998) Evolution of Enzymatic Activities in the Enolase Superfamily: Characterization of the (D)-Glucarate/Galactarate Catabolic Pathway in Escherichia coli Biochemistry, 37 (41) 14369-14375

John F. Rakus, Alexander A. Fedorov, Elena V. Fedorov, Margaret E. Glasner, Brian K. Hubbard, Joseph D. Delli, Patricia C. Babbitt, Steven C. Almo and John A. Gerlt, (2008) Evolution of Enzymatic Activities in the Enolase Superfamily: l-Rhamnonate Dehydratase, Biochemistry, 47 (38), 9944-9954

Robert Belshaw and Aris Katzourakis (2005) Blast to Align: a program that uses blast to align problematic nucleotide sequences, Bioinformatics 21(1):122-123

Dmitry Lupyan, Alejandra Leo-Macias and Angel R. Ortiz (2005) A new progressive-iterative algorithm for multiple structure alignment Bioinformatics Volume 21:15 3255-3263

Doron Betel and Christopher WV Hogue, Kangaroo (2002) A pattern-matching program for biological sequences, BMC Bioinformatics, 1186/1471-2105-3-20

Ofir Goldenberg, Elana Erez, Guy Nimrod, and Nir Ben-Tal (2009) The ConSurf-DB: pre-calculated evolutionary conservation profiles of protein structures Nucleic Acids Res. D323-D327.

Changwon Keum and Dongsup Kim (2006) Protein function prediction via ligand interface residue match, World Congress on Medical Physics and Biomedical Engineering 2006, August 27 - September 1, COEX Seoul, Korea "Imaging the Future Medicine"

LevelErik Lindahl and Arne Elofsson, (2000) Identification of Related Proteins on Family, Superfamily and Fold Journal of MolecularBiology 295: 3, 613-625

Neidhart DJ, Kenyon GL, Gerlt JA, Petsko GA (1990) Mandelate racemase and muconate lactonizing enzyme are mechanistically distinct and structurally homologous. Nature. 347(6294):692-4.

Contributions of Structure Comparison Methods to the Protein Structure Prediction Field

David Piedra[1], Marco d'Abramo[2] and Xavier de la Cruz[1,3]
[1]*IBMB-CSIC*
[2]*CNIO*
[3]*ICREA*
Spain

1. Introduction

Since their development, structure comparison methods have contributed to advance our understanding of protein structure and evolution (Greene et al, 2007; Hasegawa & Holm, 2009), to help the development of structural genomics projects (Pearl et al, 2005), to improve protein function annotations (D. A. Lee et al), etc, thus becoming an essential tool in structural bioinformatics. In recent years, their application range has grown to include the protein structure prediction field, were they are used to evaluate overall prediction quality (Jauch et al, 2007; Venclovas et al, 2001; Vincent et al, 2005; G. Wang et al, 2005), to identify a protein's fold from low-resolution models (Bonneau et al, 2002; de la Cruz et al, 2002), etc. In this chapter, after briefly reviewing some of these applications, we show how structure comparison methods can also be used for local quality assessment of low-resolution models and how this information can help refine/improve them.

Quality assessment is becoming an important research topic in structural bioinformatics because model quality determines the applicability of structure predictions (Cozzetto et al, 2007). Also, because prediction technology is now easily available and potential end-users of prediction methods, from template-based (comparative modeling and threading) to *de novo* methods, are no longer specialized structural bioinformaticians. Quality assessment methods have been routinely used for many years in structural biology in the evaluation of experimental models. These methods focus on several features of the protein structure (see (Laskowski et al, 1998) and (Kleywegt, 2000) and references therein). Because a number of quality issues are common to both experimental and predicted models, the use of these methods has been naturally extended to the evaluation of structure predictions. For example, in the case of homology modeling, a widely used structure prediction technique, evaluation of models with PROCHECK (Laskowski et al, 1993), WHAT-CHECK (Hooft et al, 1997), PROSA (Sippl, 1993), and others (see (Marti-Renom et al, 2000) and references therein) is part of the standard prediction protocol; WHATIF (Vriend, 1990) and PROSA (Sippl, 1993) have also been used in the CASP experiment to assess comparative models (Venclovas, 2001; Williams et al, 2001); etc.

Some quality assessment problems are unique to the structure prediction field, given the specific characteristics of computational models, and have led to the development of methods aimed at: the recognition of near-native predictions from a set of decoys (Jones & Thornton, 1996; Lazaridis & Karplus, 2000; Sippl, 1995); identification of a target's protein family (Bonneau et al, 2002; de la Cruz et al, 2002); overall quality assessment of predictions (Archie et al, 2009; Benkert et al, 2009; Cheng et al, 2009; Larsson et al, 2009; Lundstrom et al, 2001; McGuffin, 2009; Mereghetti et al, 2008; Wallner & Elofsson, 2003; 2005; Z. Wang et al, 2009; Zhou & Skolnick, 2008); and, more recently, residue-level quality assessment (Benkert et al, 2009; Cheng et al, 2009; Larsson et al, 2009; McGuffin, 2009; Wallner & Elofsson, 2006; 2007; Z. Wang et al, 2009). However, in spite of these promising efforts, quality assessment of protein structure predictions remains an open issue(Cozzetto et al, 2009).

Here we focus on the problem of local quality assessment, which consists on the identification of correctly modeled regions in predicted structures (Wallner & Elofsson, 2006; 2007), or, as stated by Wallner and Elofsson(Wallner & Elofsson, 2007): "The real value of local quality prediction is when the method is able to distinguish between high and low quality regions.". In many cases, global and local quality estimates are produced simultaneously (Benkert et al, 2009; Cheng et al, 2009; Larsson et al, 2009; McGuffin, 2009). However, in this chapter we separate these two issues by assuming that, irrespective of its quality, a structure prediction with the native fold of the corresponding protein is available. From a structural point of view this is a natural requirement, as a correct local feature (particularly if it is one which, like a β-strand (Chou et al, 1983), is stabilized by long-range interactions) in an otherwise wrong structure can hardly be understood. From a practical point of view, successful identification of correct parts within incorrect models may lead to costly errors. For example, identifying a correctly modeled binding site within a structurally incorrect context should not be used for drug design: it would surely have incorrect dynamics; the long-range terms of the interaction potential, like electrostatics, would be meaningless; false neighboring residues could create unwanted steric clashes with the substrate, thus hampering its docking; or, on the contrary, absence of the true neighbors could lead to unrealistic docking solutions; etc. In the remaining of the chapter we describe how structure comparison methods can be applied to obtain local quality estimates for low-resolution models and how these estimates can be used to improve the model quality.

2. A simple protocol for local quality assessment with structure comparison methods

As mentioned before, an important goal in local quality assessment(Wallner & Elofsson, 2006; 2007) is to partition the residues from a structure prediction in two quality classes: high and low. This can be done combining several predictions; however, in the last two rounds of the CASP experiment -a large, blind prediction experiment performed every two years(Kryshtafovych et al, 2009)- evaluators of the Quality Assessment category stressed that methods aimed to assess single predictions are needed(Cozzetto et al, 2007; Cozzetto et al, 2009). These methods are particularly important for users that generate their protein models with *de novo* prediction tools, which are still computationally costly(Jauch et al, 2007), particularly for large proteins.

Here we describe a single-molecule approach, based on the use of structure comparison methods, that allows to partition model residues in two sets, of high and low quality respectively. In this approach (Fig. 1), the user's model of the target is first structurally aligned with a target's homolog. This alignment, which constitutes the core of the procedure, is then used to separate the target's residues in two groups: aligned and unaligned. The main assumption of this approach is that aligned residues are of higher quality than the average. The validity of this assumption is tested in the next section. In section 3 we discuss the conditions that determine/limit the applicability and usefulness of the method.

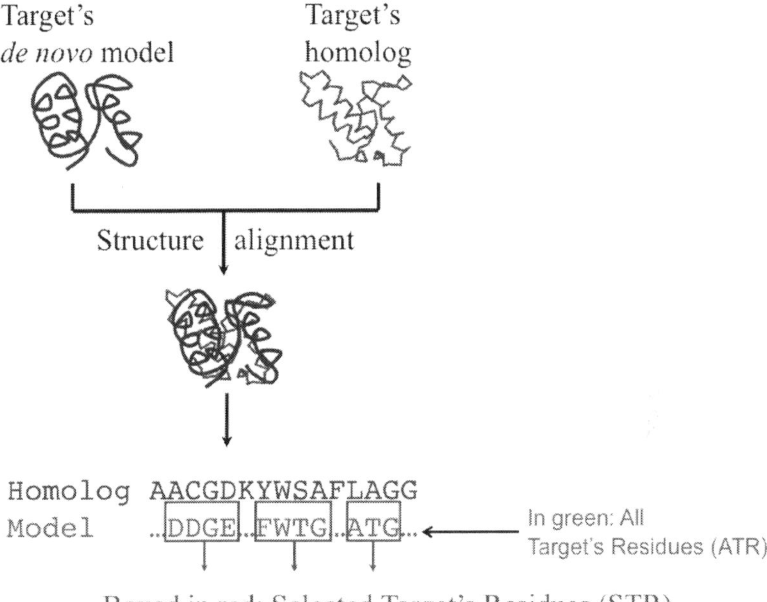

Fig. 1. Use of structure comparison methods for local quality assessment of structure predictions

2.1 Performance of structure comparison methods in local quality assessment

To show that structurally aligned residues are usually of higher quality we used a set of *de novo* predictions with medium/low to very low resolution, obtained with Rosetta(Simons et al, 1997). Although several *de novo* prediction programs have shown promising results in the CASP experiment(Jauch et al, 2007; Vincent et al, 2005), we used Rosetta predictions because: (i) Rosetta is a well known *de novo* prediction program that has ranked first in successive CASP rounds (Jauch et al, 2007; Vincent et al, 2005); (ii) many Rosetta predictions are available at the server of Baker's group, thus allowing a consistent test of our approach, with predictions from the same source; and (iii) the program is available for interested users (http://www.rosettacommons.org/software/).

We downloaded the protein structure predictions from the server of Baker's laboratory (http://depts.washington.edu/bakerpg/drupal/). This set was constituted by 999 *de novo* models generated with Rosetta(Simons et al, 1997) for 85 proteins, i.e. a total of 84915 models. Application of the protocol studied here (Fig. 1) requires that the structure of a homolog of the target is available, and that the predictions used have the fold of the target.

The former was enforced by keeping only those proteins with a CATH relative at the T-level(Pearl et al, 2005) (homologs with the same fold of the target protein, regardless of their sequence similarity). The second condition was required to focus only on the local quality assessment problem, and was implemented by excluding those models not having the fold of their target protein. Technically, this meant that we only kept those models structurally similar to any member of the target's structural family: that is, those models giving a score higher than 5.25 for the model-homolog structure comparison done with MAMMOTH (Ortiz et al, 2002), for at least one homolog of the target protein. This step was computationally costly, as it involved 7,493,499 structure comparisons, and could only be carried using MAMMOTH(Ortiz et al, 2002); the 5.25 score threshold was taken from MAMMOTH's article (Ortiz et al, 2002). The final dataset was constituted by 68 target proteins and a total of 17180 models.

The properties of the selected target's residues (STR; to avoid meaningless results we only considered STR sets larger than 20 residues) were characterized with four parameters: two structure-based, and two sequence-based. The former were used to check if STR really were of better quality, comparing their parameters' values with those obtained for the set of all the target residues (ATR), i.e. the whole model structure. It has to be noted that: (i) STR and ATR sets are constituted by residues from the target protein, more precisely STR is a subset of ATR; and (ii) three possible STR sets were produced, because we checked our procedure using three structure comparison methods (MAMMOTH (Ortiz et al, 2002), SSAP(Orengo & Taylor, 1990) and LGA(Zemla, 2003)). The sequence-based properties were utilized to describe how STR spread along the sequence of the target, which helps to assess the usefulness of the protocol. Below we provide a brief description of each parameter, together with the results obtained from their use.

2.1.1 Structural quality: rmsd

Rmsd(Kabsch, 1976) is a quality measure widely employed to assess structure models: it corresponds to the average distance between model atoms and their equivalent in the native structure. Small rmsd values correspond to higher quality predictions than larger values.

In Fig. 2 we see the STR and ATR rmsd distributions. Regardless of the structure comparison method used (MAMMOTH (Ortiz et al, 2002), SSAP(Orengo & Taylor, 1990) and LGA(Zemla, 2003) in blue, yellow and red, respectively), STR distributions are shifted towards lower rmsd values relative to ATR distributions (in grey). This confirms the starting assumption: it shows that model residues structurally aligned to the protein's homolog usually have a higher structural quality. A consensus alignment (in black), which combined the results of the three structure comparison methods, gave better results at the price of including fewer residues; for this reason we excluded the consensus approach from subsequent analyses.

An interesting feature of STR rmsd distributions was that their maxima were between 3.5 Å and 6.5 Å, and that a majority of individual values were between 3 Å and 8 Å, and below 10 Å. To further explore this issue, we plotted the values of rmsd for STR against ATR (Fig. 3, grey boxes). In accordance with the histogram results, STR rmsd tended to be smaller than ATR rmsd. We distinguished two regions in the graph: in the first region (ATR rmsd between 0 Å and 6-8 Å) there was a roughly linear relationship between ATR and STR rmsds; however, for ATR rmsd values beyond 8 Å, STR rmsd reached a plateau. This plateau is at the origin of the thresholds observed in the histograms (Fig. 2), and confirms that structure alignments can be used to identify subsets of model residues with better rmsd than the rest.

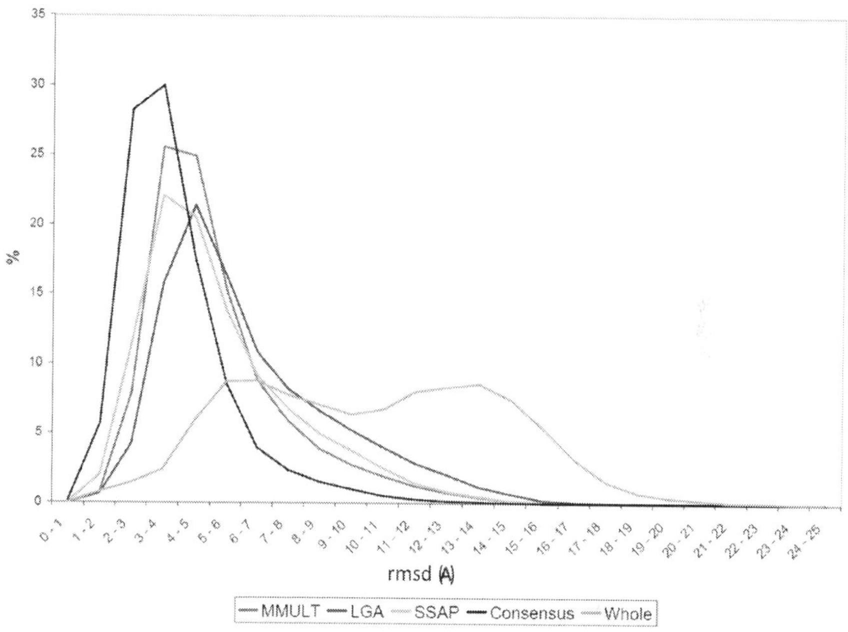

Fig. 2. Quality of structurally aligned regions vs. whole model, rmsd frequency histogram.

As a performance reference we used the PROSA program (Sippl, 1993) (white boxes) which provides a residue-by-residue, energy-based quality assessment, and is a single model method, therefore comparable to the approach presented here. PROSA was executed with default parameters, and we took as high quality residues those having energies below zero. In Fig. 3 we see that for good models, i.e. those with low ATR values, PROSA results (in white) were as good as those obtained with structure comparison methods (in grey). However, as models became poorer, PROSA results became worse, particularly after structure comparison methods reached their plateau. This indicates that when dealing with poor predictions use of structure alignments can improve/complement other quality assessment methods.

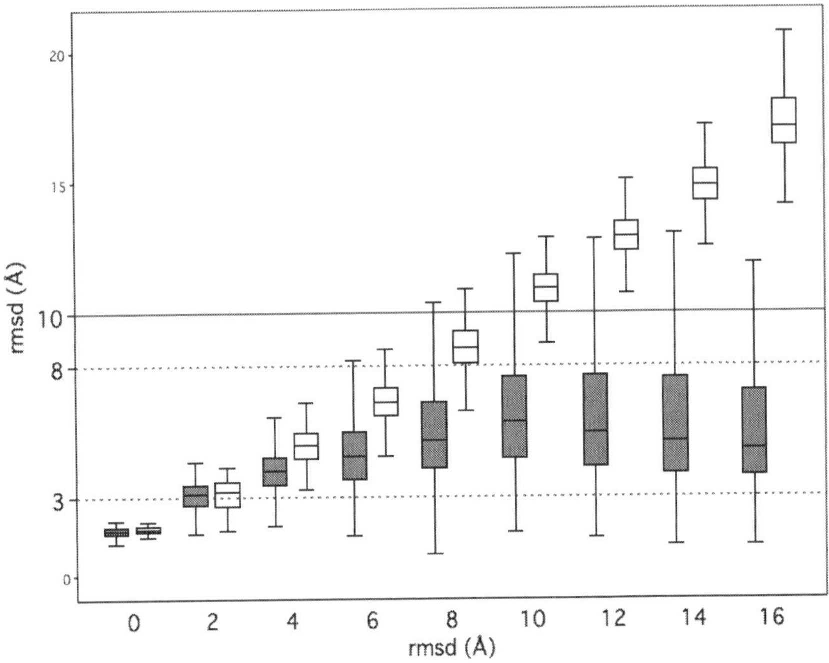

Fig. 3. Quality of structurally aligned (obtained with MAMMOTH (Ortiz et al, 2002)) regions vs. whole model, rmsd of selected residues vs. all-residues. Grey: structure comparison-based protocol (Fig. 1); white: PROSA(Sippl, 1993) results.

2.1.2 Structural quality: GDT_TS

GDT_TS is a quality measure routinely utilised by evaluator teams in the CASP community experiment (Jauch et al, 2007; Vincent et al, 2005): it is equal to the average of the percentages of model residues at less than 1 Å, 2 Å, 4 Å and 8 Å from their location in the correct structure. It was computed following the procedure described by Zemla(Zemla, 2003), using Cα atoms to compute residue-residue distances. GDT_TS varies between 0 and 100, with values approaching 100 as models become better.

We found that STR GDT_TS was in general better than ATR GDT_TS (Fig. 4); this was particularly true when the latter was below 40-50. Overall, this shows that STR is enriched in good quality sub-structures relative to ATR, particularly for poor models.

Consistency with rmsd analysis was observed when comparing the performance of structure comparison-based quality assessment (in grey) with that of PROSA (in white): for good models (GDT_TS values above 60-70) both approaches had a similar behavior; however, as model quality decreased, use of structure alignments showed increasingly better performance than PROSA at pinpointing correct substructures.

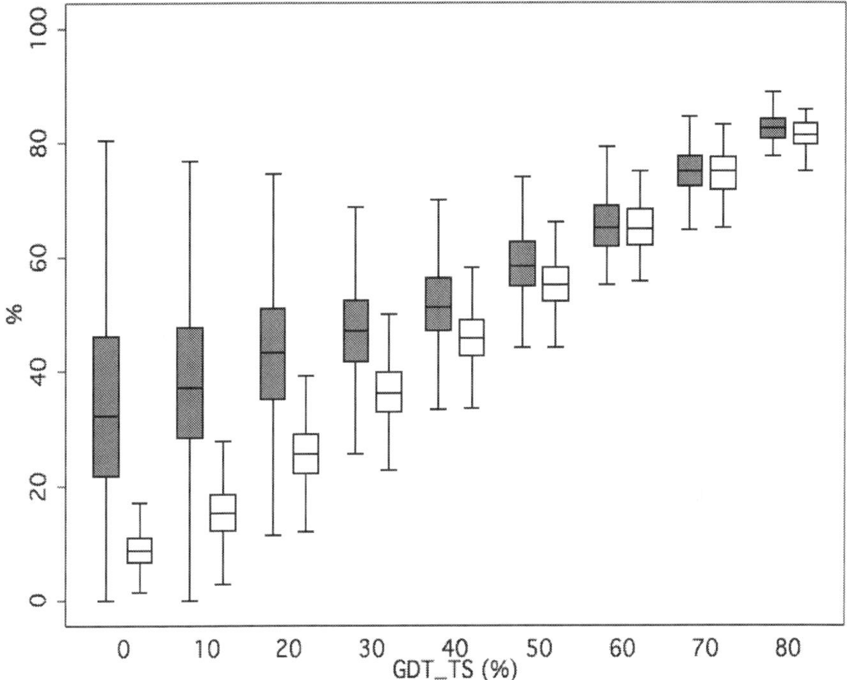

Fig. 4. Quality of structurally aligned (obtained with MAMMOTH (Ortiz et al, 2002)) regions vs. whole model, GDT_TS of the selected residues vs. all-residues GDT_TS. Grey: structure comparison-based protocol (Fig. 1); white: PROSA(Sippl, 1993) results.

2.1.3 Distribution of high quality residues along the protein sequence

Usually, STR do not form a continuous block, they tend to scatter along the sequence. The nature of this distribution is of interest for some applications of quality assessment methods (like model refinement) for which STR sets may be of little value if the involved residues are either too close in sequence, or contain too many orphan residues.

To characterize the distribution of STR along the sequence we used two measures: maximum distance (MD) between STR runs and normalized size distribution of STR runs (SAS). Both are based on the fact that, for a given model, STR sets are constituted by residue runs of varying size. MD corresponds to the largest sequence distance between STR runs (i.e. the number of residues between the rightmost and leftmost STR runs), divided by whole sequence length. MD values near 1 indicate that STR runs are spread over the whole protein, while smaller values point to a tighter residue clustering. SAS corresponds to the normalized (again by whole sequence length) size distribution for all runs constituting STR sets. SAS gives a view of how the sequence coverage is done: either by large sequence chunks, by small residue clusters, or by a mixture of both. When the alignment is constituted by small, evenly distributed residue clusters the SAS distribution will approach zero.

Our results showed that MD values are more frequent above 0.5, and more than 50% of them were higher than 0.8 (Fig. 5). The three structure comparison methods showed similar distributions, although LGA was slightly nearer to 1. This indicates that STR spread over a substantial part of the predicted protein.

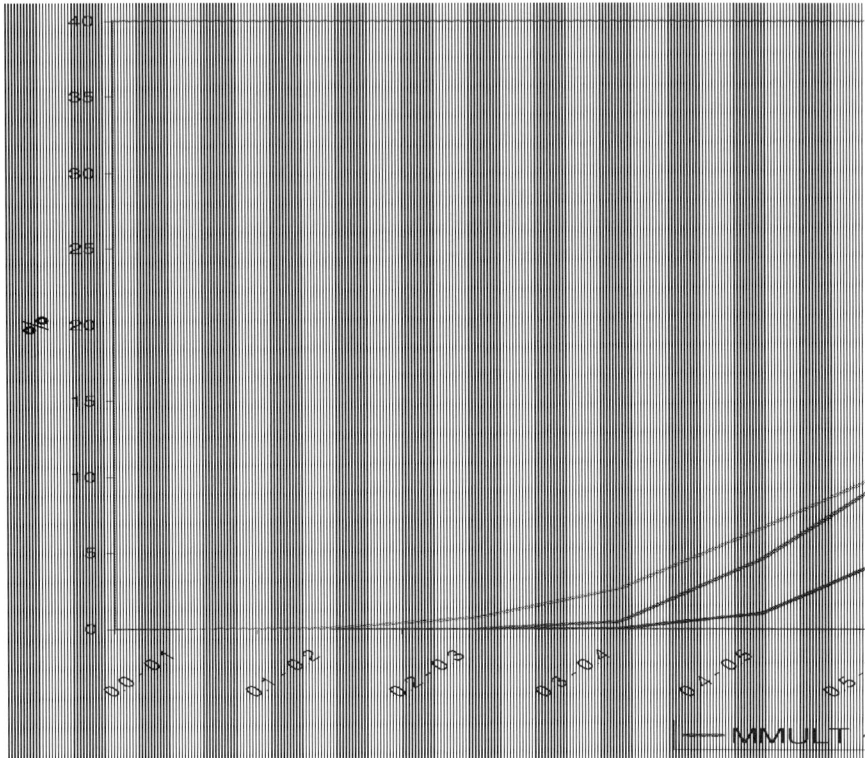

Fig. 5. Frequency distribution of the selected residues along the target sequence: normalized maximum distance between STR runs (unitless parameter).

Results for SAS (Fig. 6) showed that while ~50 % of STR formed clusters of size lower than 10 % of the whole sequence (i.e. SAS values below 0.1), the remaining residues were grouped in medium to large stretches. This means that for a 100 residue protein, clusters of more than 10 residues (which is roughly the size of an average α-helix) are frequent. In addition, for 95 % of the cases, the largest run of adjacent residues was above 30 % of the target length.

The picture arising from MD and SAS distributions is that STR usually extend over the protein length. Although STR sets are constituted by somewhat heterogeneous runs they do not contain too many orphan residues, as they include one large run (the size of a supersecondary structure motif, or larger) and several, smaller runs (the size of secondary structure elements).

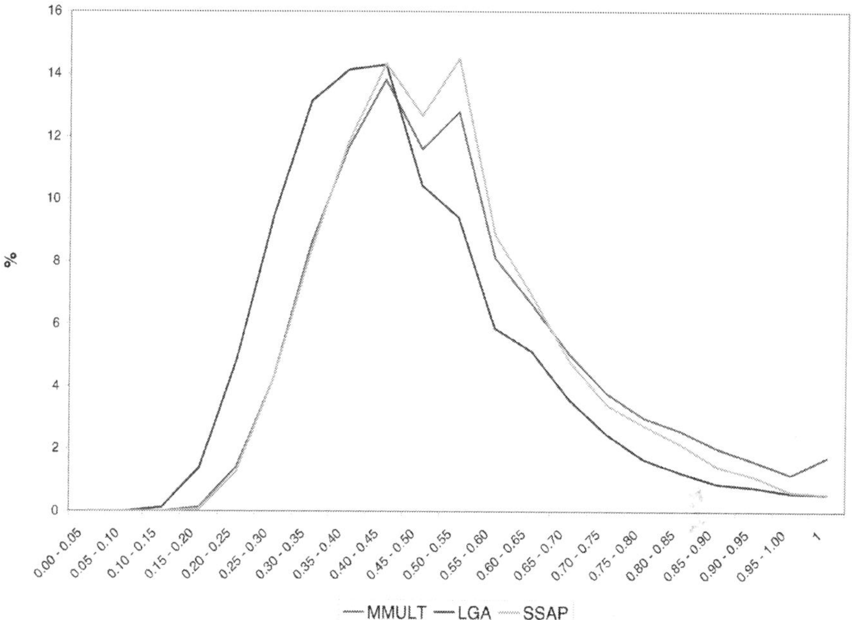

Fig. 6. Frequency distribution of the selected residues along the target sequence: normalized maximum distance between STR runs (unitless parameter).

3. Applicability range of structure comparison methods in local quality assessment

The approach described here is easy to use and has few computational requirements; however, it cannot be arbitrarily applied to any model or in any prediction scenario. In this section we describe which are its limits regarding prediction methods, target proteins and protein model nature.

3.1 Prediction methods

As far as the target protein has a homolog of known structure, model-homolog structure alignments can be computed and the quality assessment protocol (Fig. 1) can be applied, regardless of the prediction method originating the model. However, the approach presented here reaches its maximum utility when models are obtained with *de novo* structure prediction methods (methods originally devised to work using only the target's sequence and a physico-chemical/statistical potential, irrespective of the availability of homologs). This may seem somewhat contradictory, as one can think that the existence of target's homologs favors the use of comparative modeling methods instead of *de novo* methods. However, this is not the case: while *de novo* methods were initially developed with the *de novo* scenario in mind (only sequence information is available for the target protein), this situation is changing rapidly. Actually, when prediction problems become difficult, or a given method gives an unclear answer, using more than one technique is considered a good approach within the prediction community, as the evaluators of the *de novo* section in the CASP6 experiment explain (Vincent et al, 2005): "Many predicting groups now use both de novo and homology modeling/fold recognition techniques to predict structures in all categories". In addition, it has been shown that de novo methods can compete with template-based methods in the prediction of difficult targets (Jauch et al, 2007; Raman et al, 2009; Vincent et al, 2005). In this situation, which implies the existence of target's homologs, our method can be used to score the local quality of *de novo* predictions.

In addition, a completely new field of application for *de novo* methods has been unveiled by the growing interest in knowing the structure of alternative splicing isoforms (C. Lee & Wang, 2005). Due to the very localized nature of sequence changes (Talavera et al, 2007), structure prediction of alternative splicing variants seems a trivial exercise in comparative modeling. However, template-based methods fail to reproduce the structure changes introduced by alternative splicing (Davletov & Jimenez, 2004). *De novo* approaches with their ability to combine first principles with deep conformational searches are ideal candidates to tackle this problem; in this case, availability of the structure of only one isoform would allow the application of our method.

3.2 Target proteins

Proteins to which our approach can be applied must have a homolog of known structure. The number of these proteins is increasing due to: (i) the progress of structural genomics projects (Todd et al, 2005) (this will increase the number of both easy/medium and hard targets); (ii) the growing number of alternative splicing variants of unknown structure (C. Lee & Wang, 2005).

3.3 Protein models

The approach proposed (Fig. 1) is a local, not a global, quality assessment method and should only be applied to models that have the native fold of the target (see above). Present *de novo* methods still cannot consistently produce models with a native-like fold (Moult et al, 2009). Therefore, researchers must ascertain that the model's fold is correct (irrespective of its resolution). This can be done using global quality assessment methods like PROSA (Sippl, 1993), the Elofsson's suite of programs (Wallner & Elofsson, 2007), etc.

4. Applications

Once available, local quality information can be used with different purposes. For example, it may help to identify those parts of a theoretical model that are more reliable for mutant design, or to interpret the results of mutagenesis experiments; it may be used for *in sillico* docking experiments involving *de novo* models, to decide which parts of the models must be employed preferentially; etc. One the most promising applications of quality assessment methods is the refinement of low-resolution models (Wallner & Elofsson, 2007). In this section we illustrate how the results of the procedure here described can be used for this purpose.

Among the possible options available for model refinement, we propose to use the alignment resulting from the structural superimposition between a *de novo* model and the target's homolog (Fig. 1) as input to a comparative modeling program. We applied this strategy to 15 proteins (five from each of the three main CATH structural classes: alpha, beta and alpha/beta) from our initial dataset. These 15 proteins contributed a total of 2693 *de novo* models that resulted in 8033 model-homolog alignments (obtained with MAMMOTH (Ortiz et al, 2002)). These alignments were subsequently used as input to the standard homology modeling program MODELLER (Marti-Renom et al, 2000), which was run with default parameters. For the aligned regions we found (Fig. 7) that most of the refined models had lower model-native rmsd than the starting *de novo* models, i.e. they were closer to the native structure. A similar, although milder, trend was also observed when considering the whole set of protein residues (i.e. aligned as well as unaligned residues) (Fig. 8). These results show that this simple, computationally cheap model refinement protocol, based on the use of structure comparison local quality analysis, clearly helps to refine/improve low-resolution *de novo* models to an accuracy determined by the closest homolog of the target.

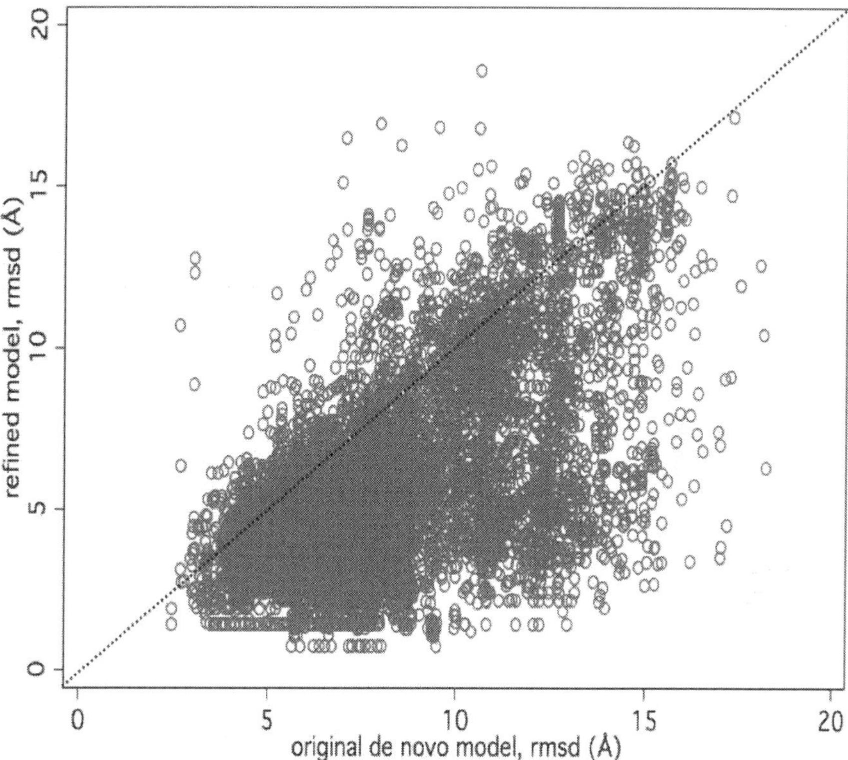

Fig. 7. Model refinement using structure comparison-based local quality assessment: rmsd of refined models vs. rmsd of original *de novo* models, subset of aligned residues. Points below the dotted line correspond to refinement-improved models.

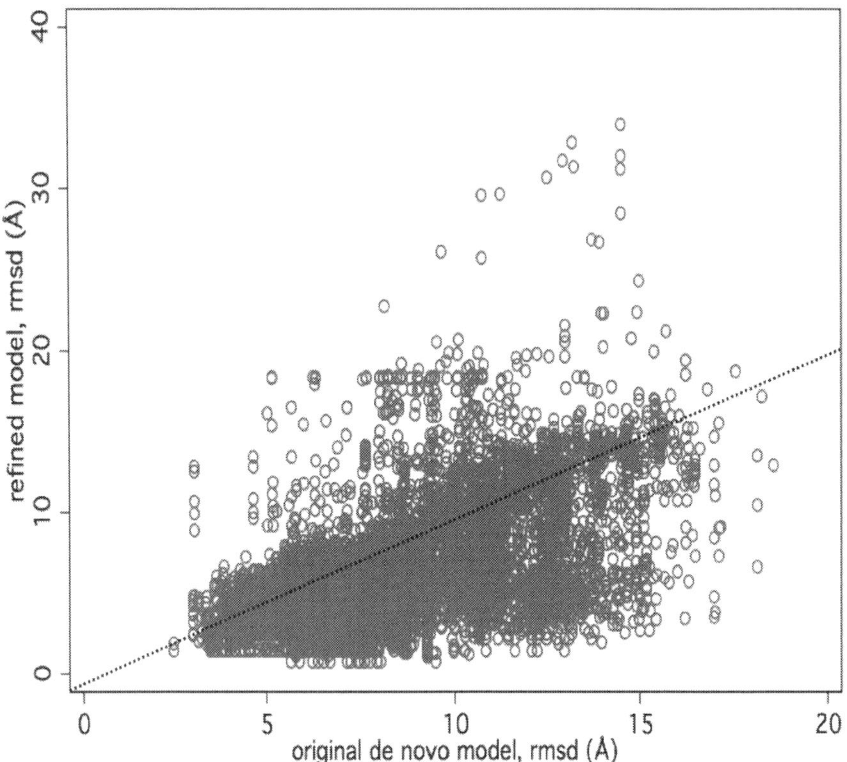

Fig. 8. Model refinement using structure comparison-based local quality assessment: rmsd of refined models vs. rmsd of original *de novo* models, all protein residues. Points below the dotted line correspond to refinement-improved models.

5. Conclusions

In this chapter we have described and tested a protocol for local quality assessment of low-resolution predictions based on the use of structure comparison methods. The testing was carried with *de novo* predictions, and the results showed that structure comparison methods allow the partitioning of the model's residues in two sets of high and low quality, respectively. This result holds even when only remote homologs of the target protein are available. The simplicity of the approach leaves room for future improvements and fruitful combination with other quality assessment methods. Two conditions determine the application range of this approach: the target protein must have at least one homolog of known structure, and models reproducing the fold of the target are required. However, results indicating that we may be near a full coverage of the proteins' fold space, together with advances in overall quality scoring indicate that these two problems are likely to become minor issues in the near future. Finally, our procedure suggests a simple refinement strategy based on the use of comparative modeling programs that may be used to improve low-resolution *de novo* models.

6. Acknowledgment

This work is dedicated to the memory of Angel Ramírez Ortíz, leading bioinformatician and designer of the MAMMOTH program for structure comparison. The authors wish to thank the CATH team for their support. Xavier de la Cruz acknowledges funding from the Spanish government (Grants BIO2006-15557 and BFU2009-11527). David Piedra acknowledges economical support from the Government of Catalonia and the Spanish *Ministerio de Educación y Ciencia*.

7. References

Archie, J.G.; Paluszewski M. & Karplus K. (2009) Applying Undertaker to quality assessment. Proteins Vol.77 Suppl 9191-195, ISSN 0887-3585.

Benkert, P.; Tosatto S.C. & Schwede T. (2009) Global and local model quality estimation at CASP8 using the scoring functions QMEAN and QMEANclust. Proteins Vol.77 Suppl 9173-180, ISSN 0887-3585.

Bonneau, R.; Strauss C.E.; Rohl C.A.; Chivian D.; Bradley P.; Malmstrom L.; Robertson T. & Baker D. (2002) De novo prediction of three-dimensional structures for major protein families. J Mol Biol Vol.322, No.1, pp. 65-78, ISSN 0022-2836.

Cheng, J.; Wang Z.; Tegge A.N. & Eickholt J. (2009) Prediction of global and local quality of CASP8 models by MULTICOM series. Proteins Vol.77 Suppl 9181-184, ISSN 0887-3585.

Chou, K.C.; Nemethy G. & Scheraga H.A. (1983) Role of interchain interactions in the stabilization of the right-handed twist of beta-sheets. J Mol Biol Vol.168, No.2, pp. 389-407, ISSN 0022-2836.

Cozzetto, D.; Kryshtafovych A.; Ceriani M. & Tramontano A. (2007) Assessment of predictions in the model quality assessment category. Proteins Vol.69 Suppl 8175-183, ISSN 0887-3585.

Cozzetto, D.; Kryshtafovych A. & Tramontano A. (2009) Evaluation of CASP8 model quality predictions. Proteins Vol.77 Suppl 9157-166, ISSN 0887-3585.

Davletov, B. & Jimenez J.L. (2004) Sculpting a domain by splicing. Nat Struct Mol Biol Vol.11, No.1, pp. 4-5,

de la Cruz, X.; Sillitoe I. & Orengo C. (2002) Use of structure comparison methods for the refinement of protein structure predictions. I. Identifying the structural family of a protein from low-resolution models. Proteins Vol.46, No.1, pp. 72-84, ISSN 0887-3585.

Greene, L.H.; Lewis T.E.; Addou S.; Cuff A.; Dallman T.; Dibley M.; Redfern O.; Pearl F.; Nambudiry R.; Reid A.; Sillitoe I.; Yeats C.; Thornton J.M. & Orengo C.A. (2007) The CATH domain structure database: new protocols and classification levels give a more comprehensive resource for exploring evolution. Nucleic Acids Res Vol.35, No.Database issue, pp. D291-297, ISSN 0305-1048.

Hasegawa, H. & Holm L. (2009) Advances and pitfalls of protein structural alignment. Curr Opin Struct Biol Vol.19, No.3, pp. 341-348, ISSN 0959-440X.

Hooft, R.W.; Sander C. & Vriend G. (1997) Objectively judging the quality of a protein structure from a Ramachandran plot. Comput Appl Biosci Vol.13, No.4, pp. 425-430, ISSN 0266-7061.

Jauch, R.; Yeo H.C.; Kolatkar P.R. & Clarke N.D. (2007) Assessment of CASP7 structure predictions for template free targets. Proteins Vol.69 Suppl 857-67, ISSN 0887-3585.

Jones, D.T. & Thornton J.M. (1996) Potential energy functions for threading. Curr Opin Struct Biol Vol.6, No.2, pp. 210-216, ISSN 0959-440X.

Kabsch, W.A. (1976) A solution for the best rotation to relate two sets of vectors. Acta Crystallogr A Vol.A32922-923,

Kleywegt, G.J. (2000) Validation of protein crystal structures. Acta Crystallogr D Biol Crystallogr Vol.56, No.Pt 3, pp. 249-265, ISSN 1399-0047.

Kryshtafovych, A.; Fidelis K. & Moult J. (2009) CASP8 results in context of previous experiments. Proteins Vol.77 Suppl 9217-228, ISSN 0887-3585.

Larsson, P.; Skwark M.J.; Wallner B. & Elofsson A. (2009) Assessment of global and local model quality in CASP8 using Pcons and ProQ. Proteins Vol.77 Suppl 9167-172, ISSN 0887-3585.

Laskowski, R.A.; MacArthur M.W.; Moss D.S. & Thornton J.M. (1993) PROCHECK: a program to check the stereochemical quality of protein structures. J Appl Crystallogr Vol.26283-291, ISSN 1600-5767.

Laskowski, R.A.; MacArthur M.W. & Thornton J.M. (1998) Validation of protein models derived from experiment. Curr Opin Struct Biol Vol.8, No.5, pp. 631-639, ISSN 0959-440X.

Lazaridis, T. & Karplus M. (2000) Effective energy functions for protein structure prediction. Curr Opin Struct Biol Vol.10, No.2, pp. 139-145, ISSN 0959-440X.

Lee, C. & Wang Q. (2005) Bioinformatics analysis of alternative splicing. Brief Bioinform Vol.6, No.1, pp. 23-33,

Lee, D.A.; Rentzsch R. & Orengo C. (2010) GeMMA: functional subfamily classification within superfamilies of predicted protein structural domains. Nucleic Acids Res Vol.38, No.3, pp. 720-737, ISSN 0305-1048.

Lundstrom, J.; Rychlewski L.; Bujnicki J. & Elofsson A. (2001) Pcons: a neural-network-based consensus predictor that improves fold recognition. Protein Sci Vol.10, No.11, pp. 2354-2362,

Marti-Renom, M.A.; Stuart A.C.; Fiser A.; Sanchez R.; Melo F. & Sali A. (2000) Comparative protein structure modeling of genes and genomes. Annu Rev Biophys Biomol Struct Vol.29291-325, ISSN 1056-8700.

McGuffin, L.J. (2009) Prediction of global and local model quality in CASP8 using the ModFOLD server. Proteins Vol.77 Suppl 9185-190, ISSN 0887-3585.

Mereghetti, P.; Ganadu M.L.; Papaleo E.; Fantucci P. & De Gioia L. (2008) Validation of protein models by a neural network approach. BMC Bioinformatics Vol.966, ISSN 1471-2105.

Moult, J.; Fidelis K.; Kryshtafovych A.; Rost B. & Tramontano A. (2009) Critical assessment of methods of protein structure prediction - Round VIII. Proteins Vol.77 Suppl 91-4, ISSN 0887-3585.

Orengo, C.A. & Taylor W.R. (1990) A rapid method of protein structure alignment. J Theor Biol Vol.147, No.4, pp. 517-551,

Ortiz, A.R.; Strauss C.E. & Olmea O. (2002) MAMMOTH (matching molecular models obtained from theory): an automated method for model comparison. Protein Sci Vol.11, No.11, pp. 2606-2621,

Pearl, F.; Todd A.; Sillitoe I.; Dibley M.; Redfern O.; Lewis T.; Bennett C.; Marsden R.; Grant A.; Lee D.; Akpor A.; Maibaum M.; Harrison A.; Dallman T.; Reeves G.; Diboun I.; Addou S.; Lise S.; Johnston C.; Sillero A.; Thornton J. & Orengo C. (2005) The CATH Domain Structure Database and related resources Gene3D and DHS provide comprehensive domain family information for genome analysis. Nucleic Acids Res Vol.33, No.Database issue, pp. D247-251,

Raman, S.; Vernon R.; Thompson J.; Tyka M.; Sadreyev R.; Pei J.; Kim D.; Kellogg E.; DiMaio F.; Lange O.; Kinch L.; Sheffler W.; Kim B.H.; Das R.; Grishin N.V. & Baker D. (2009) Structure prediction for CASP8 with all-atom refinement using Rosetta. Proteins Vol.77 Suppl 989-99, ISSN 0887-3585.

Simons, K.T.; Kooperberg C.; Huang E. & Baker D. (1997) Assembly of protein tertiary structures from fragments with similar local sequences using simulated annealing and Bayesian scoring functions. J Mol Biol Vol.268, No.1, pp. 209-225,

Sippl, M.J. (1993) Recognition of errors in three-dimensional structures of proteins. Proteins Vol.17, No.4, pp. 355-362, ISSN 0887-3585.

Sippl, M.J. (1995) Knowledge-based potentials for proteins. Curr Opin Struct Biol Vol.5, No.2, pp. 229-235, ISSN 0959-440X.

Talavera, D.; Vogel C.; Orozco M.; Teichmann S.A. & de la Cruz X. (2007) The (in)dependence of alternative splicing and gene duplication. PLoS Comput Biol Vol.3, No.3, pp. e33,

Todd, A.E.; Marsden R.L.; Thornton J.M. & Orengo C.A. (2005) Progress of structural genomics initiatives: an analysis of solved target structures. J Mol Biol Vol.348, No.5, pp. 1235-1260, ISSN 0022-2836.

Venclovas, C. (2001) Comparative modeling of CASP4 target proteins: combining results of sequence search with three-dimensional structure assessment. Proteins Vol.Suppl 547-54, ISSN 0887-3585.

Venclovas, C.; Zemla A.; Fidelis K. & Moult J. (2001) Comparison of performance in successive CASP experiments. Proteins Vol.Suppl 5163-170, ISSN 0887-3585.

Vincent, J.J.; Tai C.H.; Sathyanarayana B.K. & Lee B. (2005) Assessment of CASP6 predictions for new and nearly new fold targets. Proteins Vol.61 Suppl 767-83, ISSN 0887-3585.

Vriend, G. (1990) WHAT IF: a molecular modeling and drug design program. J Mol Graph Vol.8, No.1, pp. 52-56, 29, ISSN 0263-7855.

Wallner, B. & Elofsson A. (2003) Can correct protein models be identified? Protein Sci Vol.12, No.5, pp. 1073-1086, ISSN 0961-8368.

Wallner, B. & Elofsson A. (2005) Pcons5: combining consensus, structural evaluation and fold recognition scores. Bioinformatics Vol.21, No.23, pp. 4248-4254, ISSN 1367-4803.

Wallner, B. & Elofsson A. (2006) Identification of correct regions in protein models using structural, alignment, and consensus information. Protein Sci Vol.15, No.4, pp. 900-913, ISSN 0961-8368.

Wallner, B. & Elofsson A. (2007) Prediction of global and local model quality in CASP7 using Pcons and ProQ. Proteins Vol.69 Suppl 8184-193, ISSN 0887-3585.

Wang, G.; Jin Y. & Dunbrack R.L., Jr. (2005) Assessment of fold recognition predictions in CASP6. Proteins Vol.61 Suppl 746-66, ISSN 0887-3585.

Wang, Z.; Tegge A.N. & Cheng J. (2009) Evaluating the absolute quality of a single protein model using structural features and support vector machines. Proteins Vol.75, No.3, pp. 638-647, ISSN 0887-3585.

Williams, M.G.; Shirai H.; Shi J.; Nagendra H.G.; Mueller J.; Mizuguchi K.; Miguel R.N.; Lovell S.C.; Innis C.A.; Deane C.M.; Chen L.; Campillo N.; Burke D.F.; Blundell T.L. & de Bakker P.I. (2001) Sequence-structure homology recognition by iterative alignment refinement and comparative modeling. Proteins Vol.Suppl 592-97, ISSN 0887-3585.

Zemla, A. (2003) LGA: A method for finding 3D similarities in protein structures. Nucleic Acids Res Vol.31, No.13, pp. 3370-3374,

Zhou, H. & Skolnick J. (2008) Protein model quality assessment prediction by combining fragment comparisons and a consensus C(alpha) contact potential. Proteins Vol.71, No.3, pp. 1211-1218, ISSN 0887-3585.

18

Functional Analysis of Intergenic Regions for Gene Discovery

Li M. Fu
Pacific Tuberculosis and Cancer Research Organization, Anaheim, CA USA

1. Introduction

Gene finding can be defined as a problem of identifying a stretch of the genomic DNA sequence that is biologically functional. Such a genomic DNA sequence is known as a gene. A gene performs a function like protein coding or regulation at the molecular level and plays a biological role, such as growth, metabolism, and intelligence. Traditionally, gene finding relies on numerous biological experiments and statistical analysis to pinpoint the location of a new gene in a genetic map. With the advent of bioinformatics, gene finding has largely become a computational problem. Genes are predictable based on the genomic sequence alone. However, the determination of the specific function and biological role of a gene would still demand *in vivo* experimentation, which is hoped to be reduced or even replaced by new bioinformatics algorithms in the future.

A newly sequenced genome is annotated thoroughly so that the information it carries can be utilized. In essence, genome annotation is to identify the locations of genes and all of the coding regions in a genome, and determine their protein products as well as functions. Hundreds of bacterial genome sequences are publicly available and the number will soon reach a new milestone. Gene annotation by hand is almost impossible to handle the deluge of new genome sequences appearing at this pace. The need for automated, large-scale, high-throughput genome annotation is imminent (Overbeek, Begley et al. 2005; Van Domselaar, Stothard et al. 2005; Stothard and Wishart 2006). The basic level of genome annotation is the use of BLAST (Altschul, Gish et al. 1990) for finding similarities between related genomic sequences. Integration with other sources of information and experimental data is a trend in genome annotation.

A recent study indicates that many genomes could be either over-annotated (too many genes) or under-annotated (too few genes), and a large percentage of genes may have been assigned a wrong start codon (Nielsen and Krogh 2005). The fact that the original genome annotation is accurate and complete upon submission does not guarantee that it will not be changed, as new experimental evidence and knowledge would continue to arrive and constant updates would be inevitable. However, re-annotation of the whole genome is not very fruitful, as most of the genes have been identified in the first annotation. For example, the re-annotation of the H37Rv genome resulted in about 2% of new protein-coding sequences (CDS) added to the genome. The result reflects the limitation with current genome annotation technology. To address the issue, we developed a new method for gene finding in an annotated genome. We select the genome of *Mycobacterium tuberculosis*, the causative pathogen of tuberculosis, as the experimental genome for this study. The availability of the complete genome sequence of *M. tuberculosis* H37Rv (Cole, Brosch et al. 1998) has led to a better understanding of the biology and pathogenicity of the organism, and new molecular targets for diagnostics and therapeutics can be invented at a fast pace by focusing on genes with important functions.

In our previous studies, we found that some intergenic sequences in *M. tuberculosis* genome exhibited expression signals, as detected by the Affymetrix GeneChip (Fu 2006; Fu and Fu-Liu 2007; Fu and Shinnick 2007). The same observation has been made for other bacteria, such as *Bacillus subtilis* (Lee, Zhang et al. 2001), and also holds true in the eukaryotic system (Zheng, Zhang et al. 2005). At present, it is not clear whether or how intergenic expression represents gene activity. Here, we presented our research work concerning gene discovery in the intergenic sequences based on transcription activity. In this work, new protein-coding genes were identified by the bioinformatics criteria based on the gene structure, protein coding potential, and ortholog evidence, in conjunction with microarray-based transcriptional evidence.

2. Research methods and design

The developed method of gene finding in the intergenic sequences proceeds as follows:

1. Transcription analysis to identify intergenic regions exhibiting significant gene expression activity.
2. Coding potential and gene structure analysis on active intergenic elements identified based on transcription evidence.
3. Protein domain search to identify functional domains in each active intergenic element with significant transcription activity and coding potential.
4. Homology search based on BLAST to seek homologue evidence.

The flowchart of the method is displayed in Figure 1.

The method was applied to the originally annotated *M. tuberculosis* H37Rv genome (Cole, Brosch et al. 1998). The genes discovered in the intergenic sequences were validated against recent findings in the literature. The research protocols (Fu and Shinnick 2007) were described below.

2.1 RNA isolation

M. tuberculosis strain H37Rv was obtained from the culture collection of the Mycobacteriology Laboratory Branch, Centers for Disease Control and Prevention at Atlanta. Bacterial lysis and RNA isolation were performed following the procedure at the CDC lab, Atlanta (Fisher, Plikaytis et al. 2002). Briefly, cultures were mixed with an equal volume of RNALater™ (Ambion, Austin, TX) and the bacteria harvested by centrifugation (1 min, 25000g, 8°C) and transferred to Fast Prep tubes (Bio 101, Vista, CA) containing Trizol (Life Technologies, Gaithersburg, MD). Mycobacteria were mechanically disrupted in a Fast Prep apparatus. The aqueous phase was recovered, treated with Cleanascite (CPG, Lincoln Park, NJ), and extracted with chloroform-isoamyl alcohol (24:1 v/v). Nucleic acids were ethanol precipitated. DNAase I (Ambion) treatment to digest contaminating DNA was performed in the presence of Prime RNase inhibitor (5′-3′, Boulder, CO). The RNA sample was precipitated and washed in ethanol, and redissolved to make a final concentration of 1 mg/ml. The purity of RNA was estimated by the ratio of the readings at 260 nm and 280 nm (A260/A280) in the UV. 20 ul

RNA samples were sent to the UCI DNA core and further checked through a quality and quantity test based on electrophoresis before microarray hybridization.

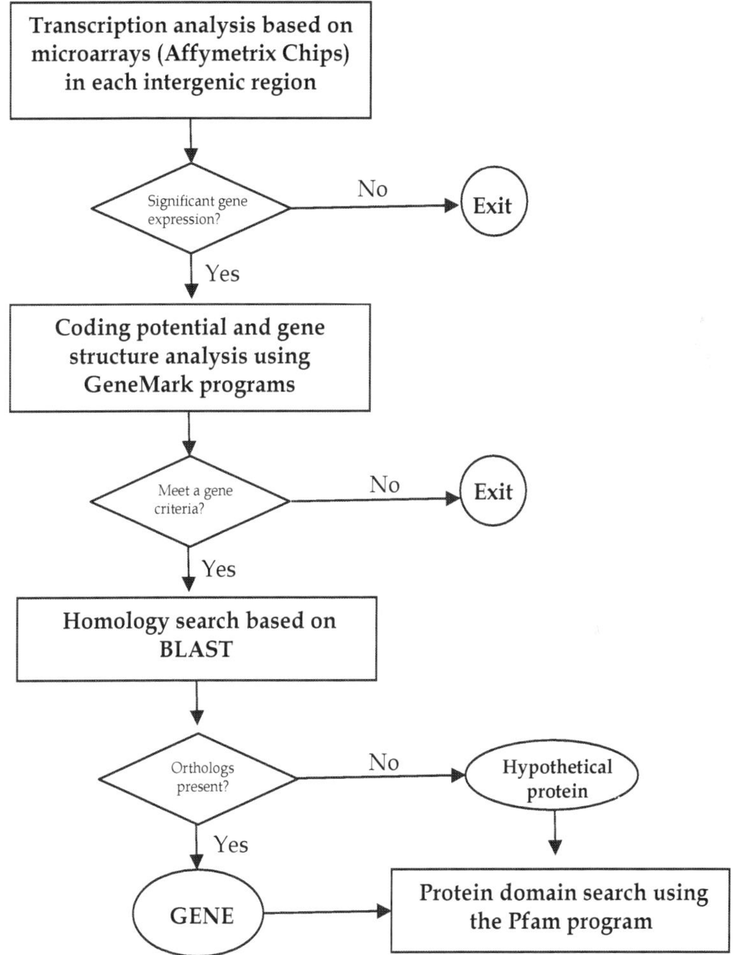

Fig. 1. The bioinformatics method with a flowchart developed for finding genes in intergenic regions.

2.2 Microarray hybridization

In this study, we used the anti-sense Affymetrix *M. tuberculosis* genome array (GeneChip). The probe selection was based on the genome sequence of *M. tuberculosis* H37Rv (Cole, Brosch et al. 1998). Each annotated ORF (Open Reading Frame) or IG (Intergenic Region) was interrogated with oligonucleotide probe pairs. An IG referred to the region between two consecutive ORFs. The gene chip represented all 3924 ORFs and 740 intergenic regions of H37Rv. The selection of these IGs in the original design was based on the sequence length. Twenty 25-mer probes were selected within each ORF or IG. These probes were called PM (Perfect-Match) probes. The sequence of each PM probe was perturbed with a single substitution at the middle base. They were called MM (Mismatch) probes. A PM probe and its respective MM probe constituted a probe pair. The MM probe served as a negative control for the PM probe in hybridization.

Microarray hybridization followed the Affymetrix protocol. In brief, the assay utilized reverse transcriptase and random hexamer primers to produce DNA complementary to the RNA. The cDNA products were then fragmented by DNAase and labeled with terminal transferase and biotinylated GeneChip DNA Labeling Reagent at the 3' terminal.

Each RNA sample underwent hybridization with one gene array to produce the expression data of all genes on the array. We performed eleven independent bacterial cultures and RNA extractions at different times, and collected eleven sets of microarray data for this study. A global normalization scheme was applied so that each array's median value was adjusted to a predefine value (500). The scale factor for achieving this transformed median value for an array was uniformly applied to all the probe set values on a specific array to result in the determined signal value for all the probe sets on the array. In this manner, corresponding probe sets can be directly compared across arrays.

2.3 Gene expression analysis

The gene expression data were analyzed by the program GCOS (GeneChip Operating Software) version 1.4. In the program, the Detection algorithm determined whether a measured transcript was detected (P Call) or not detected (A Call) on a single array according to the Detection p-value that was computed by applying the one-sided Wilcoxon's signed rank test to test the Discrimination scores (R) against a predefined adjustable threshold τ. The parameter τ controlled the sensitivity and specificity of the analysis, and was set to a typical value of 0.015, and the Detection p-value cutoffs, α_1 and α_2, set to their typical values, 0.04 and 0.06, respectively, according to the Affymetrix system.

2.4 Gene prediction

Protein-coding region identification and gene prediction were performed by the programs, GeneMark and GeneMark.hmm (Lukashin and Borodovsky 1998; Besemer and Borodovsky 2005) (http://exon.gatech.edu/GeneMark/), respectively. The prokaryotic version and the *M. tuberculosis* H37Rv genome were selected. Both programs use inhomogeneous Markov chain models for coding DNA and homogeneous Markov chain models for non-coding DNA. GeneMark adopts Bayesian formalism, while GeneMark.hmm uses a Hidden Markov Model (HMM).

2.5 Protein domain search

The Pfam program version 20.0 (Finn, Mistry et al. 2006) (http://pfam.wustl.edu/) was employed to conduct protein domain search after the input DNA sequence was translated into a protein sequence in six possible frames. The search mode was set to "global and local alignments merged", and the cut-off E-value set to 0.001, which was more stringent than the default value of 1.0. Pfam maintained a comprehensive collection of multiple sequence alignments and hidden Markov models for 8296 common protein families based on the Swissprot 48.9 and SP-TrEMBL 31.9 protein sequence databases.

2.6 Homology search

The BLASTx program (Altschul, Gish et al. 1990) (http://www.ncbi.nlm.nih.gov/BLAST/) was used to identify high-scoring homologous sequences. The program first translated the input DNA sequence into a protein sequence in six possible frames, and then matched it against the non-redundant protein sequence database (nr) in the GenBank and calculated the statistical significance of the matches. The default cut-off E-value was 10.0 but we set it to 1.0×10^{-10}. Orthologs refer to homologs in different strains of the same species. Orthologs provide critical evidence for gene finding and characterization in a new genome sequence. A typical prokaryotic gene has the following structure: the promoter, transcription initiation, the 5' untranslated region, translation initiation, the coding region, translation stop, the 3' untranslated region, and transcription stop.

3. Results

In our previous research, we conducted a genome-wide expression analysis on intergenic regions using the Affymetrix GeneChip (Fu and Shinnick 2007). The transcriptional activity of intergenic regions was measured based on a set of eleven independent RNA samples extracted from *M. tuberculosis* culture. Each RNA sample contained the information of genome-wide expression of genes and intergenic elements. The Affymetrix GeneChip was uniquely suited to this study since it had the advantage of encoding both genes and intergenic sequences whereas other types of microarray like the cDNA array could not profile intergenic sequences. As an additional strength, the Affymetrix array was designed to minimize cross-hybridization by using unique oligonucleotide probes and the pair of PM (Perfect-Match) and MM (Mismatch) probes. The cross-hybridization of related or overlapping gene sequences often contributed to false positive signals, especially in the case when long cDNA sequences were used as probes. A study demonstrated that the Affymetrix GeneChip produced more reliable results in detecting changes in gene expression than cDNA microarrays (Li, Pankratz et al. 2002).

In this work, genes in the intergenic sequences were recognized based on transcriptional activity, structural patterns, and coding potential, and subsequently validated through sequence comparison with orthologs from other *M. tuberculosis* strains.

3.1 Transcriptional analysis

An intergenic region was assumed to transcribe if there existed transcripts in the RNA sample that were bound to the probes encoding that intergenic sequence. The presence or absence of a given transcript was determined in accordance with the Detection algorithm of the Affymetrix system. In this study, a gene or intergenic region was determined to express (transcriptionally active) only if the derived mRNA was present (P-call) in more than 90% of the collected RNA samples with a Detection p-value < 0.001. The status of active transcription assigned to an intergenic sequence signified the possible presence of a gene within that sequence. We focused on finding protein-coding genes and neglected regulatory genes that transcribed into a regulatory RNA instead of mRNA. Furthermore, it was not clear how much cross-hybridization would occur between genes and intergenic sequences. As a result, the functional criterion based on expression activity was strengthened by structural analysis for minimizing false positives in gene identification.

3.2 Sequence-based gene prediction

In the sequence-based approach to gene prediction, gene structure and coding potential are the two mutually supportive elements. The GeneMark algorithm was applied to an intergenic sequence for checking whether it contained a probable coding region, and the GeneMark.hmm algorithm was used for predicting a gene within the sequence. The criteria based on the predefined transcriptional evidence, coding potential, and gene structure yielded 65 candidate genes in the intergenic regions of *M. tuberculosis* H37Rv.

3.3 Protein domain search

The biological function of a gene is determined using *in vivo* experimentation in a traditional approach. Recently, the wealth of bioinformatics knowledge in the functional domains of proteins has enabled the function of a molecular sequence to be characterized directly, subject to *in vivo* validation. Thus, the "candidate" genes within the intergenic sequences that satisfied the criteria based on transcription activity, gene structure, and coding potential were further examined for embedded functional domains. To this end, Pfam was applied to search on the protein sequences of the candidate genes. Twelve of them were found to have a known domain (Tables 1); a found domain was generally located within the predicted gene sequence, but there were a few exceptions (i.e., IG398 and IG1140) in which a domain was found within the intergenic sequence but outside the predicted gene sequence. The biological function and role of a gene is deducible from its associated functional domains; yet sufficient evidence from homology or biochemistry would serve to corroborate it.

3.4 Homologue evidence

In evolutionary biology, a reliable means for predicting the function of an unknown gene sequence is based on homologs or orthologs. BLAST is a bioinformatics program for database search, allowing functional and evolutionary inference between sequences. In this study, BLAST was employed to retrieve from sequence databases all proteins that produced statistically significant alignment with a given intergenic sequence under consideration. The sequences retrieved by BLAST were homologous to the query sequence. It turned out that the highest-scoring homologous sequences with ≥98% identity were consistently those belonging to the same strain (H37Rv) or different strains of *Mycobacterium tuberculosis* (e.g., CDC1551, F11, and C). These sequences are coding sequences described in the currently annotated genome of *M. tuberculosis*.

A homologous sequence found in different strains of the same species often represents an ortholog that shares similar function, whereas a homologous sequence found in the same organism is a paralog (which is produced via gene duplication within a genome) that tends to have a different function. No evidence suggested paralogs in our analysis, as argued based on the following observation. We noted that, given an intergenic sequence, when BLAST returned a homologous sequence pertaining to the H37Rv strain, it was apparently the same protein-coding sequence contained in the intergenic sequence. This is because the intergenic sequence used as a query and its homologous sequence returned by BLAST occupied the same physical location within the genome, as inferred from information given by the Affymetrix Genechip. This coincidence was further explained by the fact that the intergenic sequence was named according to the early version of the H37Rv genome annotation while the homologous sequence was retrieved from the GenBank which contained all up-to-date genes. The results are significant. First, we demonstrated that our method was able to identify protein-coding genes in intergenic regions previously considered as non-coding sequences. Secondly, our method based on bioinformatics and transcriptional evidence correctly predicted these changes on a high-throughput, genomic scale. The changes refer to

- IG1061 → (containing) Rv1322A
- IG499 → Rv0634B
- IG617 → Rv0787A
- IG1741 → Rv2219A
- IG2500 → Rv3198A
- IG2053 → Rv2631
- IG1179 → Rv1489A
- IG2522 → Rv3224B
- IG1291 → Rv1638A
- IG398 → Rv0500A
- IG2870 → Rv3678A
- IG188 → Rv0236A
- IG2498 → Rv3196A,
- IG2591 → Rv3312A
- IG595 → Rv0755A
- IG1814 → Rv2309A
- IG1030 → Rv1290A
- IG2141 → Rv2737A

In the above findings, each intergenic region contained an independent gene/CDS with the only exception that part of IG2053 was incorporated in its left-flanking CDS. The presence of a gene structure in an IG and its lack of functional correlation with its adjacent genes suggested that it was not a run-away segment from adjacent genes.

In our analysis, predicted genes located within intergenic sequences that met the criteria defined based on protein-coding potential, structural patterns, and transcription evidence, were called "candidate" genes. If a candidate gene of unknown function was homologous to another gene of know function, the candidate gene was assigned the function associated with its homologous gene. Nonetheless, the strategy of inferring the function of an uncharacterized sequence from its orthologs had limited value in analyzing intergenic data mainly because most of the orthologs found in this study were hypothetical proteins with unknown function. We did not assign a specific function to a candidate gene until it had an ortholog of known function, whether or not the candidate gene carried a known functional domain. In the absence of a specific function assigned, a CDS was termed a hypothetical protein rather than a gene in our system.

In this work, six intergenic sequences were identified that met the criteria we defined, including protein coding, structural patterns, transcription, and ortholog evidence: IG499,

IG617, IG1741, IG2500, IG1567, and IG2229, among which four genes had been reported in the *M. tuberculosis* H37Rv genome (Table 1). A hypothetical protein was found in 52 intergenic sequences and 14 among them had been reported in the H37Rv genome. Overall, this research discovered two genes with a specific function and 38 hypothetical proteins that had not been reported in the H37Rv genome (Fu and Shinnick 2007). The two new genes discovered were a DNA-binding protein in the CopG family and a nickel binding GTPase, located in IG1567 and IG2229, respectively (Figure 2). It was worth noting that 4.3% of intergenic regions exhibiting transcriptional activity contained a gene described in the re-annotated H37Rv genome, compared with 1.0% of intergenic regions in the absence transcriptional activity. The four-fold increase in likelihood suggested that microarray-based transcriptional analysis would facilitate genome-wide gene finding.

IG	Lt Flank	Rt Flank	Domain ID	Re-annotated H37Rv Gene
IG1061	Rv1322	Rv1323	Glyoxalase	Rv1322A*
IG499	Rv0634c	Rv0635	Ribosomal_L33	Rv0634B
IG617	Rv0787	Rv0788	PurS	Rv0787A
IG398	Rv0500	Rv0501	DUF1713	Rv0500A*
IG1741	Rv2219	Rv2220	RDD	Rv2219A
IG2500	Rv3198c	Rv3199c	Glutaredoxin	Rv3198A
IG2053	Rv2631	Rv2632c	UPF0027	Rv2631*
IG1179	Rv1489c	Rv1490	MM_CoA_mutase	Rv1489A*
IG1140	Rv1438	Rv1439c	TetR_N	None
IG2522	Rv3224	Rv3225c	YbaK	Rv3224B*
IG1567	Rv1991c	Rv1992c	RHH_1	None
IG2229	Rv2856	Rv2857c	cobW	None

*: Hypothetical protein

Table 1. The functional domains of the predicted genes located within the intergenic sequences of *M. tuberculosis* H37Rv genome. Each intergenic sequence shown is characterized by its flanking genes or ORFs and the functional domain identified in the translated protein sequence. Most of IGs with a functional domain contain a gene in the re-annotated H37Rv genome (Fu and Shinnick 2007).

4. Discussion

Computational algorithms for gene prediction are divided in two classes: One is based on sequence similarity and the other based on gene structure and signal. The latter is known as *ab initio* prediction. The first class of algorithm, represented by BLAST (Altschul, Gish et al. 1990), finds sequences (DNA, protein, or ESTs) in the database that match the given sequence, whereas the second class of algorithm, such as Hidden Markov Model (Burge and Karlin 1997; Lukashin and Borodovsky 1998; Besemer and Borodovsky 2005), builds a model of gene structure from empirical data. They both have limitations. For instance, the sequence-based approach is not applicable if no homology is found, whereas the model-based approach is not workable if no adequate training data is available for model parameter estimation. To explore an alternative in a different perspective, the method developed in our research combined sequence alignment, transcriptional evidence, and homology. In particular, the transcriptional activity of a piece of DNA is direct evidence that it is functioning. This is important because a gene means a piece of genomic DNA that is functional. In the absence of functional evidence, any gene computed by whatever algorithms will remain hypothetical.

New gene 1:
[Location]: Between Rv1991c and Rv1992c
[Product]: DNA-binding protein, CopG family
[Nucleotide Sequence]:
atcgtccatggtttctagcacgcggtatgcgttggccacggcgagggcctccgcttcgtcggtgccatggatgctctctagagccctgtcgatctggcccgtgagcaattggg
cgtccagctcgtgcaggtagcgctgcgcagccttcgtgaagaactcggaccgactcatgccgagctcactcgcacgccgcgatacccgatcgaacgtctcatccggcagag
aaatagctgtcttcat
[Protein Sequence]:
mktaislpdetfdrvsrraselgmsrsefftkaaqrylheldaqlltgqidralesihgtdeaealavanayrvletmdd

New gene 2:
[Location]: Between Rv2856 and Rv2857c
[Product]: Nickle binding GTPase involved in regulation of expression urease and hydrogenase
[Nucleotide Sequence]:
atggtctcctcggtcaccgagggcaaggacaagccgctgatgtacccggcgacgttccgctcgagggatgtagtgctgctcgacaagatcgacttggtgccctttctggac
gccgacgtggacgcgtatatcgcgcatgtccgcgaggtcaacgcagccgcgacgatcctgccgaccagcacgcgcaccggagccggcatggggtcctggtcatga
[Protein Sequence]:
mvssvtegkdkplmypatfrsrdvvlldkidlvpfldadvdayiahvrevnaaatilptstrtgagmgsws

Fig. 2. Examples of new genes with a predicted function found in the genome of *M. tuberculosis* H37Rv (Fu and Shinnick 2007).

The whole *M. tuberculosis* H37Rv genome has been sequenced and annotated comprehensively (Cole, Brosch et al. 1998). Transcriptional analysis of intergenic regions is a means of exploring unknown genes. Our idea capitalized on transcription analysis in gene finding, which was useful especially when applied to an annotated genome. Current genome annotation technology allowed all genes to be identified by a computational algorithm, and it was unlikely to add new genes through re-annotation at the same time unless using a different algorithm. Thus, it was within expectation that the number of new protein-coding sequences due to re-annotation was merely 2% of that in the original submission of *M. tuberculosis* genome (Camus, Pryor et al. 2002). Through homology and pattern-based search, most protein-coding sequences with a predicted function have been reported. Yet, transcriptional evidence could quickly hint at potential protein-coding genes in the intergenic regions. It is encouraging that we are still able to find new genes in this study given the fact that the current knowledge concerning *M. tuberculosis* genes is derived from intensive research in the field involving *in vivo* biological experiments, such as gene deletion and complementation. Thus the integration of the sequence- and function-based analyses would be a useful approach to not just gene characterization but also gene prediction. As the experiment was based on the standard in vitro growth condition of *M.*

tuberculosis, silent genes under this condition were not under examination in this study, but the same idea should be applicable to other genomes under other conditions, and contribute to the improvements of current gene databases.
The methods presented here did not address the issue of genes that did not code proteins. There are a number of regulatory, non-coding RNAs assuming a distinct role from mRNA, rRNA and tRNA. Many such RNAs have been identified and characterized both in prokaryotes and eukaryotes and their main functions are posttranscriptional regulation of gene expression and RNA-directed DNA methylation (Erdmann, Barciszewska et al. 2001; Pickford and Cogoni 2003). A non-coding RNA has neither a long open reading frame nor a gene structure. The DNA sequence that encodes a non-coding RNA is called a gene if its regulatory function can be defined. Thus it is possible that an isolated expression element lacking a gene structure is a non-coding, regulatory RNA. However, it was confirmed that the potential protein-coding genes found in this study did not match any RNA family published in the RNA-families database (www.sanger.ac.uk/ Software/Rfam/).

5. Conclusion

High-throughput gene finding on a newly sequenced genome is enabled through advanced computational genome annotation software. However, genome annotation does not guarantee all genes to be identified since knowledge and concepts about what constitutes a gene are evolving and yet to be perfected. Genome re-annotation using the same kind of computational heuristics offers limited help, unless supported with new *in vivo* experimental evidence, but such evidence often slowly arrives. We developed a method that integrated sequence-based and transcriptional information for gene finding in the intergenic regions of an annotated genome. In the experiment with the *M. tuberculosis* H37Rv genome, the method discovered genes with a specific function, such as a DNA-binding protein in the CopG family and a nickel binding GTPase, as well as hypothetical proteins that have not been reported in the *M. tuberculosis* H37Rv genome. This work has demonstrated that microarray-based transcriptional analysis could play an important role in gene finding on the genomic scale.

6. Acknowledgments

We would like to thank Dr. Thomas Shinnick at CDC for collaboration and the use of the facilities, and thank UCI for providing service for microarray hybridization. We thank Thomas R. Gingeras at Affymetrix, Inc. for designing *Mycobacterium tuberculosis* GeneChip. Bacterial culture and RNA isolation were performed by Pramod Aryal.

7. References

Altschul, S. F., W. Gish, W. Miller, E. W. Myers and D. J. Lipman (1990). "Basic local alignment search tool." J Mol Biol 215(3): 403-10.

Besemer, J. and M. Borodovsky (2005). "GeneMark: web software for gene finding in prokaryotes, eukaryotes and viruses." Nucleic Acids Res 33(Web Server issue): W451-4.

Burge, C. and S. Karlin (1997). "Prediction of complete gene structures in human genomic DNA." J Mol Biol 268(1): 78-94.

Camus, J. C., M. J. Pryor, C. Medigue and S. T. Cole (2002). "Re-annotation of the genome sequence of Mycobacterium tuberculosis H37Rv." Microbiology 148(Pt 10): 2967-73.

Cole, S. T., R. Brosch, J. Parkhill, T. Garnier, C. Churcher, D. Harris, et al. (1998). "Deciphering the biology of Mycobacterium tuberculosis from the complete genome sequence." Nature 393(6685): 537-44.

Erdmann, V. A., M. Z. Barciszewska, A. Hochberg, N. de Groot and J. Barciszewski (2001). "Regulatory RNAs." Cell Mol Life Sci 58(7): 960-77.

Finn, R. D., J. Mistry, B. Schuster-Bockler, S. Griffiths-Jones, V. Hollich, T. Lassmann, et al. (2006). "Pfam: clans, web tools and services." Nucleic Acids Res 34(Database issue): D247-51.

Fisher, M. A., B. B. Plikaytis and T. M. Shinnick (2002). "Microarray analysis of the Mycobacterium tuberculosis transcriptional response to the acidic conditions found in phagosomes." J Bacteriol 184(14): 4025-32.

Fu, L. M. (2006). "Exploring drug action on Mycobacterium tuberculosis using affymetrix oligonucleotide genechips." Tuberculosis (Edinb) 86(2): 134-43.

Fu, L. M. and C. S. Fu-Liu (2007). "The gene expression data of Mycobacterium tuberculosis based on Affymetrix gene chips provide insight into regulatory and hypothetical genes." BMC Microbiol 7: 37.

Fu, L. M. and T. M. Shinnick (2007). "Genome-Wide Analysis of Intergenic Regions of Mycobacterium tuberculosis H37Rv Using Affymetrix GeneChips." EURASIP J Bioinform Syst Biol: 23054.

Fu, L. M. and T. M. Shinnick (2007). "Understanding the action of INH on a highly INH-resistant Mycobacterium tuberculosis strain using Genechips." Tuberculosis (Edinb) 87(1): 63-70.

Lee, J. M., S. Zhang, S. Saha, S. Santa Anna, C. Jiang and J. Perkins (2001). "RNA expression analysis using an antisense Bacillus subtilis genome array." J Bacteriol 183(24): 7371-80.

Li, J., M. Pankratz and J. A. Johnson (2002). "Differential gene expression patterns revealed by oligonucleotide versus long cDNA arrays." Toxicol Sci 69(2): 383-90.

Lukashin, A. V. and M. Borodovsky (1998). "GeneMark.hmm: new solutions for gene finding." Nucleic Acids Res 26(4): 1107-15.

Nielsen, P. and A. Krogh (2005). "Large-scale prokaryotic gene prediction and comparison to genome annotation." Bioinformatics 21(24): 4322-9.

Overbeek, R., T. Begley, R. M. Butler, J. V. Choudhuri, H. Y. Chuang, M. Cohoon, et al. (2005). "The subsystems approach to genome annotation and its use in the project to annotate 1000 genomes." Nucleic Acids Res 33(17): 5691-702.

Pickford, A. S. and C. Cogoni (2003). "RNA-mediated gene silencing." Cell Mol Life Sci 60(5): 871-82.

Stothard, P. and D. S. Wishart (2006). "Automated bacterial genome analysis and annotation." Curr Opin Microbiol 9(5): 505-10.

Van Domselaar, G. H., P. Stothard, S. Shrivastava, J. A. Cruz, A. Guo, X. Dong, et al. (2005). "BASys: a web server for automated bacterial genome annotation." Nucleic Acids Res 33(Web Server issue): W455-9.

Zheng, D., Z. Zhang, P. M. Harrison, J. Karro, N. Carriero and M. Gerstein (2005). "Integrated pseudogene annotation for human chromosome 22: evidence for transcription." J Mol Biol 349(1): 27-45.

19

Prediction of Transcriptional Regulatory Networks for Retinal Development

Ying Li[1], Haiyan Huang[2] and Li Cai[1]
[1]*Department of Biomedical Engineering, Rutgers University, NJ*
[2]*Department of Statistics, University of California, Berkeley, CA*
USA

1. Introduction

The formation of the neatly layered retina during embryonic development is dictated by a series of complicated transcription factor interactions. Retina-specific expression of these transcription factors is an essential step in establishing retinal progenitor cells (RPCs) from embryonic stem cells. The transcriptional control of gene expression is largely mediated by the combinatorial interactions between *cis*-regulatory DNA elements and *trans*-acting transcription factors, which cooperate/interact with each other to form a transcription regulatory network during this developmental process. Such regulatory networks are essential in regulating tissue/cell-specific gene expression, e.g., in cell fate determination and differentiation during embryonic retinal development (Hu et al., 2010; Kumar, 2009; Swaroop et al., 2010). Many genes, which involved in transcriptional networks for the specification of a certain retinal cell lineage, have already been identified and characterized (Corbo et al., 2010; Kim et al., 2008b; Tang et al., 2010). The transcriptional regulatory networks for specific retinal cell lineages, e.g., the photoreceptors (Corbo et al., 2010; Hsiau et al., 2007) and bipolar neurons (Kim et al., 2008b), were established in recent years. However, the transcriptional regulatory network that governs the entire neural retinal development is still elusive.

Identifying tissue/cell-specific *cis*-regulatory elements, *trans*-acting factor binding sites (TFBSs), and their binding transcription factors (TFs) represent key steps towards understanding tissue/cell-specific gene expression and further successful reconstruction of transcriptional regulatory networks. These steps also present major challenges in both fields of experimental biology and computational biology.

Currently, the prevailing method of studying TFBSs and transcriptional regulatory networks is to determine the function of tissue-specific trans-acting factors based on data from genome-wide gene expression profiling and chromatin immunoprecipitation (ChIP). ChIP is often used to investigate protein-DNA interactions in a cell. Coupled with massive parallel sequencing, ChIP-seq is capable of mapping the genome-wide protein-DNA interaction at a finer resolution (Valouev et al., 2008) to identify candidate enhancer sequences (Visel et al., 2009). Thus, a regulatory cascade can be recognized via consequential analysis of the factors involved.

Here, we present a new method for the computational analysis of TFBSs and transcriptional regulatory networks utilizing genome-wide sequencing, expression, and enhancer data. In

contrast to the traditional method, which mainly focuses on factor expression analysis, we emphasize the sequence elements with tissue-specific enhancer activity. Our hypothesis is that enhancers, non-coding sequences that direct gene expression in a cell-/tissue-specific manner, contain common TFBSs that allow the key protein factors (important for the development of that cell/tissue type) to bind. Experimentally verified tissue-specific enhancer elements selected from enhancer databases were carefully screened for common trans-acting factor binding sites to predict potential sequence-factor interacting networks. DNA-binding protein factors that associate with multiple enhancers can be analyzed using experimental methods.
As proof-of-principle, simple transcriptional regulatory networks of embryonic retinal development were assembled based on common/key factors and their interacting genes, as determined by literature search. These resulting networks provide a general view of embryonic retinal development and a new hypothesis for further experimentation.

2. Methods and results

In this study, we aimed to develop a method for the identification of regulatory networks for cell/tissue-specific gene expression. To test our hypothesis of the existence of common TFBSs on the enhancers of cell/tissue specific genes, we employ the mouse developing retina as a model system. Enhancers that direct retina-specific gene expression were selected and their sequences were thoroughly screened for common TFBSs and *trans*-acting factors (protein factors) using open-access software such as TESS, JASPA, MatInspector, and MEME, etc. For common TFBS selection, we developed a Matlab program. These common protein factors were further studied for their expression profiles. Thereafter, we were able to construct transcriptional regulatory networks based on the known function of the enhancer-binding *trans*-acting factors using a network construction tool, BioTapestry (Longabaugh et al., 2009).

2.1 Retina-specific enhancers

To determine the transcriptional regulatory networks that govern retinal development, we identified and selected enhancer elements that direct gene expression in the retina by searching the VISTA Enhancer Browser. The VISTA Enhancer Browser is a central resource for experimentally validated human and mouse non-coding DNA fragments with enhancer activity as assessed in transgenic mice. Most of these non-coding elements were selected for testing based on their extreme conservation in other vertebrates or epigenomic evidence (ChIP-Seq) of putative enhancer marks. The results of this *in vivo* enhancer screen are provided through this publicly available website (Frazer et al., 2004). Of the 1503 non-coding DNA elements from human and mouse genomes that were tested for their ability to direct gene expression on embryonic day 11.5 (E11.5) using a reporter assay in transgenic mouse, a total of 47 elements were shown to possess enhancer activity in the retina (as of February 10, 2011). Of the 47 elements, 25 show enhancer activities in both the retina and other tissues, e.g., heart, limb, brain and spinal cord, etc. These 25 elements were separated for further analysis as we focused on retina-specific sequence elements. In addition, among the 22 remaining elements, 17 elements were active enhancers in the retina in at least half of the tested transgenic embryos (Supporting data 1). To further increase the possibility of identifying the key enhancer elements that are critical for retinal development, we applied more stringent selection criteria based on the following two properties: 1) the reporter expression is restricted in the retina and not in any other regions of the CNS; 2) there is expression of at least one of the flanking genes in the retina at E11.5 (or Theiler Stage 18-21). The second category was applied because a majority of known enhancers were found in the immediate up- or downstream region of their target genes, and thus the regulatory activities of enhancers were considered to be most likely associated with their flanking genes. Any enhancer elements that did not fit into at least one of the two categories was thus eliminated (details in Supporting data 5). Based on the above criteria, 8 enhancer elements were identified (Table 1).

Group	Enhancer ID	Length (bp)	Reporter expression pattern derived from enhancer activity	Annotation of reporter expression	Flanking genes of enhancers and their endogenous expression in mouse embryos	
					Upstream	Downstream
1	**hs27**	1113	eye(4/6), limb(4/6)	Retina + non-CNS	**Irx5: E11-19 retina**	**Irx6: E11-19 retina**
	hs258	1487	eye(3/5), limb,	Retina + non-CNS	Ccdc39: E14.5	**Fxr1: E12-14, E19 retina**
	hs546	1753	eye(7/7), limb, nose	Retina + non-CNS	**Nr2f1: E11-19 in retina**	Arrdc3: E14-19
	hs1170	1288	eye(8/8)	Retina	**Nr2f1: E11-19 in retina**	Arrdc3: E14-19
2	**hs932**	775	eye(6/9), limb, nose, branchial arch	Retina + non-CNS	AA408296: unknown	Irf6: not in retina
	mm165	926	eye(4/5), heart	Retina + non-CNS	Lao1: unknown	Slc2a1: not in retina
3	**hs1122**	1218	eye(6/7)	Retina+ Spinal cord	**Ascl1: E11-19 in retina (MGI)**	Pah: not in retina
	mm269	760	eye(5/5), heart, other	Retina+ Spinal cord	**Zfand5** (intragenic): **E11-13**/E13-19 in retina	

Table 1. A list of eight retina-specific enhancers. The 8 retina-specific enhancers selected from the VISTA Enhancer Browser are listed above. They are grouped into 3 sub-groups according to the reporter expression in mouse embryos derived from these enhancers and the expression pattern of their flanking genes. The enhancer IDs, their expression pattern and the flanking gene names were retrieved directly from VISTA Enhancer Browser, with an expression pattern described as 'tissue type (positive sample number/total sample number)'. Flanking genes with expression in the retina are shown in **bold**.

2.2 *Trans*-acting factor binding sites on retina-specific enhancers

The binding of trans-acting factors (e.g., transcription factors) to non-coding regulatory DNA (e.g., promoters, enhancers, etc.) is an essential process in the control of gene expression. This protein-DNA interaction helps recruit the DNA polymerase complex and co-activators to form the transcription machinery. The binding of these protein factors can also act as repressors to prevent transcription. Identification of a TFBS in the enhancer and promoter for a gene may indicate the possibility that the corresponding factors play a role in the regulation of that gene. Importantly, the ability of an enhancer to direct cell/tissue-specific gene expression is achieved via the binding of tissue-specific *trans*-acting protein factors. To determine what protein factors regulate retina-specific gene expression, it is important to determine what TFBSs are located on the retina-specific enhancers. We thus searched DNA sequences of the 8 retina-specific enhancer elements for their known TFBSs using TESS (Schug, 2002), JASPA (Portales-Casamar et al., 2010; Sandelin et al., 2004) and MatInspector (Cartharius et al., 2005). For example, TESS (Transcription Element Search System - http://www.cbil.upenn.edu/cgi-bin/tess) is a web tool for predicting TFBSs in DNA sequences. It can identify TFBSs using site or consensus strings and positional weight matrices mainly from the TRANSFAC (Knuppel et al., 1994). TRANSFAC contains data on transcription factors, their experimentally-proven binding sites, and regulated genes. Its broad compilation of binding sites allows the derivation of positional weight matrices (Knuppel et al., 1994). The following search parameters were set when searching TESS: a minimum string length of 6, a maximum allowable string mismatch of 10%, a minimum log-likelihood ratio score of 12, and organism selection of *Mus musculus* (the house mouse). Our search results show that there are approximately 150 TFBSs for each of the 8 enhancer sequences (Supporting data 2). Similar results were reported by JASPA and MatInspector. The corresponding protein factors of these TFBSs were considered to be capable of binding with the 8 retina-specific enhancers, and thus they are important in activating/suppressing gene expression in the retina.

2.3 A motif containing Pou3f2 binding sites

Since all 8 enhancers possess the ability to direct retina-specific gene expression, there may be key TFBSs shared amongst these retina-specific sequence elements. To test this hypothesis, we sorted and screened the TFBSs of each of the 8 enhancers to identify common ones using a Matlab program that we developed for this study (Supporting data 6). This Matlab program for common TFBS selection was designed to compare the TFBSs on each of the retina-specific enhancer elements predicted by TESS. TFBSs for two or three different enhancers can be sequentially compared. A "model" character was used as the comparison category instead of the binding site name in both TESS and our Matlab program. As defined in TESS, a model is "the site string or weight matrix used to pick this site" (Schug, 2002), and thus describes the nature of a binding site. One factor may have multiple models, and one model may be shared by multiple factors. The model character is the only necessary parameter to characterize the transcription factors depending on their binding site property. With this sorting/searching program, we identified a TFBS for Pou3f2 (also known as Brn2) that was present in all 8 retina-specific enhancers (Fig. 1A). Previous studies have demonstrated that the Pou3f2 transcription factor plays an important role in the development of neural progenitor cells (Catena et al., 2004; Kim et al., 2008b; McEvilly et al., 2002; Sugitani et al., 2002). Furthermore, the literature reports that this motif was first discovered as a *cis*-element in the Chx10 enhancer, which can drive reporter expression in intermediate and late RPCs (Rowan and Cepko, 2005). In this study, Pou3f2 was also shown to affect bipolar interneuron fate determination through interactions with Chx10 and Nestin. We thus speculate that this Pou3f2 binding site may exist in regulatory sequences among genes important for the development of neural retinal progenitor cells (RPCs). Therefore, the *cis*-elements of Chx10, Cyclin D1, Pax6, Rax, and Six3 were examined because these genes are known for their role in regulating RPC development and retinal cell differentiation (Conte et al., 2010; Martinez-de Luna et al., 2010; Oliver et al., 1995; Rowan and Cepko, 2005; Sicinski et al., 1995). Confirming our prediction, Pou3f2 binding sites were also present in the *cis*-elements of Chx10, Cylin D1, and Pax6 genes (sequences can be found in Supporting data 4). Next, sequence alignment analysis was performed to identify a common motif for Pou3f2 binding sites using MEME (Multiple Em for Motif Elicitation, http://meme.sdsc.edu/meme4_4_0/cgi-bin/meme.cgi) (Bailey and Elkan, 1994). It is commonly believed that there are dependencies among different positions in a motif. MEME may ignore such kind of dependency. Sequence alignment reveals a 22 bp motif containing two Pou3f2 binding sites among all 8 enhancer elements (with a stringent E-value of 7.2e-20 and p-value < 3.02e-6) and also in the *cis*-elements of RPC-specific genes, e.g., Chx10, CyclinD1, and Pax6, from multiple vertebrate species (Fig. 1B). In addition, a line of evidence indicates that Pou3f2 binds to the Rax enhancer to regulate the expression of Rax in RPCs (Martinez-de Luna et al., 2010). Such prevalent existence of repeated Pou3f2 binding sites among retina-specific enhancer elements and *cis*-elements of RPC-specific genes suggests that Pou3f2 is a key regulatory factor in the embryonic retinal development.

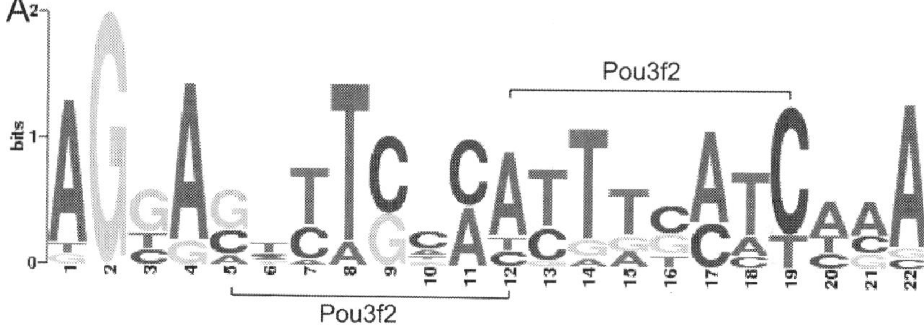

B

Name	Strand	Start	p-value	Sites		
cis-element_of_r_CyclinD1	+	0	5.31e-10		AGGACTTTGCAACTTCAACAAA	
cis-element_of_m_CyclinD1	+	0	5.31e-10		AGGACTTTGCAACTTCAACAAA	
cis-element_of_h_CyclinD1	+	0	5.31e-10		AGGACTTTGCAACTTCAACAAA	
cis-element_of_h_Chx10	+	0	6.16e-09		AGGAGATTCCCCTTTTATCCAA	
cis-element_of_m_Chx10	+	0	9.18e-09		AGGAGATTCCCTTTTTATCCAA	
mm269.chr19:21,348,574-21,349,333	+	295	1.10e-07	GACAGGCTGA	AGCGCCCTCCCATGTGATCTAA	GCTTCAGAGA
hs546.chr13:78,463,126-78,464,875	+	567	2.55e-07	CCTTCTCCCC	AGCAGAATGCAATTACCTCTAA	AGCCCGTGGT
hs1122.chr10:86,963,003-86,964,413	-	1377	2.78e-07	ACAGAAGAGC	AGGAGCCTCACATTGGCTCTGG	CTCTGCAGCT
hs932.chr1:194,967,242-194,967,917	-	343	4.61e-07	TGGAAGCCCT	GGGAGTTTCGCTCTGGCTCAGA	GCAGACATGC
hs258.chr3:33,781,318-33,782,778	+	517	7.37e-07	GAGAGATAGG	AGGACTTTCTAATTTGCTTTCC	CTCTATTTTA
hs1170.chr13:78,817,206-78,818,434	+	709	1.23e-06	ATAAACCAAG	TGTAGGCTGGCATTGTATCAAA	CTGGCTTTAT
hs27.chr8:94,973,252-94,974,310	+	916	1.61e-06	GGAGAAGGGG	AGGGGGCTGCCAGGTGCCCAGA	GGCCCTTGTG
cis-element_of_h_Pax6	+	0	2.68e-06		AGTAATTACAAATTACATTACA	
cis-element_of_q_Pax6	+	0	2.68e-06		AGTAATTACAAATTACATTACA	
mm165.chr4:118,691,756-118,692,681	-	304	3.02e-06	GGGGGAGGGT	AGCAGCCTCTCCCATCACCCCA	GGCTTAGTGA

Fig. 1. A 22 bp motif is present in all 8 retina-specific enhancers and *cis*-elements of retinal progenitor gene Chx10, CyclinD1 and Pax6. Sequence alignment was performed using MEME and the output shows a motif exists among all above 15 sequences (p-value < 3e-6). A. a SeqLog presentation of the 22bp motif. Dual binding sites of Pou3f2 were located next to each other forms Pou3f2 binding site repeats. B. sequence alignment reveals the 22bp motif among the 8 enhancers and *cis*-elements of RPC-specific genes (e.g., Chx10, CyclinD1 and Pax6) (Rowan and Cepko, 2005). Abbreviation: r, rat; m, mouse; h, human; q, quail. RPC, retinal progenitor cell.

2.4 Key *trans*-acting factors involved in transcriptional regulatory networks of retinal development

It is unlikely that only one factor (i.e., Pou3f2) is involved in regulating retina-specific gene expression. *Trans*-acting factors often do not function alone but rather in a cooperative manner. To identify other key protein factors for retina-specific gene expression, we applied the following assumptions to the enhancer elements and their binding *trans*-acting factors:

1. Key TFBSs should be common to all or a subset of retina-specific enhancer elements;
2. The flanking genes of these enhancer elements should be expressed in the retina during early retinal development, because an enhancer often regulates the expression of its flanking gene(s);
3. The binding factors should have a known function in retinal development, or
4. If the binding factors have an unknown function in retinal development, they should be at least expressed in the retina during retinal development. In this case, the expression of the factor provides novel hypothesis for their function in retinal development, which needs to be tested by functional studies.

Based on the above assumptions, the information on the expression of the flanking genes of enhancer elements in the developing retina is necessary for the TFBS analysis. Five databases of gene expression (see Table 2) were searched. The information on the expression of these flanking genes were retrieved from these databases (Tables 1, 4 and supporting data 3). The factors that do not express in the retina during embryonic retinal development, e.g., around E11.5 (when enhancer elements were active), were set aside for further analysis of retinal transcriptional regulatory networks. We then searched common TFBSs among subsets of the 8 retina-specific enhancers. The TFBSs common to individual different sub-groups were combined. In addition to Pou3f2, five other factors (i.e., Crx, Hes1, Meis1, Pbx2, and Tcf3) were identified (Table 3). Four of the 6 factors have known functions in retinal development, which is consistent with our hypothesis. The last two factors do not have known functions in the retina. However, the prediction of their binding with groups of enhancers suggests they play a role during retinogenesis. As the binding sites of these 6 factors were shared among a subset of retina-specific enhancer elements, these 6 binding *trans*-acting factors were predicted as the key factors that participate in regulating retina-specific gene expression during embryonic development.

Database	Source	Reference
Gene expression database (emage) of Edinburg Mouse Atlas Project (EMAP, v5.0_3.3)	http://www.emouseatlas.org/emage/	(Richardson et al., 2010)
Gene Expression Database in Mouse Genome Informatics (MGI, version 4.4)	http://www.informatics.jax.org	(Finger et al., 2011)
Eurexpress	http://www.eurexpress.org	(Diez-Roux et al., 2011)
VisiGene Image Browser	http://genome.ucsc.edu/cgi-bin/hgVisiGene	(Kent et al., 2002)
Genome-scale mouse brain transcription factor expression analysis	Supplementary data S4 and S6	(Gray et al., 2004)

Table 2. A list of gene expression databases used in this study.

Factor	Binding site	Known function	Expression pattern	Presence in enhancer element
Pou3f2	ATTTGCAT	Induce Bipolar cells	E10.5-14.5 in retina; diencephalon, future midbrain, future SC, rhombencephalon	All 8 elements
Crx	tgaggGGATCA Acagact	Induce Photoreceptors	E11-adult in retina	hs27, hs258, hs546, hs1170
Hes1	CTTGTG	Repress Amacrine, Horizontal, and Ganglion cells; Induce Photoreceptors	E11-13 in retina, thalamus, hypothalamus, striatum, olfactory epithelial	hs27, hs258, hs1170
Meis1	CTGTCActaaga tgaca	retinal cell fate determination	E10.5-14.5 in retina, lens vesicle, diencephalon, future sc, hindbrain	hs27, hs258, hs546, hs1170, hs932, mm165
Pbx-2	cacctgagagTGA CAGaaggaaggc agggag	No function known in retina	E10.5-14.5 in retina, thalamus, midbrain, hindbrain, sc, ear	hs27, hs258, hs546, hs1170, hs932, mm165
Tcf3	ccaccagCACCT Gtc	No function known in retina	E13.5 in retina (MGI)	hs27, hs258, hs546, hs1170

Table 3. A list of binding factors that show their temporal and spatial co-localization of expression with each group of enhancers. For each factor, elements shown in the last column indicate the enhancer elements which share a potential common binding with it. The 'Expression pattern' column shows available evidence of the co-localization with enhancers. Corresponding databases are noted because different databases recorded different expression pattern for the corresponding factor. The general function of each factor is also included for future reference. Abbreviation: RPC, retinal progenitor cell; B, bipolar cell; A, amacrine cell; H, horizontal cell; G, ganglion cell; PR, photoreceptor cell; sc, spinal cord.

Gene	Function (related to retina development) and reference
Ascl1	With Mash3, regulate the neuron/glia fate determination (Hatakeyama et al., 2001); with Mahs3 and Chx10, specify Biopolar cell identity (Satow et al., 2001).
Irx5	Off circuit subsets of bipolar interneuron (Cheng et al., 2005; Cohen et al., 2000; Kerschensteiner et al., 2008).
Irx6	No known clear function in retina. But It expresses in the in the area lining the lumen of the otic vesicle including the region giving rise to ganglion complex of CN VII/VIII at E11.5 through E16.5 and overlaps with Mash1 (Cohen et al., 2000; Mummenhoff et al., 2001)
Fxr1	Retina pigmentation(de Diego Otero et al., 2000); other function not known.
Nr2f1	Amacrine development, may involve in cone differentiation; express in a unique gradient in retina along D/V axis (Inoue et al., 2010).
Zfand5	No known function in retina.

Table 4. A list of flanking genes with their function and references

Interestingly, three common TFBSs (i.e., Pou3f2, Crx, and Meis1) were present among enhancer elements hs27, hs258, and hs1170 (see Tables 1, 3). Sequence alignment of the three enhancer elements and *cis*-elements of RPC-specific genes (e.g., Chx10, Cyclin D1, and Pax6) revealed another 22 bp motif (Fig. 2). Crx binding sites were present on enhancer elements hs546 and hs1170, while Hes1 binding sites were present on enhancer element hs546. Since these two enhancer elements (hs1170 and hs546) were both located in the non-coding region between Nr2f1 and Arrdc3, and since Arrdc3 was not active at E11.5 in the retina, the binding of Crx and Hes1 may participate in regulating the expression of Nr2f1. This result is supported by the finding that both Crx (Peng et al., 2005) and Nr2f1 (Inoue et al., 2010; Satoh et al., 2009) play a role in inducing photoreceptor cell fate, though at different stages. In addition, Crx has been shown to be expressed in bipolar cells, paired with Otx2 and Pou3f2, in binding with a 164bp Chx10 enhancer (Kim et al., 2008a). Hes1 has been shown to be active during early eye formation. By suppressing Math5, Hes1 was shown to be involved in the development of cone photoreceptors, amacrine, horizontal and ganglion cells from the RPCs (Le et al., 2006; Lee et al., 2005).

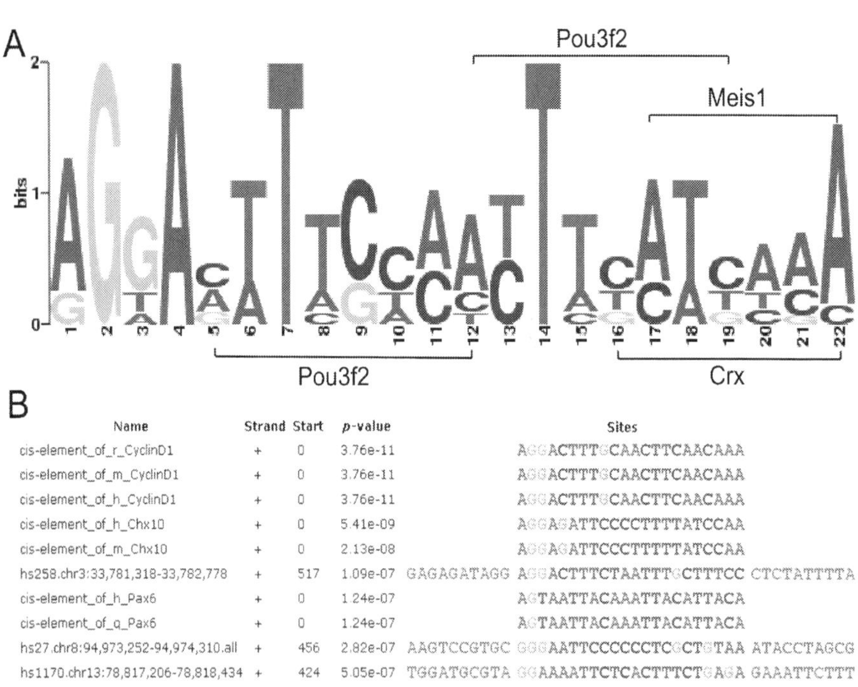

Name	Strand	Start	*p*-value	Sites		
cis-element_of_r_CyclinD1	+	0	3.76e-11		AGGACTTTGCAACTTCAACAAA	
cis-element_of_m_CyclinD1	+	0	3.76e-11		AGGACTTTGCAACTTCAACAAA	
cis-element_of_h_CyclinD1	+	0	3.76e-11		AGGACTTTGCAACTTCAACAAA	
cis-element_of_h_Chx10	+	0	5.41e-09		AGGAGATTCCCCTTTTATCCAA	
cis-element_of_m_Chx10	+	0	2.13e-08		AGGAGATTCCCTTTTTATCCAA	
hs258.chr3:33,781,318-33,782,778	+	517	1.09e-07	GAGAGATAGG	AGGACTTTCTAATTTGCTTTCC	CTCTATTTTA
cis-element_of_h_Pax6	+	0	1.24e-07		AGTAATTACAAATTACATTACA	
cis-element_of_q_Pax6	+	0	1.24e-07		AGTAATTACAAATTACATTACA	
hs27.chr8:94,973,252-94,974,310.all	+	456	2.82e-07	AAGTCCGTGC	GGGAATTCCCCCCTCGCTGTAA	ATACCTAGCG
hs1170.chr13:78,817,206-78,818,434	+	424	5.05e-07	TGGATGCGTA	GGAAAATTCTCACTTTCTGAGA	GAAATTCTTT

Fig. 2. A 22 bp motif is present in a subset of enhancer elements and *cis*-elements of RPC-specific genes, (e.g., Chx10, Cyclin D1, Pax6). Sequence alignment was performed using MEME among enhance elements (hs258, hs546, and hs1170) and *cis*-elements of genes Chx10, Cyclin D1, Pax6. A. a SeqLog presentation of the 22bp motif . This 22 bp motif contains binding sites for 3 factors: Pou3f2, Meis1 and Crx. B. Sequence alignment between elements mentioned above. Abbreviation: r, rat; m, mouse; h, human; q, quail.

Meis1 together with Meis2, as members of the TALE-homeodomain protein Homothorax (Hth) related protein family, were known to be expressed in the RPCs of mouse and chick (Heine et al., 2008). Meis1 was expressed in RPCs throughout the entire neurogenesis period, and Meis2 was expressed more specifically in RPCs before the initiation of retina differentiation. Together, they function to maintain the RPCs in a rapid proliferating state and control the expression of other ocular genes, e.g., Pax6, CyclinD1, Six3 and Chx10 (Bessa et al., 2008; Heine et al., 2008). Since Meis1 binding sites are present in a subset of retina-specific enhancers, Meis1 may function as an RPC-specific factor. Since the onset of mouse retina neurogenesis is approximately at E10.5 when the ganglion cells first appear (Leo M. Chalupa, 2008). By E11.5, RPCs of all six cell types are highly active. Therefore, binding of Meis1 with enhancers might influence the cell fate of these RPCs.

The presence of common Pbx2 binding sites may indicate a novel functional role of Pbx2 in RPCs, since the function of Pbx2 in retinal development has not been documented. Previous studies have shown that Pbx2 is expressed in the zebrafish retina and tectum (French et al., 2007) together with Pbx1 and Meis1, and down-regulation in their expression caused by the deficiency of Prep, the prolyl endopeptidase will lead to eye anomalies (Deflorian et al., 2004; Ferretti et al., 2006). Pbx and Meis proteins are major DNA-binding partners that form abundant complexes (Chang et al., 1997). Thus, there is a possibility that Pbx2 may function in the development of RPCs via the interaction with Meis1 and also regulate other RPC-specific genes (e.g., Irx5, Nr2f1, etc) through enhancer binding (Table 1).

Tcf3 is not yet known to have a function in embryonic retinal development. However, since Tcf3 binding sites are present among the retina-specific enhancer elements, and Tcf3 is expressed in the retina during embryogenesis, its specific function in retinal development needs to be confirmed.

2.5 Generation of transcriptional regulatory network for early retinal development

Based on the available expression data from VISTA Enhancer Browser and gene expression databases (Table 2), it is known that these 8 enhancer elements and their common binding *trans*-acting factors are active during embryonic development in the retina. Among the 8 retina-specific enhancer elements, we have identified 6 common *trans*-acting factors. These 6 predicted factors are experimentally verified key protein factors known to be involved in regulating gene expression and cell differentiation of progenitor cells during embryonic retinal development (Table 3).

Retina-specific gene expression is most likely determined by two kinds of interactions: (1) the enhancers with their binding protein factors, and (2) the protein factors with their interacting partners. The information about these interactions was used to generate the transcriptional regulatory networks important for retinal development. Therefore, transcriptional regulatory networks of embryonic retina were predicted based on these 6 common/key *trans*-acting factors (Table 3) and their known interacting partners (Table 5).

To construct retinal transcriptional regulatory networks, a java-based software program named BioTapestry (Longabaugh et al., 2009) (http://www.biotapestry.org/, version 5.0.2) was used to organize the factors and their known interacting partners. BioTapestry is a network facilitating software program designed for dealing with systems that exhibit increasingly complex over time, such as genetic regulatory networks. Its unique annotation system allows the illustration of enhancer-regulated gene expression and connection between factors. Experimental evidence can also be added to network elements after the network was built, as a proof of particular interactions. We only used the presenting function of BioTapestry here to show the networks of retinal development during early neurogenesis. For better illustration, the network was mapped according to the 3-layer structure of the mouse retina.

Based on the published information on the interacting factors of these 6 *trans-acting factors* and their known functions in retinal cell development, we were able to build transcriptional regulatory networks for all six major retinal cell types (Fig. 3). In RPCs, Meis1 is regulated by Prep1 since it has been shown that insufficient Prep1 expression leads to a decrease in Meis1 expression (Deflorian et al., 2004). Meis1 itself can regulate or interact with 4 other factors (e.g., Pax6, CyclinD1, Six3 and Chx10) important to retinal cell differentiation (Bessa et al., 2008; Heine et al., 2008). Crx, interacting with other factors (e.g, Otx2, Nrl and Nr2e3) plays an important role in both cone and rod photoreceptor cell fate determination (Peng et al., 2005). Crx can also influence Chx10 and Irx5 expression to affect the bipolar cell formation (Cheng et al., 2005; Kerschensteiner et al., 2008; Kim et al., 2008a), which is similar to the function of Pou3f2 (Kim et al., 2008a; Rowan and Cepko, 2005) and Otx2 (Kim et al., 2008a). Another important factor is Hes1. Hes1 functions in maintaining RPC proliferation (Wall et al., 2009) and regulating Math5, a critical factor for the generation of amacrine, horizontal, and ganglion cells (Lee et al., 2005). In addition, Hes1 is regulated by Hes6 (Bae et al., 2000). The Nr2f1 gene expresses in the retina in a gradient along the dorsal-ventral axis and has been found to influence the development of amacrine cells. At the same time it can affect cone and rod photoreceptors differentiation (Inoue et al., 2010). The references of all genes/factors in the networks are listed in Table 5.

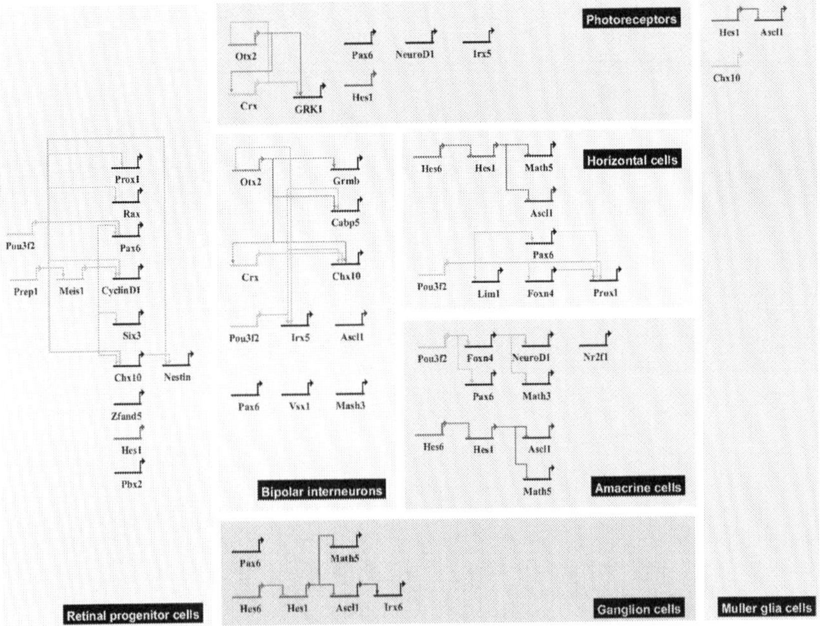

Fig. 3. Examples of transcriptional regulatory networks of embryonic retinal development. Genes and factors are connected with arrows or bars to indicate the promoting and suppressing relationship, respectively.

Factor name	Reference cited
Cabp5	(Kim et al., 2008a)
Chx10	(Hatakeyama et al., 2001)
CyclinD1	(Bessa et al., 2008; Heine et al., 2008)
Foxn4	(Shengguo Li, 2004)
Grk1	(Young and Young, 2007)
Grmb	(Kim et al., 2008a)
Hes6	(Bae et al., 2000)
Mash3	(Hatakeyama et al., 2001; Satow et al., 2001)
Math5	(Lee et al., 2005)
Nestin	(Rowan and Cepko, 2005)
NeuroD1	(Conte et al., 2010)
Nr2f1	(Inoue et al., 2010; Satoh et al., 2009)
Otx2	(Kim et al., 2008a; Young and Young, 2007)
Pax6	(Ferretti et al., 2006; Lee et al., 2005; Oliver et al., 1995)
Prep1	(Deflorian et al., 2004; Ferretti et al., 2006)
Rax	(Heine et al., 2008; Martinez-de Luna et al., 2010)
Six3	(Oliver et al., 1995)
Six6	(Conte et al., 2010)

Table 5. A list of protein factors that interact with the 6 key *trans*-acting factors in a network model

In summary, the computational method we developed in this study can be described as following. First, experimentally verified enhancer elements can be selected from enhancer databases, e.g., the Vista Enhancer Browser, based on tissue/cell-specific expression patterns derived from the enhancer element and its flanking gene (Fig. 4A-B). These tissue-specific enhancer elements can be located in the non-coding regions in inter- or intra-genetic sequences. Then, the *trans*-acting factor binding sites (TFBS) of all tissue-specific enhancer elements can be predicted using TFBS search tools such as TESS and MatInspector. Common binding sites shared by all tissue-specific enhancer elements or subsets of elements are later determined as shown in (Fig. 4C). Those common factors can be further analyzed according to their expression pattern. Only the ones with spatio-temporal expression patterns can be selected as key factors for constructing a tissue-specific transcriptional regulatory network. In addition, factors that interact with these key factors can also be found from the literature. Finally, these enhancers, genes, factors and interactions can be pieced together to build the network (Fig. 4D).

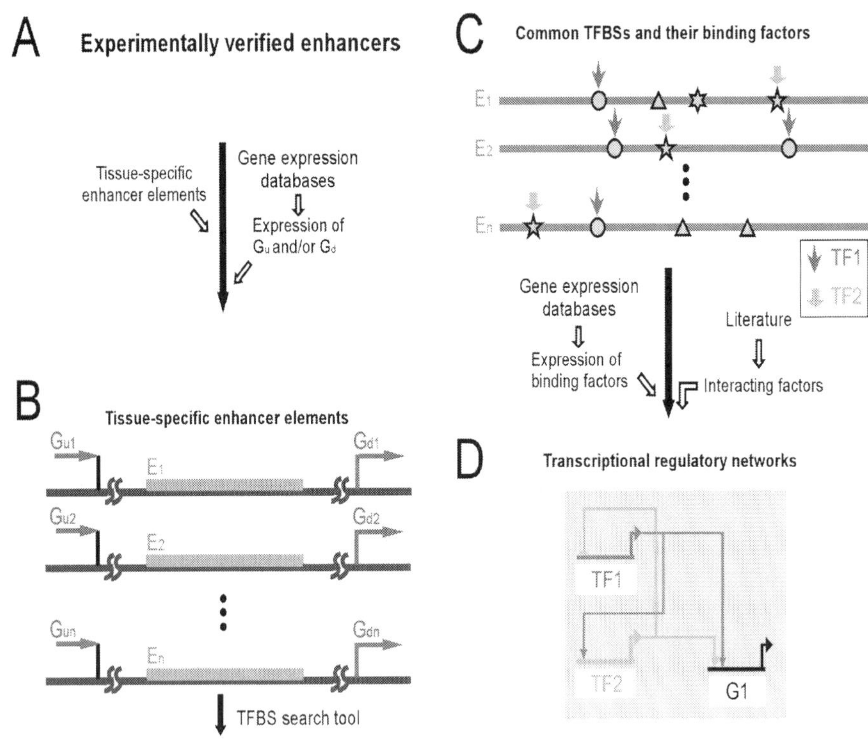

Fig. 4. Computational analysis of TFBSs and transcriptional regulatory networks for tissue/cell-specific gene expression. A. Selection of experimentally verified enhancer elements from enhancer databases. B. Depiction of enhancer elements and their flanking

genes. Orange boxes represent tissue-specific enhancer elements. Red bars represent transcript start sites of downstream flanking genes. Black bars represent the end of the last exon of upstream flanking genes. Enhancer elements in the intronic region of a gene were not illustrated. C. Comparison of the *trans*-acting factor binding sites (TFBSs). Two examples of common TFBS-TF pairs are illustrated by pairs of circle/azure arrows and pentagon/cyan arrows. Other polygons on the enhancer elements (orange bars) represent binding sites that are not common to all members. D. A simple example of a tissue-specific transcriptional regulatory network generated with two key factors (TF1 and TF2) and their interacting genes (G). The thick horizontal lines on top of factor and gene names (TF1, TF2 and G1) represent the *cis*-regulatory region of the corresponding genes. Arrows and bars terminating on this line illustrate the input from other factors, promoters or repressors, respectively. The arrow derived from the thick line represents the regulatory outputs. Abbreviation: E, enhancer element; G, gene; Gd, downstream gene; Gu, upstream gene; TF, *trans*-acting (e.g., transcription) factor; TFBS, transcription factor binding site.

3. Conclusion

In this study we have explored a new way of using existing TFBS-finding methods to predict *trans*-acting factors and networks that regulate tissue-specific gene expression (Fig. 4). As a proof-of-principle study, using this method, we have identified experimentally verified transcription factors (Pou3f2, Crx, Meis1, and Hes1) and their transcriptional regulatory networks that are known to be important for mouse embryonic retinal development. This not only provides a general idea about the development of the RPCs and retinal cell differentiation, but also validates the effectiveness of our newly developed method. Furthermore, we predicted that two other factors (Pbx2 and Tcf3, expressed in retina) may have a novel functional role in embryonic retinal development, which generates new hypotheses for future experimentation, i.e., to test the involvement of the two factors, Pbx2 and Tcf3, in the development of RPCs. This method is based on the assumption that non-coding DNA sequences contain signals regulating tissue-specific gene expression. The feasibility of this method relies on the existing data of experimentally verified tissue-specific enhancers (Visel et al., 2007) and available information of tissue-specific gene expression (Finger et al., 2011; Richardson et al., 2010). The identification of retina-specific *trans*-acting factors and networks indicates that our method is useful for the analysis of TFBSs and transcriptional regulatory networks. Since our method utilizes the existing/known information about tissue-specific enhancer elements and gene expression, we will not be able to identify novel factors that are involved in the transcriptional regulatory networks. In order to reconstruct a more sophisticated network in future studies, we will need to include all of the retina-specific enhancer elements. Finally, this method can be applied to the analysis of TFBSs and transcriptional regulatory networks in other tissue types.

4. Acknowledgement

This work is supported in part by grants (EY018738 and EY019094) from the National Institute of Health, the New Jersey Commission on Spinal Cord Research (10A-003-SCR1 and 08-3074-SCR-E-0), and Busch Biomedical Research Awards (6-49121). The authors thank the Cai lab members for helpful discussions and for proof-reading the manuscript.

5. References

Bae, S., Bessho, Y., Hojo, M., and Kageyama, R. (2000). The bHLH gene Hes6, an inhibitor of Hes1, promotes neuronal differentiation. *Development 127*, 2933-2943.

Bailey, T.L., and Elkan, C. (1994). Fitting a mixture model by expectation maximization to discover motifs in biopolymers. *Proc Int Conf Intell Syst Mol Biol 2*, 28-36.

Bessa, J., Tavares, M.J., Santos, J., Kikuta, H., Laplante, M., Becker, T.S., Gómez-Skarmeta, J.L., and Casares, F. (2008). meis1 regulates cyclin D1 and c-myc expression, and controls the proliferation of the multipotent cells in the early developing zebrafish eye. *Development 135*, 799-803.

Cartharius, K., Frech, K., Grote, K., Klocke, B., Haltmeier, M., Klingenhoff, A., Frisch, M., Bayerlein, M., and Werner, T. (2005). MatInspector and beyond: promoter analysis based on transcription factor binding sites. *Bioinformatics 21*, 2933-2942.

Catena, R., Tiveron, C., Ronchi, A., Porta, S., Ferri, A., Tatangelo, L., Cavallaro, M., Favaro, R., Ottolenghi, S., Reinbold, R., *et al.* (2004). Conserved POU binding DNA sites in the Sox2 upstream enhancer regulate gene expression in embryonic and neural stem cells. *J Biol Chem 279*, 41846-41857.

Chang, C., Jacobs, Y., Nakamura, T., Jenkins, N., Copeland, N., and Cleary, M. (1997). Meis proteins are major in vivo DNA binding partners for wild-type but not chimeric Pbx proteins. *Mol Cell Biol 17*, 5679-5687.

Cheng, C.W., Chow, R.L., Lebel, M., Sakuma, R., Cheung, H.O.-L., Thanabalasingham, V., Zhang, X., Bruneau, B.G., Birch, D.G., Hui, C.-c., *et al.* (2005). The Iroquois homeobox gene, Irx5, is required for retinal cone bipolar cell development. *Developmental Biology 287*, 48-60.

Cohen, D.R., Cheng, C.W., Cheng, S.H., and Hui, C.-c. (2000). Expression of two novel mouse Iroquois homeobox genes during neurogenesis. *Mechanisms of Development 91*, 317-321.

Conte, I., Marco-Ferreres, R., Beccari, L., Cisneros, E., Ruiz, J.M., Tabanera, N., and Bovolenta, P. (2010). Proper differentiation of photoreceptors and amacrine cells depends on a regulatory loop between NeuroD and Six6. *Development 137*, 2307-2317.

Corbo, J.C., Lawrence, K.A., Karlstetter, M., Myers, C.A., Abdelaziz, M., Dirkes, W., Weigelt, K., Seifert, M., Benes, V., Fritsche, L.G., *et al.* (2010). CRX ChIP-seq reveals the cis-regulatory architecture of mouse photoreceptors. *Genome Res 20*, 1512-1525.

de Diego Otero, Y., Bakker, C., Raghoe, P., Severijnen, L.A., Hoogeveen, A., Oostra, B., and Willemsen, R. (2000). Immunocytochemical characterization of FMRP, FXR1P and FXR2P during embryonic development in the mouse. *Gene Function & Disease 1*, 28-37.

Deflorian, G., Tiso, N., Ferretti, E., Meyer, D., Blasi, F., Bortolussi, M., and Argenton, F. (2004). Prep1.1 has essential genetic functions in hindbrain development and cranial neural crest cell differentiation. *Development 131*, 613-627.

Diez-Roux, G., Banfi, S., Sultan, M., Geffers, L., Anand, S., Rozado, D., Magen, A., Canidio, E., Pagani, M., Peluso, I., *et al.* (2011). A high-resolution anatomical atlas of the transcriptome in the mouse embryo. *PLoS biology 9*, e1000582.

Ferretti, E., Villaescusa, J.C., Di Rosa, P., Fernandez-Diaz, L.C., Longobardi, E., Mazzieri, R., Miccio, A., Micali, N., Selleri, L., Ferrari, G., *et al.* (2006). Hypomorphic Mutation of the TALE Gene Prep1 (pKnox1) Causes a Major Reduction of Pbx and Meis Proteins and a Pleiotropic Embryonic Phenotype. *Mol Cell Biol 26*, 5650-5662.

Finger, J.H., Smith, C.M., Hayamizu, T.F., McCright, I.J., Eppig, J.T., Kadin, J.A., Richardson, J.E., and Ringwald, M. (2011). The mouse Gene Expression Database (GXD): 2011 update. *Nucleic Acids Research 39*, D835-D841.

Frazer, K.A., Pachter, L., Poliakov, A., Rubin, E.M., and Dubchak, I. (2004). VISTA: computational tools for comparative genomics. *Nucleic Acids Research 32*, W273-W279.

French, C.R., Erickson, T., Callander, D., Berry, K.M., Koss, R., Hagey, D.W., Stout, J., Wuennenberg-Stapleton, K., Ngai, J., Moens, C.B., *et al.* (2007). Pbx homeodomain proteins pattern both the zebrafish retina and tectum. *BMC Dev Biol 7*, 85.

Gray, P.A., Fu, H., Luo, P., Zhao, Q., Yu, J., Ferrari, A., Tenzen, T., Yuk, D.I., Tsung, E.F., Cai, Z., *et al.* (2004). Mouse brain organization revealed through direct genome-scale TF expression analysis. *Science 306*, 2255-2257.

Hatakeyama, J., Tomita, K., Inoue, T., and Kageyama, R. (2001). Roles of homeobox and bHLH genes in specification of a retinal cell type. *Development 128*, 1313-1322.

Heine, P., Dohle, E., Bumsted-O'Brien, K., Engelkamp, D., and Schulte, D. (2008). Evidence for an evolutionary conserved role of homothorax/Meis1/2 during vertebrate retina development. *Development 135*, 805-811.

Hsiau, T.H., Diaconu, C., Myers, C.A., Lee, J., Cepko, C.L., and Corbo, J.C. (2007). The cis-regulatory logic of the mammalian photoreceptor transcriptional network. *PLoS One 2*, e643.

Hu, J., Wan, J., Hackler, L., Jr., Zack, D.J., and Qian, J. (2010). Computational analysis of tissue-specific gene networks: application to murine retinal functional studies. *Bioinformatics 26*, 2289-2297.

Inoue, M., Iida, A., Satoh, S., Kodama, T., and Watanabe, S. (2010). COUP-TFI and -TFII nuclear receptors are expressed in amacrine cells and play roles in regulating the differentiation of retinal progenitor cells. *Experimental Eye Research 90*, 49-56.

Kent, W.J., Sugnet, C.W., Furey, T.S., Roskin, K.M., Pringle, T.H., Zahler, A.M., and Haussler, D. (2002). The human genome browser at UCSC. *Genome Res 12*, 996-1006.

Kerschensteiner, D., Liu, H., Cheng, C.W., Demas, J., Cheng, S.H., Hui, C.-c., Chow, R.L., and Wong, R.O.L. (2008). Genetic Control of Circuit Function: Vsx1 and Irx5 Transcription Factors Regulate Contrast Adaptation in the Mouse Retina. *J Neurosci 28*, 2342-2352.

Kim, D.S., Matsuda, T., and Cepko, C.L. (2008a). A Core Paired-Type and POU Homeodomain-Containing Transcription Factor Program Drives Retinal Bipolar Cell Gene Expression. *J Neurosci 28*, 7748-7764.

Kim, D.S., Matsuda, T., and Cepko, C.L. (2008b). A core paired-type and POU homeodomain-containing transcription factor program drives retinal bipolar cell gene expression. *J Neurosci 28*, 7748-7764.

Knuppel, R., Dietze, P., Lehnberg, W., Frech, K., and Wingender, E. (1994). TRANSFAC retrieval program: a network model database of eukaryotic transcription regulating sequences and proteins. *Journal of computational biology : a journal of computational molecular cell biology 1*, 191-198.

Kumar, J.P. (2009). The molecular circuitry governing retinal determination. *Biochim Biophys Acta 1789*, 306-314.

Le, T.T., Wroblewski, E., Patel, S., Riesenberg, A.N., and Brown, N.L. (2006). Math5 is required for both early retinal neuron differentiation and cell cycle progression. *Developmental Biology 295*, 764-778.

Lee, H.Y., Wroblewski, E., Philips, G.T., Stair, C.N., Conley, K., Reedy, M., Mastick, G.S., and Brown, N.L. (2005). Multiple requirements for Hes1 during early eye formation. *Developmental Biology 284*, 464-478.

Leo M. Chalupa, R.W.W., ed. (2008). *Eye, Retina, and Visual System of the Mouse* (The MIT Press).

Longabaugh, W.J.R., Davidson, E.H., and Bolouri, H. (2009). Visualization, documentation, analysis, and communication of large-scale gene regulatory networks. *Biochimica et Biophysica Acta (BBA) - Gene Regulatory Mechanisms 1789*, 363-374.

Martinez-de Luna, R.I., Moose, H.E., Kelly, L.E., Nekkalapudi, S., and El-Hodiri, H.M. (2010). Regulation of retinal homeobox gene transcription by cooperative activity among cis-elements. *Gene 467*, 13-24.

McEvilly, R.J., de Diaz, M.O., Schonemann, M.D., Hooshmand, F., and Rosenfeld, M.G. (2002). Transcriptional regulation of cortical neuron migration by POU domain factors. *Science 295*, 1528-1532.

Mummenhoff, J., Houweling, A.C., Peters, T., Christoffels, V.M., and Rüther, U. (2001). Expression of Irx6 during mouse morphogenesis. *Mechanisms of Development 103*, 193-195.

Oliver, G., Mailhos, A., Wehr, R., Copeland, N.G., Jenkins, N.A., and Gruss, P. (1995). Six3, a murine homologue of the sine oculis gene, demarcates the most anterior border of the developing neural plate and is expressed during eye development. *Development 121*, 4045-4055.

Peng, G.-H., Ahmad, O., Ahmad, F., Liu, J., and Chen, S. (2005). The photoreceptor-specific nuclear receptor Nr2e3 interacts with Crx and exerts opposing effects on the transcription of rod versus cone genes. *Human Molecular Genetics 14*, 747-764.

Portales-Casamar, E., Thongjuea, S., Kwon, A.T., Arenillas, D., Zhao, X., Valen, E., Yusuf, D., Lenhard, B., Wasserman, W.W., and Sandelin, A. (2010). JASPAR 2010: the greatly expanded open-access database of transcription factor binding profiles. *Nucleic Acids Research 38*, D105-110.

Richardson, L., Venkataraman, S., Stevenson, P., Yang, Y., Burton, N., Rao, J., Fisher, M., Baldock, R.A., Davidson, D.R., and Christiansen, J.H. (2010). EMAGE mouse embryo spatial gene expression database: 2010 update. *Nucleic Acids Research 38*, D703-D709.

Rowan, S., and Cepko, C.L. (2005). A POU factor binding site upstream of the Chx10 homeobox gene is required for Chx10 expression in subsets of retinal progenitor cells and bipolar cells. *Developmental Biology 281*, 240-255.

Sandelin, A., Alkema, W., Engstrom, P., Wasserman, W.W., and Lenhard, B. (2004). JASPAR: an open-access database for eukaryotic transcription factor binding profiles. *Nucleic Acids Research 32*, D91-94.

Satoh, S., Tang, K., Iida, A., Inoue, M., Kodama, T., Tsai, S.Y., Tsai, M.-J., Furuta, Y., and Watanabe, S. (2009). The Spatial Patterning of Mouse Cone Opsin Expression Is Regulated by Bone Morphogenetic Protein Signaling through Downstream Effector COUP-TF Nuclear Receptors. *J Neurosci 29*, 12401-12411.

Satow, T., Bae, S.-K., Inoue, T., Inoue, C., Miyoshi, G., Tomita, K., Bessho, Y., Hashimoto, N., and Kageyama, R. (2001). The Basic Helix-Loop-Helix Gene hesr2 Promotes Gliogenesis in Mouse Retina. *J Neurosci 21*, 1265-1273.

Schug, J. (2002). Using TESS to Predict Transcription Factor Binding Sites in DNA Sequence (John Wiley & Sons, Inc.).

Shengguo Li, Z.M., Xuejie Yang, (2004). Foxn4 Controls the Genesis of Amacrine and Horizontal Cells by Retinal Progenitors. *Neuron 43*, 795-807.

Sicinski, P., Donaher, J.L., Parker, S.B., Li, T., Fazeli, A., Gardner, H., Haslam, S.Z., Bronson, R.T., Elledge, S.J., and Weinberg, R.A. (1995). Cyclin D1 provides a link between development and oncogenesis in the retina and breast. *Cell 82*, 621-630.

Sugitani, Y., Nakai, S., Minowa, O., Nishi, M., Jishage, K., Kawano, H., Mori, K., Ogawa, M., and Noda, T. (2002). Brn-1 and Brn-2 share crucial roles in the production and positioning of mouse neocortical neurons. *Genes Dev 16*, 1760-1765.

Swaroop, A., Kim, D., and Forrest, D. (2010). Transcriptional regulation of photoreceptor development and homeostasis in the mammalian retina. *Nat Rev Neurosci 11*, 563-576.

Tang, K., Xie, X., Park, J.I., Jamrich, M., Tsai, S., and Tsai, M.J. (2010). COUP-TFs regulate eye development by controlling factors essential for optic vesicle morphogenesis. *Development 137*, 725-734.

Valouev, A., Johnson, D.S., Sundquist, A., Medina, C., Anton, E., Batzoglou, S., Myers, R.M., and Sidow, A. (2008). Genome-wide analysis of transcription factor binding sites based on ChIP-Seq data. *Nat Meth 5*, 829-834.

Visel, A., Blow, M.J., Li, Z., Zhang, T., Akiyama, J.A., Holt, A., Plajzer-Frick, I., Shoukry, M., Wright, C., Chen, F., *et al.* (2009). ChIP-seq accurately predicts tissue-specific activity of enhancers. *Nature 457*, 854-858.

Visel, A., Minovitsky, S., Dubchak, I., and Pennacchio, L.A. (2007). VISTA Enhancer Browser—a database of tissue-specific human enhancers. *Nucleic Acids Research 35*, D88-D92.

Wall, D.S., Mears, A.J., McNeill, B., Mazerolle, C., Thurig, S., Wang, Y., Kageyama, R., and Wallace, V.A. (2009). Progenitor cell proliferation in the retina is dependent on Notch-independent Sonic hedgehog/Hes1 activity. *J Cell Biol 184*, 101-112.

Young, J.E., Kasperek, E.M., Vogt, T.M., Lis, A., and Khani, S.C. (2007). Conserved interactions of a compact highly active enhancer/promoter upstream of the rhodopsin kinase (GRK1) gene. *Genomics 90*, 236-248.

20

The Use of Functional Genomics in Synthetic Promoter Design

Michael L. Roberts
Synpromics Ltd
United Kingdom

1. Introduction

The scope of this chapter is to examine how advances in the field of Bioinformatics can be applied in the development of improved therapeutic strategies. In particular, we focus on how algorithms designed to unravel complex gene regulatory networks can then be used in the design of synthetic gene promoters that can be subsequently incorporated in novel gene transfer vectors to promote safer and more efficient expression of therapeutic genes for the treatment of various pathological conditions.

2. Development of synthetic promoters: a historical perspective

A synthetic promoter is a sequence of DNA that does not exist in nature and which has been designed to control gene expression of a target gene. *Cis*-regulatory sequences derived from naturally-occurring promoter elements are used to construct these synthetic promoters using a building block approach; which can either be carried out by rational design or by random ligation (illustrated in figure 1A). The result is a sequence of DNA composed of several distinct *cis*-regulatory elements in a completely novel orientation that can act as a promoter enhancer; typically to initiate RNA polymerase II-mediated transcription.
Construction of synthetic promoters is possible because of the modular nature of naturally-occurring gene regulatory regions. This was cleverly demonstrated by a group that used synthetic promoters to evaluate the role of the TATA box in the regulation of transcription (Mogno et al., 2010). The authors looked at the role of the TATA box in dictating the strength of gene expression. They found that the TATA box is a modular component in that its strength of binding to the RNA polymerase II complex and the resultant strength of transcription that it mediates is independent of the *cis*-regulatory element enhancers upstream. Importantly, they also found that the TATA box does not add noise to transcription, i.e. it acts as a simple amplifier without altering specificity of gene expression dictated by the upstream enhancer elements. Thus implying that any combination of *cis*-regulatory enhancers could be coupled to a TATA box and it would be the enhancers that would mediate specificity without any interference from the TATA box. The implications from this study suggest that it should be possible to construct any type of synthetic promoter that is specifically engineered to display a highly restrictive pattern of gene regulation.

Synthetic promoters have been used in the study of gene regulation for more than two decades. In one of the first examples a synthetic promoter derived from the lipoprotein gene in *E. Coli* was used to efficiently drive the expression of a number of tRNA genes (Masson et al., 1986). In the years that followed a technique was developed that enabled the mutation of prokaryotic sequences flanking the essential -10 and -35 promoter elements (illustrated in figure 1B) and thus the efficient construction of synthetic promoters for use in bacteria (Jacquet et al., 1986). This approach was successfully used to produce promoters with much higher activity compared to naturally occurring sequences and it was immediately realised that such an approach would have important applications in the biotech industry, particularly in the enhanced production of biopharmaceuticals (Trumble et al., 1992).

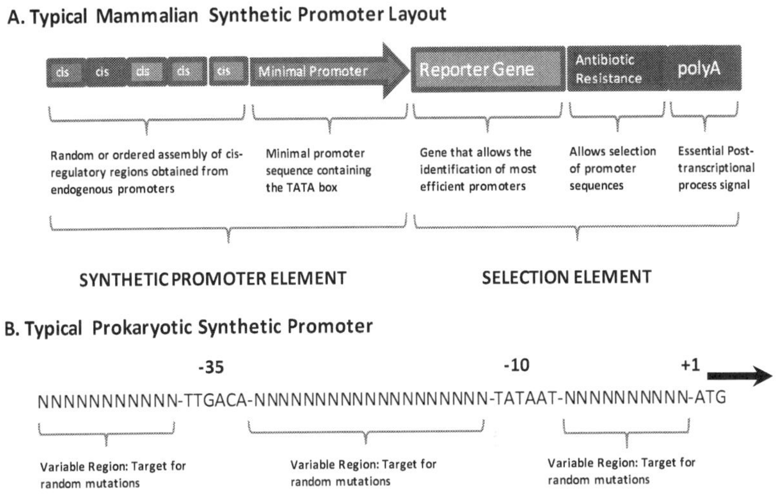

A. Typical Mammalian Synthetic Promoter Layout

cis cis cis cis cis Minimal Promoter Reporter Gene Antibiotic Resistance polyA

Random or ordered assembly of cis-regulatory regions obtained from endogenous promoters

Minimal promoter sequence containing the TATA box

Gene that allows the identification of most efficient promoters

Allows selection of promoter sequences

Essential Post-transcriptional process signal

SYNTHETIC PROMOTER ELEMENT

SELECTION ELEMENT

B. Typical Prokaryotic Synthetic Promoter

-35 -10 +1

NNNNNNNNNNN-TTGACA-NNNNNNNNNNNNNNNNNNN-TATAAT-NNNNNNNNNN-ATG

Variable Region: Target for random mutations

Variable Region: Target for random mutations

Variable Region: Target for random mutations

Fig. 1. Typical synthetic promoter layouts for prokaryotes and eukaryotes

Most of these studies were initially undertaken with a view to establish the important structural features of prokaryotic or eukaryotic promoters so that essential elements could be identified. In one example, the role of the Tat protein in the regulation of HIV gene expression was studied using synthetic promoters (Kamine et al., 1991). In this study a series of minimal promoters containing Sp1- binding sites and a TATA box were constructed and analysed to see if the Tat protein from HIV could activate them. The results demonstrated that Tat could only activate the synthetic promoters containing Sp1 sites and not promoters with the TATA box alone. The observations enabled the authors to propose that *in vivo* the Tat protein is brought to the promoter site by TAR RNA and then interacts with Sp1 to drive gene expression. In recent years more sophisticated studies using synthetic promoters have been undertaken to evaluate the important factors driving transcription factor binding to their corresponding *cis*-regulatory elements (Gertz et al., 2009a) and to thermodynamically model *trans*-factor and *cis*-element interactions (Gertz et al., 2009b).

As alluded to above, it was soon realised that synthetic promoter technology had direct implications in the improvement of the efficiency of gene expression. Indeed, one of the most widely used eukaryotic promoters employed for research purposes today is actually a synthetic promoter. The steroid-inducible Glucocorticoid Receptor Element (GRE) is a naturally occurring sequence that regulates the expression of a plethora of genes that are responsive to glucocorticoids. In a relatively early study several of these elements were linked together in order to construct a promoter with enhanced responsiveness to these steroids (Mader et al., 1993). This study detailed the construction of a 5 x GRE synthetic promoter linked to the Adenovirus type 2 major late promoter TATA region that displayed 50-fold more expression levels in response to steroid hormones when compared to the natural promoter sequence. This synthetic promoter is now a widely used constituent of a number of reporter constructs adopted in a variety of different research applications.

Finally, synthetic promoters have also been used in prokaryotic systems to reveal that regulation of gene expression follows boolean logic (Kinkhabwala et al., 2008). In this prototypical study the authors found that two transcription repressors generate a NOR logic; i.e. a OR b (on OR off), while one repressor plus one activator determines an ANDN logic; i.e. a AND NOT b (on AND NOT off). This idea was later expanded on to demonstrate that various combinations of synthetic promoters could combine to generate 12 out of 16 boolean logic terms (Hunziker et al., 2010). Most interestingly the results from these studies demonstrated that if a promoter does not follow a specific logic it is more likely to be leaky, in that it will drive gene expression under conditions where it is not expected to.

In this chapter we describe the evolution of synthetic promoter technology, its application in the development of improved tissue-specific promoters and its potential use for the development of effective disease-specific gene regulators; thus enabling the development of safer and more effective gene therapies.

3. Recent advances in the design of the synthetic promoter

In recent years some efforts have been made to construct synthetic promoters for tissue specific transcription based on the linking of short oligonucleotide promoter and enhancer elements in a random (Li et al., 1999; Edelman et al., 2000) or ordered (Chow et al., 1997; Ramon et al., 2010) fashion.

In what can be described as one of the first attempts to rationally design a tissue-specific synthetic promoter, Chow et al. describe the rearrangement of the cytokeratin K18 locus to construct a promoter mediating a highly restrictive pattern of gene expression in the lung epithelium (Chow et al., 1997). In this study the authors describe the generation of transgenic mice with this construct and demonstrate expression only in the lung. They also generated CMV (Cytomegalovirus) and SV40 (Sarcoma Virus 40) promoter based constructs and found lack of specificity and no expression in the lung epithelia. This study had important implications for researchers developing lung-based gene therapies, i.e. if CMV, one of the most widely used promoters, could not regulate gene expression in the lung epithelia then it is necessary to identify (or develop) new promoters that can efficiently regulate gene expression in this location. Indeed, it is now becoming increasingly apparent that traditional virus-derived promoters like CMV and RSV (Rous Sarcoma Virus) will have limited application in the development of modern gene therapeutics.

The random assembly of *cis*-regulatory elements has shown particular success as a means to develop synthetic promoters. In one such approach, which aimed to identify synthetic promoters for muscle-specific expression, duplex oligonucleotides from the binding sites of muscle-specific and non-specific transcription factors were randomly ligated and cloned upstream of a minimal muscle promoter driving luciferase (Li et al. 1999). Approximately

1000 plasmid clones were individually tested by transient transfection into muscle cells and luciferase activity was determined in 96-well format by luminometry. By this approach several highly active and muscle specific promoters were identified that displayed comparable strength to the most commonly used viral promoters such as CMV.

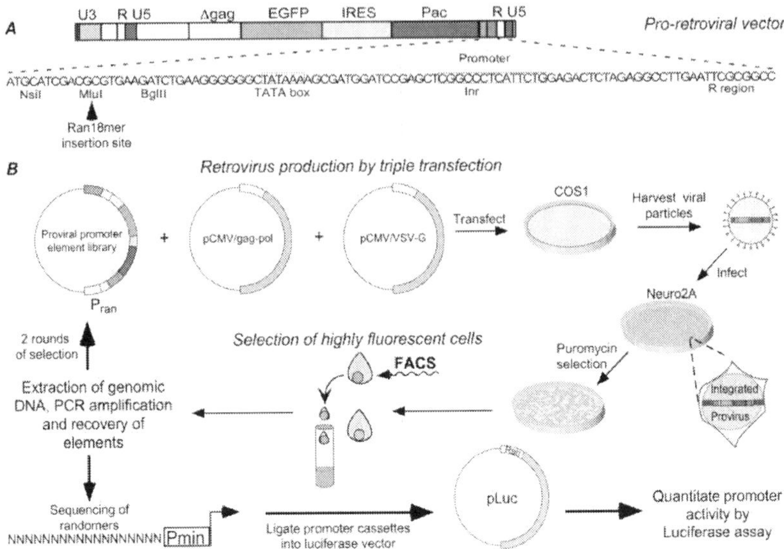

Fig. 2. Typical procedure for generation of synthetic mammalian promoters (reproduced from PNAS, Vol. 97, No. 7, pp. 3038-3043 copyright (c) 2000 by the National Academy of Sciences, USA)

Retroviral vectors have also been used to screen for synthetic promoters in eukaryotic cells (Edelman et al., 2000). This study was the first description of a retroviral library approach using antibiotic resistance and FACS selection to isolate promoter sequences (illustrated in figure 2). The libraries generated using random oligonucleotides in an effort to identify new sequences as well as examining the effects of combinations of known elements and for uncovering new transcriptional regulatory elements. After preparing a Ran18 promoter library comprises random 18mer oligonucleotides, the authors analysed the sequences of the generated synthetic promoters by searching for known transcription factor binding motifs. They found that the highest promoter activities were associated with an increased number of known motifs. They examined eight of the best known motifs; AP2, CEBp, gre, ebox, ets, creb, ap1 AND sP1/maz. Interestingly, several of the promoter sequences contained none of these motifs and the author's looked for new transcription factors.

In a similar effort employed to examine one million clones, Sutton and co-workers adopted the FACS screening approach based on the establishment of a lentiviral vector-based library (Dai et al., 2004). In this study duplex oligonucleotides from binding sites of endothelial cell-specific and non-specific transcription factors were cloned in a random manner upstream of a minimal promoter driving expression of eGFP in a HIV self-inactivating expression vector. A pool of one million clones was then transfected into endothelial cells and the highest expressers were selected by FACS sorting. Synthetic promoters were then rescued from stable transfectants by PCR from the genomic DNA where the HIV vectors had integrated. The results from this study also demonstrated the possibility of isolating several highly active endothelial cell-specific synthetic promoter elements from a random screen.

Synthetic promoters active only in the liver have also been developed (Lemken et al., 2005). In this study transcriptional units from ApoB and OTC genes were used in a controlled, non-random construction procedure to generate a series of multimeric synthetic promoters. Specifically, 2x, 4x, 8x and 16x repeats of the ApoB and OTC promoter elements were ligated together and promoter activity analysed. The results indicated that the promoter based on 4xApoB elements gave the optimal levels of gene expression and that 8x and 16x elements gave reduced levels of expression, thus demonstrating the limitations of simply ligating known promoter elements together in a repeat fashion to achieve enhanced expression.

When adopting this type of methodology in the design of synthetic tissue-specific promoters it is important to use well-designed duplex oligonucleotides. For example, each element has to be spaced in such a way that the regulatory elements appear on the same side of the DNA helix when reassembled, relevant minimal promoter elements have to be employed so that the screen produces promoters capable of expressing efficiently only in the tissue of interest and there must be some sort of mechanism, such as the addition of Sp1 sites, for the protection against promoter silencing through methylation.

In addition to tissue-specific promoters, cell-type synthetic promoters have also been developed. In one study, researchers designed a synthetic promoter to be active in nonadrenergic (NA) neurones (Hwang et al., 2001). They authors randomly ligated *cis*-regulatory elements that were identified from the human dopamine beta-hydroxylase (hDBH) gene and constructed promoters with up to 50-fold higher activity than the original promoter. Specifically, two elements from the promoter were used to generate a multimeric synthetic promoter; PRS1 and PRS2 which are bound to by the Phox2a transcription factor. The results demonstrated that the PRS2 was responsible for higher levels of gene expression as it had higher affinity to Phox-2a. It was also found that eight copies of PRS2 in the same orientation yielded maximum activity.

In a similar type of study a synthetic promoter was constructed that was specifically active in myeloid cells (He et al., 2006). The promoter comprised myeloid-specific elements for PU.1, C/EBPalpha, AML-1 and myeloid-associated elements for Sp1 and AP-1, which were randomly inserted upstream of the p47-phox minimal promoter. Synthetic promoters constructed showed very high activity. Haematopoietic Stem Cells (HSC) were initially transduced then the expression in differentiated cells was examined; only myeloid cells were found to express the reporter construct. To test therapeutic applicability of these promoters apoE-/- mice were transplanted with HSC transduced with a lentiviral vector expressing apoE from CMV and synthetic promoters. Even though transduced cells containing CMV and synthetic promoters both corrected the artherosclerotic phenotype, the cells derived from lentiviral vectors harbouring the synthetic promoter did so with less variability. Thus highlighting the improved safety features when using synthetic promoters for gene therapy applications.

In addition to tissue- and cell type-specific constitutive promoters, inducible synthetic promoters can also be constructed. One group describe a synthetic promoter constructed by placing the EPO enhancer region upstream of the SV40 promoter. The result is a strong promoter that is active only under ischaemic conditions. The authors tested this promoter by developing Neural Stem Cells (NSC) responsive to hypoxia and proposed that this system could be used to deliver therapeutic stem cells to treat ischaemic events. The authors were able to demonstrate that transplantation of NSC modified with a hypoxia-sensitive synthetic promoter resulted in specific expression of the luciferase reporter gene in response to ischaemic events *in vivo* (Liu et al., 2010).

4. Applications of synthetic promoter technology

Synthetic promoters have direct applications in large-scale industrial processes where enzymatic pathways are used in the production of biological and chemical-based products (reviewed in Hammer et al., 2006). One of the most important limitations in industrial-scale processes that synthetic promoter technology addresses is the inherent genetic instability in synthetically engineered biological systems. For instance, in prokaryotic organisms designed to express two or more enzymes, mutations will invariably arise in very few generations resulting in the termination of gene expression. This is because there is the lack of evolutionary pressure keeping all the components intact. The result is that mutations accrue over generations resulting in the deactivation of the circuit. Homologous recombination in natural promoters driving high levels of gene expression is the main reason why this circuitry fails (Sleight et al., 2010). Therefore, the use of synthetic promoters in these systems should serve to lower gene expression to result in more genetic stability, allow the avoidance of repeat sequences to prevent recombination and allow the use inducible promoters (a feature that also reduces genetic instability). In summary, the use of synthetic promoter technology in complex genetically engineered synthetic organisms expressing a variety of components should serve to increase genetic stability and improve the efficiency of the processes that the components control.

One interesting therapeutic application for synthetic promoter technology that has been described is the generation of a class of replication-competent viruses that enable tumour cell-specific killing by specifically replicating in cancer cells. In this study a replication competent retrovirus was developed to selectively kill tumour cells (Logg et al., 2002). The authors added a level of transcriptional targeting by incorporating the prostate-specific probasin (PB) promoter into the retroviral LTR and designed more efficient synthetic promoters based on the PB promoter to increase the efficiency of retroviral replication in prostate cancer cells. The result was a retrovirus that could efficiency transduce and replicate only in cancer cells. This is an attractive therapeutic strategy for the treatment of cancer, as tumour virotherapy has actually been examined as a potential therapeutic strategy for several decades.

Synthetic promoters that are active only in cycling endothelial cells would be another attractive tool for the development of cancer gene therapies. The rationale being that by targeting new blood vessels growing into tumours we would be able to develop a cancer gene therapy that could cut off supply of nutrients to the growing cancer. In a study that adopted this approach the cdc6 gene promoter was identified as a candidate promoter active only in cycling cells and was coupled to the endothelin enhancer element to construct a promoter active in dividing endothelial cells (Szymanski et al., 2006). Four endothelin elements conjugated to the cdc6 promoter gave the optimal results *in vitro*. When introduced into tumour models *in vivo*, the synthetic promoter was more efficient at driving gene expression in cancerous tissues, when compared to a CMV promoter.

Perhaps one of the most impressive applications of synthetic promoter technology thus far was the development of a liver-specific promoter that could be used to essentially cure diabetes in a transgenic mouse model (Han et al., 2010). In this study a synthetic promoter active in liver cells in response to insulin was constructed. The authors designed 3-, 6- & 9-element promoters based on random combinations of HNF-1, E/EBP and GIRE *cis*-elements. In the 3-element promoters all 27 combinations of the three were tested and the highest activity promoters were used to generate the 6-element promoter and so on. Using this technique promoters with activity up to 25% of CMV were identified. Finally, the optimal promoter was chosen depending on its responsiveness to glucose. This promoter showed highest specificity to liver cells and in response to Glucose and yielded expression levels 21% that of CMV. Adenoviral vectors containing this promoter driving expression of insulin were injected into a mouse diabetic model. Injection with the highest dose of virus resulted in protection against hyperglycaemia for 50 days. Importantly, injection with adenovirus expressing insulin from a CMV promoter resulted in death of the animals due to hypoglycaemia, thus illustrating the importance of regulated expression in gene therapy. Importantly, the results from this study excellently illustrated why the clever design of synthetic promoters controlling restricted gene expression will be essential in the development of safe gene therapy.

Synthetic promoters are increasingly being used in gene therapy type of studies. In one recent study their potential application to the gene therapy of Chronic Granulomatous Disease (X-CGD; an X-linked disorder resulting from mutations in gp91-phox, whose activity in myeloid cells is important in mounting an effective immune response) was examined (Santilli et al., 2011). The authors cite a clinical trial using a retroviral vector, which was successful at correcting the phenotype, but expression was short-lived due to promoter inactivation. In order to address this issue a chimeric promoter was constructed that was a fusion of Cathepsin G and c-Fes minimal promoter sequences, which are specifically active in cells of the myeloid lineage. This promoter was used to drive the expression of gp91-phox in myeloid cells in mice using a SIN lentiviral vector and the results show effective restricted expression to monocytes and subsequent introduction of gp91 results in high levels of expression in target cells and restoration of wild type phenotype *in vitro*. X-CGD cells were then transduced with the lentiviral vector and grafted into a mouse model of CGD. The vector was able to sustain long-term expression of gp91-phox, resulting in levels of expression that could correct the phenotype. Expression was specifically seen in granulocytes and monocytes, and not B- and T-cells.

These studies serve to highlight the potential application of synthetic promoter technology in gene therapy. They particularly highlight the importance of achieving cell-type specific gene expression and address the common issue of promoter shutdown that is seen when using stronger viral promoters like those derived from the CMV and RSV. If gene therapy is to be a success in the clinic it will be imperative to develop promoters that are highly specific and which display a restrictive and predictable expression profile. Thus, synthetic promoter technology represents the ideal solution to achieve this goal and its use is likely to become an increasingly popular approach adopted by researchers developing gene therapeutics.

5. Bioinformatic tools and synthetic promoter development

We first described how functional genomics experimentation and bioinformatics tools could be applied in the design of synthetic promoters for therapeutic and diagnostic applications several years ago (Roberts, 2007). Since then a number of scientists have also realised that this approach can be broadly applied across the biotech industry (Venter et al., 2007). In this section we discuss some of the tools that we use to analyse data obtained from large-scale gene expression analyses, which is subsequently used in the smart design of synthetic promoters conveying highly-specific regulation of gene expression.

To design a synthetic promoter it is essential to identify an appropriate number of *cis*-regulatory elements that can specifically bind to the *trans* factors that enhance gene transcription. This is where the importance of a number of bioinformatic algorithms becomes apparent. Over the past several years a number of databases and programs have been developed in order to identify transcription factor biding sites (TFBSs) on a variety of genomes. Below we introduce the most extensively used resources and discuss their application to the design of synthetic promoters, we pay particular attention to the identification of transcription networks active in cancer and how this information can be used to design cancer-specific promoters that can be used in the design of safer and more effective tumour-targeted gene therapies.

There is now a growing trend for researchers to analyse microarray data in terms of 'gene modules' instead of the presentation of differentially regulated gene lists. By grouping genes into functionally related modules it is possible to identify subtle changes in gene expression that may be biologically (if not statistically significantly) important, to more easily interpret molecular pathways that mediate a particular response and to compare many different microarray experiments from different disease states in an effort to uncover the commonalities and differences in multiple clinical conditions. Therefore, we are moving into a new era of functional genomics, where the large datasets generated by the evaluation of global gene expression studies can be more fully interpreted by improvements in computational methods. The advances in functional genomics made in recent years have resulted in the identification of many more *cis*-regulatory elements that can be directly related to the increased transcription of specific genes. Indeed, the ability to use bioinformatics to unravel complex transcriptional pathways active in diseased cells can actually serve to facilitate the process of choosing suitable *cis*-elements that can be used to design synthetic promoters specifically active in complex pathologies such as cancer.

In cancer the changes in the gene expression profile are often the result of alterations in the cell's transcription machinery induced by aberrant activation of signalling pathways that control growth, proliferation and migration. Such changes result in the activation of transcription regulatory networks that are not found in normal cells and provide us with an opportunity to design synthetic promoters that should only be active in cancerous cells. If microarray technology is to truly result in the design of tailored therapies to individual cancers or even patients, as has been heralded, it is important that the functional genomics methodology that was designed for the identification of signalling and transcription networks be applied to the design of cancer-specific promoters so that effective gene therapeutic strategies can be formulated (Roberts & Kottaridis, 2007). The development of bioinformatics algorithms for the analysis of microarray datasets has largely been applied in order to unravel the transcription networks operative under different disease and environmental conditions. To this date there has been no effort to use this type of approach to design synthetic promoters that are operative only under these certain disease or environmental conditions.

The regulation of gene expression in eukaryotes is highly complex and often occurs through the coordinated action of multiple transcription factors. The use of *trans*-factor combinations in the control of gene expression allows a cell to employ a relatively small number of transcription factors in the regulation of disparate biological processes. As discussed herein, a number of tools have been developed that allow us to utilise microarray data to identify novel *cis*-regulatory elements. It is also possible to use this information to decipher the transcriptional networks that are active in cells under different environmental conditions. In yeast, the importance of the combinatorial nature of transcriptional regulation was established by specifically examining clusters of upregulated genes for the presence of combinations of *cis*-elements. By examining microarray data from yeast exposed to a variety of conditions the authors were able to construct a network of transcription revealing the functional associations between different regulatory elements. This approach resulted in the identification of key motifs with many interactions, suggesting that some factors serve as facilitator proteins assisting their gene-specific partners in their function. The idea that a core number of transcription factors mediate such a vast array of biological responses by adopting multiple configurations implies that it may be possible to hijack the transcriptional programs that have gone awry in multifactorial diseases in an effort to develop disease-specific regulatory elements. For instance, the meta-analyses of cancer datasets has permitted the identification of gene modules, allowing for the reduction of complex cancer signatures to small numbers of activated transcription programs and even to the identification of common programs that are active in most types of cancer. This type of analysis can also help to identify specific transcription factors whose deregulation plays a key role in tumour development. In one such study, the importance of aberrant E2F activity in cancer was reaffirmed during a search for the regulatory programs linking transcription factors to the target genes found upregulated in specific cancer types (Rhodes et al., 2005). It was shown that E2F target genes were disproportionately upregulated in more than half of the gene expression profiles examined, which were obtained from a multitude of different cancer types. It was thus proposed that integrative bioinformatics analyses have the potential to generate new hypotheses about cancer progression.

Different bioinformatics tools, examples of which are given in table 1, may be used to screen for *cis*-regulatory elements. In general, such tools function by comparing gene expression profiles between differentially regulated genes and examining upstream sequences, available through genome sequence resources. For the phylogenetic footprinting tools, the untranslated regions of specific genes are compared between species and the most highly conserved sequences are returned and proposed to be potential *cis*-elements. A combination of all available approaches may be employed in order to identify regulatory sequences that predominate in the profile of specific cell or tissue types. The most common sequences identified are then used as the building blocks employed in the design of synthetic promoters.

The ability to use gene expression data to identify gene modules, which mediate specific responses to environmental stimuli (or to a diseased state) and to correlate their regulation to the *cis*-regulatory elements present upstream of the genes in each module, has transformed the way in which we interpret microarray data. For instance, by using the modular approach it is possible to examine whether particular gene modules are active in a variety of different cancers, or whether individual cancers require the function of unique gene modules. This has allowed us to look for transcriptional commonalities between different cancers, which should aid in the design of widely applicable anti-cancer therapeutic strategies. In one early study, gene expression data from 1975 microarrays, spanning 22 different cancers was used to identify gene modules that were activated or deactivated in specific types of cancer (Segal et al., 2004). Using this approach the authors found that a bone osteoblastic module was active in a number of cancers whose primary metastatic site is known to be the bone. Thus, a common mechanism of bone metastasis between varieties of different cancers was identified, which could be targeted in the development of novel anticancer therapies.

It is also possible to identify the higher-level regulator that controls the expression of the genes in each module (Segal et al., 2003). Examination of the upstream regulatory sequences of each gene in a module may reveal the presence of common *cis*-regulatory elements that are known to be the target of the module's regulator. Therefore, by identifying specific regulatory proteins that control the activation of gene modules in different cancers, it should be possible to extrapolate the important *cis*-elements that mediate transcription in the transformed cell. Thereby, allowing us to design and construct novel tumour-specific promoters based on the most active *cis*-regulatory elements in a number of tumour-specific gene modules. The ability to identify specific transcriptional elements in the human genome that control the expression of functionally related genes is transforming the application of functional genomics. Until recently the interpretation of data from microarray analysis has been limited to the identification of genes whose function may be important in a single pathway or response. How this related to global changes in the cellular phenotype had been largely ignored, as the necessary tools to examine this simply did not exist. With the advancement of bioinformatics we are now in a position to utilise all the data that is obtained from large-scale gene expression analysis and combine it with knowledge of the completed sequence of the human genome and with transcription factor, gene ontology and molecular function databases, thereby more fully utilising the large datasets that are generated by global gene expression studies.

For nearly two decades scientists have been compiling databases that catalogue the *trans*-factors and *cis*-elements that are responsible for gene regulation (Wingender et al., 1988). This has primarily been done in an effort to elucidate the various transcription programs that are activated in response to different biological stimuli in a range of organisms. The result is the emergence of useful tools that can be used to identify transcription factors and their corresponding *cis*-regulatory sequences that are useful in the design of synthetic promoters. In the remaining part of this chapter we briefly discuss each resource, indicating the unique aspect of its functionality.

TRANSFAC is perhaps the most comprehensive TFBS database available and indexes transcription factors and their target sequences based solely on experimental data (Matys et al., 2003). It is maintained as a relational database, from which public releases are made available via the web. The release consists of six flat files. At the core of the database is the interaction of transcription factors (FACTOR) with their DNA-binding sites (SITE) through which they regulate their target genes (GENE). Apart from genomic sites, 'artificial' sites which are synthesized in the laboratory without any known connection to a gene, e.g., random oligonucleotides, and IUPAC consensus sequences are also stored in the SITE table. Sites must be experimentally proven for their inclusion in the database. Experimental evidence for the interaction with a factor is given in the SITE entry in form of the method that was used (gel shift, footprinting analysis, etc.) and the cell from which the factor was derived (factor source). The latter contains a link to the respective entry in the CELL table. On the basis of those, method and cell, a quality value is given to describe the 'confidence' with which an observed DNA- binding activity could be assigned to a specific factor. From a collection of binding sites for a factor nucleotide weight matrices are derived (MATRIX). These matrices are used by the tool Match™ to find potential binding sites in uncharacterized sequences, while the program Patch™ uses the single site sequences, which are stored in the SITE table. According to their DNA-binding domain transcription factors are assigned to a certain class (CLASS). In addition to the more 'planar' CLASS table a hierarchical factor classification system is also used.

TRANSCompel® originates from COMPEL, and functions to emphasize the key role of specific interactions between transcription factors binding to their target *cis*-regulatory elements; whilst providing specific features of gene regulation in a particular cellular content (Kel-Margoulis et al., 2002). Information about the structure of known *trans* factor and *cis* sequence interactions, and specific gene regulation achieved through these interactions, is extremely useful for promoter prediction. In the TRANSCompel database, each entry corresponds to an individual *trans/cis* interaction within the context of a particular gene and thus contains information about two binding sites, two corresponding transcription factors and experiments confirming cooperative action between transcription factors.

ABS is a public database of known *cis*-regulatory binding sites identified in promoters of orthologous vertebrate genes that have been manually collated from the scientific literature (Blanco et al., 2006). In this database some 650 experimental binding sites from 68 transcription factors and 100 orthologous target genes in human, mouse, rat or chicken genome sequences have been documented. This tool allows computational predictions and promoter alignment information for each entry and is accessed through a simple and easy-to-use web interface; facilitating data retrieval and allowing different views of the information. One of the key features of this software is the inclusion of a customizable generator of artificial datasets based on the known sites contained in the whole collection and an evaluation tool to aid during the training and the assessment of various motif-finding programs.

JASPAR is an open-access database of annotated, high-quality, matrix-based TFBS profiles for multi-cellular eukaryotic organisms (Sandelin et al., 2004). The profiles were derived exclusively from sets of nucleotide sequences that were experimentally demonstrated to bind transcription factors. The database is accessible via a web-interface for browsing, searching and subset selection. The interface also includes an online sequence analysis utility and a suite of tools for genome-wide and comparative genome analysis of regulatory regions.

HTPSELEX is a public database providing access to primary and derived data from high-throughput SELEX experiments that were specifically designed in order to characterize the binding specificity of transcription factors (Jagannathan et al., 2006). The resource is primarily intended to serve computational biologists interested in building models of TFBSs from large sets of *cis*-regulatory sequences. For each experiment detailed in the database accurate information is provided about the protein material used, details of the wet lab protocol, an archive of sequencing trace files, assembled clone sequences and complete sets of *in vitro* selected protein-binding tags.

TRED is a database that stores both *cis*- and *trans*-regulatory elements and was designed to facilitate easy data access and to allow for the analysis of single-gene-based and genome-scale studies (Zhao et al., 2005). Distinguishing features of *TRED* include: relatively complete genome-wide promoter annotation for human, mouse and rat; availability of gene transcriptional regulation information including TFBSs and experimental evidence; data accuracy is ensured by hand curation; efficient user interface for easy and flexible data retrieval; and implementation of on-the-fly sequence analysis tools. *TRED* can provide good training datasets for further genome-wide *cis*-regulatory element prediction and annotation; assist detailed functional studies and facilitate the deciphering of gene regulatory networks.

Databases of known TFBSs can be used to detect the presence of protein-recognition elements in a given promoter, but only when the binding site of the relevant DNA-binding protein and its tolerance to mismatches *in vivo* is already known. Because this knowledge is currently limited to a small subset of transcription factors, much effort has been devoted to the discovery of regulatory motifs by comparative analysis of the DNA sequences of promoters. By finding conserved regions between multiple promoters, motifs can be identified with no prior knowledge of TFBSs. A number of models have emerged that achieve this by statistical overrepresentation. These algorithms function by aligning multiple untranslated regions from the entire genome and identifying sequences that are statistically significantly overrepresented in comparison to what it expected by random.

YMF is a program developed to identify novel TFBSs (not necessarily associated with a specific factor) in yeast by searching for statistically overrepresented motifs (Sinha et al., 2003; Sinha & Tompa, 2002). More specifically, *YMF* enumerates all motifs in the search space and is guaranteed to produce those motifs with the greatest z-scores.

SCORE is a computational method for identifying transcriptional *cis*-regulatory modules based on the observation that they often contain, in statistically improbable concentrations, multiple binding sites for the same transcription factor (Rebeiz et al., 2002). Using this method the authors conducted a genome-wide inventory of predicted binding sites for the Notch-regulated transcription factor Suppressor of Hairless, Su(H), in drosophila and found that the fly genome contains highly non-random clusters of Su(H) sites over a broad range of sequence intervals. They found that the most statistically significant clusters were very heavily enriched in both known and logical targets of Su(H) binding and regulation. The utility of the *SCORE* approach was validated by *in vivo* experiments showing that proper expression of the novel gene *Him* in adult muscle precursor cells depends both on Su(H) gene activity and sequences that include a previously unstudied cluster of four Su(H) sites, indicating that *Him* is a likely direct target of Su(H).

At present these tools are mainly applied in the study of lower eukaryotes where the genome is less complex and regulatory elements are easier to identify, extending these algorithms to the human genome has proven somewhat more difficult. In order to redress this issue a number of groups have shown that it is possible to mine the genome of higher eukaryotes by searching for conserved regulatory elements adjacent to transcription start site motifs such as TATA and CAAT boxes, e.g. as catalogued in the *DBTSS* resource (Suzuki et al. 2004; Suzuki et al., 2002), or one can search for putative *cis*-elements in CpG rich regions that are present in higher proportions in promoter sequences (Davuluri et al., 2001). Alternatively, with the co-emergence of microarray technology and the complete sequence of the human genome, it is now possible to search for potential TFBSs by comparing the upstream non-coding regions of multiple genes that show similar expression profiles under certain conditions. Gene sets for comparative analysis can be chosen based on clustering, e.g. hierarchical and k-means (Roth et al., 1998), from simple expression ratio (Bussemaker et al., 2001) or functional analysis of gene products (Jensen et al., 2000). This provides scientists with the opportunity to identify promoter elements that are responsive to certain environmental conditions, or those that play a key role in mediating the differentiation of certain tissues or those that may be particularly active in mediating pathologic phenotypes.

Phylogenetic footprinting, or comparative genomics, is now being applied to identify novel promoter elements by comparing the evolutionary conserved untranslated elements proximal to known genes from a variety of organisms. The availability of genome sequences between species has notably advanced comparative genomics and the understanding of evolutionary biology in general. The neutral theory of molecular evolution provides a framework for the identification of DNA sequences in genomes of different species. Its central hypothesis is that the vast majority of mutations in the genome are neutral with respect to the fitness of an organism. Whilst deleterious mutations are rapidly removed by selection, neutral mutations persist and follow a stochastic process of genetic drift through a population. Therefore, non-neutral DNA sequences (functional DNA sequences) must be conserved during evolution, whereas neutral mutations accumulate. Initial studies sufficiently demonstrated that the human genome could be adequately compared to the genomes of other organisms allowing for the efficient identification of homologous regions in functional DNA sequences.

Subsequently, a number of bioinformatics tools have emerged that operate by comparing non-coding regulatory sequences between the genomes of various organisms to enable the identification of conserved TFBSs that are significantly enriched in promoters of candidate genes or from clusters identified by microarray analysis; examples of these software suites are discussed below. Typically these tools work by aligning the upstream sequences of target genes between species thus identifying conserved regions that could potentially function as *cis*-regulatory elements and have consequently been applied in the elucidation of transcription regulatory networks in a variety of models.

TRAFAC is a Web-based application for analysis and display of a pair of DNA sequences with an emphasis on the detection of conserved TFBSs (Jegga et al., 2002). A number of programs are used to analyze the sequences and identify various genomic features (for example, exons, repeats, conserved regions, TFBSs). Repeat elements are masked out using *RepeatMasker* and the sequences are aligned using the *PipMaker-BLASTZ* algorithm. *MatInspector Professional* or *Match* (BioBase) is run to scan the sequences for TFBSs. *TRAFAC* then integrates analysis results from these applications and generates graphical outputs; termed the *Regulogram* and *Trafacgram.*

CORG comprises a catalogue of conserved non-coding sequence blocks that were initially computed based on statistically significant local suboptimal alignments of 15kb regions upstream of the translation start sites of some 10793 pairs of orthologous genes (Dieterich et al., 2003). The resulting conserved non-coding blocks were annotated with EST matches for easier detection of non-coding mRNA and with hits to known TFBSs. CORG data are accessible from the ENSEMBL web site via a DAS service as well as a specially developed web service for query and interactive visualization of the conserved blocks and their annotation.

CONSITE is a flexible suite of methods for the identification and visualization of conserved TFBSs (Lenhard et al., 2003). The system reports those putative TFBSs that are both situated in conserved regions and located as pairs of sites in equivalent positions in alignments between two orthologous sequences. An underlying collection of metazoan transcription-factor-binding profiles was assembled to facilitate the study. This approach results in a significant improvement in the detection of TFBSs because of an increased signal-to-noise ratio, as as the authors demonstrated with two sets of promoter sequences.

CONFAC enables the high-throughput identification of conserved TFBSs in the regulatory regions of hundreds of genes at a time (Karanam et al., (2004). The *CONFAC* software compares non-coding regulatory sequences between human and mouse genomes to enable identification of conserved TFBSs that are significantly enriched in promoters of gene clusters from microarray analyses compared to sets of unchanging control genes using a Mann–Whitney statistical test. The authors analysed random gene sets and demonstrated that using this approach, over 98% of TFBSs had false positive rates below 5%. As a proof-of-principle, the *CONFAC* software was validated using gene sets from four separate microarray studies and TFBSs were identified that are known to be functionally important for regulation of each of the four gene sets.

VAMP is a graphical user interface for both visualization and primary level analysis of molecular profiles obtained from functional genomics experimentation (La Rosa et al., 2006). It can be applied to datasets generated from Comparative Genomic Hybridisation (CGH) arrays, transcriptome arrays, Single Nucleotide Polymorphism arrays, loss of heterozygosity analysis (LOH), and Chromatin Immunoprecipitation experiments (ChIP-on-chip). The interface allows users to collate the results from these different types of studies and to view it in a convenient way. Several views are available, such as the classical CGH karyotype view or genome-wide multi-tumour comparisons. Many of the functionalities required for the analysis of CGH data are provided by the interface; including searches for recurrent regions of alterations, comparison to transcriptome data, correlation to clinical information, and the ability to view gene clusters in the context of genome structure.

CisMols Analyser allows for the filtering of candidate *cis*-element clusters based on phylogenetic conservation across multiple gene sets (Jegga et al., 2005). It was previously possible to achieve this for individual orthologue gene pairs, but combining data from *cis*-conservation and coordinate expression across multiple genes proved a more difficult task. To address this issue, the authors extended an orthologue gene pair database with additional analytical architecture to allow for the analysis and identification of maximal numbers of compositionally similar and phylogenetically conserved *cis*-regulatory element clusters from a list of user-selected genes. The system has been successfully tested with a series of functionally related and microarray profile-based co-expressed orthologue pairs of promoters and genes using known regulatory regions as training sets and co-expressed genes in the olfactory and immunohematologic systems as test sets. A significant amount of effort has been dedicated to the cataloguing of transcription factors and their corresponding *cis*-elements. More recently, these databases have been compiled with the aim to utilise them to unravel regulatory networks active in response to diverse stimuli.

PreMod was developed in an effort to identify *cis*-regulatory modules (CRM) active under specific environmental conditions (Blanchette et al., 2006; Ferretti et al., 2007). Starting from a set of predicted binding sites for more than 200 transcription factor families documented in the Transfac database (described above), the authors describe an algorithm relying on the principle that *cis*-regulatory modules (CRMs) generally contain several phylogenetically conserved binding sites for a small variety of transcription factors. The method allowed the prediction of more than 118,000 CRMs within the human genome. During this analysis, it was revealed that CRM density varies widely across the genome, with CRM-rich regions often being located near genes encoding transcription factors involved in development. Interestingly, in addition to showing enrichment near the 3′ end of genes, predicted CRMs were present in other regions more distant from genes. In this database, the tendency for certain transcription factors to bind modules located in specific regions was documented with respect to their target genes, and a number of transcription factors likely to be involved in tissue-specific regulation were identified.

CisView was developed to facilitate the analysis of gene regulatory regions of the mouse genome (Sharov et al., 2006). Its user interface is a browser and database of genome-wide potential TFBSs that were identified using 134 position-weight matrices and 219 sequence patterns from various sources. The output is presented with information about sequence conservation, neighbouring genes and their structures, GO annotations, protein domains, DNA repeats and CpG islands. The authors used this tool to analyse the distribution of TFBSs and revealed that many TFBSs were over-represented near transcription start sites. In the initial paper presenting the tool they also identified potential *cis*-regulatory modules defined as clusters of conserved TFBSs in the entire mouse genome. Out of 739,074 CRMs identified, 157,442 had a significantly higher regulatory potential score than semi-random sequences. The *CisView* browser provides a user-friendly computer environment for studying transcription regulation on a whole-genome scale and can also be used for interpreting microarray experiments and identifying putative targets of transcription factors.

BEARR is web browser software designed to assist biologists in efficiently carrying out the analysis of microarray data from studies of specific transcription factors (Vega et al., 2004). Batch Extraction and Analysis of *cis*-Regulatory Regions, or *BEARR*, accepts gene identifier lists from microarray data analysis tools and facilitates identification, extraction and analysis of regulatory regions from the large amount of data that is typically generated in these types of studies.

VISTA is a family of computational tools that was built to assist in the comparative analysis of DNA sequences (Dubchak & Ryaboy, 2006). These tools allow for the alignment of DNA sequences to facilitate the visualization of conservation levels and thus allow for the identification of highly conserved regions between species. Specifically, sequences can be analysed by browsing through pre-computed whole-genome alignments of vertebrates and other groups of organisms. Submission of sequences to *Genome VISTA* enables the user to align them to other whole genomes; whereas submission of two or more sequences to *mVISTA* allows for direct alignment. Submission of sequences to *Regulatory VISTA* is also possible and enables the predication of potential TFBSs (based on conservation within sequence alignments). All *VISTA* tools use standard algorithms for visualization and conservation analysis to make comparison of results from different programs more straightforward.

PromAn is a modular web-based tool dedicated to promoter analysis that integrates a number of different complementary databases, methods and programs (Lardenois et al., 2006). *PromAn* provides automatic analysis of a genomic region with minimal prior knowledge of the genomic sequence. Prediction programs and experimental databases are combined to locate the transcription start site (TSS) and the promoter region within a large genomic input sequence. TFBSs can be predicted using several public databases and user-defined motifs. Also, a phylogenetic footprinting strategy, combining multiple alignments of large genomic sequences and assignment of various scores reflecting the evolutionary selection pressure, allows for evaluation and ranking of TFBS predictions. *PromAn* results can be displayed in an interactive graphical user interface. It integrates all of this information to highlight active promoter regions, to identify among the huge number of TFBS predictions those which are the most likely to be potentially functional and to facilitate user refined analysis. Such an integrative approach is essential in the face of a growing number of tools dedicated to promoter analysis in order to propose hypotheses to direct further experimental validations.

CRSD is a comprehensive web server that can be applied in investigating complex regulatory behaviours involving gene expression signatures, microRNA regulatory signatures and transcription factor regulatory signatures (Liu et al., 2006). Six well-known and large-scale databases, including the human *UniGene*, mature microRNAs, putative promoter, *TRANSFAC*, *pathway* and *Gene Ontology* (GO) databases, were integrated to provide the comprehensive analysis in *CRSD*. Two new genome-wide databases, of microRNA and transcription factor regulatory signatures were also constructed and further integrated into *CRSD*. To accomplish the microarray data analysis at one go, several methods, including microarray data pre-treatment, statistical and clustering analysis, iterative enrichment analysis and motif discovery, were closely integrated in the web server.

MPromDb is a database that integrates gene promoters with experimentally supported annotation of transcription start sites, *cis*-regulatory elements, CpG islands and chromatin immunoprecipitation microarray (ChIP-chip) experimental results within an intuitively designed interface (Sun et al., 2006). Its initial release contained information on 36,407 promoters and first exons, 3,739 TFBSs and 224 transcription factors; with links to *PubMed* and *GenBank* references. Target promoters of transcription factors that have been identified by ChIP-chip assay are also integrated into the database and thus serving as a portal for genome-wide promoter analysis of data generated by ChIP-chip experimental studies.

A comprehensive list of the all the databases described above with a summary of their features and a reference to the original citation are shown in table 1.

Each of the aforementioned databases can be used when searching for potential regulatory sequences for inclusion in the design of synthetic promoters. Indeed, these resources can be used in order to identify *cis*-regulatory elements that may play a role in the formation of a particular cellular phenotype, or those that may be important in driving differentiation in developing organs. Synpromics, an emerging synthetic biology company recently incorporated in the United Kingdom, has cleverly utilised these tools in developing a proprietary method of synthetic promoter production where identified elements are incorporated into the design of promoters that are able to specifically regulate gene expression in a particular cellular phenotype. This method harnesses a cell's gene expression profile in order to facilitate the design of highly specific and efficient promoters. The result is a range of promoters that are inducible, tissue (or cell)-specific, active in response to a particular pathogen, chemical or biological agent and even able to mediate gene expression only under certain pathological conditions, such as cancer. Indeed, Synpromics has successfully generated a range of synthetic promoters that specifically drive high levels of gene expression in colorectal cancer and are looking to apply these promoters in the development of safer gene therapies (*manuscript in preparation*).

Resource	Description	Citation
DBTSS	Database of transcriptional start sites	Suzuki et al., (2002)
TRAFAC	Conserved *cis*-element search tool	Jegga et al., (2002)
TRANSCompel	Database of composite regulatory elements	Kel-Margoulis et al., (2002)
TRANSFAC	Eukaryotic transcription factor database	Matys et al., (2003)
Phylofoot	Tools for phylogenetic footprinting purposes	Lenhard et al., (2003)
CORG	Multi-species DNA comparison and annotation	Dieterich et al., (2003)
CONSITE	Explores *trans*-factor binding sites from two species	Lenhard et al., (2003)
CONFAC	Conserved TFBS finder	Karanam et al., (2004)
CisMols	Identifies *cis*-regulatory modules from inputed data	Jegga et al., (2005)
TRED	Catalogue of transcription regulatory elements	Zhao et al., (2005)
Oncomine	Repository and analysis of cancer microarray data	Rhodes et al., (2005)
ABS	Database of regulatory elements	Blanco et al., (2006)
JASPAR	Database of regulatory elements	Sandelin et al., (2004)
HTPSELEX	Database of composite regulatory elements	Jagannathan et al., (2006)
PReMod	Database of transcriptional regulatory modules in the human genome	Blanchette et al., (2006)
CisView	Browser of regulatory motifs and regions in the genome	Sharov et al., (2006)
BEARR	Batch extraction algorithm for microarray data analysis	Vega et al., 2004)
VISTA	Align and compare sequences from multiple species	Dubchak et al., (2006)
PromAn	Promoter analysis by integrating a variety of databases	Lardenois et al., (2006)

Table 1. Bioinformatics tools used for the identification of *cis*-regulatory elements

Importantly, synthetic promoters often mediate a level of gene expression with much greater efficiency than that seen with viral promoters, such as CMV, or compared to naturally occurring promoters within the genome. Given that the entire Biotech industry is centred on the regulation of gene expression, it is likely that synthetic promoters will eventually replace all naturally-occurring sequences in use today and help drive the growth of the synthetic biology sector in the coming decades.

6. Conclusion

In summary, synthetic promoters have emerged over the past two decades as excellent tools facilitating the identification of important structural features in naturally occurring

promoter sequences and allowing enhanced and more restrictive regulation of gene expression. A number of early studies revealed that it was possible to combine the *cis*-regulatory elements from promoters of a few tissue-specific genes and use these as building blocks to generate shorter, more efficient tissue-specific promoters. Several simple methodologies to achieve this emerged and have been applied in a multitude of organisms; including plant, bacteria, yeast, viral and mammalian systems.
Recent advances in bioinformatics and the emergence of a plethora of tools specifically designed at unravelling transcription programs has also facilitated the design of highly-specific synthetic promoters that can drive efficient gene expression in a tightly regulated manner. Changes in a cell's gene expression profile can be monitored and the transcription programs underpinning that change delineated and the corresponding *cis*-regulatory modules can be used to construct synthetic promoters whose activity is restricted to individual cell types, or to single cells subject to particular environmental conditions. This has allowed researchers to design promoters that are active in diseased cells or in tissues treated with a particular biological or chemical agent; or active in cells infected with distinct pathogens.
A number of institutions, such as Synpromics, have taken advantage of these advances and are now working to apply synthetic promoter technology to the enhanced production of biologics for use in biopharmaceutical, greentech and agricultural applications; the development of new gene therapies; and in the design of a novel class of molecular diagnostics. As the synthetic biology field continues to develop into a multi-billion dollar industry, synthetic promoter technology is likely to remain at the heart of this ever-expanding and exciting arena.

7. References

Blanchette, M., Bataille, A. R., Chen, X., Poitras, C., Laganière, J., Lefèbvre, C., et al. (2006). Genome-wide computational prediction of transcriptional regulatory modules reveals new insights into human gene expression. *Genome Research*, Vol. 16, No. 5, May 2006, pp. 656-668. ISSN 1088-9051

Blanco, E., Farré, D., Albà, M. M., Messeguer, X., & Guigó, R. (2006). ABS: a database of Annotated regulatory Binding Sites from orthologous promoters. *Nucleic Acids Research*, Vol. 34, January 2006, pp. D63-67. ISSN 0305-1048

Bussemaker, H. J., Li, H., & Siggia, E D. (2001). Regulatory element detection using correlation with expression. *Nature Genetics*, Vol. 27, No. 2, February 2001, pp. 167-71. ISSN 1061-4036

Chow, Y. H., O'Brodovich, H., Plumb, J., Wen, Y., Sohn, K. J., Lu, Z., et al. (1997). Development of an epithelium-specific expression cassette with human DNA regulatory elements for transgene expression in lung airways. *Proceedings of the National Academy of Sciences of the United States of America*, Vol. 94, No. 26, December 1997, pp. 14695-14700. ISSN 1091-6490

Dai, C., McAninch, R. E., & Sutton, R. E. (2004). Identification of synthetic endothelial cell-specific promoters by use of a high-throughput screen. *Journal of Virology*, Vol. 78, No. 12, June 2004, pp. 6209-6221. ISSN 0022-538X

Davuluri, R V, Grosse, I., & Zhang, M Q. (2001). Computational identification of promoters and first exons in the human genome. *Nature Genetics*, Vol. 29, No. 4, December 2001, pp. 412-417. ISSN 1061-4036

Dieterich, C., Wang, H., Rateitschak, K., Luz, H., & Vingron, M. (2003). CORG: a database for COmparative Regulatory Genomics. *Nucleic Acids Research*, Vol. 31, No. 1, January 2003, pp. 55-57. ISSN 0305-1048

Dubchak, I., & Ryaboy, D. V. (2006). VISTA family of computational tools for comparative analysis of DNA sequences and whole genomes. *Methods in Molecular Biology*, Vol. 338, April 2006, pp. 69-89. ISSN 1064-3745

Edelman, G. M., Meech, R., Owens, G. C., & Jones, F. S. (2000). Synthetic promoter elements obtained by nucleotide sequence variation and selection for activity. *Proceedings of the National Academy of Sciences of the United States of America*, Vol. 97, No. 7, March 2000, pp. 3038-3043. ISSN 1091-6490

Ferretti, V., Poitras, C., Bergeron, D., Coulombe, B., Robert, F., & Blanchette, M. (2007). PReMod: a database of genome-wide mammalian *cis*-regulatory module predictions. *Nucleic Acids Research*, Vol. 35, January 2007, pp. D122-126. ISSN 0305-1048

Gertz, J., & Cohen, B. A. (2009). Environment-specific combinatorial *cis*-regulation in synthetic promoters. *Molecular Systems Biology*, Vol. 5, February 2009, pp. 244. ISSN 1744-4292

Gertz, J., Siggia, Eric D, & Cohen, B. A. (2009). Analysis of combinatorial *cis*-regulation in synthetic and genomic promoters. *Nature*, Vol. 457, No. 7226, July 2009, pp. 215-218. ISSN: 0028-0836

Hammer, K., Mijakovic, I., & Jensen, P. R. (2006). Synthetic promoter libraries--tuning of gene expression. *Trends in Biotechnology*, Vol. 24, No. 2, February 2006, pp. 53-55. ISSN 0167-7799

Han, J., McLane, B., Kim, E.-H., Yoon, J.-W., & Jun, H.-S. (2010). Remission of Diabetes by Insulin Gene Therapy Using a Hepatocyte-specific and Glucose-responsive Synthetic Promoter. *Molecular Therapy*, Vol. 19, No. 3, March 2010, pp.470-478. ISSN 1525-0016

He, W., Qiang, M., Ma, W., Valente, A. J., Quinones, M. P., Wang, W., et al. (2006). Development of a synthetic promoter for macrophage gene therapy. *Human Gene Therapy*, Vol. 17, No. 9, September 2006, pp. 949-959. ISSN 1043-0342

Hunziker, A., Tuboly, C., Horváth, P., Krishna, S., & Semsey, S. (2010). Genetic flexibility of regulatory networks. *Proceedings of the National Academy of Sciences of the United States of America*, Vol. 107, No. 29, July 2010, pp. 12998-13003. ISSN 1091-6490

Hwang, D. Y., Carlezon, W. A., Isacson, O., & Kim, K. S. (2001). A high-efficiency synthetic promoter that drives transgene expression selectively in noradrenergic neurons. *Human Gene Therapy*, Vol. 12, No. 14, September 2001, pp.1731-1740. ISSN 1043-0342

Jacquet, M. A., Ehrlich, R., & Reiss, C. (1989). In vivo gene expression directed by synthetic promoter constructions restricted to the -10 and -35 consensus hexamers of E. coli. *Nucleic Acids Research*, Vol. 17, No. 8, April 1989, pp. 2933-2945. ISSN 0305-1048

Jagannathan, V., Roulet, E., Delorenzi, M., & Bucher, P. (2006). HTPSELEX--a database of high-throughput SELEX libraries for TFBSs. *Nucleic Acids Research*, Vol. 34, January 2006, pp. D90-94. ISSN 0305-1048

Jegga, A. G., Sherwood, S. P., Carman, J. W., Pinski, A. T., Phillips, J. L., Pestian, J. P., et al. (2002). Detection and visualization of compositionally similar *cis*-regulatory element clusters in orthologous and coordinately controlled genes. *Genome Research*, Vol. 12, No. 9, September 2002, pp. 1408-1417. ISSN 1088-9051

Jegga, A. G., Gupta, A., Gowrisankar, S., Deshmukh, M. A., Connolly, S., Finley, K., et al. (2005). CisMols Analyzer: identification of compositionally similar *cis*-element clusters in ortholog conserved regions of coordinately expressed genes. *Nucleic Acids Research*, Vol. 33, July 2005, pp. W408-411. ISSN 0305-1048

Jensen, L. J., & Knudsen, S. (2000). Automatic discovery of regulatory patterns in promoter regions based on whole cell expression data and functional annotation. *Bioinformatics*, Vol. 16, No. 4, April 2000, pp. 326-333. ISSN 1460-2059

Kamine, J., Subramanian, T., & Chinnadurai, G. (1991). Sp1-dependent activation of a synthetic promoter by human immunodeficiency virus type 1 Tat protein. *Proceedings of the National Academy of Sciences of the United States of America*, Vol. 88, No. 19, October 1991, pp. 8510-8514. ISSN 1091-6490

Karanam, S., & Moreno, C. S. (2004). CONFAC: automated application of comparative genomic promoter analysis to DNA microarray datasets. *Nucleic Acids Research*, Vol. 32, July 2004, pp. W475-484. ISSN 0305-1048

Kel-Margoulis, Olga V, Kel, Alexander E, Reuter, Ingmar, Deineko, I. V., & Wingender, Edgar. (2002). TRANSCompel: a database on composite regulatory elements in eukaryotic genes. *Nucleic Acids Research*, Vol. 30, No. 1, January 2002, pp. 332-334. ISSN 0305-1048

Kinkhabwala, A., & Guet, C. C. (2008). Uncovering *cis* regulatory codes using synthetic promoter shuffling. *PloS One*, Vol. 3, No. 4, April 2008, pp. e2030. ISSN 1932-6203

La Rosa, P., Viara, E., Hupé, P., Pierron, G., Liva, S., Neuvial, P., et al. (2006). VAMP: visualization and analysis of array-CGH, transcriptome and other molecular profiles. *Bioinformatics*, Vol. 22, No. 17, September 2006, pp. 2066-2073. ISSN 1460-2059

Lardenois, A., Chalmel, F., Bianchetti, L., Sahel, J.-A., Léveillard, T., & Poch, O. (2006). PromAn: an integrated knowledge-based web server dedicated to promoter analysis. *Nucleic Acids Research*, Vol. 34, July 2006, pp. W578-83. ISSN 0305-1048

Lemken, M.-L., Wybranietz, W.-A., Schmidt, U., Graepler, F., Armeanu, S., Bitzer, M., et al. (2005). Expression liver-directed genes by employing synthetic transcriptional control units. *World Journal of Gastroenterology*, Vol. 11, No. 34, September 2005, pp. 5295-5302. ISSN 1007-9327

Lenhard, B., Sandelin, A., Mendoza, L., Engström, P., Jareborg, N., & Wasserman, W. W. (2003). Identification of conserved regulatory elements by comparative genome analysis. *Journal of Biology*, Vol. 2, No. 2, May 2003, pp. 13. ISSN 1475-4924

Li, X., Eastman, E. M., Schwartz, R. J., & Draghia-Akli, R. (1999). Synthetic muscle promoters: activities exceeding naturally occurring regulatory sequences. *Nature Biotechnology*, Vol. 17, No. 3, January 1999, pp. 241-245. ISSN: 1087-0156

Liu, C.-C., Lin, C.-C., Chen, W.-S. E., Chen, H.-Y., Chang, P.-C., Chen, J. J. W., et al. (2006). CRSD: a comprehensive web server for composite regulatory signature discovery. *Nucleic Acids Research*, Vol. 34, July 2006, pp. W571-7. ISSN 0305-1048

Liu, M.-L., Oh, J. S., An, S. S., Pennant, W. A., Kim, H. J., Gwak, S.-J., et al. (2010). Controlled nonviral gene delivery and expression using stable neural stem cell line transfected with a hypoxia-inducible gene expression system. *The Journal of Gene Medicine*, Vol. 12, No. 12, December 2010, pp. 990-1001. ISSN 1521-2254

Logg, C. R., Logg, A., Matusik, R. J., Bochner, B. H., & Kasahara, N. (2002). Tissue-specific transcriptional targeting of a replication-competent retroviral vector. *Journal of Virology*, Vol. 76, No. 24, December 2002, pp. 12783-12791. ISSN 0022-538X

Mader, S., & White, J. H. (1993). A steroid-inducible promoter for the controlled overexpression of cloned genes in eukaryotic cells. *Proceedings of the National Academy of Sciences of the United States of America*, Vol. 90, No. 12, June 1993, pp. 5603-5607. ISSN 1091-6490

Masson, J. M., & Miller, J. H. (1986). Expression of synthetic suppressor tRNA genes under the control of a synthetic promoter. *Gene*, Vol. 47, No. 2-3, February 1986, pp.179-83. ISSN 0378-1119

Matys, V., Fricke, E., Geffers, R., Gössling, E., Haubrock, M., Hehl, R., et al. (2003). TRANSFAC: transcriptional regulation, from patterns to profiles. *Nucleic Acids Research*, Vol. 31, No. 1, January 2003, pp. 374-378. ISSN 0305-1048

Mogno, I., Vallania, F., Mitra, R. D., & Cohen, B. A. (2010). TATA is a modular component of synthetic promoters. *Genome Research*, Vol. 20, No. 10, October 2010, pp. 1391-1397. ISSN 1088-9051

Ramon, A., & Smith, H. O. (2010). Single-step linker-based combinatorial assembly of promoter and gene cassettes for pathway engineering. *Biotechnology Letters*, Vol 33, No. 3, March 2010, pp. 549-555. ISSN 0141-5492

Rebeiz, M., Reeves, N. L., & Posakony, J. W. (2002). SCORE: a computational approach to the identification of *cis*-regulatory modules and target genes in whole-genome sequence data. Site clustering over random expectation. *Proceedings of the National Academy of Sciences of the United States of America*, Vol. 99, No. 15, July 2002, pp. 9888-9893. ISSN 1091-6490

Roberts, M. L. (2007). A method for the construction of cancer-specific promoters using functional genomics. WIPO WO/2008/107725.

Roberts, M. L., & Kottaridis S. D. (2007). Interpreting microarray data: towards the complete bioinformatics toolkit for cancer. Cancer Genomics and Proteomics, Vol 4, No. 4, July-August 2007, pp 301-308. ISSN: 1109-6535.

Roth, F. P., Hughes, J. D., Estep, P. W., & Church, G. M. (1998). Finding DNA regulatory motifs within unaligned noncoding sequences clustered by whole-genome mRNA quantitation. *Nature Biotechnology*, Vol. 16, No. 10, October 1998, pp. 939-945. ISSN: 1087-0156

Sandelin, A., Alkema, W., Engström, P., Wasserman, W. W., & Lenhard, B. (2004). JASPAR: an open-access database for eukaryotic transcription factor binding profiles. *Nucleic Acids Research*, Vol. 32, January 2004, pp. D91-94. ISSN 0305-1048

Santilli, G., Almarza, E., Brendel, C., Choi, U., Beilin, C., Blundell, M. P., et al. (2011). Biochemical correction of X-CGD by a novel chimeric promoter regulating high levels of transgene expression in myeloid cells. *Molecular Therapy*, Vol. 19, No. 1, January 2011, pp. 122-132. ISSN 1525-0016

Segal, E., Friedman, N., Koller, D., & Regev, A. (2004). A module map showing conditional activity of expression modules in cancer. *Nature Genetics*, Vol. 36, No. 10, October 2004, pp. 1090-1098. ISSN 1061-4036

Segal, E., Shapira, M., Regev, A., Pe'er, D., Botstein, D., Koller, D., et al. (2003). Module networks: identifying regulatory modules and their condition-specific regulators from gene expression data. *Nature Genetics*, Vol. 34, No. 2, June 2003, pp. 166-176. ISSN 1061-4036

Sharov, A. A., Dudekula, D. B., & Ko, M. S. H. (2006). CisView: a browser and database of *cis*-regulatory modules predicted in the mouse genome. *DNA Research*, Vol. 13, No. 3, June 2006, pp. 123-134. ISSN 1340-2838

Sinha, S., & Tompa, M. (2002). Discovery of novel TFBSs by statistical overrepresentation. *Nucleic Acids Research*, Vol. 30, No. 24, December 2002, pp. 5549-5560. ISSN 0305-1048

Sinha, S., & Tompa, M. (2003). YMF: A program for discovery of novel TFBSs by statistical overrepresentation. *Nucleic Acids Research*, Vol. 31, No. 13, July 2003, pp. 3586-3588. ISSN 0305-1048

Sleight, S. C., Bartley, B. A., Lieviant, J. A., & Sauro, H. M. (2010). Designing and engineering evolutionary robust genetic circuits. *Journal of Biological Engineering*, Vol. 4, No. 1, November 2010, pp. 12. ISSN 1754-1611

Sun, H., Palaniswamy, S. K., Pohar, T. T., Jin, V. X., Huang, T. H.-M., & Davuluri, Ramana V. (2006). MPromDb: an integrated resource for annotation and visualization of mammalian gene promoters and ChIP-chip experimental data. *Nucleic Acids Research*, Vol. 34, January 2006, pp. D98-103. ISSN 0305-1048

Suzuki, Y., Yamashita, R., Nakai, K., & Sugano, S. (2002). DBTSS: DataBase of human Transcriptional Start Sites and full-length cDNAs. *Nucleic Acids Research*, Vol. 30, No. 1, January 2002, pp. 328-331. ISSN 0305-1048

Suzuki, Y., Yamashita, R., Sugano, S., & Nakai, K. (2004). DBTSS, DataBase of Transcriptional Start Sites: progress report 2004. *Nucleic Acids Research*, Vol. 32, January 2004, pp. D78-81. ISSN 0305-1048

Szymanski, P., Anwer, K., & Sullivan, S. M. (2006). Development and characterization of a synthetic promoter for selective expression in proliferating endothelial cells. *The Journal of Gene Medicine*, Vol. 8, No. 4, April 2006, pp. 514-523. ISSN 1521-2254

Trumble, W. R., Sherf, B. A., Reasoner, J. L., Seward, P. D., Denovan, B. A., Douthart, R. J., et al. (1992). Protein expression from an Escherichia coli/Bacillus subtilis multifunctional shuttle plasmid with synthetic promoter sequences. *Protein Expression and Purification*, Vol. 3, No. 3, June 1992, pp. 169-177. ISSN 1046-5928

Vega, V. B., Bangarusamy, D. K., Miller, L. D., Liu, E. T., & Lin, C.-Y. (2004). BEARR: Batch Extraction and Analysis of *cis*-Regulatory Regions. *Nucleic Acids Research*, Vol 32, July 2004, pp. W257-260. ISSN 0305-1048

Venter, M. (2007). Synthetic promoters: genetic control through *cis* engineering. *Trends in Plant Science*, Vol. 12, No. 3, March 2007, pp. 118-124. ISSN 1360-1385

Wingender, E. (1988). Compilation of transcription regulating proteins. *Nucleic Acids Research*, Vol. 16, No. 5, March 1988, pp. 1879-1902. ISSN 0305-1048

Zhao, F., Xuan, Z., Liu, L., & Zhang, Michael Q. (2005). TRED: a Transcriptional Regulatory Element Database and a platform for in silico gene regulation studies. *Nucleic Acids Research*, Vol. 33, January 2005, pp. D103-107. ISSN 0305-1048

21

Analysis of Transcriptomic and Proteomic Data in Immune-Mediated Diseases

Sergey Bruskin[1], Alex Ishkin[2], Yuri Nikolsky[2], Tatiana Nikolskaya[2] and Eleonora Piruzian[1]
[1]*Vavilov Institute of General Genetics, Russian Academy of Sciences, Moscow*
[2]*Thomson Reuters, Professional, Saint Joseph, MI*
[1]*Russia*
[2]*USA*

1. Introduction

Psoriasis (a skin disease) and Crohn's disease (a disease of the intestinal epithelium) are multifactorial diseases caused by abnormalities in genetic machinery regulation. Both pathologies disturb the immune system, and the pathological processes are triggered by environmental factors. In the case of psoriasis, these factors are psychoemotional stresses, infections (group A streptococci and Staphylococcus aureus), drugs (lithium-containing, antimalarial, and antituberculous agents and Novocain), smoking, and mechanical damages (the so-called Koebner phenomenon) [Bowcock A et al., 2004]. Psoriasis vulgaris is one of the most prevalent chronic inflammatory skin diseases affecting approximately 2% of individuals in Western societies, and found worldwide in all populations. Psoriasis is a complex disease affecting cellular, gene and protein levels and presented as skin lesions. The skin lesions are characterized by abnormal keratinocyte differentiation, hyperproliferation of keratinocytes, and infiltration of inflammatory cells [Boehncke WH et al. 1996; Ortonne JP, 1996]. The factors triggering Crohn's disease include psychoemotional stresses, infections (Mycobacterium avium ssp. paratuberculosis and invasive Escherichia coli variants), drugs (antibiotics and nonsteroid antiimflammatory agents), smoking, and nutritional regimen [Sartor R., 2006]. Crohn's disease known only since the 1920s [Crohn B et al., 1932] and now affecting up to 0.15% of the northwest European and North American population [Binder V., 2005].

Both psoriasis and Crohn's disease are now regarded as incurable, and the goal of their therapy is to extend the remission periods and decrease the severity of the disease. These two diseases are tightly related at the genetic level, as over five genetic loci are involved in the development of both psoriasis and Crohn's disease.

The mechanisms of both psoriasis and Crohn's disease are complex and involve genetic and environmental factors. As we gain more knowledge about molecular pathways implicated in diseases, novel therapies emerge (such as etanercept and infliximab that target TNF-α or CD11a- mediated pathways [Pastore S et al., 2008; Gisondi P et al., 2007]).

We have studied earlier the components of AP-1 transcription factor as psoriasis candidate genes. This study was performed by bioinformatics analysis of the transcription data using the GEO DataSets database (http://www.ncbi.nlm.nih.gov/geo/) [Piruzian ES et al., 2007].

The same approach was used by other researchers to detect potential therapeutic targets for psoriasis [Yao Y., et al., 2008]. In next step, we performed a comparative analysis of the molecular processes involved in the pathogenesis of two diseases, psoriasis and Crohn's disease [Piruzian ES et al., 2009].

Despite the fact that psoriasis and Crohn`s disease affect completely different body systems (skin and intestine), they are much more similar that it may seem at first glance. Both skin and intestinal epithelium are barrier organs, that are the first to resist the environmental factors, including microorganisms. Both pathologies are immune-mediated inflammatory diseases, that is also marked by the same drug therapies. Finally, they have a lot of common susceptibility loci (Fig. 1).

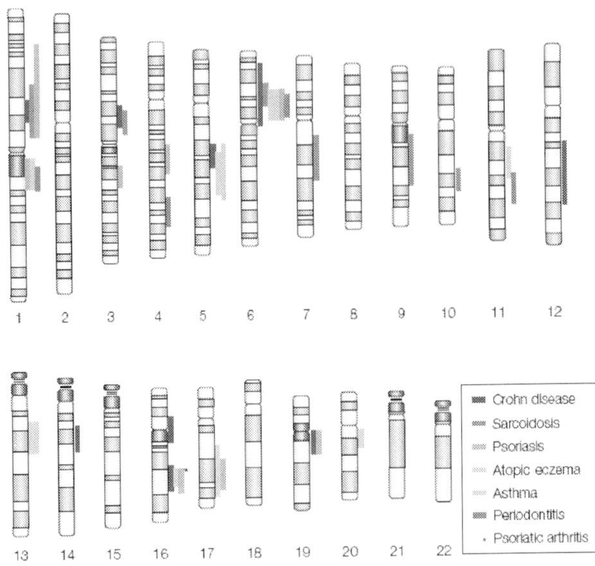

Fig. 1. Localization of various linkage regions for barrier diseases on human chromosomes map [Schreiber S et al., 2005])

In recent years, microarray mRNA expression profiling [Oestreicher JL et al., 2001; Bowcock AM et al., 2001; Zhou X et al., 2003; Quekenborn-Trinquet V et al., 2005] of lesional psoriatic skin revealed over 1,300 differentially expressed genes. Enrichment analysis (EA) showed that these genes encode proteins involved in regeneration, hyperkeratosis, metabolic function, immune response, and inflammation and revealed a number of modulating signaling pathways. These efforts may help to develop new-generation drugs. However, enrichment analysis limits our understanding of altered molecular interactions in psoriasis as it provides a relative ranking based on ontology terms resulting in the representation of fragmented and disconnected perturbed pathways. Furthermore, analysis of gene expression alone is not sufficient for understanding the whole variety of pathological changes at different levels of cellular organization. Indeed, new methodologies have been applied to the analysis of OMICs data in complex diseases that include algorithm-based biological network analysis [Nikolskaya T, et al., 2009; Nikolsky Y et al., 2005; Bhavnani SK et al., 2009; Ideker T et al., 2008; Chuang HY et al., 2007] and meta-analysis of multiple datasets of different types [Cox B et al., 2005; Wise LH et al., 1999; Ghosh D et al., 2003; Warnat P et al., 2005; Hack CJ, 2004; Menezes R et al., 2009]. Here, we applied several techniques of network and meta-analysis to reveal the similarities and differences between transcriptomics- and proteomics-level perturbations in psoriasis lesions. We particularly focused on revealing novel regulatory pathways playing a role in psoriasis development and progression.

2. Transcriptomic and proteomic data, network analysis

Data preparation. The data deposited with the public database of microarray experiments, GEO (http://www.ncbi.nlm.nih.gov/geo/), were analyzed. The expression data on psoriasis were contained in entry GDS1391, and on Crohn's disease, in entry GDS1330. Since these data were obtained using different microarrays and experimental schemes, analysis was individually performed for each disease with subsequent comparison of the lists of genes with altered expression for each case.

Two sets were selected from the overall data on psoriasis, namely, four experiments with gene expression in psoriatic skin lesions, and four, with gene expression in the healthy skin of the same patients. The selected data for Crohn's disease were also represented by two sets: 10 experiments on expression in intestinal epithelial lesions, and 11, on expression in the intestinal tissue of healthy individuals. The data were prepared for analysis using the GeneSpring GX (http://www.chem.agilent.com/scripts/pds.asp?lpage=27881) software package. This processing comprised discarding of the genes with poorly detectable expression and normalization of the remaining data. In addition to the values of expression, the so-called absent call flags were added for psoriasis cases; these flags characterize the significance of the difference in expression of a particular gene from the background noise. The genes displaying the flag value of A (i.e., absent, which means that the expression of a particular gene in experiment is undetectable) in over 50% of experiments were discarded from further analysis. This information was unavailable for Crohn's disease; therefore, this step was omitted. The results were normalized by the median gene expression in the corresponding experiment to make them comparable with one another.

Detection of the genes with altered expression. Differentially expressed genes were sought using Welch's t-test [Welch B.L., 1947]. This test does not require that the distribution variances for the compared samples be equal; therefore, it is more convenient for analyzing expression data than a simple t-test. FDR algorithm [Benjamini Y et al., 1995] with a significance threshold of 0.1 was used to control the type I errors in finding differentially expressed genes; in this case, the threshold determined the expected rate of false positive predictions in the final set of genes after statistical control.

Detection of common biological processes. The resulting gene lists were compared, and the molecular processes mediated by the genes displaying altered expressions in both diseases were sought using the MetaCore (GeneGo Inc., www.genego.com) program. The significance of the biological processes where the genes displaying altered expressions in both diseases was assessed according to the degree to which overlapping between the list of differentially expressed genes and the list of genes ascribed to the process exceeded random overlapping. Hypergeometric distribution [Draghici S et al., 2007] was used as a model of random overlapping between the gene lists. The measure of signifi- cance for the input gene list, the p value, in this distribution is calculated as

$$pVal(r,n,R,N) = \sum_{i=\max(r,R+n-N)}^{\min(n,R)} P(i,n,R,N)$$
$$= \frac{R!n!(N-R)!(N-n)!}{N!}$$
$$\times \sum_{i=\max(r,R+n-N)}^{\min(n,R)} \frac{1}{i!(R-i)!(n-i)!(N-R-n+i)!}'$$

where N is the number of genes in the MetaCore database; R, the number of genes ascribed to a particular process; n, the size of the input gene list; and r, the number of genes from the input list related to this process.

Three ontologies of biological processes were used in this work: GO (www.gene-ontology.org) and two ontologies included in the MetaCore, Canonical pathways and GeneGo process networks. The processes contained in the MetaCore ontologies are gene networks, which reflect the interaction of proteins involved in a particular biological regulatory or metabolic pathway. The processes for all three ontologies were prioritized by the negative logarithm of p value.

The common molecular biological pathways were determined based on the analysis of significant biological processes and expressions of the genes involved in these processes. The MetaCore contains the algorithms providing for detection in the total network of gene interactions the particular regulatory pathways and subnetworks saturated with the objects of research interest, in this case, the genes with altered expression. The resulting assumptions on the pattern of common biological pathways were visualized as a gene network using the MetaCore.

Skin biopsies. Acquisition of the human tissue was approved by the Vavilov Institute of General Genetics of Russian Academy of Sciences review board and the study was conducted after patient's consent and according to the Declaration of Helsinki Principles. A total of 6 paired nonlesional and lesional (all were plaque-type) skin biopsies from 3 psoriatic patients were profiled using 2D electrophoresis. All the donors who gave biopsy tissue (both healthy controls and individuals with psoriasis) provided a written informed consent for the tissue to be taken and used in this study. Clinical data for all patients are listed in Table 3. Full-thickness punch biopsies were taken from uninvolved skin (at least 2 cm distant from any psoriatic lesion; 6 mm diameter) and from the involved margin of a psoriatic plaque (6 mm diameter) from every patient.

Sample preparation, two-dimensional electrophoresis, gel image analysis and massspectrometry was carried out using the standard procedure [Gravel P & Golaz O, 1996; Mortz E, et al., 2001].

Microarray data analysis. We used recently published data set [Yao Y, et al., 2008] from GEO data base (http://www.ncbi.nlm.nih.gov/geo/; accession number GSE14095). We compared 28 pairs of samples (in each pair there was a sample of lesional skin and a sample of healthy skin from the same patient). Values for each sample were normalized by sample median value in order to unify distributions of expression signals. For assessment of differential expression we used paired Welch ttest with FDR correction [Benjamini Y et al.,

1995]. Probe set was considered as differentially expressed if its average fold change exceeded 2.5 and FDR corrected *p*-value was less than 0.01.

Overconnection analysis. All network-based analyses were conducted with MetaCore software suite http://www.genego.com. This software employs a dense and manually curated database of interactions between biological objects and variety of tools for functional analysis of high-throughput data. We defined a gene as overconnected with the gene set of interest if the corresponding node had more direct interactions with the nodes of interest than it would be expected by chance. Significance of overconnection was estimated using hypergeometric distribution with parameters r - number of interactions between examined node and the list of interest; R - degree of examined node, n - sum of interactions involving genes of interest and N - total number of interactions in the database:

$$pVal(r,n,R,N) = \sum_{i=\max(r,R+n-N)}^{\min(n,R)} P(i,n,R,N)$$

Hidden nodes analysis. In addition to direct interacting objects, we also used objects that may not interact directly with objects of interest but are important upstream regulators of those [Dezso Z et al., 2009]. The approach is generally the same as described above, but the shortest paths instead of direct links are taken into account. As we were interested in transcriptional regulation, we defined a transcriptional activation shortest path as the preferred shortest path from any object in the MetaCore database to the transcription factor target object from the data set. We added an additional condition to include the uneven number of inhibiting interactions in the path (that's required for the path to have activating effect). If the number of such paths containing examined gene and leading to one of objects of interest were higher than expected by chance, this gene was considered as significant hidden regulator. The significance of a node's importance was estimated using hypergeometric distribution with parameters r - number of shortest paths between containing currently examined gene; R - total number of shortest paths leading to a gene of interest through transcriptional factor, n - total number of transcription activation shortest paths containing examined gene and N - total number of transcription activation shortest paths in the database.

Rank aggregation. Both topology significance approaches produced lists of genes significantly linked to a gene or protein set of interest, ranked by corresponding p-values. To combine results of these two approaches, we used a weighted rank aggregation method described in [Pihur V et al., 2009]. Weighted Spearman distance was used as distance measure and the genetic algorithm was employed to select the optimal aggregated list of size 20. This part of work was accomplished in R 2.8.1 http://www.r-project.org.

Network analysis. In addition to topology analysis, we examined overexpressed genes and proteins using various algorithms for selecting connected biologically meaningful subnetworks enriched with objects of interest. Significance of enrichment is estimated using hypergeometric distribution. We first used an algorithm intended to find regulatory pathways that are presumably activated under pathological conditions. It defines a set of transcription factors that are directly regulating genes of interest and a set of receptors whose ligands are in the list of interest and then constructs series of networks; one for each receptor. Each network contains all shortest paths from a receptor to the selected transcriptional factors and their targets. This approach allows us to reveal the most important areas of regulatory machinery affected under the investigated pathological condition. Networks are sorted by enrichment p-value. The second applied algorithm used was aimed to define the most influential transcription factors. It considers a transcriptional factor from the data base and gradually expands the subnetwork around it until it reaches a predefined threshold size (we used networks of 50 nodes). Networks are sorted by enrichment p-value.

3. A comparative analysis of the molecular genetic processes in the pathogenesis of psoriasis and Crohn's disease

Constructing List of the Genes with Altered Expression in Both Pathologies We detected the lists of differentially expressed genes separately in each dataset and compared these lists at the system level. This approach to analysis was dictated by the properties of expression data in general (a high noise level and a large volume of analyzed data) and individual properties of datasets selected for analysis, which were obtained using different microarrays with different numbers of probes. That is why the datasets were incomparable in a direct fashion. The dataset on psoriasis initially contained information on the expression levels of 12626 probes from eight experiments (four specimens of skin lesions, and four of the healthy skin from the same patients). After discarding the probes with poorly detectable expression (see Materials and Methods), the set was reduced to 5076 probes. The list of the probes with statistically significant differences in expression between the lesion and healthy tissue contained 410 items at a significance level of 0.1.

The dataset on Crohn's disease contained information on the expression level of 24016 probes from 21 experiments (11 specimens of epithelial lesions and 10 specimens of healthy epithelium). The list of probes displaying statistically significant differences in expression between the lesion and healthy tissue contained 3850 probes at a significance level of 0.1. This pronounced difference in the sizes of gene lists result from the fact that the algorithm used for controlling type I errors (FDR) depends on the input set. The larger the initial gene list, the larger number of genes will pass the FDR control at a similar p-value distribution; in our case, the number of analyzed probes in the dataset for Crohn's disease is five times larger than that in the dataset for psoriasis.

The lists of differentially expressed genes were input into the MetaCore program. Because microarrays contained not only gene probes, but also a large number of ESTs with unidentified functions, the size of gene lists at this stage changed because not all the probes had the corresponding gene in the MetaCore database and because some probes corresponded to more than one gene. The lists of recognized genes comprised 425 and 2033 items for psoriasis and Crohn's disease, respectively.

The common part for the compared lists comprised 49 genes, which is a significant overlapping (p value = 4.94×10^{-2}). The significance was estimated using Fisher's test. The complete set contained 9017 genes present in both studied datasets (this set was identified by comparing the complete lists of genes for both microarrays in MetaCore). The lists of genes with altered expression were reduced to the subset of genes present in both datasets. Thus, these particular 49 genes were selected for further analysis (Table 1).

It was of interest to determine the molecular processes with which the genes common to psoriasis and Crohn's disease are associated. Table 2 consolidates the most probable cell processes with involvement of the genes listed in Table 1, as determined by the MetaCore software tools. These processes (Table 2) fall into two main groups—related to inflammation and cell cycle. Indeed, the pathological lesions in both psoriasis and Crohn's disease are inflammatory foci. The cell cycle is also considerably affected in both pathologies.

An increased proliferation of keratinocytes is observed in the psoriatic skin, an inflammatory focus.

GNA15	SFPQ	IFI35	IER2	OAS2	RFK	UBE2L6
CBX3	CG018	CSNK1D	SYNCRIP	PSME2	CTSC	CASP4
GPM6B	UGT1A4	STAT3	S100A8	FOXC1	SOSTDC1	ETS2
UGT1A6	VKORC1	TRIM22	RARG	TRAK2	SERPINB5	MECP2
IFI44	H2AFY	TXNDC1	ARMET	ZNF207	KIAA1033	QPCT
DEGS1	MIB1	IRF9	DDOST	DNAJC7	RBPMS	JUNB
LONRF1	HMGN1	MRPL9	FGFR2	CDC42EP1	S100A9	PHGDH

Table 1. Genes displaying altered expression in both psoriasis and Crohn's disease.

Process	***p* value**
Inflammation: interferon signaling pathways	2.19E-03
Signal transduction: Wnt signaling pathways	1.20E-02
Regulation of translation initiation	5.66E-02
Morphogenesis of blood vessels	9.76E-02
DNA repair	1.17E-01
Inflammation: amphoterin signaling pathways	1.19E-01
Proteolysis determined by the cell cycle and apoptosis	1.29E-01
Interleukin regulation of the cell cycle in G1-S phase	1.29E-01
Signal transduction: androgen receptor signaling pathways	1.34E-01

Table 2. Cell processes common to psoriasis and Crohn's disease.

For a more detailed description of the inflammatory response and cell cycle in the parts of them most tightly related to the genes listed in Table 1, we constructed gene networks, which are fragments of the larger gene networks describing the inflammatory response (Fig. 2) and cell cycle control (Fig. 3). Figure 2 shows that the inflammatory response is initiated by such well-known cytokines as TNF-α, IFN-γ, IL-2, IL-6, IL-17, and IL-23. Then protein kinases activate the transcription factors AP-1, STAT3, C/EBP, NF-κB, ISGF3, and others.
Figure 3 shows that the key cell cycle regulators that changed gene expression are the transcription factors AP-1, c-Myc, and STAT3. It is also evident that the genes encoding AP-1 transcription factor components are involved in both the inflammatory response and cell cycle control. It is known that the genes depending on AP-1 play an important role in regulation of proliferation, morphogenesis, apoptosis, and cell differentiation. Induction of cell differentiation activates transcription of the genes encoding the components of AP-1 complex [Turpaev K.T., 2006]. We assume that the genes of AP-1 transcription factor are the candidate genes involved in the pathogenesis of both psoriasis and Crohn's disease; moreover, this hypothesis is particularly based on the bioinformatics analysis of microarray data. Therefore, it was interesting to compare our data with the available information about the chromosome localization of the loci associated with psoriasis and Crohn's disease.

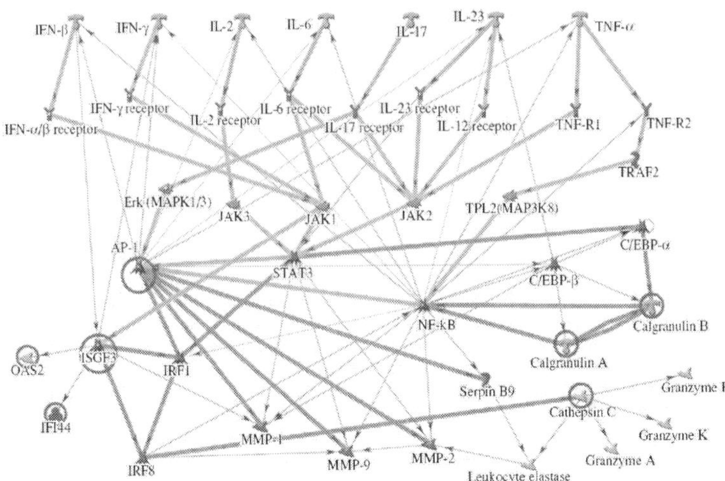

Fig. 2. Detail of the gene network describing inflammatory response. Green arrows indicate activation of the corresponding network elements, from the level of cytokines to transcription factors; light blue arrows, the activation of effector genes by transcription factors; and red circles, genes from the list.

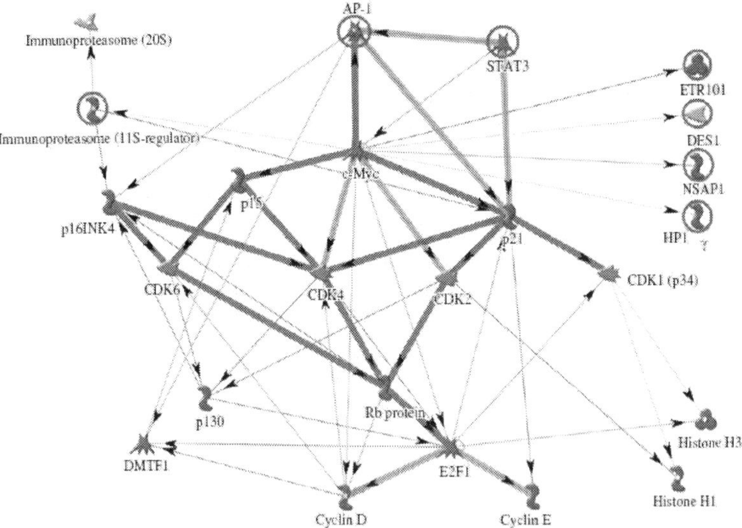

Fig. 3. Detail of the gene network describing cell cycle control. Green arrows indicate activation of the corresponding elements; red arrows, inhibition; and red circles, genes from the list.

4. Integrated network analysis of transcriptomic and proteomic data in psoriasis

Differentially abundant proteins. Protein abundance was determined by densitometric quantification of the protein spots on 2D-electophoresis gel (Figure 4) followed by MALDI-TOF mass spectrometry. Total of 10 proteins were over-abundant at least 2-fold in lesional skin compared with uninvolved skin: Keratin 14, Keratin 16, Keratin 17, Squamous cell carcinoma antigen, Squamous cell carcinoma antigen-2, Enolase 1, Superoxide dismutase [Mn], Galectin-7, S100 calcium-binding protein A9 and S100 calcium-binding protein A7.

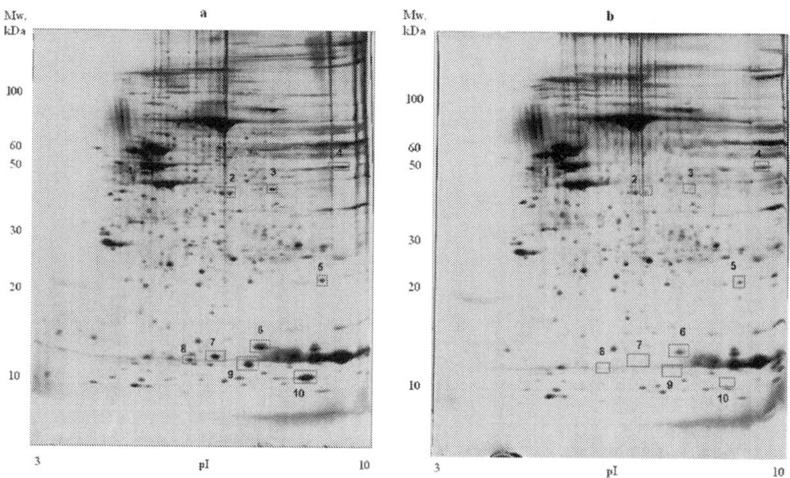

Fig. 4. Representative silver-stained 2DE gel images of lesional and uninvolved skin biopsy lysates. a) - gel image of lesional skin biopsy lysate; b) - gel image of uninvolved skin biopsy lysate. Spots corresponding to proteins overexpressed in lesions are marked with red rectangles and numbered. Spot 1 correspond to 3 proteins of keratin family, spot 2 - SCCA2, spot 3 - SCCA1, spot 4 - enolase 1, spot 5 - SOD2, spot 6 - galectin-7. S100A7 is found in spots 7 and 8 and S100A9 corresponds to 9 th and 10 th spots.

Several of these proteins were previously reported to be over-abundant in psoriatic plaques [Leigh IM et al., 1995; Madsen P et al., 1991; Vorum H et al., 1996; Takeda A et al.., 2002]. The proteins belonged to a diverse set of pathways and processes.

We attempted to connect the proteins into a network using a collection of over 300,000 manually curated protein interactions and several variants of "shortest path" algorithms applied in MetaCore suite [Nikolsky Y et al, 2009] (Figure 5). The genes encoding overabundant proteins were found to be regulated by several common transcription factors (TFs) including members of the NFkB and AP-1 complexes, STAT1, STAT3, c-Myc and SP1. Moreover, the upstream pathways activating these TFs were initiated by the overabundant S100A9 through its receptor RAGE [Ghavami S et al., 2008] and signal transduction kinases (JAK2, ERK, p38 MAPK). This network also included a positive feedback loop as S100A9 expression was determined to be controlled by NF-kB [Schreiber J et al., 2006]. The topology of this proteomics-derived network was confirmed by several transcriptomics studies

[Tsuruta D, 2009; Sano S et al., 2008; Ghoreschi K et al., 2003; Piruzian ES et al., 2009; Gandarillas A & Watt FM, 1997; Arnold I & Watt FM, 2001] which showed overexpression of these TFs in psoriasis lesions. Transiently expressed TFs normally have low protein level and, therefore, usually fail to be detected by proteomics methods.

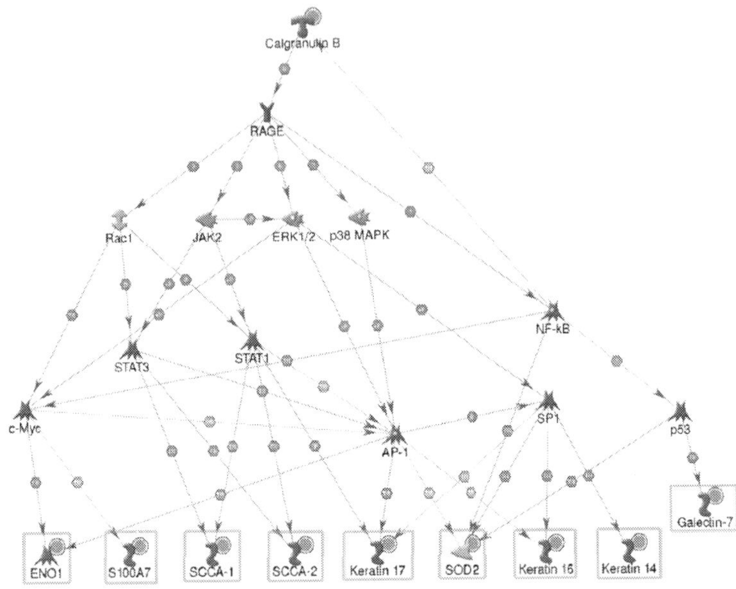

Fig. 5. Network illustrating regulatory pathways leading to transcription activation of proteomics markers. Red circles denote upregulated proteins.

RAGE receptor is clearly the key regulator on this network and plays the major role in orchestrating observed changes of protein abundance. This protein is abundant in both keratinocytes and leukocytes, though normally its expression is low [Lohwasser C et al., 2006]. RAGE participates in a range of processes in these cell types, including inflammation. It is being investigated as a drug target for treatment of various inflammatory disorders [Santilli F et al., 2009]. Thus, we may propose that RAGE can also play significant role in psoriasis.

We used Affymetrix gene expression data set from the recent study [Yao Y et al., 2008] involving 33 psoriasis patients. Originally, more than 1300 probe sets were found to be upregulated in lesions as compared with unlesional skin of the same people. We identified 451 genes overexpressed in lesional skin under more stringent statistical criteria (28 samples of lesional skin were matched with their nonlesional counterparts from the same patients in order to exclude individual expression variations, genes with fold change >2.5 and FDR-adjusted p-value < 0.01 were considered as upregulated). The genes encoding 7 out of 10 proteomic markers were overexpressed, well consistent with proteomics data. Expression of Enolase 1, Keratin 14 and Galectin 7 was not altered.

Despite good consistency between the proteomics and expression datasets, the two orders of magnitude difference in list size make direct correlation analysis difficult. Therefore, we applied interactome methods for the analysis of common upstream regulation of the two datasets at the level of transcription factors. First, we defined the sets of the most influential transcription factors using two recently developed methods of interactome analysis [Nikolsky Y et al., 2008] and the "hidden nodes" algorithm [Nikolskaya T et al., 2009]. The former method ranks TFs based on their one-step overconnectivity with the dataset of interest compared to randomly expected number of interactions. The latter approach takes into account direct and more distant regulation, calculating the p-values for local subnetworks by an aggregation algorithm [Nikolskaya T et al., 2009]. We calculated and ranked the top 20 TFs for each data type and added several TFs identified by network analysis approaches (data not shown). The TFs common for both data types were taken as set of 'important pathological signal transducers' (Figure 6). Noticeably, they closely resemble the set of TFs regulating the protein network on Figure 5.

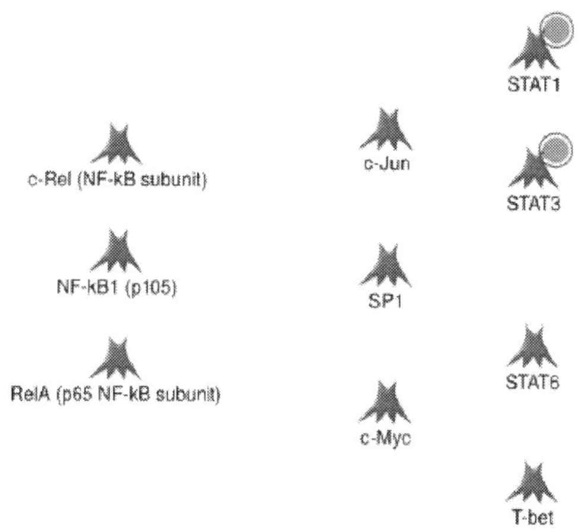

Fig. 6. Common transcriptional factors important for regulation of objects at both transcriptomics and proteomic levels. Objects in MetaCore database representing transcriptional factors found to be important regulators of pathology-related genes. Red circles denote that corresponding gene is upregulated in psoriatic lesion.

In the next step, we applied "hidden nodes" algorithm to identify the most influential receptors that could trigger maximal possible transcriptional response. In total, we found 226 membrane receptors significantly involved into regulation of 462 differentially expressed genes ('hidden nodes' p-value < 0.05). Assuming that topological significance alone does not necessarily prove that all receptors are involved in real signaling or are even expressed in the sample; we filtered this list by expression performance. The receptors used were those whose encoding genes or corresponding ligands were overexpressed greater than 2.5 fold. We assumed that the pathways initiated by over-expressed receptors and ligands are more likely to be activated in psoriasis. Here we assumed that expression alterations and protein abundance are at least collinear. An additional criterion was that the candidate receptors had to participate in the same signaling pathways with at least one of the common TFs. No receptor was rejected based on this criterion. In total, 44 receptors passed the transcription cut-off. Of these 24 receptor genes were overexpressed; 23 had overexpressed ligands and 3 cases had overexpression of both ligands and receptors (IL2RB, IL8RA and CCR5; see Figures 7 and 8). Interestingly, for several receptors, more than one ligand was overexpressed (Figure 7). Several receptors are composed of several subunits, only one of which was upregulated (for example, IL-2 receptor has only gamma subunit gene significantly upregulated). Out of 44 receptors we identified by topology analysis, 21 were previously reported as psoriasis markers (they are listed in Table 3 with corresponding references). The other 23 receptors were not reported to be linked to psoriasis or known to be implicated in other inflammatory diseases. These receptors belong to different cellular processes (development, cell adhesion, chemotaxis, apoptosis and immune response) (Table 6).

Gene	Connection to psoriasis	Gene	Connection to psoriasis
EPHA2	No	AGER	Yes [Foell D et al., 2003]
EPHB2	No	CCR1	Yes [Horuk R, 2005]
FCER1G	No	CCR2	Yes [Vestergaard C et al., 2004]
INSR	No	CCR3	Yes [Rottman JB et al., 2001]
LTBR	No	CCR5	Yes [de Groot M et al., 2007]
PLAUR	No	CD2	Yes [Ellis CN & Krueger GG., 2001]
TNFRSF10A	No	CD27	Yes [De Rie MA et al., 1996]
TNFRSF10B	No	CD36	Yes [Prens E et al., 1996]
CD44	Possible [Reichrath J et al, 1997]	CD3D	Yes [Haider AS, et al., 2007]
CSF2RB	Possible [Kelly R et al., 1993]	EGFR	Yes [Castelijns FA et al., 1999]
CXCR4	Possible [Gu J et al., 2002]	IL17RA	Yes [Johansen C et al., 2009]
FZD4	Possible[Reischl J et al., 2007]	IL1R1	Yes [Debets R et al., 1997]
GABBR1	Possible[Shiina T et al., 2009]	IL8RA	Yes [Schulz BS et al., 1993]
IL10RA	Possible [Asadullah K et al., 1998]	IL8RB	Yes [Schulz BS et al., 1993]
IL13RA1	Possible [Cancino-Diaz JC et al., 2002]	ITGAL	Yes [Guttman-Yassky E et al., 2008]
IL2RB	Possible [Pietrzak A et al., 2008]	ITGB2	Yes [Sjogren F et al., 1999]
IL2RG	Possible [Pietrzak A et al., 2008]	LRP1	Yes [Curry JL et al., 2003]
IL4R	Possible [Martin R, 2003]	PTPRC	Yes [Vissers WH et al., 2004]
LILRB2	Possible [Penna G et al., 2005]	SDC3	Yes [Patterson AM et al., 2008]
LRP2	Possible [Fu X et al., 2009]	SELE	Yes [Wakita H &Takigawa M, 1994]
LRP8	Possible [Fu X et al., 2009]	SELPLG	Yes [Chu A et al., 1999]
ROR2	Possible [Reischl J et al., 2007]	TLR4	Yes [Seung NR et al., 2007]

Table 3. Receptors identified in our study and not yet studied in connection to psoriasis ('Possible' term was used if protein name co-occurred with psoriasis in articles, but no clear evidence of its implication was shown. In some cases, ligands are associated with psoriasis (i.e, IL-10)).

Meta-analysis of multiple OMICs data types and studies is becoming an important research tool in understanding complex diseases. Several methods were developed for correlation analysis between the datasets of different type, such as mRNA and proteomics [Hack CJ, 2004; Le Naour F et al., 2001; Steiling K et al., 2009; Conway JP & Kinter M, 2005; Di Pietro C et al., 2009]. However, there are many technological challenges to resolve, including mismatching protein IDs and mRNA probes, fundamental differences in OMICs technologies, differences in experimental set-ups in studies done by different groups etc [Mijalski T et al., 2005]. Moreover, biological reasons such as differences in RNA and protein degradation processes also contribute to variability of different data types. As a result, transcriptome and proteome datasets usually show only weak positive correlation although were considered as complimentary. More recent studies focused on functional similarities and differences observed for different levels of cellular organization and reflected in different types of OMICs data [Habermann JK et al., 2007; Chen YR et al., 2006; Shachaf CM et al., 2008; Zhao C et al., 2009]. For example, common interacting objects were found for distinct altered transcripts and proteins in type 2 diabetes [Gerling IC et al., 2006]. In one leukemia study [Zheng PZ et al., 2005] authors found that distinct alterations at transcriptomics and proteomic levels reflect different sides of the same deregulated cellular processes.

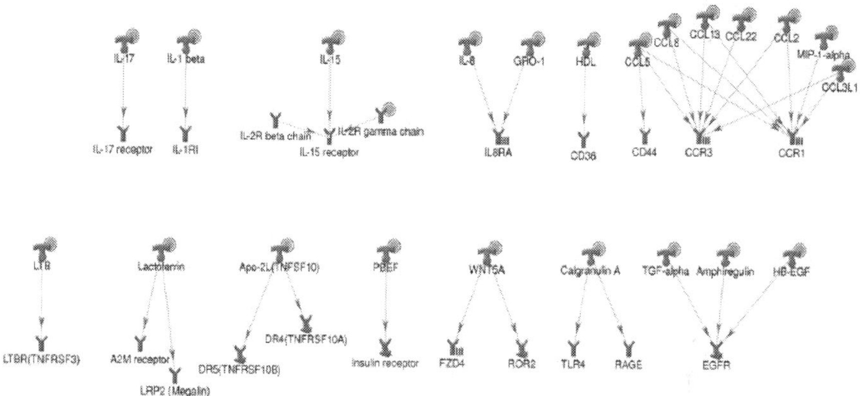

Fig. 7. Candidate receptors with their respective upregulated ligands. Initial steps of pathways presumably activated in lesions (ligands, overexpressed at transcriptional level and their corresponding receptors) Red circles denote that corresponding gene is upregulated in psoriatic lesion.

The overall concordance between mRNA and protein expression landscapes was addressed in earlier studies, although the data types were compared mostly at the gene/protein level with limited functional analysis [Cox B et al., 2005; Mijalski T et al., 2005]. Later, ontology enrichment co-examination of transcriptomics and proteomic data has shown that the two data types affect similar biological processes and are complimentary [Chen YR, et al., 2006; Zheng PZ et al., 2005; Zhao C et al., 2009]. However, the key issue of biological causality and functional consequences of distinct regulation events at both mRNA and protein levels of cellular organization were not yet specifically addressed. These issues cannot be resolved by low resolution functional methods like enrichment analysis. Instead, one has to apply more precise computational methods such as topology and biological networks, which take into consideration directed binary interactions and multi-step pathways connecting objects between the datasets of different types regardless of their direct overlap at gene/protein level [Ideker T & Sharan R, 2008; Chuang HY et al., 2007]. For example, topology methods such as "hidden nodes" [Dezso Z et al., 2009; Nikolsky Y et al., 2008] can identify and rank the upstream regulatory genes responsible for expression and protein level alterations while network tools help to uncover functional modules most affected in the datasets, identify the most influential genes/proteins within the modules and suggest how specific modules contribution to clinical phenotype [Nikolsky Y et al., 2005; Gerling IC et al., 2006].

In this study, we observed substantial direct overlap between transcriptomics and proteomics data, as 7 out of 10 over-abundant proteins in psoriasis lesions were encoded by differentially over-expressed genes. However, the two orders of magnitude difference in dataset size (462 genes versus 10 proteins) made the standard correlation methods inapplicable. Besides, proteomics datasets display a systematic bias in function of abundant proteins, favoring "effector" proteins such as structural, inflammatory, core metabolism proteins but not the transiently expressed and fast degradable signaling proteins. Therefore, we applied topological network methods to identify common regulators for two datasets such as the most influential transcription factors and receptors. We have identified some key regulators of the "proteomics" set among differentially expressed genes, including transcription factors, membrane receptors and extracellular ligands, thus reconstructing upstream signaling pathways in psoriasis. In particular, we identified 24 receptors previously not linked to psoriasis.

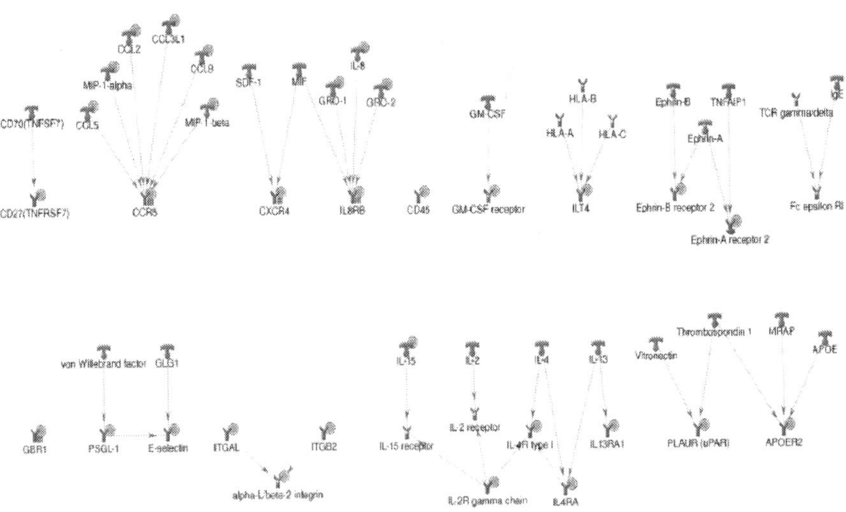

Fig. 8. Upregulated candidate receptors with their respective ligands. Initial steps of pathways presumably activated in lesions (receptors, overexpressed at transcriptional level and their corresponding ligands) Red circles denote that corresponding gene is upregulated in psoriatic lesion.

Importantly, many ligands and receptors defined as putative starts of signaling pathways were activated by transcription factors at the same pathways, clearly indicating on positive regulatory loops activated in psoriasis. The versatility and the variety of signaling pathways activated in psoriasis is also impressive, which is evident from differentially overexpression of 44 membrane receptors and ligands in skin lesions. This complexity and redundancy of psoriasis signaling likely contributes to the inefficiency of current treatments, even novel therapies such as monoclonal antibodies against TNF-α and IL-23. Thus, the key regulator, RAGE receptor, triggers multiple signaling pathways which stay activated even when certain immunological pathways are blocked. Our study suggests that combination therapy targeting multiple pathways may be more efficient for psoriasis (particularly considering feasibility for topical formulations). In addition, the 24 receptors we identified by topology analysis and previously not linked with psoriasis can be tested as potential novel targets for disease therapy. The functional machinery of psoriasis is still not complete and additional studies can be helpful in "filling the gaps" of our understanding of its molecular mechanisms. For instance, kinase activity is still unaccounted for, as signaling kinases are activated only transiently and are often missed in gene expression studies. Topological analysis methods such as "hidden nodes" [Dezso Z et al., 2004] may help to reconstruct regulatory events missing in the data. Also, the emerging phosphoproteomics methodology may prove to become a helpful and complimentary OMICs technology. The network analysis methodology is not dependent on the type of data analyzed and or any gene/protein content overlap between the studies and is well applicable for functional integration of multiple data types.

3. Conclusion

Thus, we succeeded in comparing the molecular processes characteristic of psoriasis and Crohn's disease and detecting the candidate genes involved in the processes common for both pathologies and critical for their development. Identification of the proteins encoded by these genes is an important aspect of the research performed, because the proteins are particular targets for elaborating new approaches to treating psoriasis and Crohn's disease. Our data obtained by analyzing expression of the candidate genes for psoriasis and Crohn's disease can enhance the search for new biological targets for the corresponding therapeutics. In order to gain insight into molecular machinery underlying the disease, we conducted a comprehensive meta-analysis of proteomics and transcriptomics of psoriatic lesions from independent studies. Network-based analysis revealed similarities in regulation at both proteomics and transcriptomics level. We identified a group of transcription factors responsible for overexpression of psoriasis genes and a number of previously unknown signaling pathways that may play a role in this process. We also evaluated functional synergy between transcriptomics and proteomics results.

We have successfully applied network-based methods to integrate and explore two distinct high-throughput disease data sets of different origin and size. Through identification of common regulatory machinery that is likely to cause overexpression of genes and proteins, we came to the signaling pathways that might contribute to the altered state of regulatory network in psoriatic lesion. Our approach allows easy integrative investigation of different data types and produces biologically meaningful results, leading to new potential therapy targets. We have demonstrated that pathology can be caused and maintained by a great amount of various cascades, many previously not described as implicated in psoriasis; therefore, combined therapies targeting multiple pathways might be effective in treatment.

4. Acknowledgment

This work was supported by grants 16.512.11.2049 from Ministry of Education and Science of the Russian Federation, grant from Russian Academy of Sciences ("Fundamental Sciences to Medicine" program), and grant P1309 from Ministry of Education and Science of the Russian Federation (Federal Special Program "Scientific and Educational Human Resources of Innovative Russia" for 2009 - 2013)

5. Reference

Arnold I, Watt FM: c-Myc activation in transgenic mouse epidermis results in mobilization of stem cells and differentiation of their progeny. Curr Biol 2001, 11(8):558-68.

Asadullah K, et al.: IL-10 is a key cytokine in psoriasis. Proof of principle by IL-10 therapy: a new therapeutic approach. J Clin Invest 1998, 101(4):783-94.

Benjamini Y., Hochberg Y. 1995. Controlling the false discovery rate: A practical and powerful approach to multiple testing. J. Royal Stat. Soc. Ser. B, Methodol. 57, 289–300.

Bhavnani SK, et al.: Network analysis of genes regulated in renal diseases: implications for a molecular-based classification. BMC Bioinformatics 2009, 10(Suppl 9):S3.

Binder V. 2004. Epidemiology of IBD during the twentieth century: An integrated view. Best Pract. Res. Clin. Gastroenterol. 18, 463–479.

Boehncke WH, et al.: Pulling the trigger on psoriasis. Nature 1996, 379(6568):777.

Bowcock A., Cookson W. 2004. The genetics of psoriasis, psoriatic arthritis and atopic dermatitis. Hum. Mol. Genet. 13, 43–55.

Bowcock AM, et al.: Insights into psoriasis and other inflammatory diseases from large-scale gene expression studies. Hum Mol Genet 2001, 10(17):1793-805.

Cancino-Diaz JC, et al.: Interleukin-13 receptor in psoriatic keratinocytes: overexpression of the mRNA and underexpression of the protein. J Invest Dermatol 2002, 119(5):1114-20.

Castelijns FA, et al.: The epidermal phenotype during initiation of the psoriatic lesion in the symptomless margin of relapsing psoriasis. J Am Acad Dermatol 1999, 40(6 Pt 1):901-9.

Chen YR, et al.: Quantitative proteomic and genomic profiling reveals metastasis-related protein expression patterns in gastric cancer cells. J Proteome Res 2006, 5(10):2727-42.

Chu A, et al.: Tissue specificity of E- and P-selectin ligands in Th1- mediated chronic inflammation. J Immunol 1999, 163(9):5086-93.

Chuang HY, et al.: Network-based classification of breast cancer metastasis. Mol Syst Biol 2007, 3:140.

Conway JP, Kinter M: Proteomic and transcriptomic analyses of macrophages with an increased resistance to oxidized low density lipoprotein (oxLDL)-induced cytotoxicity generated by chronic exposure to oxLDL. Mol Cell Proteomics 2005, 4(10):1522-40.

Cox B, Kislinger T, Emili A: Integrating gene and protein expression data: pattern analysis and profile mining. Methods 2005, 35(3):303-14.

Crohn B., Ginzburg L., Oppenheimer G. 1932. Regional ileitis: A pathological and clinical entity. J. Actinomycetes. Med. Assoc. 251, P73–P79 (1984).

Curry JL, et al.: Innate immune-related receptors in normal and psoriatic skin. Arch Pathol Lab Med 2003, 127(2):178-86.

de Groot M, et al.: Expression of the chemokine receptor CCR5 in psoriasis and results of a randomized placebo controlled trial with a CCR5 inhibitor. Arch Dermatol Res 2007, 299(7):305-13.

De Rie MA, et al.: Expression of the T-cell activation antigens CD27 and CD28 in normal and psoriatic skin. Clin Exp Dermatol 1996, 21(2):104-11.

Debets R, et al.: The IL-1 system in psoriatic skin: IL-1 antagonist sphere of influence in lesional psoriatic epidermis. J Immunol 1997, 158(6):2955-63.

Dezso Z, et al.: Identifying disease-specific genes based on their topological significance in protein networks. BMC Syst Biol 2009, 3:36

Di Pietro C, et al.: The apoptotic machinery as a biological complex system: analysis of its omics and evolution, identification of candidate genes for fourteen major types of cancer, and experimental validation in CML and neuroblastoma. BMC Med Genomics 2009, 2(1):20.

Draghici S., Khatri P., Tarca A.L., Amin K., Done A., Voichita C., Georgescu C., Romero R. 2007. A systems biology approach for pathway level analysis. Genome Res. 17, 1537–1545.

Ellis CN, Krueger GG: Treatment of chronic plaque psoriasis by selective targeting of memory effector T lymphocytes. N Engl J Med 2001, 345(4):248-55.

Foell D, et al.: Expression of the pro-inflammatory protein S100A12 (ENRAGE) in rheumatoid and psoriatic arthritis. Rheumatology (Oxford) 2003, 42(11):1383-9.

Fu X, et al.: Estimating accuracy of RNA-Seq and microarrays with proteomics. BMC Genomics 2009, 10(1):161.

Gandarillas A, Watt FM: c-Myc promotes differentiation of human epidermal stem cells. Genes Dev 1997, 11(21):2869-82.

Gerling IC, et al.: New data analysis and mining approaches identify unique proteome and transcriptome markers of susceptibility to autoimmune diabetes. Mol Cell Proteomics 2006, 5(2):293-305.

Ghavami S, et al.: S100A8/A9 at low concentration promotes tumor cell growth via RAGE ligation and MAP kinase-dependent pathway. Journal of leukocyte biology 2008, 83(6):1484-1492

Ghoreschi K, Mrowietz U, Rocken M: A molecule solves psoriasis? Systemic therapies for psoriasis inducing interleukin 4 and Th2 responses. J Mol Med 2003, 81(8):471-80.

Ghosh D, et al.: Statistical issues and methods for meta-analysis of microarray data: a case study in prostate cancer. Funct Integr Genomics 2003, 3(4):180-8.

Gisondi P, Girolomoni G: Biologic therapies in psoriasis: a new therapeutic approach. Autoimmun Rev 2007, 6(8):515-9.

Gravel P, Golaz O: Two-Dimensional PAGE Using Carrier Ampholyte pH Gradients in the First Dimension. The Protein Protocols Handbook 1996:127-132.

Gu J, et al.: A 588-gene microarray analysis of the peripheral blood mononuclear cells of spondyloarthropathy patients. Rheumatology (Oxford) 2002, 41(7):759-66

Guttman-Yassky E, et al.: Blockade of CD11a by efalizumab in psoriasis patients induces a unique state of T-cell hyporesponsiveness. J Invest Dermatol 2008, 128(5):1182-91.

Habermann JK, et al.: Stage-specific alterations of the genome, transcriptome, and proteome during colorectal carcinogenesis. Genes Chromosomes Cancer 2007, 46(1):10-26.

Hack CJ: Integrated transcriptome and proteome data: the challenges ahead. Brief Funct Genomic Proteomic 2004, 3(3):212-9.

Haider AS, et al.: Novel insight into the agonistic mechanism of alefacept in vivo: differentially expressed genes may serve as biomarkers of response in psoriasis patients. J Immunol 2007, 178(11):7442-9.

Horuk R: BX471: a CCR1 antagonist with anti-inflammatory activity in man. Mini Rev Med Chem 2005, 5(9):791-804.

Ideker T, Sharan R: Protein networks in disease. Genome Res 2008, 18(4):644-52.

Johansen C, et al.: Characterization of the interleukin-17 isoforms and receptors in lesional psoriatic skin. Br J Dermatol 2009, 160(2):319-24.

Kelly R, Marsden RA, Bevan D: Exacerbation of psoriasis with GM-CSF therapy. Br J Dermatol 1993, 128(4):468-9.

Le Naour F, et al.: Profiling changes in gene expression during differentiation and maturation of monocyte-derived dendritic cells using both oligonucleotide microarrays and proteomics. J Biol Chem 2001, 276(21):17920-31.

Leigh IM, et al.: Keratins (K16 and K17) as markers of keratinocyte hyperproliferation in psoriasis in vivo and in vitro. The British journal of dermatology 1995, 133(4):501-511.

Lohwasser C, et al.: The receptor for advanced glycation end products is highly expressed in the skin and upregulated by advanced glycation end products and tumor necrosis factor-alpha. J Invest Dermatol 2006, 126(2):291-9.

Madsen P, et al.: Molecular cloning, occurrence, and expression of a novel partially secreted protein "psoriasin" that is highly up-regulated in psoriatic skin. The Journal of investigative dermatology 1991, 97(4):701-712.

Martin R: Interleukin 4 treatment of psoriasis: are pleiotropic cytokines suitable therapies for autoimmune diseases? Trends Pharmacol Sci 2003, 24(12):613-6.

Menezes R, et al.: Integrated analysis of DNA copy number and gene expression microarray data using gene sets. BMC Bioinformatics 2009, 10(1):203.

Mijalski T, et al.: Identification of coexpressed gene clusters in a comparative analysis of transcriptome and proteome in mouse tissues. Proceedings of the National Academy of Sciences of the United States of America 2005, 102(24):8621-8626

Mortz E, et al.: Improved silver staining protocols for high sensitivity protein identification using matrix-assisted laser desorption/ionization-time of flight analysis. Proteomics 2001, 1(11):1359-63

Nikolskaya T, et al.: Network analysis of human glaucomatous optic nerve head astrocytes. BMC Med Genomics 2009, 2:24.

Nikolsky Y, et al.: Genome-wide functional synergy between amplified and mutated genes in human breast cancer. Cancer research 2008, 68(22):9532-9540.

Nikolsky Y, et al.: Functional analysis of OMICs data and small molecule compounds in an integrated "knowledge-based" platform. Methods in molecular biology (Clifton, N.J.) 2009, 563:177-196.

Nikolsky Y, Nikolskaya T, Bugrim A: Biological networks and analysis of experimental data in drug discovery. DrugDiscov Today 2005, 10(9):653-62.

Oestreicher JL, et al.: Molecular classification of psoriasis diseaseassociated genes through pharmacogenomic expression profiling. Pharmacogenomics J 2001, 1(4):272-87.

Ortonne JP: Aetiology and pathogenesis of psoriasis. Br J Dermatol 1996, 135(Suppl 49):1-5.

Pastore S, et al.: Biological drugs targeting the immune response in the therapy of psoriasis. Biologics 2008, 2(4):687-97.
Patterson AM, et al.: Differential expression of syndecans and glypicans in chronically inflamed synovium. Ann Rheum Dis 2008, 67(5):592-601.
Penna G, et al.: Expression of the inhibitory receptor ILT3 on dendritic cells is dispensable for induction of CD4+Foxp3+ regulatory T cells by 1,25-dihydroxyvitamin D3. Blood 2005, 106(10):3490-7.
Pietrzak A, et al.: Genes and structure of selected cytokines involved in pathogenesis of psoriasis. Folia Histochem Cytobiol 2008, 46(1):11-21.
Pihur V, Datta S: RankAggreg, an R package for weighted rank aggregation. BMC Bioinformatics 2009, 10:62.
Piruzian E.S., Nikolskaya T.A., Abdeev R.M., Bruskin S.A. 2007. Transcription factor AP-1 components as candidate genes involved in the development of psoriatic process. Mol. Biol. 41, 1069–1080.
Piruzian ES, et al.: [The comparative analysis of psoriasis and Crohn disease molecular-genetical processes under pathological conditions]. Mol Biol (Mosk) 2009, 43(1):175-9.
Prens E, et al.: Adhesion molecules and IL-1 costimulate T lymphocytes in the autologous MECLR in psoriasis. Arch Dermatol Res 1996, 288(2):68-73.
Quekenborn-Trinquet V, et al.: Gene expression profiles in psoriasis: analysis of impact of body site location and clinical severity. Br J Dermatol 2005, 152(3):489-504.
Reichrath J, et al.: Expression of integrin subunits and CD44 isoforms in psoriatic skin and effects of topical calcitriol application. J Cutan Pathol 1997, 24(8):499-506.
Reischl J, et al.: Increased expression of Wnt5a in psoriatic plaques. J Invest Dermatol 2007, 127(1):163-9.
Rottman JB, et al.: Potential role of the chemokine receptors CXCR3, CCR4, and the integrin alphaEbeta7 in the pathogenesis of psoriasis vulgaris. Lab Invest 2001, 81(3):335-47.
Sano S, Chan KS, DiGiovanni J: Impact of Stat3 activation upon skin biology: a dichotomy of its role between homeostasis and diseases. J Dermatol Sci 2008, 50(1):1-14.
Santilli F, et al.: Soluble forms of RAGE in human diseases: clinical and therapeutical implications. Curr Med Chem 2009, 16(8):940-52.
Sartor R. 2006. Mechanisms of disease: Pathogenesis of Crohn's disease and ulcerative colitis. Nature Clin. Pract. Gastroenterol. Hepatol. 3, 390–407.
Schreiber J, et al.: Coordinated binding of NF-kappaB family members in the response of human cells to lipopolysaccharide. Proc Natl Acad Sci USA 2006, 103(15):5899-904
Schreiber S., Rosenstiel P., Albrecht M., Hampe J., Krawczak M. 2005. Genetics of Crohn disease, an archetypal inflammatory barrier disease. Nature Rev. Genet. 6, 376–388.
Schulz BS, et al.: Increased expression of epidermal IL-8 receptor in psoriasis. Down-regulation by FK-506 in vitro. J Immunol 1993, 151(8):4399-406.
Seung NR, et al.: Comparison of expression of heat-shock protein 60, Toll-like receptors 2 and 4, and T-cell receptor gammadelta in plaque and guttate psoriasis. J Cutan Pathol 2007, 34(12):903-11.
Shachaf CM, et al.: Genomic and proteomic analysis reveals a threshold level of MYC required for tumor maintenance. Cancer Res 2008, 68(13):5132-42.
Shiina T, et al.: The HLA genomic loci map: expression, interaction, diversity and disease. J Hum Genet 2009, 54(1):15-39.

Sjogren F, et al.: Expression and function of beta 2 integrin CD11B/CD18 on leukocytes from patients with psoriasis. Acta Derm Venereol 1999, 79(2):105-10.

Steiling K, et al.: Comparison of proteomic and transcriptomic profiles in the bronchial airway epithelium of current and never smokers. PLoS One 2009, 4(4):e5043.

Takeda A, et al.: Overexpression of serpin squamous cell carcinoma antigens in psoriatic skin. J Invest Dermatol 2002, 118(1):147-54.

Tsuruta D: NF-kappaB links keratinocytes and lymphocytes in the pathogenesis of psoriasis. Recent Pat Inflamm Allergy Drug Discov 2009, 3(1):40-8.

Turpaev K.T. 2006. Role of transcription factor AP-1in integration of intracellular signal pathways. Mol. Biol. 40, 945–961.

Vestergaard C, et al.: Expression of CCR2 on monocytes and macrophages in chronically inflamed skin in atopic dermatitis and psoriasis. Acta Derm Venereol 2004, 84(5):353-8.

Vissers WH, et al.: Memory effector (CD45RO+) and cytotoxic (CD8+) T cells appear early in the margin zone of spreading psoriatic lesions in contrast to cells expressing natural killer receptors, which appear late. Br J Dermatol 2004, 150(5):852-9.

Vorum H, et al.: Expression and divalent cation binding properties of the novel chemotactic inflammatory protein psoriasin. Electrophoresis 1996, 17(11):1787-96.

Wakita H, Takigawa M: E-selectin and vascular cell adhesion molecule-1 are critical for initial trafficking of helper-inducer/memory T cells in psoriatic plaques. Arch Dermatol 1994, 130(4):457-63.

Warnat P, Eils R, Brors B: Cross-platform analysis of cancer microarray data improves gene expression based classification of phenotypes. BMC Bioinformatics 2005, 6:265.

Welch B.L. 1947. The generalization of "student's" problem when several different population variances are involved. Biometrika. 34, 28–35.

Wise LH, Lanchbury JS, Lewis CM: Meta-analysis of genome searches. Ann Hum Genet 1999, 63(Pt 3):263-72.

Yao Y., Richman L., Morehouse C., de los Reyes M., Higgs B.W., Boutrin A., White B., Coyle A., Krueger J., Kiener P.A., Jallal B. 2008. Type I interferon: Potential therapeutic target for psoriasis? PLoS ONE. 3, e2737.

Zhao C, et al.: Identification of Novel Functional Differences in Monocyte Subsets Using Proteomic and Transcriptomic Methods. Journal of Proteome Research 2009, 8(8):4028-4038

Zheng PZ, et al.: Systems analysis of transcriptome and proteome in retinoic acid/arsenic trioxide-induced cell differentiation/apoptosis of promyelocytic leukemia. Proc Natl Acad Sci USA 2005, 102(21):7653-8.

Zhou X, et al.: Novel mechanisms of T-cell and dendritic cell activation revealed by profiling of psoriasis on the 63,100-element oligonucleotide array. Physiol Genomics 2003, 13(1):69-78.

22

Emergence of the Diversified Short ORFeome by Mass Spectrometry-Based Proteomics

Hiroko Ao-Kondo, Hiroko Kozuka-Hata and Masaaki Oyama
Medical Proteomics Laboratory, Institute of Medical Science, University of Tokyo
Japan

1. Introduction

In proteomics analyses, protein identification by mass spectrometry (MS) is usually performed using protein sequence databases such as RefSeq (NCBI; http://www.ncbi.nlm.nih.gov/RefSeq/), UniProt (http://www.uniprot.org/) or IPI (http://www.ebi.ac.uk/IPI/IPIhelp.html). Because these databases usually target the longest (main) open reading frame (ORF) in the corresponding mRNA sequence, whether shorter ORFs on the same mRNA are actually translated still shrouds in mystery. In the first place, it had been considered that almost all eukaryotic mRNAs contains only one ORF and functions as monocistronic mRNAs. It is now known, however, that some eukaryotic mRNAs had multiple ORFs, which are recognized as polycistronic mRNAs. One of the well-known extra ORFs is an upstream ORF (uORF) and it functions as regulators of mRNA translation (Diba et al., 2001; Geballe & Morris, 1994; Morris & Geballe, 2000; Vilela & McCarthy, 2003; Zhang & Dietrich, 2005). For getting clues to the mystery of diversified short ORFs, full-length mRNA sequence databases with complete 5′-untranslated regions (5′-UTRs) were essentially needed (Morris & Geballe, 2000; Suzuki et al., 2001).
The oligo-capping method was developed to construct full-length cDNA libraries (Maruyama & Sugano, 1994) and the corresponding sequence were stored into the database called DBTSS (DataBase of Transcriptional Start Site; http://dbtss.hgc.jp/) (Suzuki et al., 1997, 2002, 2004; Tsuchihara et al., 2009; Wakaguri et al., 2008; Yamashita et al., 2006). Comparing the dataset of DBTSS with the corresponding RefSeq entries, it was found that about 50 % of the RefSeq entries had at least one upstream ATG (uATG) except the functional ATG initiator codon (Yamashita et al., 2003). Although it had been suggested that upstream AUGs (uAUGs) and uORFs play important roles for translation of the main ORF, none of the proteins from these uORFs was detected in biological experiments in vivo. Our previous proteomics analysis focused on small proteins revealed the first evidence of the existence of four novel small proteins translated from uORFs in vivo using highly sensitive nanoflow liquid chromatography (LC) coupled with the electrospray ionization-tandem mass spectrometry (ESI-MS/MS) system (Oyama et al., 2004). Large-scale analysis based on in-depth separation by two-dimensional LC also led to the identification of additional eight novel small proteins not only from uORFs but also from downstream ORFs and one of them was found to be translated from a non-AUG initiator codon (Oyama et al., 2007). Finding of these novel small proteins indicate the possibility of diverse control mechanisms of translation initiation.

In this chapter, we first introduce widely-recognized mechanism of translation initiation and functional roles of uORF in translational regulation. We then review how we identified novel small proteins with MS and lastly discuss the progress of bioinformatical analyses for elucidating the diversification of short coding regions defined by the transcriptome.

2. Translational regulation by short ORFs

It is well known that 5′-UTRs of some mRNAs contain functional elements for translational regulation defined by uAUG and uORF. In this section, we show how uAUG and uORF have biological consequences for protein synthesis on eukaryotic mRNAs.

2.1 Outline of translation initiation

Initiation of translation on eukaryotic mRNAs occurs roughly as follows (Fig. 1) (Kozak, 1989, 1991, 1999).

1. A small (40S) ribosomal subunit binds near the 5′-end of mRNA, i.e. the cap structure.
2. The 40S subunit migrates linearly downstream of the 5′-UTR until it encounters the optimum AUG initiator codon.
3. A large (60S) ribosomal subunit joins the paused 40S subunit.
4. The complete ribosomal complex (40S + 60S) starts protein synthesis.

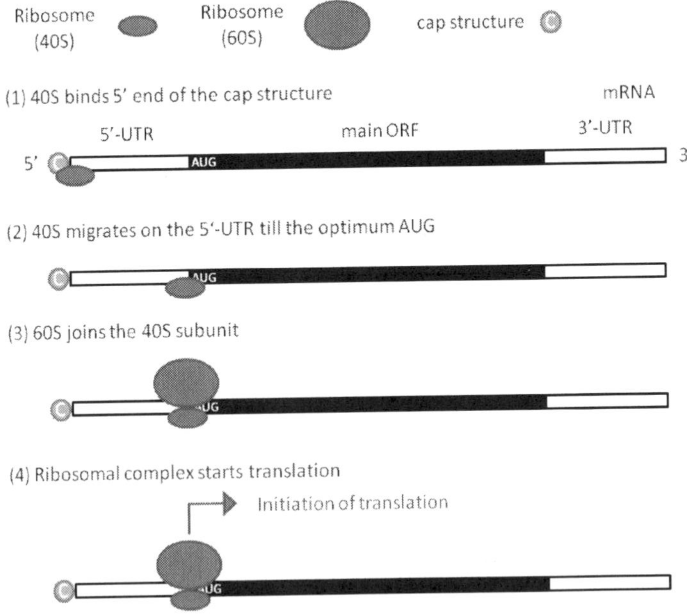

Fig. 1. The proposed procedure for initiation of translation in eukaryotes. The black region indicates the main ORF of the mRNA.

In addition to the above mechanism, initiation of translation without the step of ribosome scanning is also known. It is called "internal initiation", which depends on some particular structure on an mRNA termed internal ribosome entry site (IRES).

2.2 The relationship between uORF and main ORF

In case that an mRNA contains a uORF, two models for the initiation of translation are suggested (Fig. 2) (Hatzigeorgiou, 2002). One is called "leaky scanning" and the other is "reinitiation". If the first AUG codon is in an unfavorable sequence context defined by Kozak (see the section 3.2), a small ribosomal subunit (40S) ignores the first AUG and initiates translation from a more favorable AUG codon downstream located. This phenomenon is known as "leaky scanning" (Fig. 2-(A)). In case that a complete ribosomal complex translates a main ORF after termination of translation of the uORF on the same mRNA, it is termed "reinitiation" (Fig. 2-(B)).

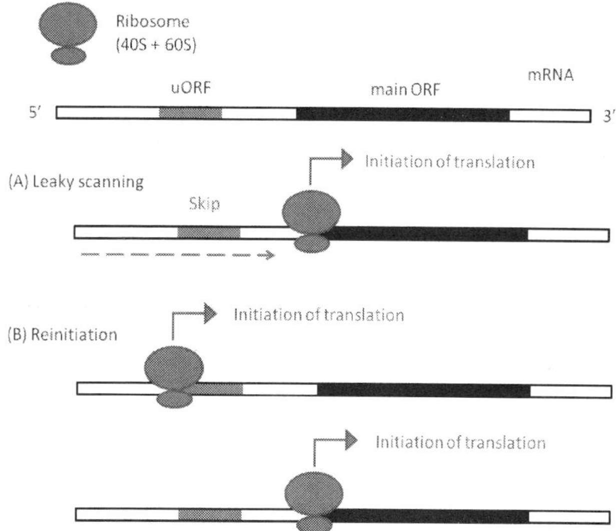

Fig. 2. The irregular models of ribosome scanning on eukaryotic mRNAs.
(A) Leaky scanning and (B) Reinitiation. Gray regions indicate uORFs on the mRNA, whereas black ones represent the main ORFs.

The relations between two ORFs are classified into three types as follows; (1) A distant type; in-frame/out-of-frame, (2) A contiguous type; in-frame and (3) An overlapped type; in-frame/out-of-frame (Fig. 3). *In-frame* means that a uORF and the main ORF are on the same frame of the mRNA sequence, whereas *out-of-frame* means that they are on the different frame. According to the previous analysis of the accumulated 5'-end sequence data, the average size of uORF was estimated at 31 amino acids and 20 % of ORFs were categorized into Type (3) (Yamashita et al., 2003).

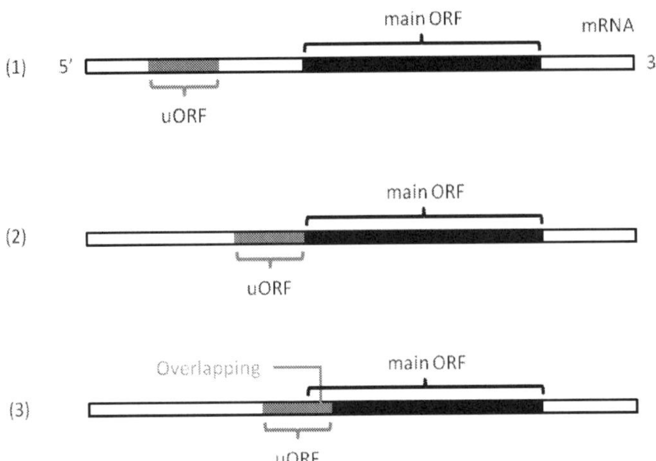

Fig. 3. The location of a uORF and the main ORF on the mRNA.
(1) A distant type, (2) A contiguous type and (3) An overlapped type. Types (1) and (3) have two subtypes based on the frames of two ORFs. One is defined by the same reading frame (in-frame) and the other is by the different one (out-of-frame). Gray and black regions indicate uORFs and the main ORFs on mRNAs, respectively, whereas a blue one represents an overlap.

These different relations might bring about different events in initiating translation. In eukaryotes, it has a tendency to increase an efficiency of reinitiation if the distance between a uORF and the main ORF is long (Kozak, 1991; Meijer & Thomas, 2002; Morris & Geballe, 2000). Therefore, the ORFs classified as Types (2) and (3) would be difficult to be regulated by reinitiation. It is also said that reinitiation occurs only when the length of uORF is short (Kozak, 1991), whereas the sequence context of an inter-ORF's region, that of upstream of uORF, uORF itself and even the main ORF can also affect reinitiation (Morris & Geballe, 2000). On the contrary, the ORFs of Type (3) might easily cause leaky scanning (Geballe & Morris, 1994; Yamashita et al., 2003). As a special case, when a termination codon of the uORF is near the AUG initiator codon of the downstream ORF, within about 50 nucleotides, ribosomes could scan backwards and reinitiate translation from the AUG codon of the downstream ORF (Peabody et al., 1986).

2.3 The role of short ORFs in translation regulation

The 5′-UTR elements such as uAUGs and uORFs are well known as important regulators for translation initiation. In case of some genes that have multiple uORFs, considerably different effects can be generated on the translation of the main ORF depending on which combination of uORFs is translated. Some uORFs seem to promote reinitiation of the main ORFs and the others seem to inhibit it. It is supposed that these effects are caused by the nucleotide sequences of the 3′ ends of the uORFs, that of uORFs or protein products encoded by uORFs. Such differential enhancement of translation are considered to be one of the responses of adaptation to the environment (Altmann & Trachsel, 1993; Diba et al., 2001;

Geballe & Morris, 1994; Hatzigeorgiou, 2002; Iacono et al., 2005; Meijer & Thomas, 2002; Morris & Geballe, 2000; Vilela & McCarthy, 2003; Wang & Rothnagel, 2004; Zhang & Dietrich, 2005). In addition to that, various factors or events are known to influence on the translational inhibition of the main ORF; the presence of arginine, a stalling of a ribosomal complex at the termination or an interaction between a ribosomal complex and the peptide encoded by the uORF, which indicates that down-regulated controls by uORFs are general (Diba et al., 2001; Geballe & Morris, 1994; Iacono et al., 2005; Meijer & Thomas, 2002; Morris & Geballe, 2000; Vilela & McCarthy, 2003; Zhang & Dietrich, 2005).
As for downstream ORFs, there is also a report that a peptide encoded in the 3′-UTR may be expressed (Rastinejad & Blau, 1993). However, whether and how the peptides control the translation initiation of the main ORF is still unknown.

3. Variability of translation start sites

How a ribosomal complex (40S + 60S) recognizes an initiator codon on the mRNA is a matter of vital importance for defining the proteome. Here we present a part of already proposed elements for regulation of translation initiation.

3.1 The first-AUG rule

Traditionally, the first-AUG rule is widely recognized for initiation of translation (Kozak, 1987, 1989, 1991). It states that ribosomes start translation from the first-AUG on the corresponding mRNA. Although this rule is not absolute, 90-95 % of vertebrate ORFs was established by the first AUG codon on the mRNA (Kozak, 1987, 1989, 1991). Our previous proteomics analysis of small proteins also indicated that about 84 % of proteins in RefSeq were translated from the first AUG of the corresponding mRNAs (Oyama et al., 2004). On the other hand, there are also many negative reports concerning the rule; 29 % of cDNA contained at least one ATG codon in their 5′-UTR (Suzuki et al., 2000); 41 % of transcripts had more than one uAUG and 24 % of genes had more than two uAUGs (Peri & Pandey, 2001); about 50 % of the RefSeq entries had at least one uAUG (Yamashita et al., 2003); about 44 % of 5′-UTRs had uAUGs and uORFs (Iacono et al., 2005). There are also some reports that the first AUG is skipped if it is too close to the cap structure, within 12 (Kozak, 1991) to 14 (Sedman et al., 1990) nucleotides (see the section 3.3). In this chapter, we cited a variety of statistical data on the UTRs. Because they are based on different versions or generations of sequence databases, the data vary widely (Meijer & Thomas, 2002), which is the point to be properly considered.

3.2 Kozak's consensus sequence

The strongest bias for initiation of translation in vertebrates is the sequence context called "Kozak's sequence", known as GCCA/GCCATGG (Kozak, 1987). The nucleotides in positions -3 (A or G) and +4 (G) are highly conserved and greatly effective for a ribosomal complex to start translation (Kozak, 1987, 2002; Matsui et al., 2007; Suzuki et al., 2001; Wang & Rothnagel, 2004). The context of an AUG codon in position -3 is the most highly conserved and functionally the most important; it is regarded as strong or optimal only when this position matches A or G, and that in position +4 is also highly conserved (Kozak, 2002). Some reports mentioned that only 0.86 % (Kozak, 1987) to 6 % (Iacono et al., 2005) of functional initiator codons lacked Kozak's sequence in positions -3 and +4, whereas 37 %

(Suzuki et al., 2000) to 46 % (Kozak, 1987) of uATGs would be skipped because of unfavorable Kozak's sequence in both of the positions. On the contrary, another report mentioned that most initiator codons were not in close agreement with Kozak's consensus sequence (Peri & Pandey, 2001).

3.3 The length of the 5'-UTR

The length of 5'-UTR is also effective when translation occurs from an AUG codon near the 5′ end of the mRNA (Kozak, 1991; Sedman et al., 1990). About half of ribosomes skip an AUG codon even in an optimal context if the length of 5′-UTR is less than 12 nucleotides (mentioned in the section 3.1) and this type of leaky scanning can be reduced if the length of 5′-UTR is more than or equal to 20 nucleotides (Kozak, 1991). In the traditional analysis based on incomplete 5′-UTR sequences, the distance from the 5' end to the AUG initiator codon in vertebrate mRNAs was generally from 20 and 100 nucleotides (Kozak, 1987). The previous analysis using RefSeq human mRNA sequences indicated that 85 % of 5′-UTR sequences less than 100 nucleotides contain no uAUGs (Peri & Pandey, 2001). The evidence convinced us that the first-AUG rule was widely supported in eukaryotes. In the recent analysis based on full-length 5′-UTR sequences, it is 125 nucleotides long on average (Suzuki et al., 2000) and transcriptional start sites (TSSs) vary widely (Carninci et al., 2006; Kimura et al., 2006; Suzuki et al., 2001). The average scattered length of 5'-UTR was more than 61.7 nucleotides, with a standard deviation of 19.5 nucleotides (Suzuki et al., 2001) and 52 % of the human RefSeq genes contained 3.1 TSS clusters on average (Kimura et al., 2006), which has an over 500 nucleotides interval (Fig. 4). In protein-coding genes, differentially regulated alternative TSSs are common (Carninci et al., 2006). Because the diversity of transcription initiation greatly affects the length of the 5'-UTR, there remain some doubts whether the length of the 5'-UTR contributes to the efficiency of translation initiation. There is also a report that the degree of leaky scanning is not affected by the length of 5′-UTR (Wang & Rothnagel, 2004).

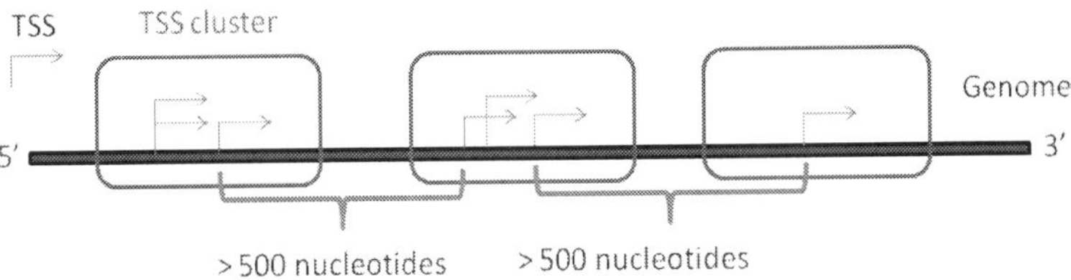

Fig. 4. The schematic representation of the 5′ ends of the TSSs.
Each TSS cluster consists of at least one TSS and has an over 500 nucleotides interval.

3.4 non-AUG initiator codon

In the general translation model, a non-AUG codon is considered to be ignored by ribosomes unless a downstream AUG codon is in a relatively weak context (Geballe & Morris, 1994; Kozak, 1999). In case that an upstream non-AUG codon, such as ACG, CUG or GUG, satisfies Kozak's consensus sequence, it possibly functions as an initiator of translation in addition to the first AUG initiator codon (Kozak, 1999, 2002). Besides Kozak's consensus sequence, downstream stem-and-loop and highly structured GC-rich context in the 5′-UTR could enhance translation initiation from a non-AUG codon (Kozak, 1991, 2002).

4. Protein identification by MS

The recent progress of proteomic methodologies based on highly sensitive liquid chromatography-tandem mass spectrometry (LC-MS/MS) technology have enabled us to identify hundreds or thousands of proteins in a single analysis.

We succeeded in the discovery of novel small proteins translated from short ORFs using direct nanoflow LC-MS/MS system (Oyama et al., 2004, 2007). Among 54 proteins less than 100 amino acids that were identified by retrieving several sequence databases with a representative search engine, Mascot (Matrix Science; http://www.matrixscience.com/), four ones were turned out to be encoded in 5′-UTRs (Oyama et al., 2004). This showed the first direct evidence of peptide products from the uORFs actually translated in human cells. In the subsequent analysis using more sophisticated two-dimensional LC system, we also discovered eight novel small proteins (Oyama et al., 2007), five of which were encoded in the 5′-UTR and three were encoded in the 3′-UTR of the corresponding mRNA. Even based on the accumulated DBTSS data, two ORFs had no putative AUG codon, which indicated the possibility that they were translated from non-AUG initiator codon. In the article above, 197 proteins less than 20 kDa were identified by Mascot. The procedure for identifying novel proteins by MS is described as follows.

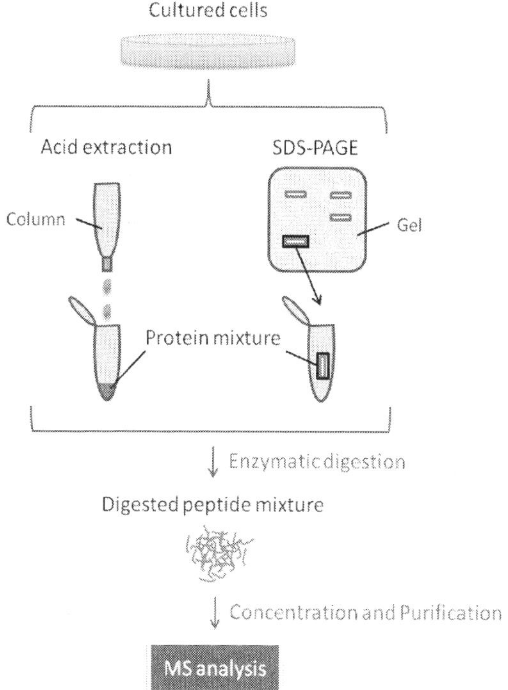

Fig. 5. The procedure for preparing samples for proteomic analyses of small proteins.

4.1 Materials and methods

The proteins included in cultured cell lysates were first separated according to their size. Small protein-enriched fraction through acid extraction and SDS-PAGE were treated with enzymes. In case of SDS-PAGE, the digested peptides were extracted from the gel. The samples were desalted and concentrated to introduce into the MS system. The schematic procedure is shown in Fig. 5.

4.2 Protein identification

The samples were analyzed using nanoflow LC-MS/MS system. The purified peptides were eluted with a linear gradient of acetonitrile and sprayed into the high-resolution tandem mass spectrometer. Acquired tandem mass (MS/MS) spectra were then converted to text files and processed against sequence databases using Mascot. Based on the principle that each peptide has a MS/MS spectrum with unique characteristics, the search engine compares measured data on precursor/product ions with those theoretically calculated from protein sequence data (Fig. 6). The MS/MS spectrum file contains mass to charge ratio (m/z) values of precursor and product ions along with their intensity. The measured spectrum lists are searched against sequence databases to identify the corresponding peptide in a statistical manner. The theoretical spectrum lists are totally dependent on the contents of sequence databases themselves.

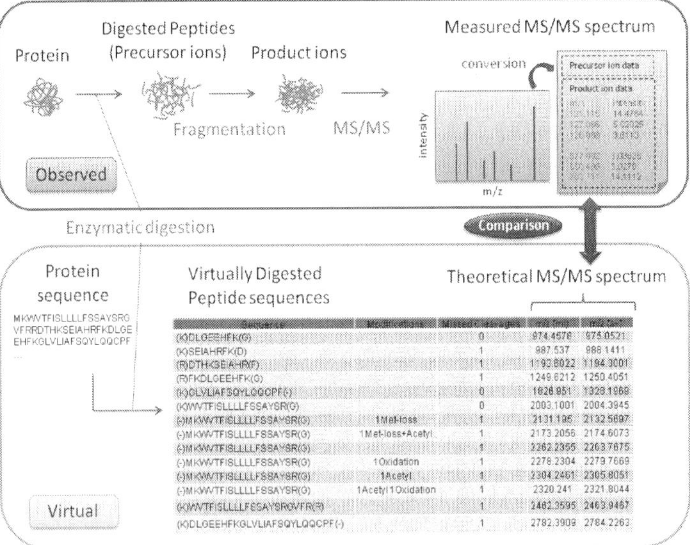

Fig. 6. The principle of protein identification.
A search engine compares measured MS/MS spectrum lists with theoretical ones. The acquired MS/MS spectra are converted into a text file that is composed of precursor ion data and product ion data, in a format defined by the search engine. Product ion data usually consist of multiple pairs of m/z and its intensity. The theoretical m/z values are calculated virtually.

4.3 Finding of novel small proteins

For exploring novel small proteins, two types of sequence databases were used; one was an artificial database computationally translated from the cDNA sequences in all the reading frames and the other was an already established protein database. In order to process the comparison of the large-scale protein identification data from the two kinds of databases, several Perl scripts have been developed based on the definition that candidates of novel small proteins were identified only in the cDNA database(s) (Fig. 7). In a result datasheet using RefSeq sequences, each protein was annotated with NM numbers for the cDNA database and with NP numbers for the protein database. The Perl scripts then exchanged NM to NP numbers and evaluated them.

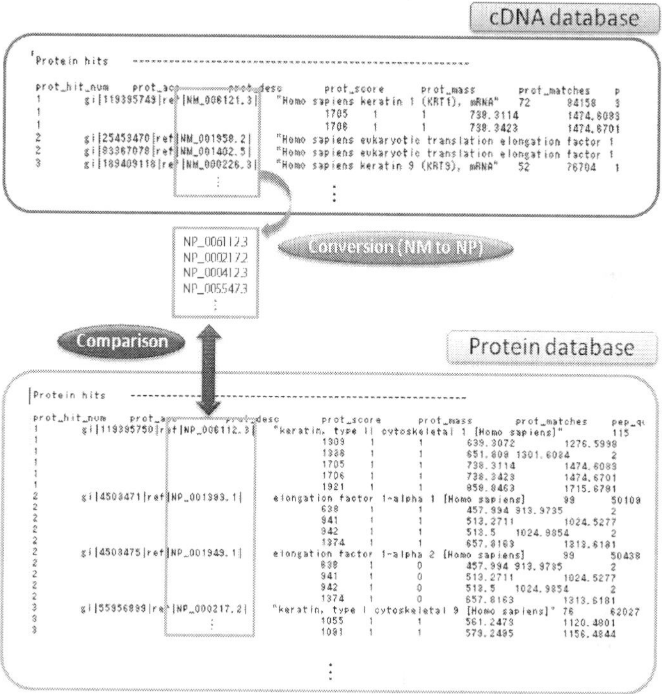

Fig. 7. The algorithm to compare the lists of search results using RefSeq cDNA and protein databases.
The proteins identified from the cDNA database are annotated with NM numbers, whereas, those from the protein database are with NP numbers. To compare these results, it is needed to exchange NM to NP numbers. The NP numbers annotated only from the cDNA database are considered to be candidates of novel proteins.

5. Bioinformatics approach

In order to forward MS-based identification of novel coding regions of mRNAs, MS systems, sequence databases and bioinformatics methodologies are required to improve together.

Regarding bioinformatics, two aspects seem to be demanded; one is for retrieving target proteins from an enormous size of database searching results, the other is for constructing platforms to predict novel coding sequences (CDSs).

5.1 Contribution of sequence databases & bioinformatics to MS-based proteomics

The recent advances in MS-based proteomics technology have enabled us to perform large-scale protein identification with high sensitivity. The accumulation of well-established sequence databases also made a great contribution to efficient identification in proteomics analyses. One of the representative databases is a specialized 5′-end cDNA database like DBTSS and the other is a series of whole genome sequence databases for various species. To investigate the mechanisms in transcriptional control, DBTSS has lately attracted considerable attention because it contains accumulated information on the transcriptional regulation of each gene (Suzuki et al., 2002, 2004; Tsuchihara et al., 2009; Wakaguri et al., 2008; Yamashita et al., 2006). Based on the accumulated data, the diverse distribution of TSSs was clearly indicated (Kimura et al., 2006; Suzuki et al., 2000, 2001). On the other hand, many whole genome sequencing projects are progressing all over the world (GOLD: Genomes Online Database; http://www.genomesonline.org/). Complement and maintenance of sequence databases for various species must help to find more novel proteins across the species. For example, there are several reports that conducted bioinformatical approaches to explore novel functional uORFs by comparing the 5'-UTR regions of orthologs based on multiple sequence alignments (Zhang & Dietrich, 2005), using ORF Finder (http://bioinformatics.org/sms/orf find.html) and a machine learning technique, inductive logic programming (ILP) with biological background knowledge (Selpi et al., 2006), or applying comparative genomics and a heuristic rule-based expert system (Cvijovic et al., 2007). Using advanced sequence databases, new protein CDSs were added as a result of the prediction by various algorithms (e.g. Hatzigeorgiou, 2002; Ota et al., 2004). Based on the well-established cDNA databases, MS could evaluate whether these CDSs are actually translated in a high-throughput manner. Construction of more detailed sequence databases will lead to detection of more novel small proteins in the presumed 5'-UTRs (Oyama et al., 2004). To make good use of those exhaustive sequence databases, bioinformatical techniques, especially data mining tools such as search engines to retrieve target proteins from an enormous size of database search results, are obviously indispensable.

5.2 Contribution of MS-based proteomics to sequence databases & bioinformatics

In addition to the technological progress of MS, sequence databases and data mining tools, development of other bioinformatical techniques called prediction tools, are also important. Ad-hoc algorithms for predicting new CDSs, as mentioned above, could be improved by using MS-based novel protein data. Those novel ones can be applied to play a role in a collection of supervised training data for machine learning, pattern recognition or rule-based manual approach. There is an interesting bioinformatical report which hypothesized that a uORF in the transcript down-regulates transcription of the corresponding RNA via RNA decay mechanisms (Matsui et al., 2007). They obtained human and mouse transcripts from RefSeq and UniGene (http://www.ncbi.nlm.nih.gov/unigene) and classified the transcripts into Level 0 (not containing uORF) and Level 1-3 (containing uORF). Then, they prepared the data of expression intensities and half-lives of mRNA transcripts mainly from SymAtlas (now linked to BioGPS; http://biogps.gnf.org/#goto=welcome) and Genome Research website (http://genome.cshlp.org/). Although they suggested that not only the

expression level but also the half-life of transcripts was obviously declined in the latter group, they did not demonstrate any interaction between uORFs and transcripts.
Advanced MS instruments can not only evaluate whether uORFs are actually translated but also quantify time-course changes of their expression levels. Stable isotope labeling with amino acids in cell culture (SILAC) technology enables us to quantify the changes regarding all the proteins in vivo (Oyama et al., 2009). Based on time-course changes of specific peptides, we could also hypothesize some regulatory interactions. In combination with the measurement of the dynamics of the corresponding mRNAs using microarray or reverse transcription-polymerase chain reaction (RT-PCR), transcriptional regulation by short ORFs will be analyzed at the system level.

6. Conclusion

Although the roles of 5′-UTR elements, especially uORFs, had been well discussed as translational regulators for the main ORFs in the biological context, whether the proteins encoded by the uORFs were translated had not been approached for a long time. We first unraveled the mystery by demonstrating the existence of novel protein products defined by these ORFs using advanced proteomics technology. Thanks to the progress of nanoLC-MS/MS-based shotgun proteomics strategies, thousands of proteins can now be identified from protein mixtures such as cell lysates. Some of the presumed UTRs are no longer "untranslated", and other noncoding transcripts are no longer "noncoding". One of the novel small proteins revealed in our analysis was indeed defined by a short transcript variant generated by utilization of the downstream alternative promoters (Oyama et al., 2007). Alternative uses of diverse transcription initiation, splicing and translation start sites could increase the complexity of short protein-coding regions and MS -based annotation of these novel small proteins will enable us to perform a more detailed analysis of the real outline of the proteome, along with the translational regulation by the diversified short ORFeome systematically.

7. References

Altmann, M. & Trachsel, H. (1993). Regulation of translation initiation and modulation of cellular physiology. *Trends in Biochemical Sciences*, Vol. 18, No. 11, pp. 429-432, Online ISSN 0167-7640; 0376-5067, Print ISSN 0968-0004.

Carninci, P., Sandelin, A., Lenhard, B., Katayama, S., Shimokawa, K., Ponjavic, J., Semple, C.A.M., Taylor, M.S., Engstrom, P.G., Frith, M.C., Forrest, A.R.R., Alkema, W.B., Tan, S.L., Plessy, C., Kodzius, R., Ravasi, T., Kasukawa, T., Fukuda, S., Kanamori-Katayama, M., Kitazume, Y., Kawaji, H., Kai, C., Nakamura, M., Konno, H., Nakano, K., Mottagui-Tabar, S., Arner, P., Chesi, A., Gustincich, S., Persichetti, F., Suzuki, H., Grimmond, S.M., Wells, C.A., Orlando, V., Wahlestedt, C., Liu, E.T., Harbers, M., Kawai, J., Bajic, V.B., Hume, D.A. & Hayashizaki, Y. (2006). Genome-wide analysis of mammalian promoter architecture and evolution. *Nature Genetics*, Vol. 38, No. 6, pp. 626-635, Online ISSN 1546-1718, Print ISSN 1061-4036.

Cvijovic, M., Dalevi, D., Bilsland, E., Kemp, G.J.L. & Sunnerhagen, P. (2007). Identification of putative regulatory upstream ORFs in the yeast genome using heuristics and evolutionary conservation. *BMC Bioinformatics*, Vol. 8, Article No. 295, Online ISSN 1471-2105.

Diba, F., Watson, C.S. & Gametchu, B. (2001). 5'UTR Sequences of the Glucocorticoid Receptor 1A Transcript Encode a Peptide Associated With Translational Regulation of the Glucocorticoid Receptor. *Journal of Cellular Biochemistry*, Vol. 81, No. 1, pp. 149-161, Online ISSN 1097-4644, Print ISSN 0730-2312.

Geballe, A.P. & Morris, D.R. (1994). Initiation codons within 5'-leaders of mRNAs as regulators of translation. *Trends in Biochemical Sciences*, Vol. 19, No. 4, pp. 159-164, Online ISSN 0167-7640; 0376-5067, Print ISSN 0968-0004.

Hatzigeorgiou, A.G. (2002). Translation initiation start prediction in human cDNAs with high accuracy. *Bioinformatics*, Vol. 18, No. 2, pp. 343-350, Online ISSN 1460-2059, Print ISSN 1367-4803.

Iacono, M., Mignone, F. & Pesole, G. (2005). uAUG and uORFs in human and rodent 5'untranslated mRNAs. *Gene*, Vol. 349, pp. 97-105, Online ISSN 1879-0038, Print ISSN 0378-1119.

Kimura, K., Wakamatsu, A., Suzuki, Y., Ota, T., Nishikawa, T., Yamashita, R., Yamamoto, J., Sekine, M., Tsuritani, K., Wakaguri, H., Ishii, S., Sugiyama, T., Saito, K., Isono, Y., Irie, R., Kushida, N., Yoneyama, T., Otsuka, R., Kanda, K., Yokoi, T., Kondo, H., Wagatsuma, M., Murakawa, K., Ishida, S., Ishibashi, T., Takahashi-Fujii, A., Tanase, T., Nagai, K., Kikuchi, H., Nakai, K., Isogai, T. & Sugano, S. (2006). Diversification of transcriptional modulation: Large-scale identification and characterization of putative alternative promoters of human genes. *Genome Research*, Vol. 16, No. 1, pp. 55-65, Online ISSN 1549-5469, Print ISSN 1088-9051.

Kozak, M. (1987). An analysis of 5'-noncoding sequences from 699 vertebrate messenger RNAs. *Nucleic Acids Research*, Vol. 15, No. 20, pp. 8125-8148, Online ISSN 1362-4962, Print ISSN 0305-1048.

Kozak, M. (1989). The Scanning Model for Translation: An Update. *The Journal of Cell Biology*, Vol. 108, No. 2, pp. 229-241, Online ISSN 1540-8140, Print ISSN 0021-9525.

Kozak, M. (1991). Structural Features in Eukaryotic mRNAs That Modulate the Initiation of Translation. *The Journal of Biological Chemistry*, Vol. 266, No. 30, pp. 19867-19870, Online ISSN 1083-351X, Print ISSN 0021-9258.

Kozak, M. (1999). Initiation of translation in prokaryotes and eukaryotes. *Gene*, Vol. 234, No. 2, pp. 187-208, Online ISSN 1879-0038, Print ISSN 0378-1119.

Kozak, M. (2002). Pushing the limits of the scanning mechanism for initiation of translation. *Gene*, Vol. 299, No. 1-2, pp. 1-34, Online ISSN 1879-0038, Print ISSN 0378-1119.

Maruyama, K. & Sugano, S. (1994). Oligo-capping: a simple method to replace the cap structure of eukaryotic mRNAs with oligoribonucleotides. *Gene*, Vol. 138, No. 1-2, pp. 171-174, Online ISSN 1879-0038, Print ISSN 0378-1119.

Matsui, M., Yachie, N., Okada, Y., Saito, R. & Tomita, M. (2007). Bioinformatic analysis of post-transcriptional regulation by uORF in human and mouse. *FEBS Letters*, Vol. 581, No. 22, pp. 4184-4188, Online ISSN 1873-3468, Print ISSN 0014-5793.

Meijer, H.A. & Thomas, A.A.M. (2002). Control of eukaryotic protein synthesis by upstream open reading frames in the 5'-untranslated region of an mRNA. *Biochemical Journal*, Vol. 367, No. 1, pp. 1-11, Online ISSN 1470-8728, Print ISSN 0264-6021.

Morris, D.R. & Geballe, A.P. (2000). Upstream Open Reading Frames as Regulators of mRNA Translation. *Molecular and Cellular Biology*, Vol. 20, No. 23, pp. 8635-8642, Online ISSN 1098-5549, Print ISSN 0270-7306.

Ota, T., Suzuki, Y., Nishikawa, T., Otsuki, T., Sugiyama, T., Irie, R., Wakamatsu, A., Hayashi, K., Sato, H., Nagai, K., Kimura, K., Makita, H., Sekine, M., Obayashi, M., Nishi, T., Shibahara, T., Tanaka, T., Ishii, S., Yamamoto, J., Saito, K., Kawai, Y., Isono, Y.,

Nakamura, Y., Nagahari, K., Murakami, K., Yasuda, T., Iwayanagi, T., Wagatsuma, M., Shiratori, A., Sudo, H., Hosoiri, T., Kaku, Y., Kodaira, H., Kondo, H., Sugawara, M., Takahashi, M., Kanda, K., Yokoi, T., Furuya, T., Kikkawa, E., Omura, Y., Abe, K., Kamihara, K., Katsuta, N., Sato, K., Tanikawa, M., Yamazaki, M., Ninomiya, K., Ishibashi, T., Yamashita, H., Murakawa, K., Fujimori, K., Tanai, H., Kimata, M., Watanabe, M., Hiraoka, S., Chiba, Y., Ishida, S., Ono, Y., Takiguchi, S., Watanabe, S., Yosida, M., Hotuta, T., Kusano, J., Kanehori, K., Takahashi-Fujii, A., Hara, H., Tanase, T., Nomura, Y., Togiya, S., Komai, F., Hara, R., Takeuchi, K., Arita, M., Imose, N., Musashino, K., Yuuki, H., Oshima, A., Sasaki, N., Aotsuka, S., Yoshikawa, Y., Matsunawa, H., Ichihara, T., Shiohata, N., Sano, S., Moriya, S., Momiyama,H., Satoh, N., Takami, S., Terashima, Y., Suzuki, O., Nakagawa, S., Senoh, A., Mizoguchi, H., Goto, Y., Shimizu, F., Wakebe, H., Hishigaki, H., Watanabe, T., Sugiyama, A., Takemoto, M., Kawakami, B., Yamazaki, M., Watanabe, K., Kumagai, A., Itakura, S., Fukuzumi, Y., Fujimori, Y., Komiyama, M., Tashiro, H., Tanigami, A., Fujiwara, T., Ono, T., Yamada, K., Fujii, Y., Ozaki, K., Hirao, M., Ohmori, Y., Kawabata, A., Hikiji, T., Kobatake, N., Inagaki, H., Ikema, Y., Okamoto, S., Okitani, R., Kawakami, T., Noguchi, S., Itoh, T., Shigeta, K., Senba, T., Matsumura, K., Nakajima, Y., Mizuno, T., Morinaga, M., Sasaki, M., Togashi, T., Oyama, M., Hata, H., Watanabe, M., Komatsu, T., Mizushima-Sugano, J., Satoh, T., Shirai, Y., Takahashi, Y., Nakagawa, K., Okumura, K., Nagase, T., Nomura, N., Kikuchi, H., Masuho, Y., Yamashita, R., Nakai, K., Yada, T., Nakamura, Y., Ohara, O., Isogai, T. & Sugano, S. (2004). Complete sequencing and characterization of 21,243 full-length human cDNAs. *Nature Genetics*, Vol. 36, No. 1, pp. 40-45, Online ISSN 1546-1718, Print ISSN 1061-4036.

Oyama, M., Itagaki, C., Hata, H., Suzuki, Y., Izumi, T., Natsume, T., Isobe, T. & Sugano, S. (2004). Analysis of Small Human Proteins Reveals the Translation of Upstream Open Reading Frames of mRNAs. *Genome Research*, Vol. 14, No. 10B, pp. 2048-2052, Online ISSN 1549-5469, Print ISSN 1088-9051.

Oyama, M., Kozuka-Hata, H., Suzuki, Y., Semba, K., Yamamoto, T. & Sugano, S. (2007). Diversity of Translation Start Sites May Define Increased Complexity of the Human Short ORFeome. *Molecular & Cellular Proteomics*, Vol. 6, No. 6, pp. 1000-1006, Online ISSN 1535-9484, Print ISSN 1535-9476.

Oyama, M., Kozuka-Hata, H., Tasaki, S., Semba, K., Hattori, S., Sugano, S., Inoue, J. & Yamamoto, T. (2009). Temporal Perturbation of Tyrosine Phosphoproteome Dynamics Reveals the System-wide Regulatory Networks. *Molecular & Cellular Proteomics*, Vol. 8, No. 2, pp. 226-231, Online ISSN 1535-9484, Print ISSN 1535-9476.

Peabody, D.S., Subramani, S. & Berg, P. (1986). Effect of Upstream Reading Frames on Translation Efficiency in Simian Virus 40 Recombinants. *Molecular and Cellular Biology*, Vol. 6, No. 7, pp. 2704-2711, Online ISSN 1098-5549, Print ISSN 0270-7306.

Peri, S. & Pandey, A. (2001). A reassessment of the translation initiation codon in vertebrates. *Trends in Genetics*, Vol. 17, No. 12, pp. 685-687, Print ISSN 0168-9525.

Rastinejad, F. & Blau, H.M. (1993). Genetic Complementation Reveals a Novel Regulatory Role for 3′ Untranslated Regions in Growth and Differentiation. *Cell*, Vol. 72, No. 6, pp. 903-917, Online ISSN 1097-4172, Print ISSN 0092-8674.

Sedman, S.A., Gelembiuk, G.W. & Mertz, J.E. (1990). Translation Initiation at a Downstream AUG Occurs with Increased Efficiency When the Upstream AUG Is Located Very Close to the 5' Cap. *Journal of Virology*, Vol. 64, No. 1, pp. 453-457, Online ISSN 1098-5514, Print ISSN 0022-538X.

Selpi, Bryant, C.H., Kemp, G.J.L. & Cvijovic, M. (2006). A First Step towards Learning which uORFs Regulate Gene Expression. *Journal of Integrative Bioinformatics*, Vol. 3, No. 2, ID. 31, Online ISSN 1613-4516.

Suzuki, Y., Yoshitomo-Nakagawa, K., Maruyama, K., Suyama, A. & Sugano, S. (1997). Construction and characterization of a full length-enriched and a 5'-end-enriched cDNA library. *Gene*, Vol. 200, No. 1-2, pp. 149-156, Online ISSN 1879-0038, Print ISSN 0378-1119.

Suzuki, Y., Ishihara, D., Sasaki, M., Nakagawa, H., Hata, H., Tsunoda, T., Watanabe, M., Komatsu, T., Ota, T., Isogai, T., Suyama, A. & Sugano, S. (2000). Statistical Analysis of the 5'Untranslated Region of Human mRNA Using "Oligo-Capped" cDNA Libraries. *Genomics*, Vol. 64, No. 3, pp. 286-297, Online ISSN 1089-8646, Print ISSN 0888-7543.

Suzuki, Y., Taira, H., Tsunoda, T., Mizushima-Sugano, J., Sese, J., Hata, H., Ota, T., Isogai, T., Tanaka, T., Morishita, S., Okubo, K., Sakaki, Y., Nakamura, Y., Suyama, A. & Sugano, S. (2001). Diverse transcriptional initiation revealed by fine, large-scale mapping of mRNA start sites. *EMBO reports*, Vol. 2, No. 5, pp. 388-393, Online ISSN 1469-3178, Print ISSN 1469-221X.

Suzuki, Y., Yamashita, R., Nakai, K. & Sugano, S. (2002). DBTSS: DataBase of human Transcriptional Start Sites and full-length cDNAs. *Nucleic Acids Research*, Vol. 30, No. 1, pp. 328-331, Online ISSN 1362-4962, Print ISSN 0305-1048.

Suzuki, Y., Yamashita, R., Sugano, S. & Nakai, K. (2004). DBTSS: DataBase of Transcriptional Start Sites: progress report 2004. *Nucleic Acids Research*, Vol. 32 (suppl 1), Database issue D78-D81, Online ISSN 1362-4962, Print ISSN 0305-1048.

Tsuchihara, K., Suzuki, Y., Wakaguri, H., Irie, T., Tanimoto, K., Hashimoto, S., Matsushima, K., Mizushima-Sugano, J., Yamashita, R., Nakai, K., Bentley, D., Esumi, H. & Sugano, S. (2009). Massive transcriptional start site analysis of human genes in hypoxia cells. *Nucleic Acids Research*, Vol. 37, No. 7, pp. 2249-2263, Online ISSN 1362-4962, Print ISSN 0305-1048.

Vilela, C. & McCarthy, J.E.G. (2003). Regulation of fungal gene expression via short open reading frames in the mRNA 5'untranslated region. *Molecular Microbiology*, Vol. 49, No. 4, pp. 859-867, Online ISSN 1365-2958, Print ISSN 0950-382X.

Wakaguri, H., Yamashita, R., Suzuki, Y., Sugano, S. & Nakai, K. (2008). DBTSS: database of transcription start sites, progress report 2008. *Nucleic Acids Research*, Vol. 36 (suppl 1), Database issue D97-D101, Online ISSN 1362-4962, Print ISSN 0305-1048.

Wang, X-Q. & Rothnagel, J.A. (2004). 5'-Untranslated regions with multiple upstream AUG codons can support low-level translation via leaky scanning and reinitiation. *Nucleic Acids Research*, Vol. 32, No. 4, pp. 1382-1391, Online ISSN 1362-4962, Print ISSN 0305-1048.

Yamashita, R., Suzuki, Y., Nakai, K. & Sugano, S. (2003). Small open reading frames in 5' untranslated regions of mRNAs. *Comptes Rendus Biologies*, Vol. 326, No. 10-11, pp. 987-991, Online ISSN 1768-3238, Print ISSN 1631-0691.

Yamashita, R., Suzuki, Y., Wakaguri, H., Tsuritani, K., Nakai, K. & Sugano, S. (2006). DBTSS: DataBase of Human Transcription Start Sites, progress report 2006. *Nucleic Acids Research*, Vol. 34 (suppl 1), Database issue D86-D89, Online ISSN 1362-4962, Print ISSN 0305-1048.

Zhang, Z. & Dietrich, F.S. (2005). Identification and characterization of upstream open reading frames (uORF) in the 5' untranslated regions (UTR) of genes in Saccharomyces cerevisiae. *Current Genetics*, Vol. 48, No. 2, pp. 77-87 Online ISSN 1432-0983, Print ISSN 0172-8083.

23

Acrylamide Binding to Its Cellular Targets: Insights from Computational Studies

Emmanuela Ferreira de Lima[1] and Paolo Carloni[2]
[1]*Scuola Internazionale Superiore di Studi Avanzati – SISSA - Statistical and Biological Physics Sector, via Bonomea, Trieste*
[2]*German Research School for Simulation Sciences, Jülich Research Center and RWTH-Aachen University,*
[1]*Italy*
[2]*Germany*
To the memory of Ernesto Illy (1925 – 2008), who suggested this project.

1. Introduction

Acrylamide (AC, CH_2=CH-$CONH_2$, Chart 1), present in heated foodstuffs [Rice, 2005], has been classified by the International Agency for Research on Cancer as "probably carcinogenic to humans" (group 2A) (IARC, 1994). The Scientific Committee on Toxicity, Ecotoxicity and the Environment (CSTEE) in the European Union (EU) demonstrated that AC exposure to humans should be controlled as low as possible because of its inherently toxic properties [Zhang et. al., 2009; Dearfield et. al., 1995]. AC is a low molecular-weight, odorless, and colorless compound. It is readily soluble in water and can rapidly polymerize from a monomer state to a polymer form [Klaunig, 2008]. AC is biotransformed *in vivo* to its epoxide glycidamide (GDE) by cytochrome P450 2E1 (CYP2E1) [Ghanayem et. al., 2005]. GDE has genotoxic properties in both *in vitro* and *in vivo* test systems [Kurebayashi & Ohno, 2008].
In spite of the possible carcinogenic nature of AC, no consistent effect of AC exposure on cancer incidence in humans could be identified [Rice, 2005]. This strikingly contrasts with AC subministration in both mice and rats, which may cause tumors at multiple sites [Besaratinia & Pfeifer, 2005; 2007]. A plausible hypothesis is this might be caused, at least partially, by the fact that AC interacts differently with the mouse and human proteins. AC may interact with cysteine residues not engaged in S-S bridges. Indeed, the double bond of conjugated vinyl compounds has strong affinity with SH groups [Friedman, 2003; Carere, 2006].

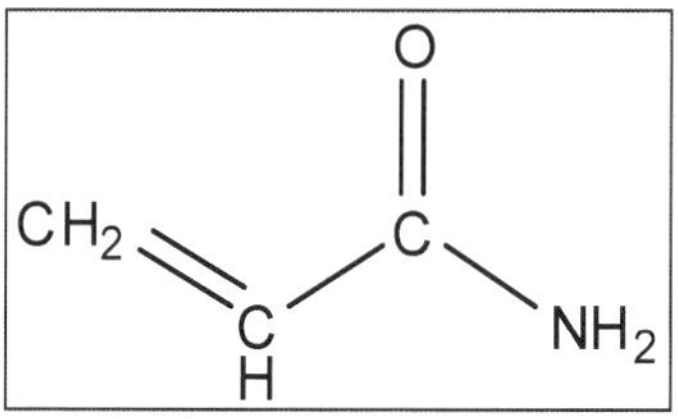

Fig. 1. AC structure.

There are six human proteins which are known to bind to AC. For all of them, structural information is available. These are five enzymes [Howland et. al., 1980; Sakamoto & Hashimoto, 1985] and the serum albumin protein [Ferguson et al., 2010] (Tab. 1). All of them are dimers. Inspection of the X-ray structures of all of these targets shows Cys residues may be present in the enzymatic active sites and/or solvent exposed (Tab. 1).

2. Objective

This work has two main goals. On one hand, using bioinformatics, molecular docking, and molecular simulation procedures, we aim at predicting the structural determinants of AC in complex with its cellular targets reported in Tab. 1. On the other hand, we aim at detecting the difference (if any) of binding these molecules to the correspondent proteins in rodents. These differences might contribute to the carcinogenic features of these molecules in rodents as opposed to humans.

Protein	PDB ID of human proteins	Sequence Identity (mouse)	Cys not engaged in disulfur bridges
Topoisomerase IIα	1ZXM	88%	**Cys216, Cys104, Cys392, Cys405,** Cys170, Cys300
Creatine Kinase (CK)	3B6R	96%	**Cys74**, Cys283*, **Cys254, Cys141, Cys146**
Aldolase	1QO5	95%	**Cys72, Cys239, Cys289, Cys338, Cys201, Cys149, Cys177, Cys134**
Serum albumin	1AO6	72%	**Cys34**
glyceraldehyde-3-phosphate dehydrogenase (GAPDH)	1ZNQ	93%	Cys152*, **Cys156, Cys247**
Enolase	2PSN	97%	Cys118, Cys356, Cys336, Cys338, **Cys398, Cys388**

Table 1. Structural information of AC targets. For human proteins, for which the experimental structures are available, the PDB codes are included. For mouse proteins, as there is no structure available, the sequence identities with human proteins are reported. The cysteines present in the active sites of the enzymes are shown with an asterisk while the solvent-exposed ones are highlighted in bold.

3. Methodology

3.1. Molecular docking

Docking methods attempt to find the "best" matching between two molecules, typically a receptor and a ligand [Halperin et. al., 2002]. Hence, they make the prediction of ligand conformation and orientation within a targeted binding site [Kitchen et. al., 2004; Halperin

et. al., 2002]. This prediction is carried out by performing a conformational space search based on an ad hoc potential energy function. [Halperin et. al., 2002].
The accuracy of the method can be investigated by docking ligands into the protein from which they were extracted (self-docking) and by docking them into its target protein in a different conformation (usually taken from different protein/ligand complexes) (cross-docking) [Kawatkar et. al., 2009; Sutherland et. al., 2007].
The increasing availability of protein three-dimensional structures combined with continuing computational advances have made docking a very important tool for small-molecule lead discovery [Campbell et. al., 2003].
Here we use the Genetic Optimisation for Ligand Docking (GOLD) program [Jones et. al., 1995; 1997]. GOLD is an automated ligand docking program that uses a genetic algorithm to explore the full range of ligand's conformational flexibility with partial flexibility of the protein. It satisfies the fundamental requirement that the ligand must displace loosely bound water on binding [Jones et. al., 1995; 1997].
GOLD uses a genetic algorithm (GA) to optimize a variety of structural parameters: (a) dihedrals of ligand's rotatable bonds; (b) ligand's ring geometries; (c) dihedrals of protein OH groups and NH_3^+ groups; and (d) the position of the ligand in the binding site [Verdonk et. at., 2003].
In GOLD, one minimizes a molecular mechanics–like function with four terms, the so called GOLD Fitness (Eq. 1) [Jones et al., 1995; 1997]:

$$\text{GOLD Fitness} = S_{hb\_ext} + S_{vdw\_ext} + S_{hb\_int} + S_{vdw\_int} \quad (1)$$

S_{hb_ext} decreases with the number of protein–ligand hydrogen-bonds; S_{vdw_ext} decreases with the van der Waals interactions; S_{hb_int} decreases with the intramolecular hydrogen bonds in the ligand; S_{vdw_int} increases with the intramolecular strain in the ligand [Verdonk et. al., 2003]. The scoring function is taken as the negative of the GOLD Fitness [Jones, 1995; 1997]. The larger the scoring function of a pose, the higher its rank.
We have performed GOLD-based docking for the human proteins in Tab. 1. These are: human topoisomerase IIα [Sciandrello et al, 2010]; human-brain creatine kinase [Lü Z-R. et al, 2009]; human aldolase [Dobryszycki et al., 1999a; 1999b]; human serum albumin [Ferguson et al., 2010]; human glyceraldehyde-3-phosphate dehydrogenase [Sakamoto & Hashimoto, 1985] and human enolase [Howland et. al., 1980]. Since alkylation by acrylamide is limited to cysteine SH groups, the best cys residues putatively binding AC will be those ones with the shortest distances between S atom and AC.
For rodents, there is no experimental structure available. Fortunately, the sequence identity (SI) between the mouse proteins and the human enzymes is always greater than 88%, except for serum albumin, for which SI= 72% (Tab. 1 and Fig. S1, SI). The structure of the latter was built by homology modeling using the server HHpred [Soding, 2005; Soding et. al., 2005; Hildebrand et. al., 2009].
The active site was defined here as a sphere with radius of 10 Å centered at the sulfur atoms of the Cys residues of Tab. 1. The cavity detection algorithm, LIGSITE [Hendlich et. al., 1997], was used to restrict the region of interest to concave, solvent-accessible surfaces. To be sure that most of the high-affinity binding modes were explored, the genetic algorithm was run 200 times for each complex examined.
The poses were ranked using the GOLD Fitness [Jones et al., 1995; 1997] as well as in terms of the number of contacts between AC and the target cysteine residues.

4. Results and discussion

Here we report the results for AC docking to all its known human targets [Friedman, 2003]. We also discuss the putative binding of AC to the correspondent mouse proteins, for which binding has not been shown. We consider here Cys residues not involved in S-S bridges (Target cysteines hereafter, see Tab. 1).
Table 2 shows the docking score values for the best AC poses in all targets, the values are in Kcal/mol. It can be seen that the best score obtained was for creatine kinase, -46.2, this result shows the great protein-ligand interaction provided by the docking calculations, this result is in good agreement with experiments, which have shown that CK is alkylated by AC.

Protein	Docking score
Topoisomerase IIα	-23.3
Creatine Kinase (CK)	-46.2
Aldolase	-21.3
Serum albumin	-22.6
glyceraldehyde-3-phosphate dehydrogenase (GAPDH)	-28.1
Enolase	-28.3

Table 2. Docking score values. The lower the score, the better the binding.

In the **topoisomerase** enzyme (Fig. 2), the cysteine(s) that react with AC were not identified experimentally. The X-ray structure [Wei et. al., 2005] features six target cysteines in each subunit, four of which are solvent-exposed (Tab. 1). Our docking procedure suggests that, in the most likely pose, AC interacts with Cys 405 (Fig. 3). In all figures, distances are reported in Angstroms.

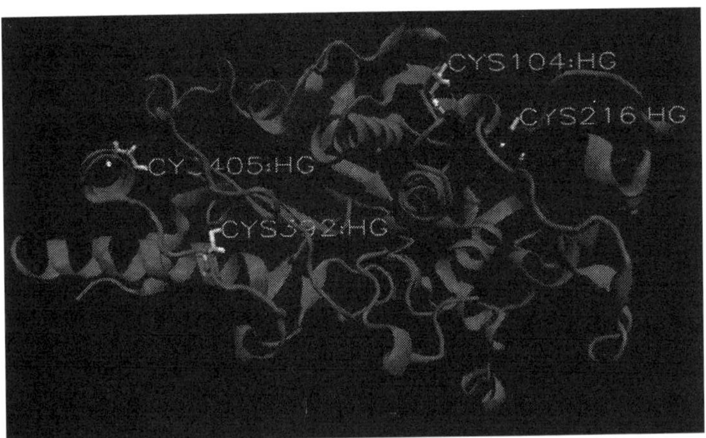

Fig. 2. 3D structure of topoisomerase enzyme [Sciandrello et al, 2010]. The four exposed cysteines (Cys104, Cys216, Cys392 and Cys405) are named and indicated in licorice representation.

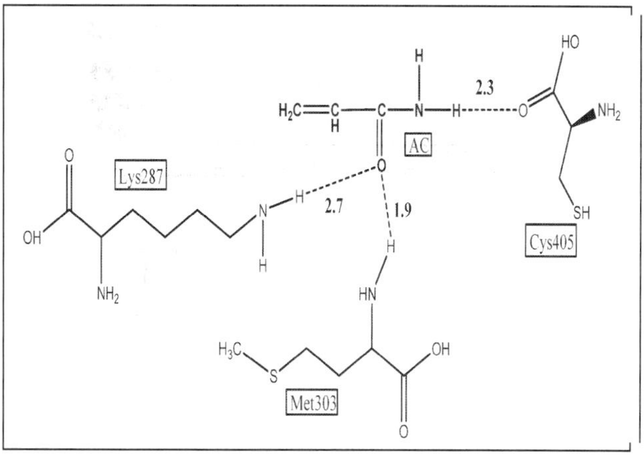

Fig. 3. AC interactions with Lys287, Met303 and Cys405 residues inside topoisomerase enzyme.

The SI between human and mouse proteins is as high as 88% (Fig. S2, SI). In particular, all of the residues surrounding the target cysteines are conserved. Inspection of the structure and of the sequence alignment in Fig. S2 strongly suggests that the chemical environment of the target cysteines in the mouse proteins is basically the same as in the human ones. The same argument applies to all of the enzymes considered here. Hence, AC might bind to mouse topoisomerase in a similar way as it does to the human protein.
The human enzyme **creatine kinase** (Fig. 4) has five target cysteines in each subunit (Tab. 1). One of them (Cys283) is located in each of the active sites (Fig. 5) and it is known experimentally to bind to AC [Matsuoka et. al., 1996; Meng et. al., 2001]. Consistently with experiment, we find that in the best pose AC interacts with Cys283 (Fig. 5). AC is not found to bind to any other target cysteine.
The SI between human and mouse proteins is considerably high (96%) and hence also in this case AC might bind to the mouse creatine kinase in a similar fashion.
The human enzyme **enolase** (Fig. 6) has 6 target cysteines in each binding site. Our docking procedure suggests that Cys 388 may bind to AC (Fig. 7). Because of the large SI between human and mouse protein (97%), AC might bind to the correspondent Cys residue (Cys 388, see Fig. S3, SI) also in the mouse protein.
The human enzyme **glyceraldehyde-3-phosphate dehydrogenase (GAPDH)** (Fig. 8) has three target cysteines in each dimer. One of them is present in each active sites, Cys152 (Fig. 9). Experiments show that Cys152 in fact interacts with AC [Campian et. Al., 2002]. Our docking is consistent with the experimental evidence (Fig. 9).
The protein human **aldolase** (Fig. 10) have eight and one target cysteine (Cys239, about 100 times more reactive than remaining exposed groups) in each subunit. Human **serum albumin** (Fig. 11) has 35 cysteine residues with 17 disulphide bonds. Sulphydryl residue in position 34 is left free to react with thiols of the environment [Candiano et al., 2009]. Our docking procedure could identify binding poses for AC in neither proteins. This suggests that conformational rearrangements, which are not taken into account in the docking procedure, might allow AC to bind to one or more Cys residues. For mouse aldolase, the SI

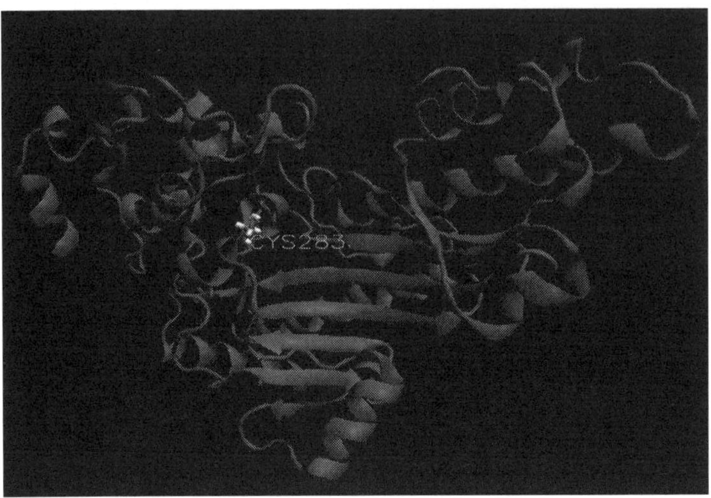

Fig. 4. 3D structure of creatine kinase enzyme [Lü Z-R. et al, 2009]. The Cys283, which is located in the active side is named and indicated in licorice representation.

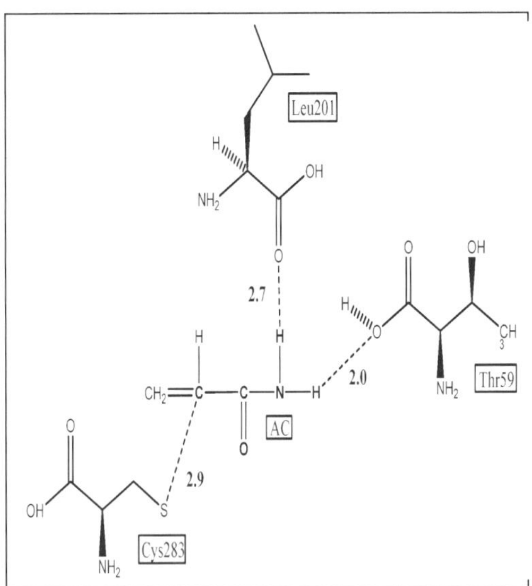

Fig. 5. AC interactions with Leu201, Thr59 and Cys283 residues inside creatine kinase active site.

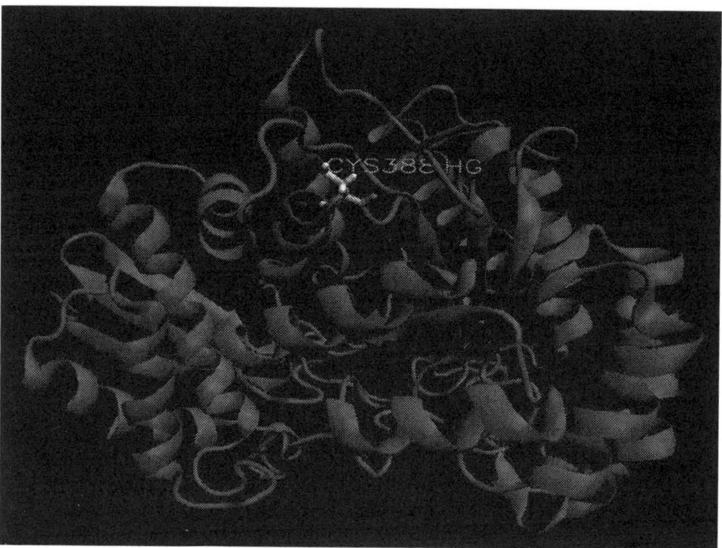

Fig. 6. 3D structure of enolase enzyme [Howland et. al., 1980]. The Cys388, which binds AC is named and indicated in licorice representation.

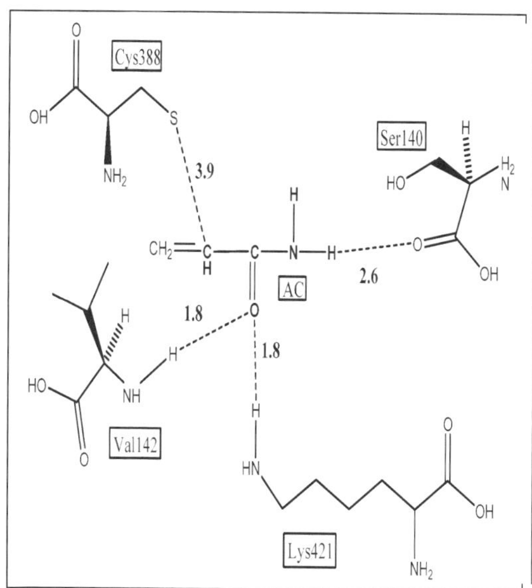

Fig. 7. AC interactions with Val142, Lys421, Ser140 and Cys388 residues, inside enolase enzyme.

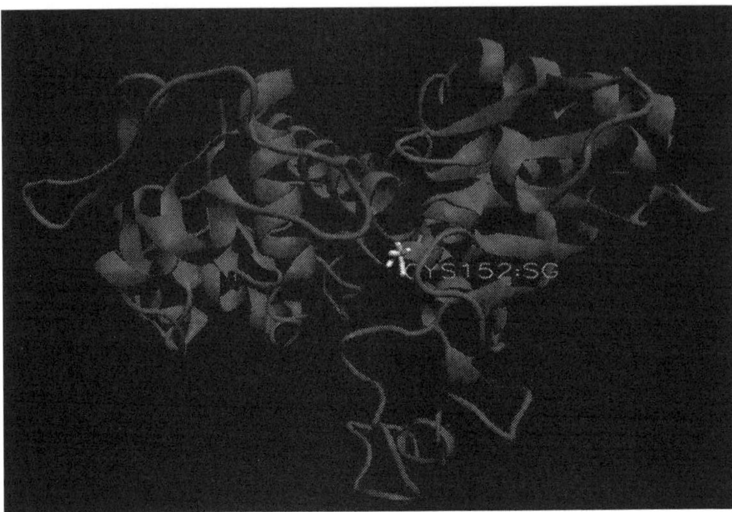

Fig. 8. 3D structure of GAPDH enzyme [Sakamoto & Hashimoto, 1985]. The Cys152, which binds AC is named and indicated in licorice representation.

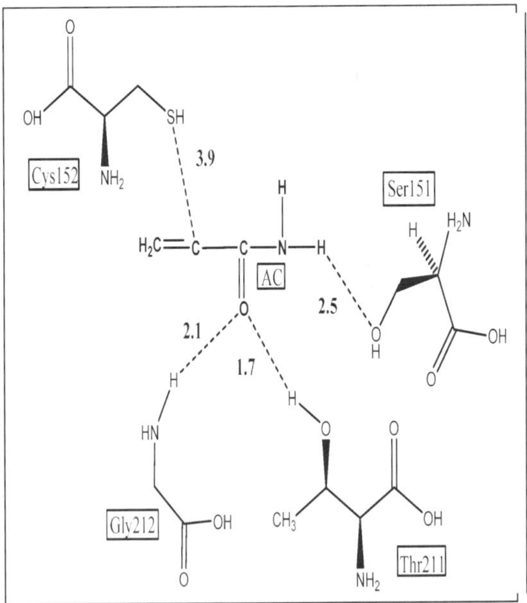

Fig. 9. AC interactions with Gly212, Thr211, Ser151 and Cys152 residues inside GAPDH active site.

is as high as 95%. This suggests similar consideration for this species. Instead, for mouse serum albumin, SI = 72% (see Fig. S1, SI). In the structure of the mouse serum albumin, which was built by homology modeling, there are two target cysteines (at position 58 and 603). Therefore, we may expect rather different binding in the two species. Because of the limitations of the docking procedure with homology models, we did not proceed to investigate AC binding poses to this protein [Leach, 2001; McGovern & Shoichet, 2003].

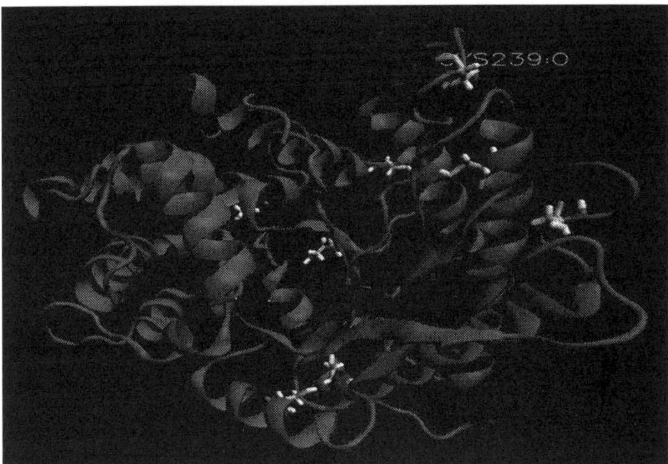

Fig. 10. 3D structure of aldolase enzyme [Dobryszycki et al., 1999a; 1999b]. All the cysteine residues are in licorice representation with indication of the most reactive one, Cys239.

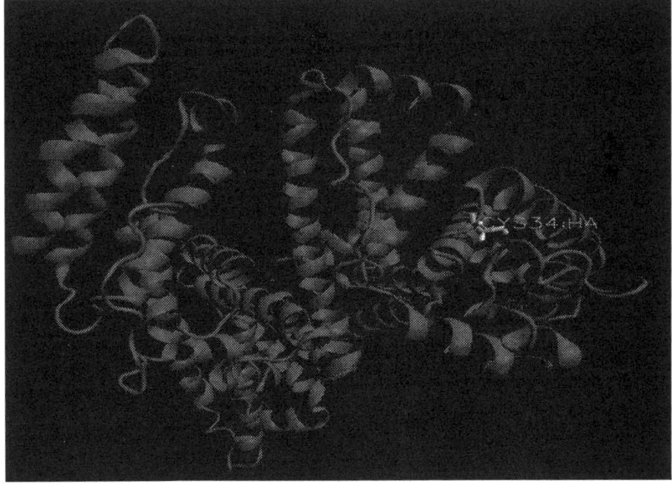

Fig. 11. 3D structure of serum albumin [Ferguson et al., 2010]. The only cysteine free to react, Cys34 is indicated in licorice representation.

5. Conclusions

By means of molecular docking methodology, we have studied the interactions between AC and its human targets. The investigation is complemented by a study of AC interactions with the mouse protein, for which binding has not been reported so far. The calculations are consistent with the available biochemical data and they provide novel information on putative cysteines to which AC could be bound in both mouse and human protein. In the case of one protein, serum albumin, binding is likely to occur at different locations in the protein. Hence, this difference could contribute to the experimentally known differences in toxic properties of AC in humans and mice.

6. Acknowledgment

One of the authors (Lima, EF) is recipient of a grant from the 'Fondazione Ernesto Illy', Trieste-Italy.

7. References

Besaratinia, A., & Pfeifer, G. P. (2005). DNA adduction and mutagenic properties of acrylamide. *Mutatational Research*, Vol. 580, pp. 31–40.

Besaratinia A., & Pfeifer G. P. (2007). A review of mechanisms of acrylamide carcinogenicity. *Carcinogenesis*, Vol. 28, No. 3, pp. 519–528.

Campbell S. J., Gold N. D., Jackson R. M. & Westhead D. R. (2003). Ligand binding: functional site location, similarity and docking, *Current Opinion in Structural Biology*, Vol. 13, pp. 389–395.

Campian E. C., Cai J., & Benz F. W. (2002). Acrylonitrile irreversibly inactivates glyceraldehyde-3-phosphate dehydrogenase by alkylating the catalytically active cysteine 149. *Chemico-Biological Interactions*, Vol. 140, pp. 279–291.

Candiano G., Petretto A., Bruschi M., Santucci L., Dimuccio V., Prunotto M., Gusmano R., Urbani A., Ghiggeri G. M. (2009). (Review) The oxido-redox potential of albumin Methodological approach and relevance to human diseases, journal of proteomics, 73, 188-195.

Carere A. (2006). Genotoxicity and carcinogenicity of acrylamide: a critical review, *Annali dell'Istituto Superiore di Sanità*, Vol. 42, No. 2, pp. 144-155.

Dearfield K. L. , Douglas G. R., Ehling U. H., Moore M. M., Sega G. A., & Brusick D. J. (1995). Acrylamide: a review of its genotoxicity and an assessment of heritable genetic risk. *Mutation Research*, Vol. 330, pp. 71-99.

Dobryszycki P., Rymarczuk M., Bulaj G., & Kochman M. (1999a). Effect of acrylamide on aldolase structure.I. Induction of intermediate states. *Biochimica et Biophysica Acta*, Vol. 1431, pp. 338-350.

Dobryszycki P., Rymarczuk M., Gapinski J., & Kochman M. (1999b). Effect of acrylamide on aldolase structure. II. Characterization of aldolase unfolding intermediates. *Biochimica et Biophysica Acta*, Vol. 1431, pp. 351-362.

Ferguson S. A., Garey J., Smith M. E., Twaddle N. C., Doerge D. R., & Paule M. G. (2010). Preweaning behaviors, developmental landmarks, and acrylamide and glycidamide levels after pre- and postnatal acrylamide treatment in rats. *Neurotoxicology and Teratology*, Vol. 32, pp. 373–382.

Friedman M. (2003). Chemistry, Biochemistry, and Safety of Acrylamide. A Review. *Journal of Agricultural and Food Chemistry*, Vol. 51, No. 16, pp. 4504-4526.

Ghanayem B. I., McDaniel L. P., Churchwell M. I., Twaddle N. C., Snyder R., Fennell T. R., & Doerge D. R. (2005). Role of CYP2E1 in the Epoxidation of Acrylamide to Glycidamide and Formation of DNA and Hemoglobin Adducts. *Toxicological Science*, Vol. 88, No. 2, pp. 311–318.

Halperin I., Ma B., Wolfson H., & Nussinov R. (2002). Principles of Docking: An Overview of Search Algorithms and a Guide to Scoring Functions. *PROTEINS: Structure, Function, and Genetics*, Vol. 47, pp. 409-443.

Hendlich M., Rippmann F., & Barnickel G. (1997). LIGSITE: Automatic and efficient detection of potential small molecule binding sites in proteins. *Journal of Molecular Graphics and Modelling*, Vol. 15, No. 6, pp. 359-363.

Hildebrand A., Remmert M., Biegert A., & Soding J. (2009). Fast and accurate automatic structure prediction with HHpred. *Proteins*, Vol. 77, pp. 128–132.

Howland R. D., Vyas I. L., Lowndes H. E., & Argentieri T. M. (1980). The etiology of toxic peripheral neuropathies: in vitro effects of acrylamide and 2,5-hexanedione on brain enolase and other glycolytic enzymes. *Brain Research*, Vol. 202, pp. 131-142.

Jones G., Willett P., & Glen R. C. (1995). Molecular recognition of receptor sites using a genetic algorithm with a description of desolvation. *Journal of Molecular Biology*, Vol. 245, pp. 43-53.

Jones G., Willett P., Glen R. C., Leach A. R., & Taylor R. (1997). Development and Validation of a Genetic Algorithm for Flexible Docking. *Journal of Molecular Biology*, Vol. 267, pp. 727-748.

Kawatkar S., Wang H., Czerminski R., & McCarthy D. J. (2009). Virtual fragment screening: an exploration of various docking and scoring protocols for fragments using Glide, *Journal of Computer-Aided Molecular Design*, Vol. 23, pp. 527–539.

Kitchen D. B., Decornez H., Furr J. R., & Bajorath J. (2004). Docking and scoring in virtual screening for drug discovery: methods and applications, *Nature Reviews – Drug Discovery*, Vol. 3, pp. 935.

Klaunig J. E. (2008). Acrylamide Carcinogenicity. *Journal of Agricultural and Food Chemistry*, Vol. 56, pp. 5984–5988.

Kurebayashi H., & Ohno Y. (2006). Metabolism of acrylamide to glycidamide and their cytotoxicity in isolated rat hepatocytes: protective effects of GSH precursors. *Archives of Toxicology*, Vol. 80, pp. 820–828.

Leach, A. R. (2nd Edition). (2001). *Molecular modelling, principles and applications*, Harlow, Essex Pearson Education Limited, ISBN 0-582-38210-6, Harlow, England.

Lü Z-R., Zou H-C., Jin P. S., Park D., Shi L., Ho O-S., Park Y-D., Bhak J., & Zou F. (2009). The effects of acrylamide on brain creatine kinase: Inhibition kinetics and computational docking simulation. *International Journal of Biological Macromolecules*, Vol. 44, pp. 128–132.

Matsuoka M., Matsumura H., & Igisu H. (1996). Creatine kinase activities in brain and blood: possible neurotoxic indicator of acrylamide intoxication. *Occupational and Environmental Medicine*, Vol. 53, pp.468-471.

McGovern S. L. & Shoichet Brian K. (2003). Information decay in molecular docking screens against Holo, Apo, and modeled conformations of enzymes. *Journal of Medicinal Chemistry*, Vol. 46, pp. 2895-2907.

Meng F-G., Zhou H-W., & Zhou H-M. (2001). Effects of acrylamide on creatine kinase from rabbit muscle. *The International Journal of Biochemistry & Cell Biology*, Vol. 33, pp. 1064–1070.

Rice, J. (2005). The carcinogenicity of acrylamide. *Mutatational Research*, Vol. 580, pp. 3–20.

Sakamoto J., & Hashimoto K. (1985). Effect of acrylamide and related compounds on glycolytic enzymes in rat sciatic nerve in vivo. *Archives of Toxicology*, Vol. 57, pp. 282-284.

Sciandrello G., Mauro M., Caradonna F., Catanzaro I., Saverini M., & Barbata G. (2010). Acrylamide catalytically inhibits topoisomerase II in V79 cells. *Toxicology in Vitro*, Vol. 24, pp. 830–834.

Soding J. (2005). Protein homology detection by HMM–HMM comparison. *Bioinformatics*, Vol. 21, No. 7, pp. 951–960.

Soding J., Biegert A., & Lupas A. N. (2005). The HHpred interactive server for protein homology detection and structure prediction. *Nucleic Acids Research*, Vol. 33, pp. W244–W248.

Sutherland J. J., Nandigam R. K., Erickson J. A., & Vieth M. (2007). Lessons in Molecular Recognition. 2. Assessing and Improving Cross-Docking Accuracy, Journal of Chemical Information and Modeling, Vol. 47, pp. 2293-2302.

Verdonk M. L., Cole J. C., Hartshorn M. J., Murray C. W., & Taylor R. D. (2003). Improved Protein–Ligand Docking Using GOLD, *Proteins: Structure, Function, and Genetics*, Vol. 52, pp. 609–623.

Wei H., Ruthenburg A. J., Bechis S. K., & Verdine G. L. (2005). Nucleotide-dependent Domain Movement in the ATPase Domain of a Human Type IIA DNA Topoisomerase. *Journal of Biological Chemistry*, Vol. *280, pp. 37041-37047.*

Zhang Y., Ren Y., & Zhang Y. (2009). New Research Developments on Acrylamide: Analytical Chemistry, Formation Mechanism, and Mitigation Recipes. *Chemical Reviews*, Vol. 109, pp. 4375–4397.